OXFORD IB STUDY GUIDES

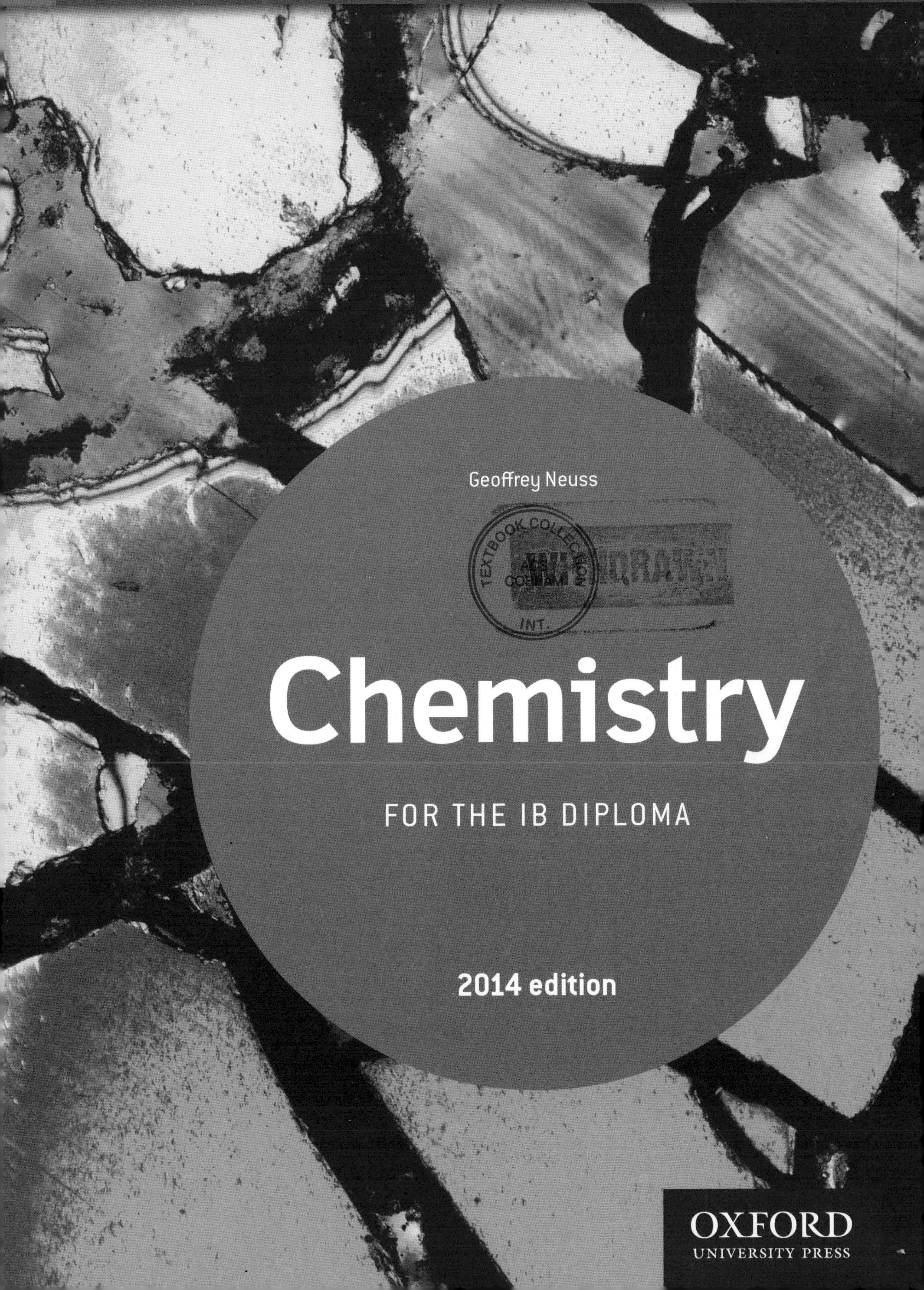

Geoffrey Neuss

Chemistry

FOR THE IB DIPLOMA

2014 edition

OXFORD
UNIVERSITY PRESS

OXFORD
UNIVERSITY PRESS

Great Clarendon Street, Oxford, OX2 6DP, United Kingdom

Oxford University Press is a department of the University of Oxford. It furthers the University's objective of excellence in research, scholarship, and education by publishing worldwide. Oxford is a registered trade mark of Oxford University Press in the UK and in certain other countries

First published in 2014

British Library Cataloguing in Publication Data
Data available

978-0-19-839353-5

15 14 13 12 11 10

Paper used in the production of this book is a natural, recyclable product made from wood grown in sustainable forests. The manufacturing process conforms to the environmental regulations of the country of origin.

Printed in Great Britain by Bell and Bain Ltd, Glasgow

Acknowledgements

Cover: ALFRED PASIEKA/SCIENCE PHOTO LIBRARY; **p105**: Oxford University Press; **p117**: Oxford University Press; **p145**: Kaspri/Dreamstime.com; **p172**: IUPAC.org; **p174**: Now Art

Artwork by Six Red Marbles and Oxford University Press.

We have tried to trace and contact all copyright holders before publication. If notified the publishers will be pleased to rectify any errors or omissions at the earliest opportunity.

MIX
Paper from responsible sources
FSC www.fsc.org FSC® C007785

Introduction and acknowledgements

This book replaces the very successful first and second editions. Like earlier editions it is written specifically for students studying Chemistry for the International Baccalaureate Diploma although many students following their own national systems will also find it helpful. It comprehensively covers the new programme that will be examined from 2016 onwards. All the information required for each topic is set out in separate boxes with clear titles that follow faithfully the layout of the syllabus. The first 11 topics cover both the Core content needed by all students and the Additional Higher Level material under the main topic headings. The difference between the two levels is clearly distinguished. Both Higher Level and Standard Level students must study one of the four options and each option stands in its own right even if this has meant repeating small amounts of some of the material. Worked examples are included where they are appropriate and at the end of each main topic there are both multiple choice and short-answer practice questions. Similarly at the end of each option I have included short-answer questions. Many of these questions are taken from past IB examination papers and I would like to thank the International Baccalaureate Organization for giving me permission to use them. The remaining questions are written to the same IB standard specifically for this book. Worked answers to the questions are provided. These answers are not necessarily full 'model' answers but they do contain all the information needed to score each possible mark. I have included one chapter on the underlying philosophy of the course to explain and give examples of the Essential Ideas, the Nature of Science, the International Dimension, and Utilization. To help you, the student, gain the highest grade possible the final chapter is devoted to giving you advice on how to study and prepare for the final examination. It also advises you on how to excel at the internally assessed practical component of the course. For those who opt for Chemistry as the subject for their Extended Essay it gives advice and guidance on how to choose the topic and write your Essay. A comprehensive Extended Essay checklist is included to help you gain bonus points towards your IB Diploma.

I have stuck rigorously to the syllabus to produce this study guide which contains all the necessary subject content required for the examination in one easily accessible and compact format. IB Chemistry is, of course, about much more than the final examination and you are also encouraged to read widely around the subject to further your knowledge and understanding as well as your enjoyment of chemistry. Many more multiple choice tests and short answer questions, as well as a blog and much more background information (e.g. videos) about each topic and option can be found on my website for IB Diploma Chemistry teachers at www.thinkib.net/chemistry.

I have been fortunate at Atlantic College to teach many highly motivated and gifted students who have often challenged me with searching questions. During my association with the International Baccalaureate, the European Baccalaureate and the United World Colleges I have been privileged to meet, work alongside and exchange ideas with many excellent Chemistry teachers who exude a real enthusiasm for their subject. Many of these have influenced me greatly – in particular, John Devonshire, a fellow teacher at Atlantic College and Jacques Furnemont, an Inspector of Chemistry in Belgium. I value greatly their advice, opinions and knowledge. I would also like to pay tribute to two former Chief Examiners for the IB, Ron Ragsdale and Arden Zipp. The high regard with which IB Chemistry is held today owes much to both of them.

Finally I should like to thank my wife Chris and my friend and colleague John for their patience and unstinting support throughout.

Dr Geoffrey Neuss

Contents

(Italics denote topics which are exclusively Higher Level.)

Particulate nature of matter

ELEMENTS

All substances are made up of one or more elements. An element cannot be broken down by any chemical process into simpler substances. There are just over 100 known elements. The smallest part of an element is called an atom.

Names of the first 20 elements

Atomic Number	Name	Symbol	Relative atomic mass
1	hydrogen	H	1.01
2	helium	He	4.00
3	lithium	Li	6.94
4	beryllium	Be	9.01
5	boron	B	10.81
6	carbon	C	12.01
7	nitrogen	N	14.01
8	oxygen	O	16.00
9	fluorine	F	19.00
10	neon	Ne	20.18
11	sodium	Na	22.99
12	magnesium	Mg	24.31
13	aluminium	Al	26.98
14	silicon	Si	28.09
15	phosphorus	P	30.97
16	sulfur	S	32.07
17	chlorine	Cl	35.45
18	argon	Ar	39.95
19	potassium	K	39.10
20	calcium	Ca	40.08

COMPOUNDS

Some substances are made up of a single element, although there may be more than one atom of the element in a particle of the substance. For example, oxygen is diatomic, that is, a molecule of oxygen contains two oxygen atoms and has the formula O_2. A compound contains more than one element combined chemically in a fixed ratio. For example, a molecule of water contains two hydrogen atoms and one oxygen atom. It has the formula H_2O. Water is a compound not an element because it can be broken down chemically into its constituent elements: hydrogen and oxygen. Compounds have different chemical and physical properties from their component elements.

MIXTURES

The components of a mixture may be elements or compounds. These components are not chemically bonded together. Because they are not chemically combined, the components of a mixture retain their individual properties. All the components of a mixture may be in the same phase, in which case the mixture is said to be **homogeneous**. Air is an example of a gaseous homogenous mixture. If the components of a mixture are in different phases the mixture is said to be **heterogeneous**. There is a physical boundary between two phases. A solid and a liquid is an example of a two-phase system. It is possible to have a single state but two phases. For example, two immiscible liquids such as oil and water form a heterogeneous mixture.

STATES OF MATTER

sublimation

SOLID STATE

- Fixed shape
- Fixed volume
- Particles held together by intermolecular forces in a fixed position
- Particles can vibrate about a fixed point but do not have translational velocity
- As heat is supplied at a certain temperature the vibration is sufficient to overcome the attractive forces holding the solid together and the solid melts

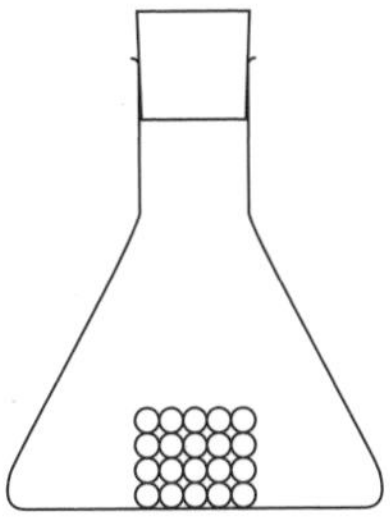

melting

freezing

LIQUID STATE

- Fixed volume
- Takes up shape of container
- Particles held closely together by intermolecular forces
- Particles have translational velocity so diffusion can occur
- As heat is supplied the liquid particles move faster. Some particles move faster than others and escape from the surface of the liquid to form a vapour. Once the pressure of the vapour is equal to the pressure above the liquid the liquid boils.

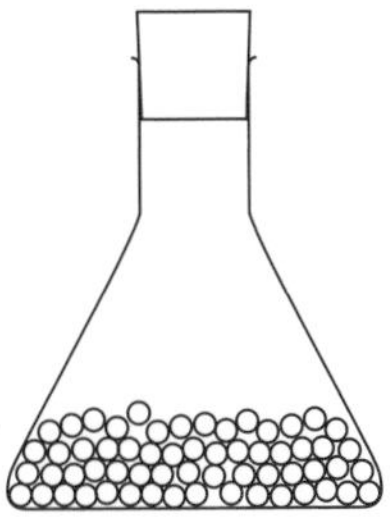

boiling/ vaporization/ evaporation

condensation

GASEOUS STATE

- Widely spaced particles that completely fill container
- Pressure of the gas due to gaseous particles colliding with the walls of the container
- Intermolecular forces between particles negligible
- Volume occupied by molecules themselves negligible compared with total volume of gas
- Particles moving with rapid, random motion so diffusion can occur

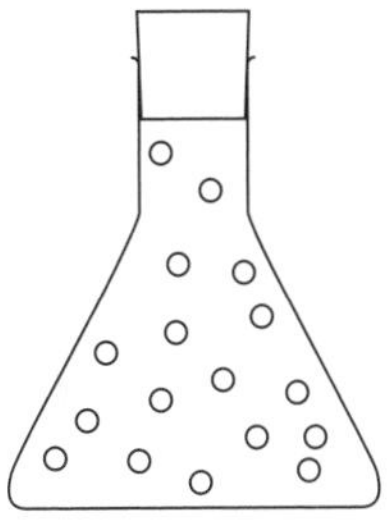

deposition

The mole concept and chemical formulas

MOLE CONCEPT AND AVOGADRO'S CONSTANT

A single atom of an element has an extremely small mass. For example, an atom of carbon-12 has a mass of 1.993×10^{-23} g. This is far too small to weigh. A more convenient amount to weigh is 12.00 g. 12.00 g of carbon-12 contains 6.02×10^{23} atoms of carbon-12. This number is known as Avogadro's constant (N_A or L).

Chemists measure amounts of substances in moles. A mole is the amount of substance that contains L particles of that substance. The mass of one mole of **any** substance is known as the **molar mass** and has the symbol M. For example, hydrogen atoms have $\frac{1}{12}$ of the mass of carbon-12 atoms so a mole of hydrogen atoms contains 6.02×10^{23} hydrogen atoms and has a mass of 1.01 g. In reality elements are made up of a mixture of isotopes.

The **relative atomic mass** of an element A_r is the weighted mean of all the naturally occurring isotopes of the element relative to carbon-12. This explains why the relative atomic masses given for the elements on page 1 are not whole numbers. The units of molar mass are g mol^{-1} but relative molar masses M_r have no units. For molecules **relative molecular mass** is used. For example, the M_r of glucose, $C_6H_{12}O_6 = (6 \times 12.01) + (12 \times 1.01) + (6 \times 16.00) = 180.18$. For ionic compounds the term **relative formula mass** is used.

Be careful to distinguish between the words **mole** and **molecule**. A molecule of hydrogen gas contains two atoms of hydrogen and has the formula H_2. A mole of hydrogen gas contains 6.02×10^{23} hydrogen molecules made up of two moles (1.20×10^{24}) of hydrogen atoms.

FORMULAS OF COMPOUNDS

Compounds can be described by different chemical formulas.

Empirical formula (literally the formula obtained by experiment)

This shows the simplest whole number ratio of atoms of each element in a particle of the substance. It can be obtained by either knowing the mass of each element in the compound or from the percentage composition by mass of the compound. The percentage composition can be converted directly into mass by assuming 100 g of the compound are taken.

Example: A compound contains 40.00% carbon, 6.73% hydrogen and 53.27% oxygen by mass, determine the empirical formula.

	Amount / mol	Ratio	
C	40.00/12.01 = 3.33	1	
H	6.73/1.01 = 6.66	2	Empirical formula $= CH_2O$
O	53.27/16.00 = 3.33	1	

Molecular formula

For molecules this is much more useful as it shows the actual number of atoms of each element in a molecule of the substance. It can be obtained from the empirical formula if the molar mass of the compound is also known.

Methanal CH_2O ($M_r = 30$), ethanoic acid $C_2H_4O_2$ ($M_r = 60$) and glucose $C_6H_{12}O_6$ ($M_r = 180$) are different substances with different molecular formulas but all with the same empirical formula CH_2O. Note that subscripts are used to show the number of atoms of each element in the compound.

Structural formula

This shows the arrangement of atoms and bonds within a molecule and is particularly useful in organic chemistry.

The three different formulas can be illustrated using ethene:

CH_2

empirical formula

C_2H_4

molecular formula

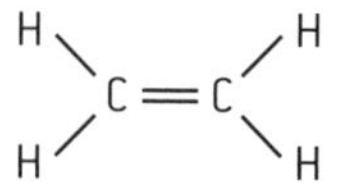

(can also be written $H_2C{=}CH_2$)

structural formula

EXPERIMENTAL DETERMINATION OF AN EMPIRICAL FORMULA

The empirical formula of magnesium oxide can be determined simply in the laboratory. A coil of magnesium ribbon about 10 cm long is placed in a pre-weighed crucible and its mass recorded. The crucible is placed on a clay triangle and heated strongly. When the magnesium ribbon starts to burn the lid is lifted slightly to allow more air to enter and the heating is continued until all the magnesium has burned. After cooling the crucible, its lid and its contents are reweighed.

Table of typical raw quantitative data:

	Mass / g (± 0.001 g)
Mass of crucible + lid	30.911
Mass of crucible + lid + magnesium	31.037
Mass of crucible + lid + magnesium oxide	31.106

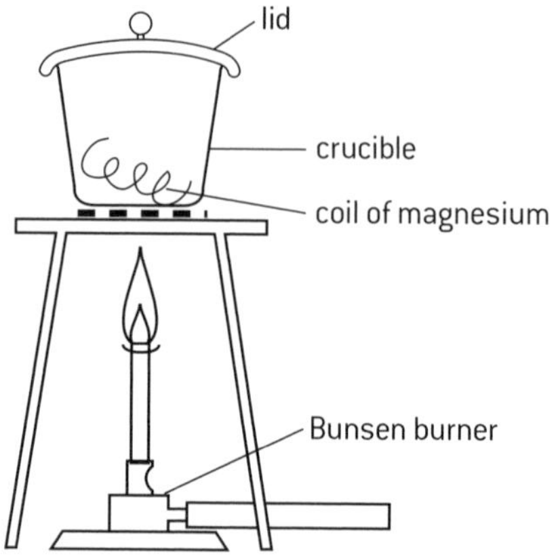

Mass of magnesium = 31.037 − 30.911 = 0.126 g

Mass of magnesium oxide = 31.106 − 30.911 = 0.195 g

Mass of oxygen combining with magnesium = 0.195 − 0.126 = 0.069 g

Amount of magnesium $= \dfrac{0.126}{24.31} = 5.2 \times 10^{-3}$ mol

Amount of oxygen $= \dfrac{0.069}{16.00} = 4.3 \times 10^{-3}$ mol

Ratio of Mg to O $= \dfrac{5.2 \times 10^{-3}}{4.3 \times 10^{-3}} = 1.2$ to 1

Convert to whole number ratio = 6 : 5

Empirical formula of magnesium oxide as determined by this experiment is Mg_6O_5.

Chemical reactions and equations

PROPERTIES OF CHEMICAL REACTIONS

In all chemical reactions:

- new substances are formed
- bonds in the reactants are broken and bonds in the products are formed resulting in an energy change between the reacting system and its surroundings
- there is a fixed relationship between the number of particles of reactants and products resulting in no overall change in mass – this is known as the stoichiometry of the reaction.

CHEMICAL EQUATIONS

Chemical reactions can be represented by chemical equations. Reactants are written on the left-hand side and products on the right-hand side. The number of moles of each element must be the same on both sides in a balanced chemical equation, e.g. the reaction of nitric acid (one of the acids present in acid rain) with calcium carbonate (the main constituent of marble statues).

$$\underbrace{\underset{\text{calcium carbonate}}{CaCO_3(s)} + \underset{\text{nitric acid}}{2HNO_3(aq)}}_{\text{REACTANTS}} \rightarrow \underbrace{\underset{\text{calcium nitrate}}{Ca(NO_3)_2(aq)} + \underset{\text{carbon dioxide}}{CO_2(g)} + \underset{\text{water}}{H_2O(l)}}_{\text{PRODUCTS}}$$

STATE SYMBOLS

Because the physical state that the reactants and products are in can affect both the rate of the reaction and the overall energy change it is good practice to include the state symbols in the equation.

(s) – solid (l) – liquid (g) – gas (aq) – in aqueous solution

$\rightarrow$ OR $\rightleftharpoons$

A single arrow $\rightarrow$ is used if the reaction goes to completion. Sometimes the reaction conditions are written on the arrow:

e.g. $$C_2H_4(g) + H_2(g) \xrightarrow{\text{Ni catalyst, 180 °C}} C_2H_6(g)$$

Reversible arrows are used for reactions where both the reactants and products are present in the equilibrium mixture:

e.g. $$3H_2(g) + N_2(g) \underset{\text{250 Atm}}{\overset{\text{Fe(s), 550 °C}}{\rightleftharpoons}} 2NH_3(g)$$

COEFFICIENTS AND MOLAR RATIO

The coefficient refers to the number in front of each reactant and product in the equation. The coefficients give information on the molar ratio. In the first example above, two moles of nitric acid react with one mole of calcium carbonate to produce one mole of calcium nitrate, one mole of carbon dioxide and one mole of water. In the reaction between hydrogen and nitrogen above, three moles of hydrogen gas react with one mole of nitrogen gas to produce two moles of ammonia gas.

IONIC EQUATIONS

Because ionic compounds are completely dissociated in solution it is sometimes better to use ionic equations. For example, when silver nitrate solution is added to sodium chloride solution a precipitate of silver chloride is formed.

$$Ag^+(aq) + NO_3^-(aq) + Na^+(aq) + Cl^-(aq) \rightarrow AgCl(s) + Na^+(aq) + NO_3^-(aq)$$

$Na^+(aq)$ and $NO_3^-(aq)$ are spectator ions and do not take part in the reaction. So the ionic equation becomes:

$$Ag^+(aq) + Cl^-(aq) \rightarrow AgCl(s)$$

From this we can deduce that any soluble silver salt will react with any soluble chloride to form a precipitate of silver chloride.

Mass and gaseous volume relationships

SOLIDS

Normally measured by weighing to obtain the mass.

1.000 kg = 1000 g

When weighing a substance the mass should be recorded to show the accuracy of the balance. For example, exactly 16 g of a substance would be recorded as 16.00 g on a balance weighing to + or − 0.01 g but as 16.000 g on a balance weighing to + or − 0.001 g.

LIQUIDS

Pure liquids may be weighed or the volume recorded.

The density of the liquid = $\frac{\text{mass}}{\text{volume}}$ and is usually expressed in g cm^{-3}.

MEASUREMENT OF MOLAR QUANTITIES

In the laboratory moles can conveniently be measured using either mass or volume depending on the substances involved.

GASES

Mass or volume may be used for gases.

SOLUTIONS

Volume is usually used for solutions.

1.000 litre = 1.000 dm^3 = 1000 cm^3

Concentration is the amount of **solute** (dissolved substance) in a known volume of **solution** (solute plus **solvent**). It is expressed either in g dm^{-3}, or, more usually in mol dm^{-3}. For very dilute solutions it is also sometimes expressed in parts per million, ppm. A solution of known concentration is known as a **standard solution**.

To prepare a 1.00 mol dm^{-3} solution of sodium hydroxide dissolve 40.00 g of solid sodium hydroxide in distilled water and then make the total volume up to 1.00 dm^3.

Concentration is often represented by square brackets, e.g.

[NaOH(aq)] = 1.00 mol dm^{-3}

A 25.0 cm^3 sample of this solution contains

$1.00 \times \frac{25.0}{1000} = 2.50 \times 10^{-2}$ mol of NaOH

CHANGING THE VARIABLES FOR A FIXED MASS OF GAS

$P \propto \frac{1}{V}$ (or PV = constant)

At constant temperature: as the volume decreases the concentration of the particles increases, resulting in more collisions with the container walls. This increase in pressure is inversely proportional to the volume, i.e. doubling the pressure halves the volume.

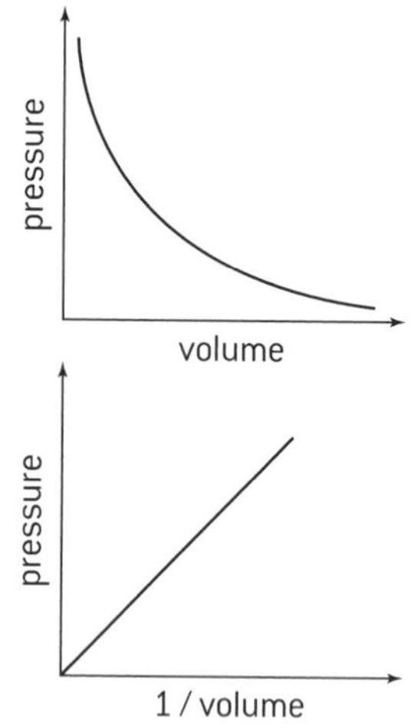

$P \propto T$ (or $\frac{P}{T}$ = constant)

At constant volume: increasing the temperature increases the average kinetic energy so the force with which the particles collide with the container walls increases. Hence pressure increases and is directly proportional to the absolute temperature, i.e. doubling the absolute temperature doubles the pressure.

IDEAL GAS EQUATION

The different variables for a gas are all related by the ideal gas equation.

$PV = nRT$

P = pressure in Pa (N m^{-2})
(1 atm = 1.013×10^5 Pa)

T = absolute temperature in K

V = volume in m^3
(1 cm^3 = 1×10^{-6} m^3)

n = number of moles

R = gas constant = 8.314 J K^{-1} mol^{-1}

V $\propto T$ (or $\frac{V}{T}$ = constant)

At constant pressure: at higher temperatures the particles have a greater average velocity so individual particles will collide with the container walls with greater force. To keep the pressure constant there must be fewer collisions per unit area so the volume of the gas must increase. The increase in volume is directly proportional to the absolute temperature, i.e. doubling the absolute temperature doubles the volume.

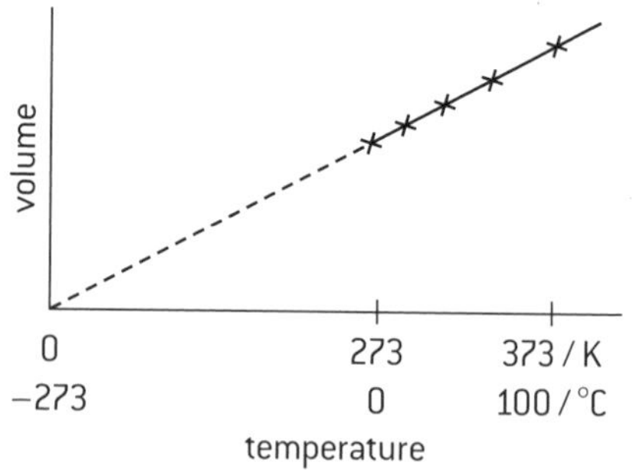

Extrapolating the graph to zero volume gives the value for absolute zero.

UNITS

The gas constant can be expressed in different units but it is easier to use SI units.

$$R = \frac{PV}{nT} = \frac{\text{N m}^{-2} \times \text{m}^3}{\text{mol} \times \text{K}} = \text{N m mol}^{-1}\text{ K}^{-1}$$

$$= \text{J K}^{-1}\text{ mol}^{-1}$$

REAL GASES

An ideal gas exactly obeys the gas laws. Real gases do have some attractive forces between the particles and the particles themselves do occupy some space so they do not exactly obey the laws. If they did they could never condense into liquids. A gas behaves most like an ideal gas at high temperatures and low pressures.

Molar volume of a gas and calculations

MOLAR VOLUME OF A GAS

The ideal gas equation depends on the amount of gas (number of moles of gas) but not on the nature of the gas. Avogadro's Law states that equal volumes of different gases at the same temperature and pressure contain the same number of moles. From this it follows that one mole of any gas will occupy the same volume at the same temperature and pressure. This is known as the molar volume of a gas. At 273 K and 1.00×10^5 Pa pressure this volume is 22.7×10^{-2} m³ (22.7 dm³ or 22 700 cm³).

When the mass of a particular gas is fixed (nR is constant) a useful expression to convert the pressure, temperature and volume under one set of conditions (1) to another set of conditions (2) is:

$$\frac{P_1V_1}{T_1} = \frac{P_2V_2}{T_2}$$

In this expression there is no need to convert to SI units as long as the same units for pressure and volume are used on both sides of the equation. However do not forget that T refers to the absolute temperature and must be in kelvin.

CALCULATIONS FROM EQUATIONS

Work methodically.

Step 1. Write down the correct formulas for all the reactants and products.

Step 2. Balance the equation to obtain the correct stoichiometry of the reaction.

Step 3. If the amounts of all reactants are known work out which are in **excess** and which one is the limiting reagent. By knowing the **limiting reagent** the maximum **yield** of any of the products can be determined.

Step 4. Work out the amount (in mol) of the substance required.

Step 5. Convert the amount (in mol) into the mass or volume.

Step 6. Express the answer to the correct number of significant figures and include the appropriate units.

WORKED EXAMPLES

(a) Calculate the volume of hydrogen gas evolved at 273 K and 1.00×10^5 Pa when 0.623 g of magnesium reacts with 27.3 cm³ of 1.25 mol dm⁻³ hydrochloric acid.

Equation:

$$Mg(s) + 2HCl(aq) \rightarrow H_2(g) + MgCl_2(aq)$$

A_r for Mg = 24.31. Amount of Mg present $= \frac{0.623}{24.31} = 2.56 \times 10^{-2}$ mol

Amount of HCl present $= 1.25 \times \frac{27.3}{1000} = 3.41 \times 10^{-2}$ mol

From the equation $2 \times 2.56 \times 10^{-2} = 5.12 \times 10^{-2}$ mol of HCl would be required to react with all of the magnesium.

Therefore the magnesium is in excess and the limiting reagent is the hydrochloric acid.

The maximum amount of hydrogen produced $= \frac{3.41 \times 10^{-2}}{2} = 1.705 \times 10^{-2}$ mol

Volume of hydrogen at 273 K, 1.00×10^5 Pa $= 1.705 \times 10^{-2} \times 22.7 = 0.387$ dm³ (or 387 cm³)

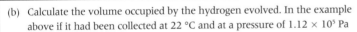

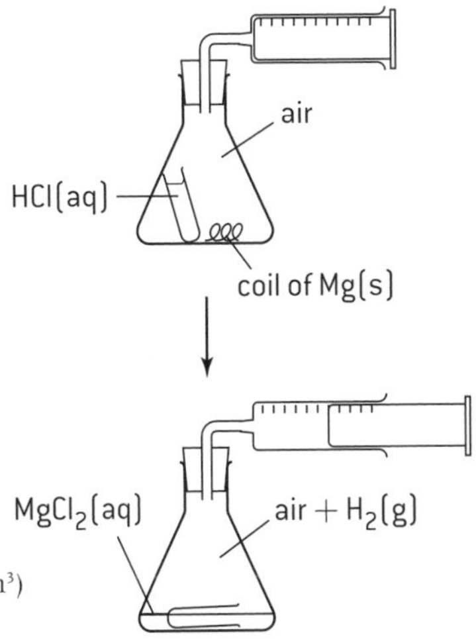

(b) Calculate the volume occupied by the hydrogen evolved. In the example above if it had been collected at 22 °C and at a pressure of 1.12×10^5 Pa

Step 1. Express the temperature as an absolute temperature
22 °C = 295 K

Step 2. Apply the ideal gas equation $pV = nRT$

$$1.12 \times 10^5 \times V = 1.705 \times 10^{-2} \times 8.314 \times 295$$

$$V = \frac{1.705 \times 10^{-2} \times 8.314 \times 295}{1.12 \times 10^5} = 3.73 \times 10^{-4}\ \text{m}^3\ (373\ \text{cm}^3)$$

This could also be solved using $\frac{P_1V_1}{T_1} = \frac{P_2V_2}{T_2}$

$$V_2 = V_1 \times \frac{P_1}{P_2} \times \frac{T_2}{T_1} = 0.387 \times \frac{1.00 \times 10^{-5}}{1.12 \times 10^{-5}} \times \frac{295}{273}$$

$$= 0.373\ \text{cm}^3\ (373\ \text{cm}^3)$$

(c) The actual volume of hydrogen collected under the conditions stated in (a) was 342 cm³. Determine the percentage yield.

Step 1. Use the mole ratio from the equation and the amounts of reactants to determine the limiting reagent and hence the theoretical maximum yield. From part (a) theoretical yield = 387 cm³

Step 2. Apply the relationship:

$$\text{Percentage yield} = \frac{\text{Experimental yield}}{\text{Theoretical yield}} \times 100$$

$$\text{Percentage yield} = \frac{342}{387} \times 100 = 88.4\%$$

Titration and atom economy

DETERMINING AN UNKNOWN CONCENTRATION BY TITRATION

Titration is a useful technique to find the concentration of a solution of unknown concentration by reacting it with a stoichiometric amount of a standard solution. A known accurate volume of one of the solutions is placed in a conical flask using a pipette. A burette is then used to add the other solution dropwise until the reaction is complete. This can be seen when one drop causes the solution to just change colour. For acid–base titrations, it is usual to add an indicator but this is not always necessary for some other types of titration, e.g. redox titrations using acidified potassium permanganate, as the reactant itself causes the colour change.

It is usual to obtain at least two accurate readings, which should be within 0.15 cm^3 of each other.

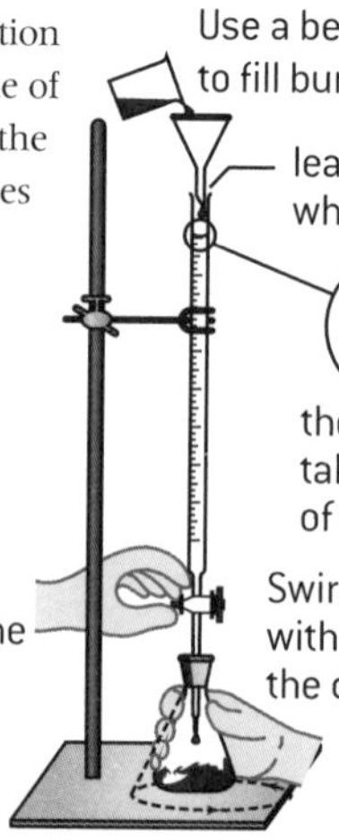

Use a beaker and funnel to fill burette.
leave air gap when filling
the burette reading is taken from the bottom of the meniscus
Use left hand to control the flow rate.
Swirl the flask with right hand while the drops are being added.

Worked examples

1. 25.00 cm^3 of a solution of sodium hydroxide of unknown concentration required 23.65 cm^3 of 0.100 mol dm^{-3} hydrochloric acid solution for complete neutralization. Calculate the concentration of the sodium hydroxide solution.

   Equation for the reaction: $NaOH(aq) + HCl(aq) \rightarrow NaCl(aq) + H_2O(l)$

   Amount of hydrochloric acid present in 23.65 cm^3 = $\left(\frac{23.65}{1000}\right) \times 0.100 = 2.365 \times 10^{-3}$ mol

   Since one mol of NaOH reacts with one mol of HCl

   Amount of sodium hydroxide present in 25.00 cm^3 = 2.365×10^{-3} mol

   Concentration of sodium hydroxide = $2.365 \times 10^{-3} \times \left(\frac{1000}{25.00}\right) = 0.0946$ mol dm^{-3}

2. 50.0 cm^3 of 1.00 mol dm^{-3} hydrochloric acid solution, HCl(aq) was added to some egg shell with a mass of 2.016 g. After all the egg shell had reacted the resulting solution was put into a 100 cm^3 volumetric flask and the volume made up to the mark with distilled water. 10.0 cm^3 of this solution required 11.40 cm^3 of 1.00×10^{-1} mol dm^{-3} sodium hydroxide solution, NaOH(aq) for complete neutralization. Calculate the percentage of calcium carbonate in the egg shell.

   Titration equation: $NaOH(aq) + HCl(aq) \rightarrow NaCl(aq) + H_2O(l)$

   Amount of sodium hydroxide present in 11.40 cm^3 = $\left(\frac{11.40}{1000}\right) \times 1.00 \times 10^{-1} = 1.140 \times 10^{-3}$ mol

   Since one mol of NaOH reacts with one mol of HCl

   Amount of diluted excess hydrochloric acid in 10.0 cm^3 = 1.140×10^{-3} mol

   Amount of excess hydrochloric acid in 100 cm^3 = $(1.140 \times 10^{-3}) \times 10 = 1.140 \times 10^{-2}$ mol

   Initial amount of hydrochloric acid added to egg shell = $\left(\frac{50.0}{1000}\right) \times 1.00 = 5.00 \times 10^{-2}$ mol

   Amount of hydrochloric acid reacting with egg shell = $(5.00 \times 10^{-2}) - (1.140 \times 10^{-2}) = 3.860 \times 10^{-2}$ mol

   Equation for reaction: $CaCO_3(s) + 2HCl(aq) \rightarrow CaCl_2(aq) + CO_2(g) + H_2O(l)$

   Amount of $CaCO_3$ that reacted with the acid = $\frac{1}{2} \times 3.860 \times 10^{-2} = 1.930 \times 10^{-2}$ mol

   $M_r\ (CaCO_3) = 40.08 + 12.01 + (3 \times 16.00) = 100.09$

   Mass of $CaCO_3$ in egg shell = $100.09 \times 1.930 \times 10^{-2} = 1.932$ g

   Percentage of calcium carbonate in the egg shell = $\left(\frac{1.932}{2.016}\right) \times 100 = 95.8\%$

ATOM ECONOMY (AN EXAMPLE OF UTILIZATION)

As well as trying to achieve high yields in industrial processes, chemists try to increase the conversion efficiency of a chemical process. This is known as atom economy. Ideally in a chemical process no atom is wasted. The atom economy is a measure of the amount of starting materials that become useful products. A high atom economy means that fewer natural resources are used and less waste is created. The atom economy can be calculated by using the following steps:

1. Write the balanced equation for the reaction taking place.
2. Calculate the relative molecular mass of each product and then the total mass of each product formed assuming molar quantities. Note that this is the same as the total mass of the reactants.
3. Calculate the relative molecular mass of each desired product and then the total mass of each desired product formed assuming molar quantities.
4. Atom economy = $\frac{\text{total mass of desired product(s)}}{\text{total mass of all products}} \times 100$

For example, consider the production of iron by the reduction of iron(III) oxide using the thermite reaction.

$$2Al(s) + Fe_2O_3(s) \rightarrow 2Fe(s) + Al_2O_3(s)$$

The total mass of products formed = $2 \times 55.85 + [(2 \times 26.98) + (3 \times 16.00)] = 213.66$ g

The total amount of iron (the desired product) formed = $2 \times 55.85 = 111.70$ g

The atom economy for this reaction is $\frac{111.70}{213.66} \times 100 = 52.3\%$

Obviously if a use can also be found for all the aluminium oxide produced then the atom economy for this reaction will increase to 100%.

MULTIPLE CHOICE QUESTIONS – STOICHIOMETRIC RELATIONSHIPS

1. How many oxygen **atoms** are in 0.100 mol of $CuSO_4.5H_2O$?

   A. 5.42×10^{22} C. 2.41×10^{23}

   B. 6.02×10^{22} D. 5.42×10^{23}

2. Which is not a true statement?

   A. One mole of methane contains four moles of hydrogen atoms

   B. One mole of ^{12}C has a mass of 12.00 g

   C. One mole of hydrogen gas contains 6.02×10^{23} atoms of hydrogen

   D. One mole of methane contains 75% of carbon by mass

3. A pure compound contains 24 g of carbon, 4 g of hydrogen and 32 g of oxygen. No other elements are present. What is the empirical formula of the compound?

   A. $C_2H_4O_2$ C. CH_4O

   B. CH_2O D. CHO

4. What is the mass in grams of one molecule of ethanoic acid CH_3COOH?

   A. 0.1 C. 1×10^{-22}

   B. 3.6×10^{25} D. 60

5. What is the relative molecular mass, M_r, of carbon dioxide, CO_2?

   A. 44.01 g mol^{-1} C. 44.01 kg mol^{-1}

   B. 44.01 mol g^{-1} D. 44.01

6. Which of the following changes of state is an exothermic process?

   A. melting C. vaporizing

   B. condensing D. boiling

7. What is the empirical formula for the compound $C_6H_5(OH)_2$?

   A. C_6H_6O C. C_6H_7O

   B. $C_6H_5O_2H_2$ D. $C_6H_7O_2$

8. Phosphorus burns in oxygen to produce phosphorus pentoxide P_4O_{10}. What is the sum of the coefficients in the balanced equation?

   $\_P_4(s) + \_O_2(g) \rightarrow \_P_4O_{10}(s)$

   A. 3 C. 6

   B. 5 D. 7

9. Magnesium reacts with hydrochloric acid according to the following equation:

   $$Mg(s) + 2HCl(aq) \rightarrow MgCl_2(aq) + H_2(g)$$

   What mass of hydrogen will be obtained if 100 cm^3 of 2.00 mol dm^{-3} HCl are added to 4.86 g of magnesium?

   A. 0.2 g C. 0.8 g

   B. 0.4 g D. 2.0 g

10. Butane burns in oxygen according to the equation below.

    $$2C_4H_{10}(g) + 13O_2(g) \rightarrow 8CO_2(g) + 10H_2O(l)$$

    If 11.6 g of butane is burned in 11.6 g of oxygen which is the limiting reagent?

    A. Butane C. Neither

    B. Oxygen D. Oxygen and butane

11. Four identical containers under the same conditions are filled with gases as shown below. Which container and contents will have the highest mass?

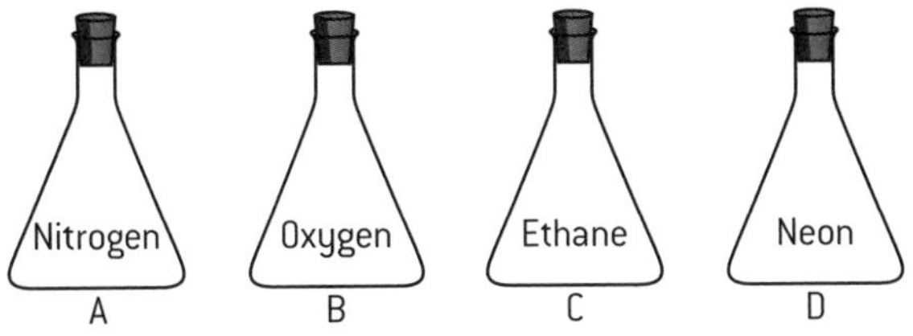

12. What is the amount, in moles, of sulfate ions in 100 cm^3 of 0.020 mol dm^{-3} $FeSO_4(aq)$?

    A. 2.0×10^{-3} C. 2.0×10^{-1}

    B. 2.0×10^{-2} D. 2.0

13. 300 cm^3 of water is added to a solution of 200 cm^3 of 0.5 mol dm^{-3} sodium chloride. What is the concentration of sodium chloride in the new solution?

    A. 0.05 mol dm^{-3} C. 0.2 mol dm^{-3}

    B. 0.1 mol dm^{-3} D. 0.3 mol dm^{-3}

14. Separate samples of two gases, each containing a pure substance, are found to have the same density under the same conditions of temperature and pressure. Which statement about these two samples must be correct?

    A. They have the same volume

    B. They have the same relative molecular mass

    C. There are equal numbers of moles of gas in the two samples

    D. They condense at the same temperature

15. The graph below represents the relationship between two variables in a fixed amount of gas.

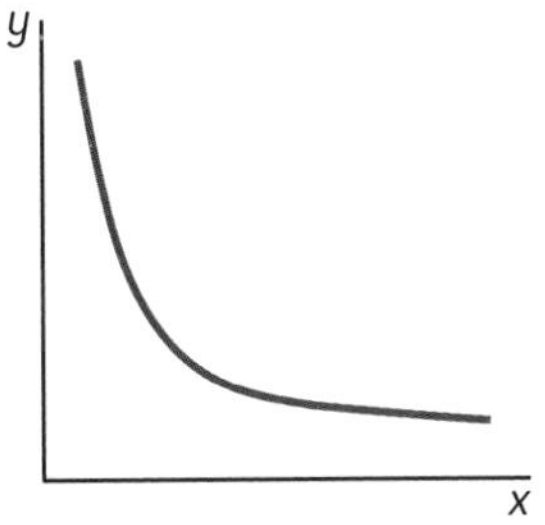

Which variables could be represented by each axis?

	x-axis	*y*-axis
A.	pressure	temperature
B.	volume	temperature
C.	pressure	volume
D.	temperature	volume

16. Sulfuric acid and sodium hydroxide react together according to the equation:

    $H_2SO_4(aq) + 2NaOH(aq) \rightarrow Na_2SO_4(aq) + 2H_2O(l)$

    What volume of 0.250 mol dm^{-3} NaOH is required to neutralize exactly 25.0 cm^3 of 0.125 mol dm^{-3} H_2SO_4?

    A. 25.0 cm^3 C. 50 cm^3

    B. 12.5 cm^3 D. 6.25 cm^3

SHORT ANSWER QUESTIONS – STOICHIOMETRIC RELATIONSHIPS

1. Aspirin, $C_9H_8O_4$, is made by reacting ethanoic anhydride, $C_4H_6O_3$ (M_r = 102.1), with 2-hydroxybenzoic acid (M_r = 138.1), according to the equation:

   $$2C_7H_6O_3 + C_4H_6O_3 \rightarrow 2C_9H_8O_4 + H_2O$$

   a) If 15.0 g 2-hydroxybenzoic acid is reacted with 15.0 g ethanoic anhydride, determine the limiting reagent in this reaction. [3]

   b) Calculate the maximum mass of aspirin that could be obtained in this reaction. [2]

   c) If the mass obtained in this experiment was 13.7 g, calculate the percentage yield of aspirin. [1]

2. 14.48 g of a metal sulfate with the formula M_2SO_4 was dissolved in water. Excess barium nitrate solution was added in order to precipitate all the sulfate ions in the form of barium sulfate. 9.336 g of precipitate was obtained.

   a) Calculate the amount of barium sulfate $BaSO_4$ precipitated. [2]

   b) Calculate the amount of sulfate ions present in the 14.48 g of M_2SO_4. [1]

   c) Deduce the relative molar mass of M_2SO_4. [1]

   d) Calculate the relative atomic mass of M and hence identify the metal. [2]

3. A student added 7.40×10^{-2} g of magnesium ribbon to 15.0 cm^3 of 2.00 mol dm^{-3} hydrochloric acid. The hydrogen gas produced was collected using a gas syringe at 20.0 °C and 1.00×10^5 Pa.

   a) State the equation for the reaction between magnesium and hydrochloric acid. [1]

   b) Determine the limiting reactant. [3]

   c) Calculate the theoretical yield of hydrogen gas:

      (i) in mol [1]

      (ii) in cm^3, under the stated conditions of temperature and pressure. [2]

   d) The actual volume of hydrogen measured was lower than the calculated theoretical volume. Suggest two reasons why the volume of hydrogen gas obtained was less. [2]

4. In 1921 Thomas Midgley discovered that the addition of a lead compound could improve the combustion of hydrocarbons in automobile (car) engines. This was the beginning of the use of leaded gasoline (petrol).

   The percentage composition, by mass, of the lead compound used by Midgley is Pb: 64.052%, C: 29.703% and H: 6.245%.

   a) (i) Determine the empirical formula of the lead compound. [3]

      (ii) Leaded gasoline has been phased out because the lead(IV) oxide, PbO_2, produced as a side product in the combustion reaction may cause brain damage in children.
      0.01 mol of Midgley's lead compound produces 0.01 mol of lead(IV) oxide. Deduce the molecular formula of Midgley's compound. [1]

      (iii) Determine the equation for the complete combustion of Midgley's compound. [2]

   b) The combustion of unleaded gasoline still produces pollution with both local and global consequences. Identify one exhaust gas that causes local pollution and one exhaust gas that causes global pollution. [2]

5. An experiment was performed to determine the percentage of iron present in a sample of iron ore. 3.682×10^{-1} g of the sample was dissolved in acid and all of the iron was converted to Fe^{2+}.

   The resulting solution was titrated with a standard solution of potassium manganate(VII), $KMnO_4$ with a concentration of 2.152×10^{-2} mol dm^{-3}. The end point was indicated when one drop caused a slight pink colour to remain. It was found that 22.50 cm^3 of the potassium manganate(VII) solution was required to reach the end point.

   In acidic solution, MnO_4^- reacts with Fe^{2+} ions to form Mn^{2+} and Fe^{3+} ions according to the following equation:

   $$MnO_4^-(aq) + 5Fe^{2+}(aq) + 8H^+(aq) \rightarrow Mn^{2+}(aq) + 5Fe^{3+}(aq) + 4H_2O(l)$$

   a) Calculate the amount (in mol) of MnO_4^- used in the titration. [2]

   b) Calculate the amount (in mol) of Fe present in the 3.682×10^{-1} g sample of iron ore. [2]

   c) Determine the percentage by mass of Fe present in the 3.682×10^{-1} g sample of iron ore. [2]

6. Copper metal may be produced by the reaction of copper(I) oxide and copper(I) sulfide according to the equation.

   $$2Cu_2O(s) + Cu_2S(s) \rightarrow 6Cu(s) + SO_2(g)$$

   A mixture of 10.0 kg of copper(I) oxide and 5.00 kg of copper(I) sulfide was heated until no further reaction occurred.

   a) Determine the limiting reagent in this reaction. [3]

   b) Calculate the maximum mass of copper that could be obtained from these masses of reactants. [2]

   c) Assuming the reaction to produce copper goes to completion according to the equation deduce the atom economy for this reaction. [3]

7. The empirical formula of magnesium oxide is MgO. Suggest four assumptions that were made in the experiment detailed on page 2 that may not be true and which might account for the wrong result being obtained. [4]

8. The percentage composition by mass of a hydrocarbon is C: 85.6% and H: 14.4%.

   a) Calculate the empirical formula of the hydrocarbon. [2]

   b) A 1.00 g sample of the hydrocarbon at a temperature of 273 K and a pressure of 1.00×10^5 Pa has a volume of 0.405 dm^3.

      (i) Calculate the molar mass of the hydrocarbon. [2]

      (ii) Deduce the molecular formula of the hydrocarbon. [2]

   c) Explain why the incomplete combustion of hydrocarbons is harmful to humans. [2]

The nuclear atom

COMPOSITION OF ATOMS

The smallest part of an element is an atom. It used to be thought that atoms are indivisible but they can be broken down into many different sub-atomic particles. All atoms, with the exception of hydrogen, are made up of three fundamental sub-atomic particles – protons, neutrons and electrons.

The hydrogen atom, the simplest atom of all, contains just one proton and one electron. The actual mass of a proton is 1.673×10^{-24} g but it is assigned a relative value of 1. The mass of a neutron is virtually identical and also has a relative mass of 1. Compared with a proton and a neutron an electron has negligible mass with a relative mass of only $\frac{1}{2000}$. Neutrons are neutral particles. An electron has a charge of 1.602×10^{-19} coulombs which is assigned a relative value of -1. A proton carries the same charge as an electron but of an opposite sign so has a relative value of $+1$. All atoms are neutral so must contain equal numbers of protons and electrons.

SUMMARY OF RELATIVE MASS AND CHARGE

Particle	Relative mass	Relative charge
proton	1	+1
neutron	1	0
electron	5×10^{-4}	−1

SIZE AND STRUCTURE OF ATOMS

Atoms have a radius in the order of 10^{-10} m. Almost all of the mass of an atom is concentrated in the nucleus which has a very small radius in the order of 10^{-14} m. All the protons and neutrons (collectively called nucleons) are located in the nucleus. The electrons are to be found in energy levels or shells surrounding the nucleus. Much of the atom is empty space.

SHORTHAND NOTATION FOR AN ATOM OR ION

$$^{A}_{Z}X^{n+/n-}$$

MASS NUMBER *A*

Equal to the number of protons and neutrons in the nucleus.

ATOMIC NUMBER *Z*

Equal to the number of protons in the nucleus and to the number of electrons in the atom. The atomic number defines which element the atom belongs to and consequently its position in the periodic table.

CHARGE

Atoms have no charge so n = 0 and this is left blank. However by losing one or more electrons atoms become positive ions, or by gaining one or more electrons atoms form negative ions.

EXAMPLES

Symbol	Atomic number	Mass number	Number of protons	Number of neutrons	Number of electrons
$^{9}_{4}Be$	4	9	4	5	4
$^{40}_{20}Ca^{2+}$	20	40	20	20	18
$^{37}_{17}Cl^{-}$	17	37	17	20	18

ISOTOPES

All atoms of the same element must contain the same number of protons, however they may contain a different number of neutrons. Such atoms are known as isotopes. Chemical properties are related to the number of electrons so isotopes of the same element have identical chemical properties. Since their mass is different their physical properties such as density and boiling point are different.

Examples of isotopes: $^{1}_{1}H$ $^{2}_{1}H$ $^{3}_{1}H$ $^{12}_{6}C$ $^{14}_{6}C$ $^{35}_{17}Cl$ $^{37}_{17}Cl$

RELATIVE ATOMIC MASS

The two isotopes of chlorine occur in the ratio of 3:1. That is, naturally occurring chlorine contains 75% $^{35}_{17}Cl$ and 25% $^{37}_{17}Cl$. The weighted mean molar mass is thus:

$$\frac{(75 \times 35) + (25 \times 37)}{100} = 35.5 \text{ g mol}^{-1}$$

and the relative atomic mass is 35.5. Accurate values to 2 d.p. for all the relative atomic masses of the elements are given in Section 6 of the IB data booklet. These are the values that must be used when performing calculations in the examinations.

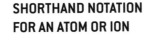

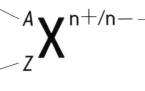

Mass spectrometer and relative atomic mass

MASS SPECTROMETER

Relative atomic masses can be determined using a mass spectrometer. A vaporized sample is injected into the instrument. Atoms of the element are ionized by being bombarded with a stream of high energy electrons in the ionization chamber. In practice the instrument is set so that only ions with a single positive charge are formed. The resulting unipositive ions pass through holes in parallel plates under the influence of an electric field where they are accelerated. The ions are then deflected by an external magnetic field.

The amount of deflection depends both on the mass of the ion and its charge. The smaller the mass and the higher the charge the greater the deflection. Ions with a particular mass/charge ratio are then recorded on a detector which measures both the mass and the relative amounts of all the ions present.

THE MASS SPECTRUM OF NATURALLY OCCURRING LEAD

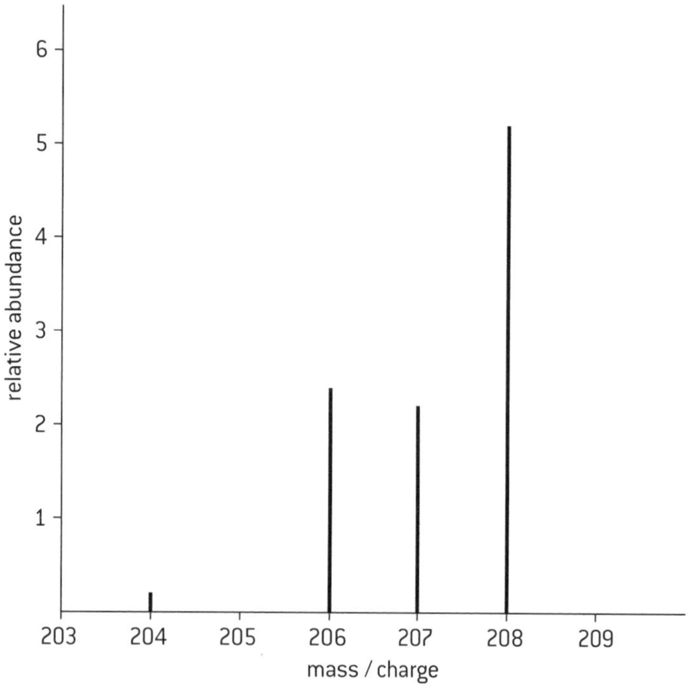

The relative atomic mass of lead can be calculated from the weighted average:

Isotopic mass	Relative abundance	% relative abundance
204	0.2	2
206	2.4	24
207	2.2	22
208	5.2	52

$$\text{relative atomic mass} = \frac{(2 \times 204) + (24 \times 206) + (22 \times 207) + (52 \times 208)}{100} = 207.2$$

USES OF RADIOACTIVE ISOTOPES

Isotopes have many uses in chemistry and beyond. Many, but by no means all, isotopes of elements are radioactive as the nuclei of these atoms break down spontaneously. When they break down these radioisotopes emit radiation which is dangerous to living things. There are three different forms of radiation. Gamma (γ) radiation is highly penetrating whereas alpha (α) radiation can be stopped by a few centimetres of air and beta (β) radiation by a thin sheet of aluminium. Radioisotopes can occur naturally or be created artificially. Their uses include nuclear power generation, the sterilization of surgical instruments in hospitals, crime detection, finding cracks and stresses in metals and the preservation of food. ${}^{14}_{6}C$ is used for carbon dating, ${}^{60}_{27}Co$ is used in radiotherapy and ${}^{131}_{53}I$ and ${}^{125}_{53}I$ are used as tracers in medicine for treating and diagnosing illness.

Emission spectra

THE ELECTROMAGNETIC SPECTRUM

Electromagnetic waves can travel through space and, depending on the wavelength, also through matter. The velocity of travel c is related to its wavelength λ and its frequency v. Velocity is measured in m s^{-1}, wavelength in m and frequency in s^{-1} so it is easy to remember the relationship between them:

$$c = \lambda \times v$$
$$(\text{m s}^{-1}) \quad (\text{m}) \quad (\text{s}^{-1})$$

Electromagnetic radiation is a form of energy. The smaller the wavelength and thus the higher the frequency the more energy the wave possesses. Electromagnetic waves have a wide range of wavelengths ranging from low energy radio waves to high energy γ-radiation. Visible light occupies a very narrow part of the spectrum.

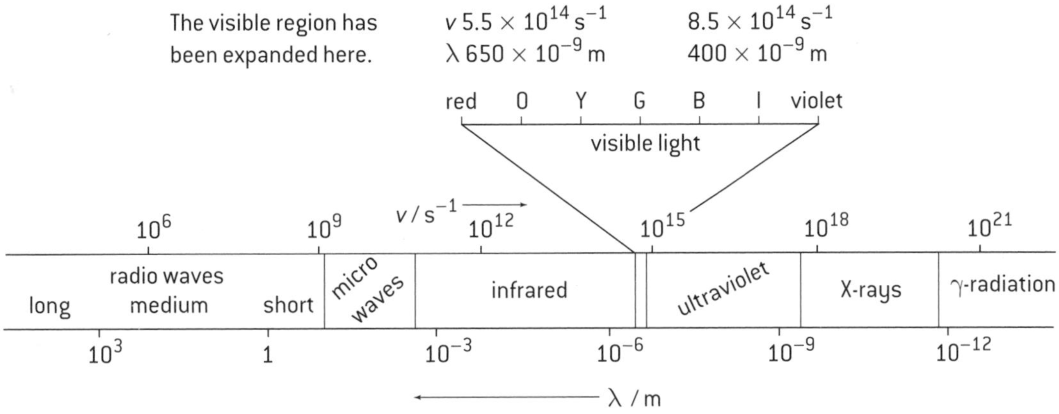

ATOMIC EMISSION SPECTRA

White light is made up of all the colours of the spectrum. When it is passed through a prism a **continuous spectrum** of all the colours can be obtained.

When energy is supplied to individual elements they emit a spectrum which only contains emissions at particular wavelengths. Each element has its own characteristic spectrum known as a **line spectrum** as it is not continuous.

The visible hydrogen spectrum

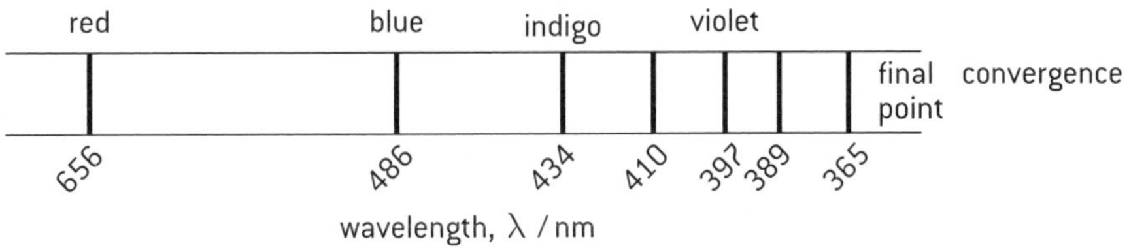

Note that the spectrum consists of discrete lines and that the lines converge towards the high energy (violet) end of the spectrum. A similar series of lines at even higher energy also occurs in the ultraviolet region of the spectrum and several other series of lines at lower energy can be found in the infrared region of the spectrum.

EXPLANATION OF EMISSION SPECTRA

When energy is supplied to an atom electrons are excited (gain energy) from their lowest (ground) state to an excited state. Electrons can only exist in certain fixed energy levels. When electrons drop from a higher level to a lower level they emit energy. This energy corresponds to a particular wavelength and shows up as a line in the spectrum. When electrons return to the first level ($n = 1$) the series of lines occurs in the ultraviolet region as this involves the largest energy change. The visible region spectrum is formed by electrons dropping back to the $n = 2$ level and the first series in the infrared is due to electrons falling to the $n = 3$ level. The lines in the spectrum converge because the energy levels themselves converge.

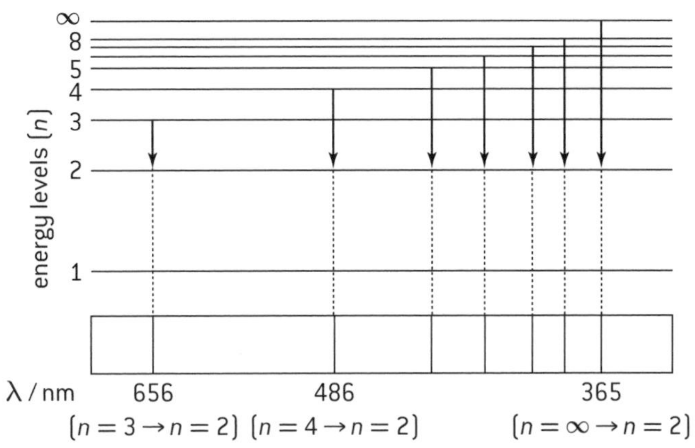

Electronic configuration

TYPES OF ORBITAL

Electrons are found in orbitals. Each orbital can contain a maximum of two electrons each with opposite spins. The first level ($n = 1$) contains just one orbital, called an s orbital. The second level ($n = 2$) contains one s orbital and three p orbitals. The 2p orbitals are all of equal energy but the sub-level made up of these three 2p orbitals is slightly higher in energy than the 2s orbital.

Principal level (shell)	**Number of each type of orbital**				**Maximum number of electrons in level ($=2n^2$)**
n	s	p	d	f	
1	1	–	–	–	2
2	1	3	–	–	8
3	1	3	5	–	18
4	1	3	5	7	32

Relative energies of sub-levels within an atom

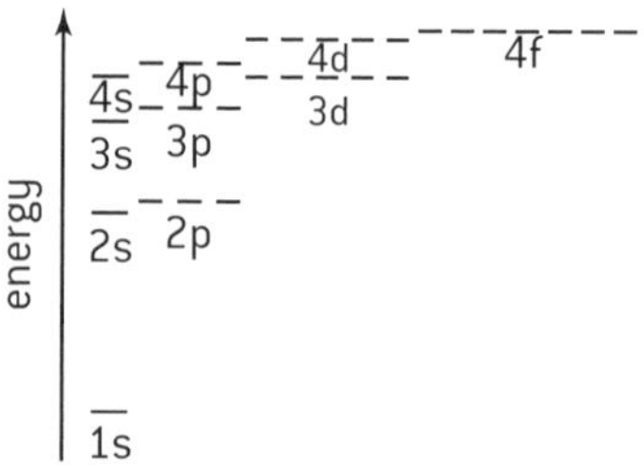

The relative position of all the sub-levels for the first four main energy levels is shown.

Note that the 4s sub-level is below the 3d sub-level. This explains why the third level is sometimes stated to hold 8 or 18 electrons.

ELECTRONIC CONFIGURATION AND AUFBAU PRINCIPLE

The electronic configuration can be determined by following the aufbau (building up) principle. The orbitals with the lowest energy are filled first. Each orbital can contain a maximum of two electrons. Orbitals within the same sub-shell are filled singly first – this is known as Hund's rule,

e.g.

H $1s^1$	Li $1s^22s^1$	Na $1s^22s^22p^63s^1$	K $1s^22s^22p^63s^23p^64s^1$
He $1s^2$	Be $1s^22s^2$	Mg $1s^22s^22p^63s^2$	Ca $1s^22s^22p^63s^23p^64s^2$
	B $1s^22s^22p^1$	Al $1s^22s^22p^63s^23p^1$	Sc $1s^22s^22p^63s^23p^64s^23d^1$
	C $1s^22s^22p^2$	Si $1s^22s^22p^63s^23p^2$	Ti $1s^22s^22p^63s^23p^64s^23d^2$
	N $1s^22s^22p^3$	P $1s^22s^22p^63s^23p^3$	V $1s^22s^22p^63s^23p^64s^23d^3$
	O $1s^22s^22p^4$	S $1s^22s^22p^63s^23p^4$	
	F $1s^22s^22p^5$	Cl $1s^22s^22p^63s^23p^5$	
	Ne $1s^22s^22p^6$	Ar $1s^22s^22p^63s^23p^6$	

To save writing out all the lower levels the configuration may be shortened by building on the last noble gas configuration. For example, continuing on from titanium, vanadium can be written [Ar]$4s^23d^3$, then Cr [Ar]$4s^13d^5$, Mn [Ar]$4s^23d^5$, Fe [Ar]$4s^23d^6$ etc.

Note that the electron configurations of the transition metals show two irregularities. When there is the possibility of the d sub-level becoming half-full or completely full it takes precedence over completely filling the 4s level first so chromium has the configuration [Ar]$4s^13d^5$ (rather than [Ar]$4s^23d^4$) and copper has the configuration [Ar]$4s^13d^{10}$ (rather than [Ar]$4s^23d^9$). When transition metals form ions the 4s electrons are removed first so Fe^{2+} has the configuration [Ar]$3d^6$. The IB requires that you can write the configuration for any element or ion up to krypton ($Z = 36$). The full electronic configuration for krypton is $1s^22s^22p^63s^23p^64s^23d^{10}4p^6$ or it can be shortened to [Ar]$4s^23d^{10}4p^6$.

(When writing electronic configurations check that for a neutral atom the sum of the superscripts adds up to the atomic number of the element.)

Sometimes boxes are used to represent orbitals so the number of unpaired electrons can easily be seen, e.g.

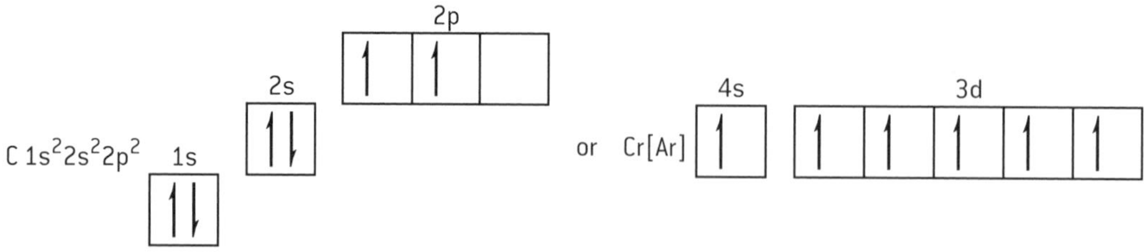

SHAPES OF ORBITALS

An electron has the properties of both a particle and a wave. Heisenberg's uncertainty principle states that it is impossible to know the exact position of an electron at a precise moment in time. An orbital describes the three-dimensional shape where there is a high probability that the electron will be located.

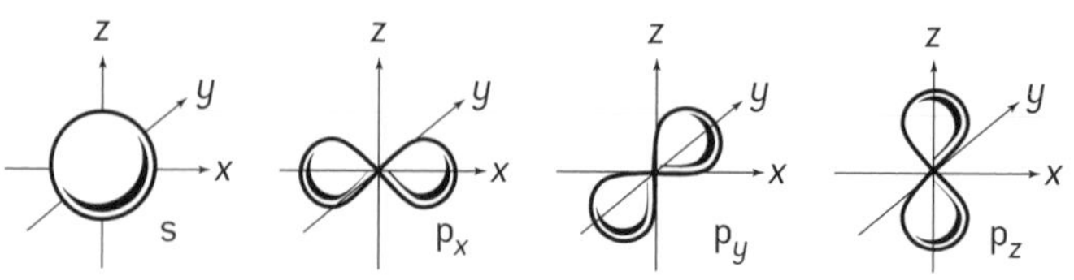

s orbitals are spherical and the three p orbitals are orthogonal (at right angles) to each other.

Evidence from ionization energies

EVIDENCE FROM IONIZATION ENERGIES

The first ionization energy of an element is defined as the energy required to remove one electron from an atom in its gaseous state. It is measured in kJ mol^{-1}.

$$X(g) \rightarrow X^{+}(g) + e^{-}$$

A graph of first ionization energies plotted against atomic number shows a repeating pattern.

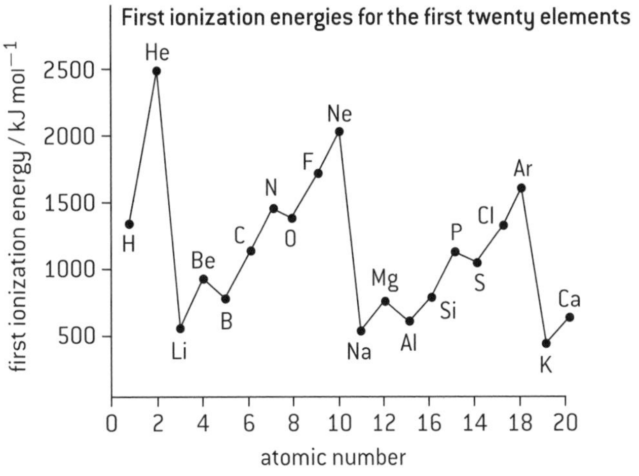

It can be seen that the highest value is for helium, an atom that contains two protons and two electrons. The two electrons are in the lowest level and are held tightly by the two protons. For lithium it is relatively easy to remove an electron, which suggests that the third electron in lithium is in a higher energy level than the first two. The value then generally increases until element 10, neon, is reached before it drops sharply for sodium. This graph provides evidence that the levels can contain different numbers of electrons before they become full.

Electrons with opposite spins tend to repel each other. When orbitals of the same energy (degenerate) are filled the electrons will go singly into each orbital first before they pair up to minimize repulsion. This explains why there is a regular increase in the first ionization energies going from B to N as the three 2p orbitals each gain one electron. Then there is a slight decrease between N and O as one of the 2p orbitals gains a second electron before a regular increase again.

EVIDENCE FOR SUB-LEVELS

The graph already shown above was for the first ionization energy for the first 20 elements. Successive ionization energies for the same element can also be measured, e.g. the second ionization energy is given by:

$$X^{+}(g) \rightarrow X^{2+}(g) + e^{-}$$

As more electrons are removed the pull of the protons holds the remaining electrons more tightly so increasingly more energy is required to remove them, hence a logarithmic scale is usually used. A graph of the successive ionization energies for potassium also provides evidence of the number of electrons in each main level.

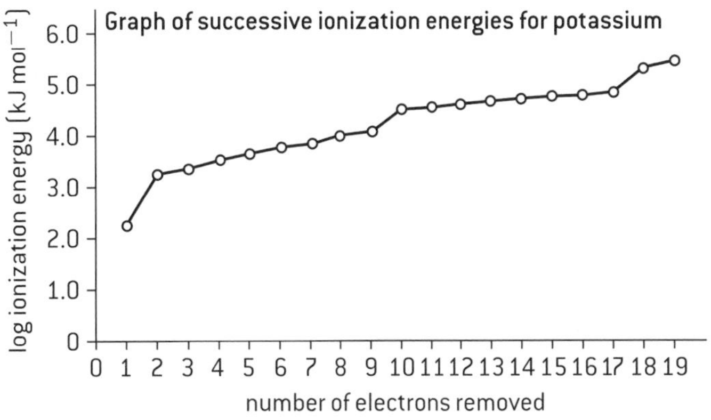

By looking to see where the first 'large jump' occurs in successive ionization energies one can determine the number of valence electrons (and hence the group in the periodic table to which the element belongs).

If the graph for first ionization energies is examined more closely then it can be seen that the graph does not increase regularly. This provides evidence that the main levels are split into sub-levels.

IONIZATION ENERGIES FROM EMISSION SPECTRA

It can be seen from the emission spectrum of hydrogen that the energy levels converge. Hydrogen contains just one electron, which will be in the lowest energy level in its ground state. If sufficient energy is supplied it can be promoted to the infinite level – that is it has been removed from the atom and the atom has become ionized to form the H^{+} ion. This amount of energy corresponds to the energy it would emit if it fell back from $n = \infty$ to $n = 1$ which produces a line in the ultraviolet region of the spectrum at a wavelength of 91.2 nm. We can use this value to calculate the energy involved. Wavelength and frequency are related by the expression $c = \lambda v$ where c is the velocity of light. Energy and frequency are related by the expression $E = hv$ where h is Planck's constant and has the value 6.63×10^{-34} J s.

The energy to remove one electron $= hv = \dfrac{hc}{\lambda} = \dfrac{6.63 \times 10^{-34}\ \text{J s} \times 3.00 \times 10^{8}\ \text{m s}^{-1}}{91.2 \times 10^{-9}\ \text{m}} = 2.18 \times 10^{-18}$ J

For one mole of electrons we need to multiply by Avogadro's constant (6.02×10^{23}) to give 1.312×10^{6} J mol^{-1} or 1312 kJ mol^{-1}. This value is exactly the same as the experimentally determined value for the first ionization energy of hydrogen.

MULTIPLE CHOICE QUESTIONS – ATOMIC STRUCTURE

1. Which of the following particles contain more neutrons than electrons?

   I. ${}^{1}_{1}H^{+}$ II. ${}^{79}_{35}Br^{-}$ III. ${}^{23}_{11}Na^{+}$

   A. I and II only C. II and III only

   B. I and III only D. I, II and III

2. Which one of the following sets represents a pair of isotopes?

   A. ${}^{31}_{15}P$ and ${}^{32}_{15}P$ C. Diamond and C_{60}

   B. ${}^{24}_{12}Mg$ and ${}^{24}_{12}Mg^{2+}$ D. ${}^{40}_{18}Ar$ and ${}^{40}_{20}Ca$

3. Which species contains 16 protons, 17 neutrons and 18 electrons?

   A. ${}^{32}S^{-}$ C. ${}^{34}S^{-}$

   B. ${}^{33}S^{2-}$ D. ${}^{35}S^{2-}$

4. Which quantities are the same for all atoms of chlorine?

   I. Number of protons

   II. Number of electrons

   III. Number of neutrons

   A. I and II only C. II and III only

   B. I and III only D. I, II and III

5. A sample of zinc has the following composition:

Isotope	% abundance
${}^{64}Zn$	55
${}^{66}Zn$	40
${}^{68}Zn$	5

   What is the relative atomic mass of the zinc in this sample?

   A. 64.5 C. 65.9

   B. 65.0 D. 66.4

6. In the electromagnetic spectrum, which will have the shortest wavelength **and** the greatest energy?

	Shortest wavelength	Greatest energy
A.	ultraviolet	ultraviolet
B.	infrared	infrared
C.	ultraviolet	infrared
D.	infrared	ultraviolet

7. Which electronic transition in a hydrogen atom releases the most energy?

   A. $n = 1 \rightarrow n = 2$

   B. $n = 7 \rightarrow n = 6$

   C. $n = 6 \rightarrow n = 7$

   D. $n = 2 \rightarrow n = 1$

8. Which shows the sub-levels in order of **increasing** energy in the fourth energy level of an atom?

   A. f < d < p < s

   B. p < d < f < s

   C. d < f < p < s

   D. s < p < d < f

9. What is the electron configuration of copper?

   A. $[Ar]4s^{2}3d^{9}$ C. $1s^{2}2s^{2}2p^{6}3s^{2}3p^{6}4s^{1}3d^{10}$

   B. $1s^{2}2s^{2}2p^{6}3s^{2}3p^{6}3d^{10}$ D. $[Ar]3d^{9}$

10. How many unpaired electrons are present in an atom of sulfur in its ground state?

    A. 1 C. 4

    B. 2 D. 6

HL

11. Which is the correct definition for the second ionization energy of carbon?

    A. $C(s) \rightarrow C(s) + 2e^{-}$ C. $C^{+}(s) \rightarrow C^{2+}(s) + 2e^{-}$

    B. $C(g) \rightarrow C(g) + 2e^{-}$ D. $C^{+}(g) \rightarrow C^{2+}(g) + 2e^{-}$

12. The first five ionization energies, in kJ mol^{-1}, for a certain element are 577, 1980, 2960, 6190 and 8700 respectively. Which group in the periodic table does this element belong to?

    A. 1 C. 3

    B. 2 D. 4

13. Which transition in the hydrogen emission spectrum corresponds to the first ionization energy of hydrogen?

    A. $n = \infty \rightarrow n = 1$ C. $n = \infty \rightarrow n = 2$

    B. $n = 2 \rightarrow n = 1$ D. $n = 4 \rightarrow n = 2$

14. Which ionization requires the most energy?

    A. $B(g) \rightarrow B^{+}(g) + e^{-}$ C. $N(g) \rightarrow N^{+}(g) + e^{-}$

    B. $C(g) \rightarrow C^{+}(g) + e^{-}$ D. $O(g) \rightarrow O^{+}(g) + e^{-}$

15. All of the following factors affect the value of the ionization energy of an atom **except** the:

    A. mass of the atom

    B. charge on the nucleus

    C. size of the atom

    D. main energy level from which the electron is removed

16. The graph below shows the first four ionization energies of four elements A, B, C and D (the letters are not their chemical symbols). Which element is magnesium?

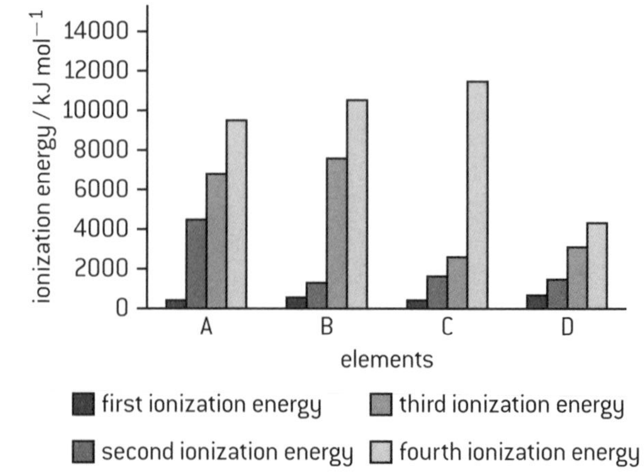

SHORT ANSWER QUESTIONS – ATOMIC STRUCTURE

1. a) Define the term relative atomic mass, A_r. [1]

   b) The relative atomic mass of naturally occurring chlorine is 35.45. Calculate the abundances of ^{35}Cl and ^{37}Cl in naturally occurring chlorine. [2]

   c) (i) State the electron configuration of chlorine. [1]

   (ii) State the electron configuration for a chloride ion, Cl^-. [1]

   d) Explain how ^{35}Cl and ^{37}Cl differ in their chemical properties. [2]

2. a) Explain why the relative atomic mass of cobalt is greater than the relative atomic mass of nickel, even though the atomic number of nickel is greater than the atomic number of cobalt. [1]

   b) Deduce the numbers of protons and electrons in the Co^{2+} ion. [1]

   c) (i) Deduce the electron configuration of the Co atom. [1]

   (ii) Deduce the electron configuration of the Co^{2+} ion. [1]

3. $^{99}_{43}Tc$ is a radioactive isotope of technetium.

   a) (i) Define the term isotope. [1]

   (ii) Determine the number of neutrons in one atom of technetium-99. [1]

   (iii) Technetium-99 is used as a tracer in medicine. Suggest a reason why it is potentially dangerous. [2]

   b) Carbon in living organisms consists of two isotopes, ^{12}C and ^{14}C, in a fixed ratio. This ratio remains constant in a living organism as the carbon is constantly being replaced through photosynthesis. Once an organism dies the ^{14}C slowly decays to ^{14}N with a half-life of 5300 years.

   (i) Identify the number of protons, neutrons and electrons in carbon-12 and in carbon-14. [2]

   (ii) Suggest the identity of the particle that is emitted when an atom of ^{14}C is converted into an atom of ^{14}N. [1]

   (iii) Discuss how the decay of carbon-14 can be used in carbon dating. [2]

4. Annotate the 2s and 2p boxes, using ↑ or ↓ to represent a spinning electron, to complete the electron configuration for an oxygen atom. [1]

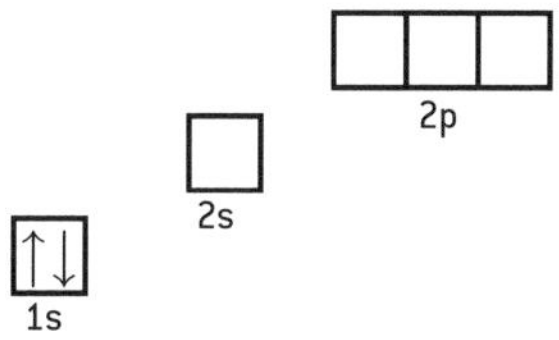

5. Draw and label an energy level diagram for the hydrogen atom. In your diagram show how the series of lines in the ultraviolet and visible regions of its emission spectrum are produced, clearly labelling each series. [4]

6. The electron configuration of chromium can be expressed as $[Ar]4s^x3d^y$.

   (i) Explain what the square brackets around argon, [Ar], represent. [1]

   (ii) State the values of x and y. [1]

   (iii) Annotate the diagram below showing the 4s and 3d orbitals for a chromium atom using an arrow, ↑ or ↓, to represent a spinning electron. [1]

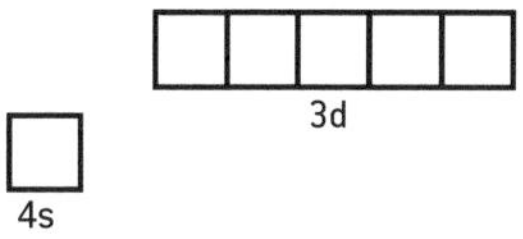

HL

7. The graph below shows the first ionization energy plotted against atomic number for the first 20 elements.

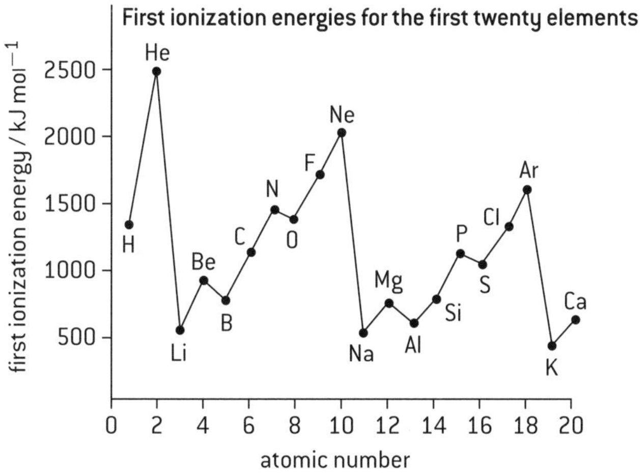

   a) Define the term *first ionization energy*. [2]

   b) Explain the following:

   (i) why there is a general increase in the value for the first ionization energy across period 2 from Li to Ne [2]

   (ii) why the first ionization energy of neon is higher than that of sodium [2]

   (iii) why the first ionization energy of beryllium is higher than that of boron [2]

   (iv) why the first ionization energy of sulfur is lower than that of phosphorus. [2]

   c) Predict how the graph for the second ionization energy plotted against atomic number for the first 20 elements differs from the graph shown above. [3]

8. Electrons are much too small to ever be 'seen'. Discuss the evidence that electrons exist in fixed energy levels and that these levels can be split into sub-levels. [5]

9. The first ionization energy of hydrogen is 1312 kJ mol^{-1}. Determine the frequency and wavelength of the convergence line in the ultraviolet emission spectrum of hydrogen. (Use information given in Sections 1 and 2 of the IB data booklet.) [3]

The periodic table

THE PERIODIC TABLE

group number: 1–18 (s-block: groups 1–2; d-block: groups 3–12; p-block: groups 13–18)

Key: atomic number / **element**

period number (n)	1	2	3	4	5	6	7	8	9	10	11	12	13	14	15	16	17	18
1	1 H																	2 He
2	3 Li	4 Be											5 B	6 C	7 N	8 O	9 F	10 Ne
3	11 Na	12 Mg											13 Al	14 Si	15 P	16 S	17 Cl	18 Ar
4	19 K	20 Ca	21 Sc	22 Ti	23 V	24 Cr	25 Mn	26 Fe	27 Co	28 Ni	29 Cu	30 Zn	31 Ga	32 Ge	33 As	34 Se	35 Br	36 Kr
5	37 Rb	38 Sr	39 Y	40 Zr	41 Nb	42 Mo	43 Tc	44 Ru	45 Rh	46 Pd	47 Ag	48 Cd	49 In	50 Sn	51 Sb	52 Te	53 I	54 Xe
6	55 Cs	56 Ba	57 * La	72 Hf	73 Ta	74 W	75 Re	76 Os	77 Ir	78 Pt	79 Au	80 Hg	81 Tl	82 Pb	83 Bi	84 Po	85 At	86 Rn
7	87 Fr	88 Ra	89 ** Ac	104 Rf	105 Db	106 Sg	107 Bh	108 Hs	109 Mt	110 Ds	111 Rg	112 Cn	113 Uut	114 Uuq	115 Uup	116 Uuh	117 Uus	118 Uuo

f-block

*	58 Ce	59 Pr	60 Nd	61 Pm	62 Sm	63 Eu	64 Gd	65 Tb	66 Dy	67 Ho	68 Er	69 Tm	70 Yb	71 Lu
**	90 Th	91 Pa	92 U	93 Np	94 Pu	95 Am	96 Cm	97 Bk	98 Cf	99 Es	100 Fm	101 Md	102 No	103 Lr

FEATURES OF THE PERIODIC TABLE

- In the periodic table elements are placed in order of increasing atomic number.
- Elements are arranged into four blocks associated with the four sub-levels – s, p, d and f.
- The period number (*n*) is the outer energy level that is occupied by electrons.
- Elements in the same vertical group contain the same number of electrons in the outer energy level.
- The number of the principal energy level and the number of the valence electrons in an atom can be deduced from its position on the periodic table.
- The periodic table shows the positions of metals, non-metals and metalloids.
- Certain groups have their own name. For example, group 1 is known as the alkali metals, group 17 as the halogens and group 18 as the noble gases.
- The d-block elements (groups 3 to 12) are known as the transition metals and the f-block elements form two distinct groups – the lanthanoids (from La to Lu) and the actinoids (from Ac to Lr).
- Metals are on the left and in the centre of the table and non-metals are on the right (distinguished by the thick line). Metalloids, such as boron, silicon and germanium have properties intermediate between those of a metal and a non-metal.

ELECTRON CONFIGURATION AND THE PERIODIC TABLE

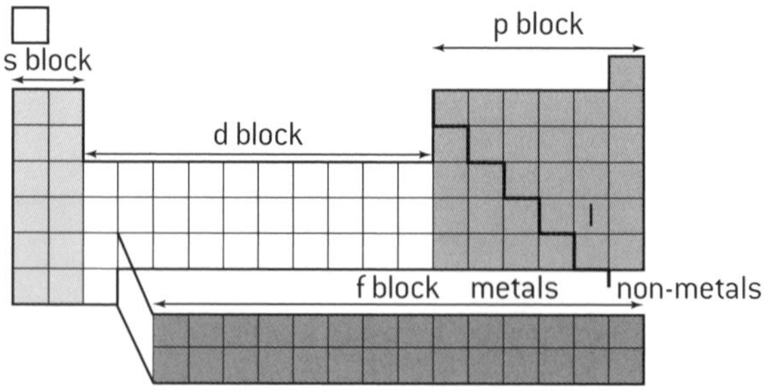

An element's position in the periodic table is related to its valence electrons so the electronic configuration of any element can be deduced from the table, e.g. iodine ($Z = 53$) is a p block element. It is in group 17 so its configuration will contain ns^2np^5. If one takes H and He as being the first period then iodine is in the fifth period so $n = 5$. The full configuration for iodine will therefore be:

$1s^22s^22p^63s^23p^64s^23d^{10}4p^65s^24d^{10}5p^5$ or [Kr] $5s^24d^{10}5p^5$

Periodic trends (1)

PERIODICITY

Elements in the same group tend to have similar chemical and physical properties. There is a change in chemical and physical properties across a period. The repeating pattern of physical and chemical properties shown by the different periods is known as **periodicity**.

These periodic trends can clearly be seen in atomic radii, ionic radii, ionization energies, electronegativities, electron affinities and melting points.

ATOMIC RADIUS

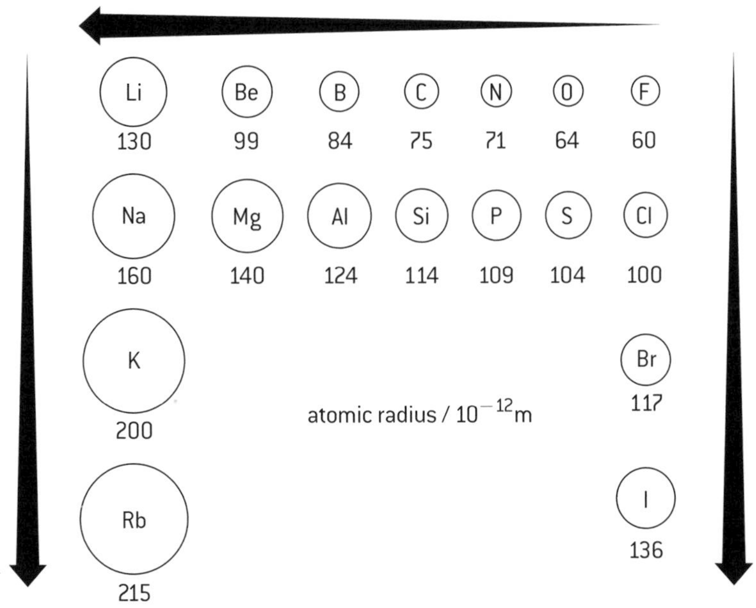

The atomic radius is the distance from the nucleus to the outermost electron. Since the position of the outermost electron can never be known precisely, the atomic radius is usually defined as half the distance between the nuclei of two bonded atoms of the same element.

As a group is descended, the outermost electron is in a higher energy level, which is further from the nucleus, so the radius increases.

Across a period electrons are being added to the same energy level, but the number of protons in the nucleus increases. This attracts the energy level closer to the nucleus and the atomic radius decreases across a period.

IONIC RADIUS

It is important to distinguish between positive ions (**cations**) and negative ions (**anions**). Both cations and anions increase in size down a group as the outer level gets further from the nucleus.

Cations contain fewer electrons than protons so the electrostatic attraction between the nucleus and the outermost electron is greater and the ion is smaller than the parent atom. It is also smaller because the number of electron shells has decreased by one. Across the period the ions contain the same number of electrons (**isoelectronic**), but an increasing number of protons, so the ionic radius decreases.

Anions contain more electrons than protons so are larger than the parent atom. Across a period the size decreases because the number of electrons remains the same but the number of protons increases.

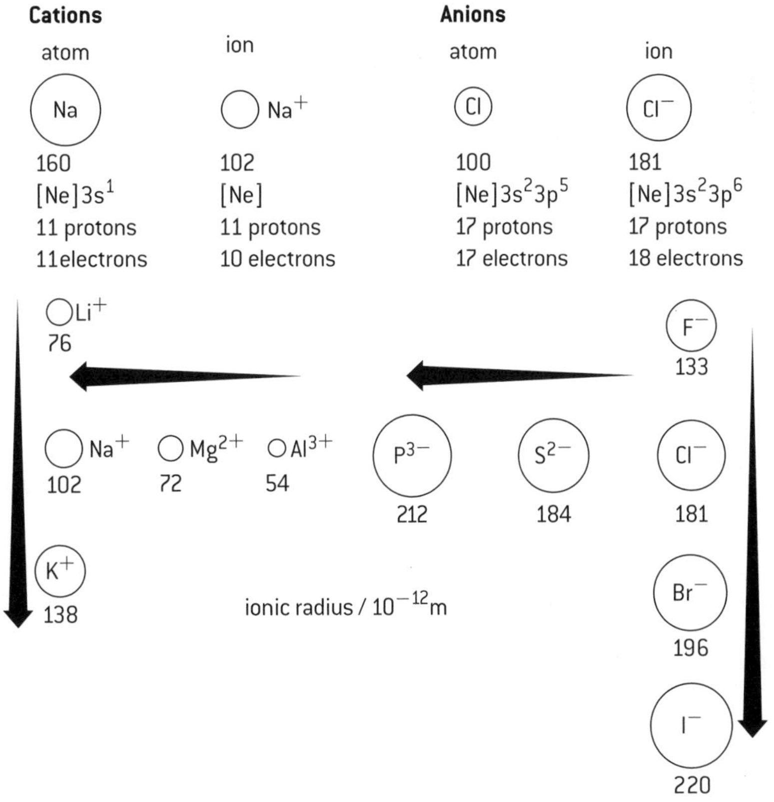

Periodic trends (2)

MELTING POINTS

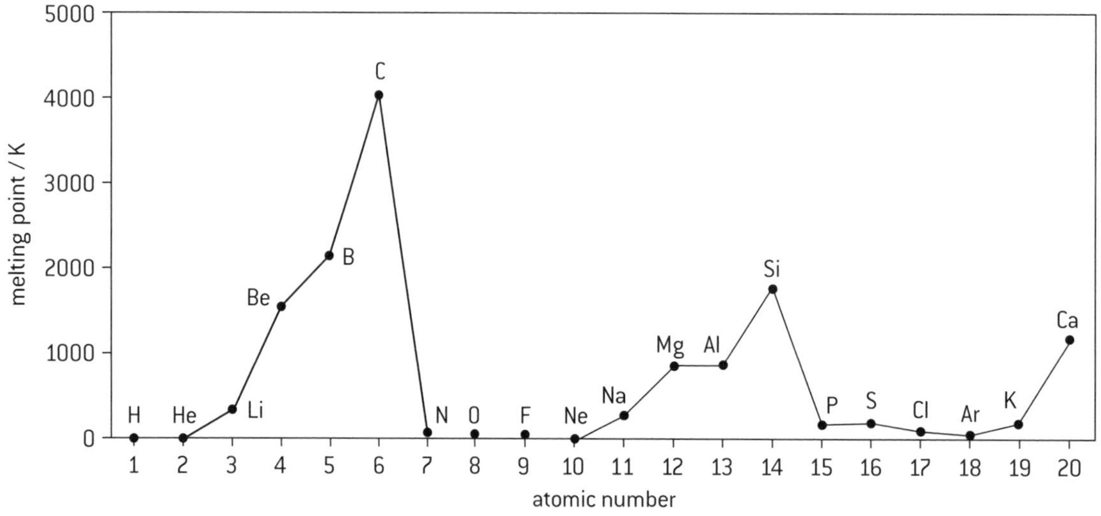

Melting points depend both on the structure of the element and on the type of attractive forces holding the atoms together. Using period 3 as an example:

- At the left of the period elements exhibit metallic bonding (Na, Mg, Al), which increases in strength as the number of valence electrons increases.
- Silicon, the metalloid, in the middle of the period has a macromolecular covalent structure with very strong bonds resulting in a very high melting point.
- Elements in groups 15, 16, and 17 (P_4, S_8, and Cl_2) show simple molecular structures with weak intermolecular forces of attraction between the molecules.
- The noble gases (Ar) exist as **monatomic molecules** (single atoms) with extremely weak forces of attraction between the atoms.

Within groups there are also clear trends:

- In group 1 the melting point decreases down the group as the atoms become larger and the strength of the metallic bond decreases.

	Li	Na	K	Rb	Cs
M. pt / K	454	371	336	312	302

- In group 17 the intermolecular attractive forces between the diatomic molecules increase down the group so the melting points increase.

	F_2	Cl_2	Br_2	I_2
M. pt / K	53	172	266	387

FIRST IONIZATION ENERGY

The definition of first ionization energy and a graph showing the values for the first 20 elements has already been given on page 13. The values decrease down each group as the outer electron is further from the nucleus and therefore less energy is required to remove it, e.g. for the group 1 elements, Li, Na and K.

Element:	Li	Na	K
Electron configuration	$1s^2 2s^1$	$[Ne]3s^1$	$[Ar]4s^1$
First ionization energy (kJ mol^{-1})	520	496	419

Generally the values increase across a period. This is because the extra electrons are filling the same energy level and the extra protons in the nucleus attract this energy level closer making it harder to remove an electron.

The values do not increase regularly across a period because new sub-levels are being filled. The p sub-level is higher in energy than the s sub-level. This explains why the value for B ($1s^2 2s^2 2p^1$) is slightly lower than the value for Be ($Is^2 2s^2$) and the value for Al ($[Ne]3s^2 3p^1$) is slightly lower than Mg ($[Ne]3s^2$). There is also a drop in value between N ($1s^2 2s^2 2p^3$) and O ($1s^2 2s^2 2p^4$) and between P ($[Ne]3s^2 3p^3$) and S ($[Ne]3s^2 3p^4$). This is because when electrons pair up in an orbital there is increased repulsion so the paired electron is easier to remove compared with when the three electrons are all unpaired, one each in the three separate p orbitals.

ELECTRONEGATIVITY

Electronegativity is a relative measure of the attraction that an atom has for a shared pair of electrons when it is covalently bonded to another atom. As the size of the atom decreases the electronegativity increases, so the value increases across a period and decreases down a group. The three most electronegative elements are F, N, and O.

ELECTRON AFFINITY

The electron affinity is the energy change when an electron is added to an isolated atom in the gaseous state, i.e.

$$X(g) + e^- \rightarrow X^-(g)$$

Atoms 'want' an extra electron so electron affinity values are negative for the addition of the first electron. However, when oxygen forms the O^{2-} ion the overall process is endothermic.

	$O(g) + e^- \rightarrow O^-(g)$	$\Delta H^\ominus = -141$ kJ mol^{-1}
	$O^-(g) + e^- \rightarrow O^{2-}(g)$	$\Delta H^\ominus = +753$ kJ mol^{-1}
overall	$O(g) + 2e^- \rightarrow O^{2-}(g)$	$\Delta H^\ominus = +612$ kJ mol^{-1}

Periodic trends (3)

CHEMICAL PROPERTIES OF ELEMENTS IN THE SAME GROUP

Group 1 – the alkali metals

Li
Na
K
Rb
Cs
(↓ increasing reactivity)

Lithium, sodium, and potassium all contain one electron in their outer shell. They are all reactive metals and are stored under liquid paraffin to prevent them reacting with air. They react by losing their outer electron to form the metal ion. Because they can readily lose an electron they are good reducing agents. The reactivity increases down the group as the outer electron is in successively higher energy levels and less energy is required to remove it.

They are called alkali metals because they all react with water to form an alkali solution of the metal hydroxide and hydrogen gas. Lithium floats and reacts quietly, sodium melts into a ball which darts around on the surface, and the heat generated from the reaction with potassium ignites the hydrogen.

$$2Li(s) + 2H_2O(l) \rightarrow 2Li^+(aq) + 2OH^-(aq) + H_2(g)$$

$$2Na(s) + 2H_2O(l) \rightarrow 2Na^+(aq) + 2OH^-(aq) + H_2(g)$$

$$2K(s) + 2H_2O(l) \rightarrow 2K^+(aq) + 2OH^-(aq) + H_2(g)$$

They all also react readily with chlorine, bromine and iodine to form ionic salts, e.g.

$$2Na(s) + Cl_2(g) \rightarrow 2Na^+Cl^-(s)$$

$$2K(s) + Br_2(l) \rightarrow 2K^+Br^-(s)$$

$$2Li(s) + I_2(g) \rightarrow 2Li^+I^-(s)$$

Group 17 – the halogens

F_2
Cl_2
Br_2
I_2
(↑ increasing reactivity)

The halogens react by gaining one more electron to form halide ions. They are good oxidizing agents. The reactivity decreases down the group as the outer shell is increasingly at higher energy levels and further from the nucleus. This, together with the fact that there are more electrons between the nucleus and the outer shell, decreases the attraction for an extra electron.

Chlorine is a stronger oxidizing agent than bromine, so can remove the electron from bromide ions in solution to form chloride ions and bromine. Similarly both chlorine and bromine can oxidize iodide ions to form iodine.

$$Cl_2(aq) + 2Br^-(aq) \rightarrow 2Cl^-(aq) + Br_2(aq)$$

$$Cl_2(aq) + 2I^-(aq) \rightarrow 2Cl^-(aq) + I_2(aq)$$

$$Br_2(aq) + 2I^-(aq) \rightarrow 2Br^-(aq) + I_2(aq)$$

Test for halide ions

The presence of halide ions in solution can be detected by adding silver nitrate solution. The silver ions react with the halide ions to form a precipitate of the silver halide. The silver halides can be distinguished by their colour. These silver halides react with light to form silver metal. This is the basis of old-fashioned film photography.

$$Ag^+(aq) + X^-(aq) \rightarrow AgX(s) \quad \text{where } X = Cl, Br, \text{ or } I$$

$$AgX(s) \xrightarrow{\text{light}} Ag(s) + \tfrac{1}{2}X_2$$

AgCl white
AgBr cream
AgI yellow

CHANGE FROM METALLIC TO NON-METALLIC NATURE OF THE ELEMENTS ACROSS PERIOD 3

Metals tend to be shiny and are good conductors of heat and electricity. Sodium, magnesium and aluminium all conduct electricity well. Silicon is a semiconductor and is called a **metalloid** as it possesses some of the properties of a metal and some of a non-metal. Phosphorus, sulfur, chlorine and argon are non-metals and do not conduct electricity. Metals can also be distinguished from non-metals by their chemical properties. Metal oxides tend to be basic, whereas non-metal oxides tend to be acidic.

Sodium oxide and magnesium oxide are both basic and react with water to form hydroxides,

e.g. $Na_2O(s) + H_2O(l) \rightarrow 2NaOH(aq)$ $\quad$ $MgO(s) + H_2O(l) \rightarrow Mg(OH)_2$

Aluminium is a metal but its oxide is amphoteric, that is, it can be either basic or acidic depending on whether it is reacting with an acid or a base.

The remaining elements in period 3 have acidic oxides. For example, sulfur trioxide reacts with water to form sulfuric acid, phosphorus pentoxide reacts with water to form phosphoric(V) acid and nitrogen(IV)oxide reacts with water to form nitric acid.

$SO_3(g) + H_2O(l) \rightarrow H_2SO_4(aq)$ $\quad$ $P_4O_{10}(s) + 6H_2O(l) \rightarrow 4H_3PO_4(aq)$ $\quad$ $3NO_2(g) + H_2O(l) \rightarrow 2HNO_3(aq) + NO(g)$

The transition metals

THE FIRST ROW TRANSITION ELEMENTS

Element	(Sc)	Ti	V	Cr	Mn	Fe	Co	Ni	Cu	(Zn)
Electron configuration [Ar]	$4s^2 3d^1$	$4s^2 3d^2$	$4s^2 3d^3$	$4s^1 3d^5$	$4s^2 3d^5$	$4s^2 3d^6$	$4s^2 3d^7$	$4s^2 3d^8$	$4s^1 3d^{10}$	$4s^2 3d^{10}$

A transition element is defined as an element that possesses an incomplete d sub-level in one or more of its oxidation states. Scandium is not a typical transition metal as its common ion Sc^{3+} has no d electrons. Zinc is not a transition metal as it contains a full d sub-level in all its oxidation states. (Note: for Cr and Cu it is more energetically favourable to half-fill and completely fill the d sub-level respectively so they contain only one 4s electron).

Transition elements have variable oxidation states, form complex ions with ligands, have coloured compounds, and display catalytic and magnetic properties.

VARIABLE OXIDATION STATES

The 3d and 4s sub-levels are very similar in energy. When transition metals lose electrons they lose the 4s electrons first. All transition metals can show an oxidation state of +2. Some of the transition metals can form the +3 or +4 ion (e.g. Fe^{3+}, Mn^{4+}) as the ionization energies are such that up to two d electrons can also be lost. The M^{4+} ion is rare and in the higher oxidation states the element is usually found not as the free metal ion but either covalently bonded or as the oxyanion, such as MnO_4^-. Some common examples of variable oxidation states in addition to +2 are:

Cr(+3)	$CrCl_3$	chromium(III) chloride
Cr(+6)	$Cr_2O_7^{2-}$	dichromate(VI) ion
Mn(+4)	MnO_2	manganese(IV) oxide
Mn(+7)	MnO_4^-	manganate(VII) ion
Fe(+3)	Fe_2O_3	iron(III) oxide
Cu(+1)	Cu_2O	copper(I) oxide

A full list of the common oxidation states can be found in Section 14 of the IB data booklet.

CATALYTIC BEHAVIOUR

Many transition elements and their compounds are very efficient catalysts, that is, they increase the rate of chemical reactions. This helps to make industrial processes, such as the production of ammonia and sulfuric acid, more efficient and economic. Platinum and palladium are used in catalytic converters fitted to cars. In the body, iron is found in haem and cobalt is found in vitamin B_{12}. Other common examples include:

Iron in the Haber process	$3H_2(g) + N_2 \xrightleftharpoons{Fe(s)} 2NH_3(g)$
Vanadium(V) oxide in the Contact process	$2SO_2(g) + O_2(g) \xrightleftharpoons{V_2O_5(s)} 2SO_3(g)$
Nickel in hydrogenation reactions	$C_2H_4(g) + H_2(g) \xrightarrow{Ni(s)} C_2H_6(g)$
Manganese(IV) oxide with hydrogen peroxide	$2H_2O_2(aq) \xrightarrow{MnO_2(s)} 2H_2O(l) + O_2(g)$

MAGNETIC PROPERTIES

Transition metals and their complexes that contain unpaired electrons can exhibit magnetism. Iron metal and some other metals (e.g. nickel and cobalt) show ferromagnetism. This is a permanent type of magnetism. In this type of magnetism the unpaired electrons align parallel to each other in domains irrespective of whether an external magnetic or electric field is present. This property of iron has been utilized for centuries to make compasses, which align with the Earth's magnetic field to point north.

Many complexes of transition metals contain unpaired electrons. Unlike paired electrons, where the spins cancel each other out, the spinning unpaired electrons create a small magnetic field and will line up in an applied electric or magnetic field to make the transition metal complex weakly magnetic when the field is applied, i.e. they reinforce the external magnetic field. This type of magnetism is known as paramagnetism. The more unpaired electrons there are in the complex the more paramagnetic the complex will be.

When all the electrons in a transition metal complex are paired up the complex is said to be diamagnetic.

Iron(II) complex ions ($[Ar]3d^6$) can be paramagnetic or diamagnetic. The five d orbitals are split by the ligands according to the spectrochemical series (see page 22). If the ligands are low in the spectrochemical series (e.g. H_2O) and split the d orbitals by only a small amount, the electrons can occupy all the d orbitals giving four units of paramagnetism as there will be four unpaired electrons. This is known as weak ligand field splitting. Ligands high in the spectrochemical series (e.g. CN^-) will cause a larger splitting. Only the lower d orbitals will be occupied and the complex will be diamagnetic as there are no unpaired electrons.

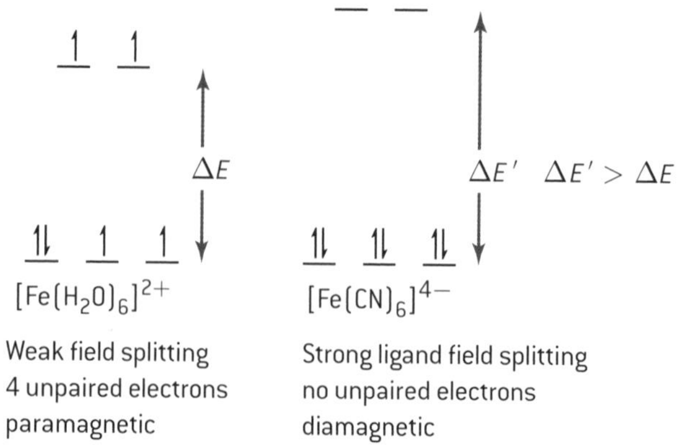

Transition metal complex ions

FORMATION OF COMPLEX IONS

Because of their small size d-block ions act as Lewis acids and attract species that are rich in electrons. Such species are known as **ligands**. Ligands are neutral molecules or anions which contain a non-bonding pair of electrons. These electron pairs can form co-ordinate covalent bonds with the metal ion to form **complex ions**.

A common ligand is water and most (but not all) transition metal ions exist as hexahydrated complex ions in aqueous solution, e.g. $[Fe(H_2O)_6]^{3+}$. Ligands can be replaced by other ligands. A typical example is the addition of ammonia to an aqueous solution of copper(II) sulfate to give the deep blue colour of the tetraaminecopper(II) ion. Similarly if concentrated hydrochloric acid is added to a solution of $Cu^{2+}(aq)$ the yellow tetrachlorocopper(II) anion is formed.

Note: in this ion the overall charge on the ion is −2 as the four ligands each have a charge of −1.

$$[CuCl_4]^{2-} \underset{H_2O}{\overset{Cl^-}{\rightleftharpoons}} [Cu(H_2O)_4]^{2+} \underset{H_2O}{\overset{NH_3}{\rightleftharpoons}} [Cu(NH_3)_4]^{2+}$$

The number of lone pairs bonded to the metal ion is known as the **coordination number**. Compounds with a coordination number of six are octahedral in shape, those with a coordination number of four are tetrahedral or square planar, whereas those with a coordination number of two are usually linear.

Coordination number	6	4	2
Examples	$[Fe(CN)_6]^{3-}$	$[CuCl_4]^{2-}$	$[Ag(NH_3)_2]^+$
	$[Fe(OH)_3(H_2O)_3]$	$[Cu(NH_3)_4]^{2+}$	

POLYDENTATE LIGANDS

Ligands such as water and cyanide ions are known as monodentate ligands as they utilize just one non-bonding pair to form a coordinate covalent bond to the metal ion. Some ligands contain more than one non-bonding pair and can form two or more coordinate bonds to the metal ion. Three common examples are ethylenediamine, $H_2NCH_2CH_2NH_2$, oxalate ions $(COO^-)_2$, both of which can use two non-bonding pairs to form bidentate ligands, and EDTA (EthyleneDiamineTetraAcetic acid) or its ion, $EDTA^{4-}$, which can act as hexadentate ligands.

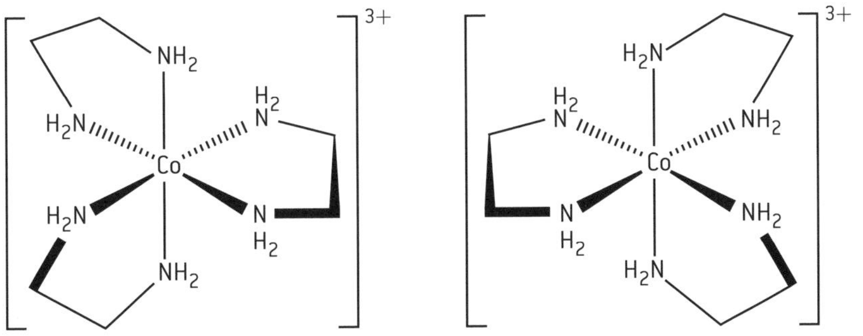

The Co^{3+} ion can form two different stereoisomers (mirror images) with the three separate ethylenediamine molecules acting as bidentate ligands. The coordination number is six as although each metal ion is surrounded by only three ligands the metal ion forms a total of six coordinate bonds with the ligands.

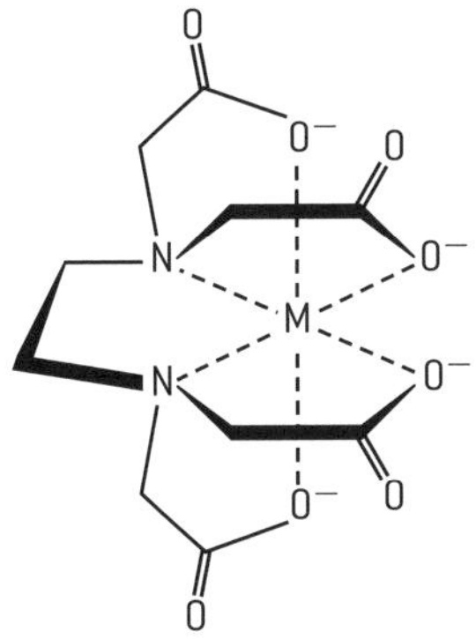

$EDTA^{4-}$ acting as a hexadentate ligand with a transition metal, M to form a complex such as $[Cu(EDTA)]^{2-}$. Note that the coordination number is still six even though only one ligand surrounds the metal ion.

FACTORS AFFECTING THE COLOUR OF TRANSITION METAL COMPLEXES

Transition metals are defined as elements having an incomplete d sub-level in one or more of their oxidation states. Compounds of Sc^{3+} which have no d electrons and of Cu^+ and Zn^{2+} which both have complete d sub-shells are colourless. This strongly suggests that the colour of transition metal complexes is related to an incomplete d level. The actual colour is determined by four different factors.

1. The nature of the transition element. For example $Mn^{2+}(aq)$ and $Fe^{3+}(aq)$ both have the configuration $[Ar]3d^5$. $Mn^{2+}(aq)$ is pink whereas $Fe^{3+}(aq)$ is yellow.
2. The oxidation state. $Fe^{2+}(aq)$ is green whereas $Fe^{3+}(aq)$ is yellow.
3. The identity of the ligand. $[Cu(H_2O)_6]^{2+}$ (sometimes shown as $[Cu(H_2O)_4]^{2+}$), is blue, $[Cu(NH_3)_4(H_2O)_2]^{2+}$ (sometimes shown as $[Cu(NH_3)_4]^{2+}$), is blue/violet whereas $[CuCl_4]^{2-}$ is yellow (green in aqueous solution).
4. The stereochemistry of the complex. The colour is also affected by the shape of the molecule or ion. In the above example $[Cu(H_2O)_6]^{2+}$ is octahedral whereas $[CuCl_4]^{2-}$ is tetrahedral. However for the IB only octahedral complexes in aqueous solution will be considered.

Colour of transition metal complexes

SPLITTING OF THE d ORBITALS

In the free ion the five d orbitals are degenerate. That is they are all of equal energy.

Note that three of the orbitals (d_{xy}, d_{yz}, and d_{zx}) lie *between* the axes whereas the other two ($d_{x^2-y^2}$ and d_{z^2}) lie *along* the axes. Ligands act as Lewis bases and donate a non-bonding pair of electrons to form a co-ordinate bond. As the ligands approach the metal along the axes to form an octahedral complex the non-bonding pairs of electrons on the ligands will repel the $d_{x^2-y^2}$ and d_{z^2} orbitals causing the five d orbitals to split, three to lower energy and two to higher energy. The difference in energy between the two levels corresponds to the wavelength of visible light.

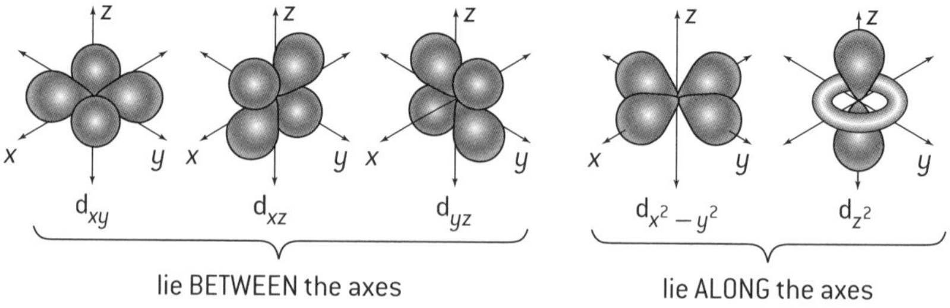

When white light falls on the aqueous solution of the complex the colour corresponding to ΔE is absorbed as an electron is promoted and the transmitted light will be the complementary colour. For example, $[Cu(H_2O)_6]^{2+}$ absorbs red light so the compound appears blue. The amount that the d orbitals are split will determine the exact colour. Changing the transition metal changes the number of protons in the nucleus which will affect the levels. Similarly changing the oxidation state will affect the splitting as the number of electrons in the level is different. Different ligands will also cause different amounts of splitting depending on their electron density.

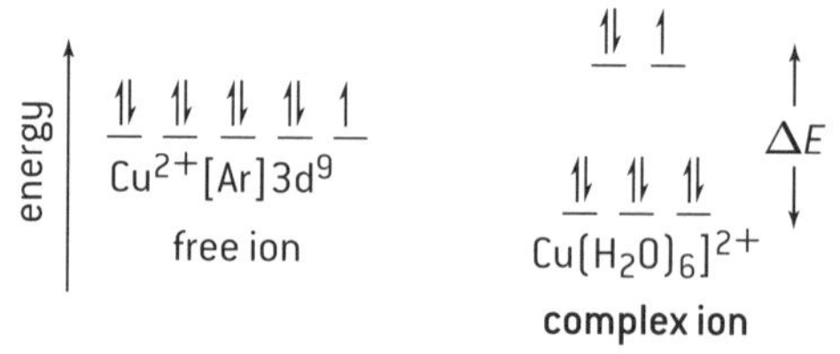

THE SPECTROCHEMICAL SERIES

Ligands can be arranged in order of their ability to split the d orbitals in octahedral complexes.

$$I^- < Br^- < S^{2-} < Cl^- < F^- < OH^- < H_2O < SCN^- < NH_3 < CN^- < CO$$

This order is known as the spectrochemical series. Iodide ions cause the smallest splitting and the carbonyl group, CO, the largest splitting. The energy of light absorbed increases when ammonia is substituted for water in Cu^{2+} complexes as the splitting increases, i.e. in going from $[Cu(H_2O)_6]^{2+}$ to $[Cu(NH_3)_4(H_2O)_2]^{2+}$. This means that the wavelength of the light absorbed decreases and this is observed in the colour of the transmitted light which changes from blue to a blue-violet (purple) colour.

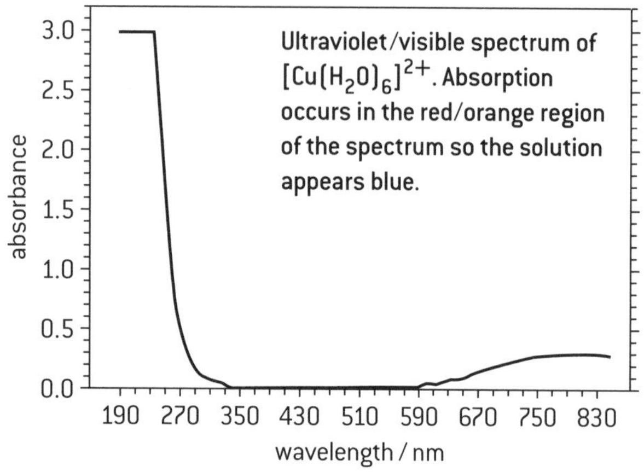

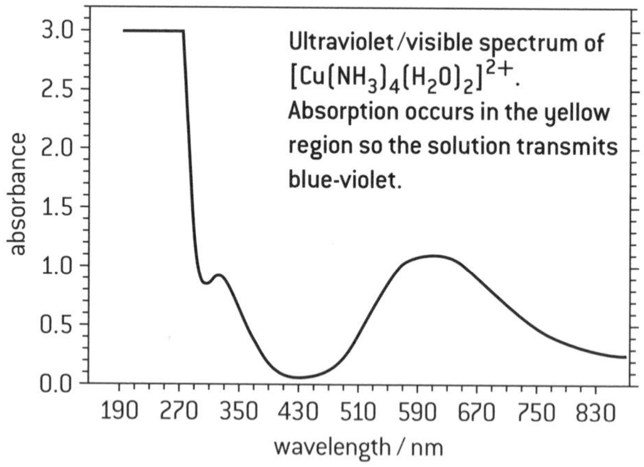

COMPLEMENTARY COLOURS

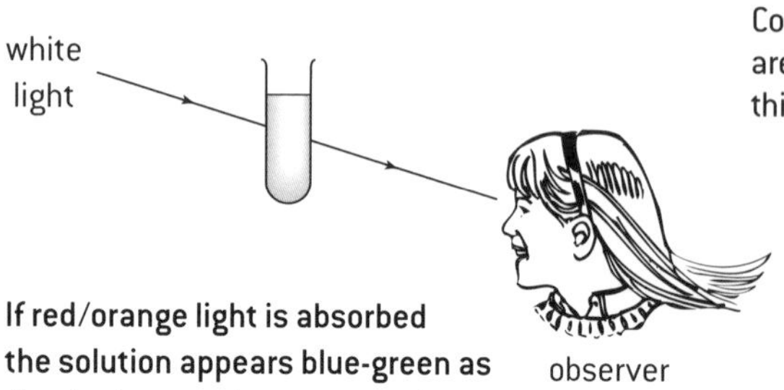

If red/orange light is absorbed the solution appears blue-green as that is the complementary colour which is transmitted.

Complementary colours are opposite each other in this 'colour wheel'.

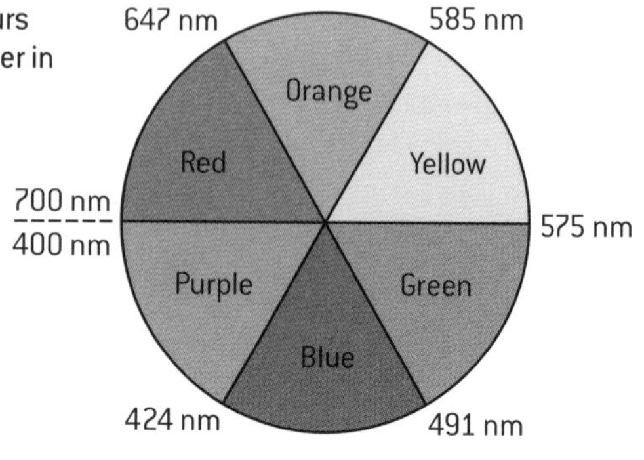

MULTIPLE CHOICE QUESTIONS – PERIODICITY

1. Where in the periodic table would the element with the electronic structure $1s^22s^22p^63s^23p^64s^23d^{10}4p^2$ be located?

   A. s block
   C. d block
   B. p block
   D. lanthanoids

2. Which property decreases with increasing atomic number in group 17 (the halogens)?

   A. Melting point
   B. Electronegativity
   C. Atomic radius
   D. Ionic radius of the negative ion

3. Which property increases with increasing atomic number for both the halogens and the alkali metals?

   A. Reactivity with water
   C. Electron affinity
   B. Electronegativity
   D. Atomic radius

4. Which one of the following series is arranged in order of increasing size?

   A. P^{3-}, S^{2-}, Cl^-
   C. Na^+, Mg^{2+}, Al^{3+}
   B. Cl, Ar, K
   D. H^+, H, H^-

5. Which element has the lowest electronegativity value?

   A. potassium
   C. iodine
   B. fluorine
   D. hydrogen

6. Which statements about the periodic table are correct?

   I. The elements Mg, Ca and Sr have similar chemical properties.
   II. Elements in the same period have the same number of main energy levels.
   III. The oxides of Na, Mg and P are basic.

   A. I and II only
   C. II and III only
   B. I and III only
   D. I, II and III

7. The *x*-axis of the graph below represents the atomic number of the elements in period 3.

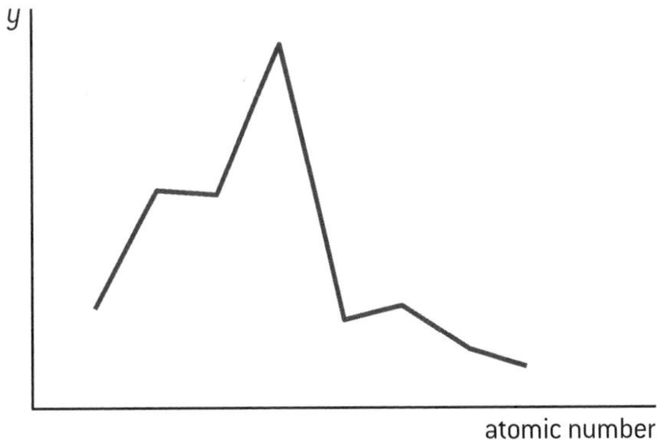

   Which variable could represent the *y*-axis?

   A. Melting point
   C. Ionic radius
   B. Electronegativity
   D. Atomic radius

8. Which is the correct definition for the electron affinity of element X?

   A. $X(s) + e^- \rightarrow X^-(s)$
   C. $X(g) + e^- \rightarrow X^-(g)$
   B. $X(s) \rightarrow X^+(s) + e^-$
   D. $X(g) \rightarrow X^+(g) + e^-$

9. Which oxide will react with water to give a solution with a pH greater than 7?

   A. NO_2
   C. MgO
   B. P_4O_{10}
   D. SiO_2

10. Which reaction does **not** occur readily?

    A. $Cl_2(aq) + 2Br^-(aq) \rightarrow Br_2(aq) + 2Cl^-(aq)$
    B. $2Na(s) + H_2O(l) \rightarrow 2NaOH(aq) + H_2(g)$
    C. $Na_2O(s) + H_2O(l) \rightarrow 2NaOH(aq)$
    D. $I_2(aq) + 2Br^-(aq) \rightarrow Br_2(aq) + 2I^-(aq)$

HL

11. Which complex ion is colourless in aqueous solution?

    A. $[Mn(H_2O)_6]^{2+}$
    C. $[Cu(NH_3)_4]^{2+}$
    B. $[Zn(H_2O)_4]^{2+}$
    D. $[CuCl_4]^{2-}$

12. A certain element has the electron configuration $1s^22s^22p^63s^23p^64s^23d^3$. What oxidation state(s) would this element most likely show?

    A. +2 only
    C. +2 and +5 only
    B. +3 only
    D. +2, +3, +4 and +5

13. What is the overall charge on the complex ion formed by Fe(III) and six CN^- ligands?

    A. +3
    C. −3
    B. −6
    D. +6

14. Which statements about transition metal complexes are correct?

    I. The colour of the complex is due to light being emitted when an electron falls from a higher split d sub-level to a lower split d sub-level.
    II. The difference in energy between the split d sub-levels depends on the nature of the surrounding ligands.
    III. The colour of the complex is influenced by the oxidation state of the metal.

    A. I and II only
    C. II and III only
    B. I and III only
    D. I, II and III

15. What is the electron configuration of the Mn^{2+} ion?

    A. $1s^22s^22p^63s^23p^63d^5$
    B. $1s^22s^22p^63s^23p^64s^23d^5$
    C. $1s^22s^22p^63s^23p^64s^23d^3$
    D. $1s^22s^22p^63s^23p^64s^13d^4$

16. Which statement is true for all polydentate ligands?

    A. They contain more than one pair of non-bonding electrons.
    B. They can only form one coordinate bond with a transition metal ion.
    C. They do not affect the size of the splitting of the d sub-level.
    D. They can only form complexes with transition metal ions.

SHORT ANSWER QUESTIONS – PERIODICITY

1. Carbon, silicon and tin belong to the same group in the periodic table.
   a) Distinguish between the terms *group* and *period* in terms of electron configuration. [2]
   b) State in which block of the periodic table (s, p, d or f) these three elements are located. [1]
   c) Explain why the first ionization energy of silicon is lower than that of carbon. [2]
   d) Describe a simple experiment to show the change in non-metallic to metallic properties as the atomic number of the elements in the group increases. [3]
2. a) The maximum number of electrons in each energy level is determined by the expression $2n^2$ where n is the number of the level. Explain why the third level is sometimes said to contain eight electrons and sometimes said to contain eighteen electrons. [2]
   b) In terms of electron configuration describe the essential difference between the lanthanoids and the actinoids. [2]
3. a) Describe how the acid–base nature of the oxides changes across period 3 ($Na \rightarrow Cl$). [3]
   b) State the equation for the reaction of water with
      (i) sodium oxide, Na_2O [1]
      (ii) phosphorus(V) oxide, P_4O_{10} [1]
      (iii) sulfur(VI) oxide, SO_3 [1]
   c) Explain why oxides of nitrogen and sulfur cause damage to many types of buildings. Include relevant equations in your explanation. [4]
4. a) State the electron configuration for the P^{3-} ion. [1]
   b) Explain why the ionic radius of the S^{2-} ion is smaller than that of the P^{3-} ion. [2]
   c) Potassium and bromine are in the same period. Explain why the atomic radius of bromine is considerably smaller than the atomic radius of potassium. [2]
   d) Suggest a reason why no value is given for the atomic radius of neon in the IB data booklet. [1]
5. The periodic table shows the relationship between electron configuration and the properties of elements and is a valuable tool for making predictions in chemistry.
   a) (i) Identify the property used to arrange elements in the periodic table. [1]
      (ii) Outline two reasons why electronegativity increases across period 3 in the periodic table and one reason why noble gases are not assigned electronegativity values. [3]
   b) (i) Define the term first ionization energy of an atom. [2]
      (ii) Explain the general increasing trend in the first ionization energies of the period 2 elements, Li to Ne. [2]
      (iii) Explain why the first ionization energy of boron is lower than that of beryllium. [2]
   c) Explain why sodium conducts electricity but phosphorus does not. [2]
6. In many cities around the world, public transport vehicles use diesel, a liquid hydrocarbon fuel, which often contains sulfur impurities and undergoes incomplete combustion. All public transport vehicles in New Delhi, India, have been converted to use compressed natural gas (CNG) as fuel. Suggest two ways in which this improves air quality, giving a reason for your answer. [3]

HL

7. Transition elements, such as iron, have many characteristic properties and uses.
   a) State and explain in terms of **acid–base properties** the type of reaction that occurs between Fe^{2+} ions and water to form $[Fe(H_2O)_6]^{2+}$. [2]
   b) Explain why the colour of $[Fe(H_2O)_6]^{2+}$ is different to that of $[Fe(H_2O)_6]^{3+}$. [2]
   c) Explain why iron metal is magnetic and why some of its complexes such as $[Fe(H_2O)_6]^{2+}$ are paramagnetic whereas others such as $[Fe(CN)_6]^{4-}$ are diamagnetic. [4]
   d) The Haber process to form ammonia is exothermic. Explore the economic significance of using an iron catalyst. [3]
8. The diagram shows the $[Cu(EDTA)]^{4-}$ ion.

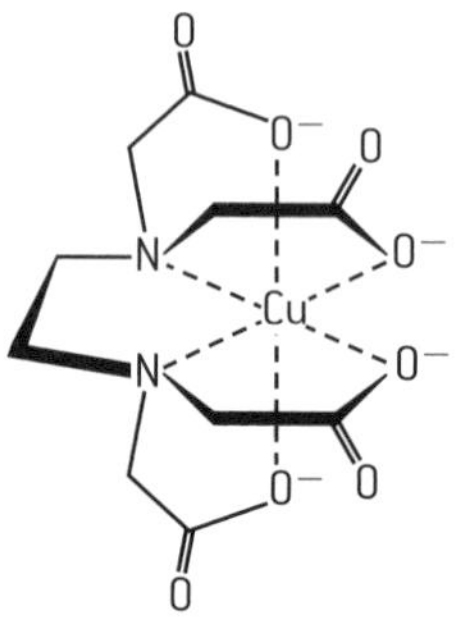

   a) Deduce the formula mass of this complex ion. [2]
   b) State the coordination number and the shape of the ion. [2]
   c) Explain how the EDTA ion is able to bond to the Cu^{2+} ion. [3]
9. Ligands can be listed in terms of their ability to split the d orbitals. This is known as the spectrochemical series.
   a) Suggest a reason why different ligands cause the d orbitals to be split by different amounts. [2]
   b) Ammonia molecules are higher in the spectrochemical series (cause greater splitting) than water molecules. Explain why the light blue colour of $[Cu(H_2O)_6]^{2+}$ changes to a more purple colour when excess ammonia is added to copper(II) sulfate solution to form the $[Cu(NH_3)_4(H_2O)_2]^{2+}$ ion. [3]
   c) Explain why aqueous solutions of scandium(III) compounds are not coloured. [2]

Ionic bonding

IONIC BOND

When atoms combine they do so by trying to achieve a noble gas configuration. Ionic compounds are formed when electrons are transferred from one atom to another to form ions with complete outer shells of electrons. In an ionic compound the positive and negative ions are attracted to each other by strong electrostatic forces, and build up into a strong lattice. Ionic compounds have high melting points as considerable energy is required to overcome these forces of attraction.

$[Ne]3s^1$	$[Ne]3s^23p^5$	$[Ne]$	$[Ar]$
11 protons	17 protons	11 protons	17 protons
11 electrons	17 electrons	10 electrons	18 electrons

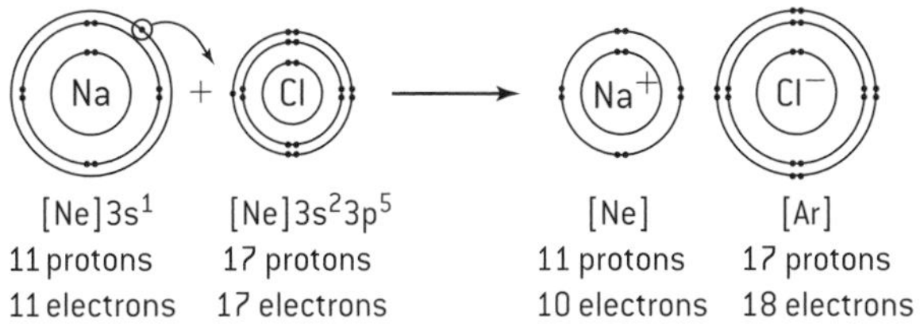

The classic example of an ionic compound is sodium chloride Na^+Cl^-, formed when sodium metal burns in chlorine. Chlorine is a covalent molecule, so each atom already has a noble gas configuration. However, the energy given out when the ionic lattice is formed is sufficient to break the bond in the chlorine molecule to give atoms of chlorine. Each sodium atom then transfers one electron to a chlorine atom to form the ions.

The charge carried by an ion depends on the number of electrons the atom needed to lose or gain to achieve a full outer shell.

Cations					Anions		
Group 1	**Group 2**	**Group 3**			**Group 15**	**Group 16**	**Group 17**
+1	+2	+3			−3	−2	−1
Li^+ Na^+ K^+	Mg^{2+} Ca^{2+}	Al^{3+}			N^{3-} P^{3-}	O^{2-} S^{2-}	F^- Cl^- Br^-

Thus, in magnesium chloride, two chlorine atoms each gain one electron from a magnesium atom to form $Mg^{2+}Cl^-_2$. In magnesium oxide two electrons are transferred from magnesium to oxygen to give $Mg^{2+}O^{2-}$. Transition metals can form more than one ion. For example, iron can form Fe^{2+} and Fe^{3+} and copper can form Cu^+ and Cu^{2+}.

FORMULAS OF IONIC COMPOUNDS

It is easy to obtain the correct formula as the overall charge of the compound must be zero.

lithium fluoride Li^+F^- magnesium chloride $Mg^{2+}Cl^-_2$ aluminium bromide $Al^{3+}Br^-_3$
sodium oxide $Na^+_2O^{2-}$ calcium sulfide $Ca^{2+}S^{2-}$ iron(III) oxide $Fe^{3+}_2O^{2-}_3$
potassium nitride $K^+_3N^{3-}$ calcium phosphide $Ca^{2+}_3P^{3-}_2$ iron(II) oxide $Fe^{2+}O^{2-}$

Note: the formulas above have been written to show the charges carried by the ions. Unless asked specifically to do this it is common practice to omit the charges and simply write LiF, $MgCl_2$, etc.

IONS CONTAINING MORE THAN ONE ELEMENT (POLYATOMIC IONS)

In ions formed from more than one element the charge is often spread (delocalized) over the whole ion. An example of a positive ion is the ammonium ion NH_4^+, in which all four N–H bonds are identical. Negative ions are sometimes known as acid radicals as they are formed when an acid loses one or more H^+ ions.

hydroxide OH^-
nitrate NO_3^- (from nitric acid, HNO_3)
sulfate SO_4^{2-} $\left.\begin{array}{l}\\ \\ \end{array}\right\}$ (from sulfuric acid, H_2SO_4)
hydrogensulfate HSO_4^-

carbonate CO_3^{2-}
hydrogencarbonate HCO_3^- $\left.\begin{array}{l}\\ \\ \end{array}\right\}$ (from carbonic acid, H_2CO_3)
ethanoate CH_3COO^- (from ethanoic acid, CH_3COOH)
phosphate PO_4^{3-} (from phosphoric acid H_3PO_4)

The formulas of the ionic compounds are obtained in exactly the same way. Note: brackets are used to show that the subscript covers all the elements in the ion e.g. sodium nitrate, $NaNO_3$, ammonium sulfate, $(NH_4)_2SO_4$ and calcium phosphate, $Ca_3(PO_4)_2$.

IONIC BOND AND PROPERTIES OF IONIC COMPOUNDS

Ionic compounds are formed between metals on the left of the periodic table and non-metals on the right of the periodic table; that is, between elements in groups 1, 2, and 3 with a low electronegativity (electropositive elements) and elements with a high electronegativity in groups 15, 16, and 17. Generally the difference between the electronegativity values needs to be greater than about 1.8 for ionic bonding to occur.

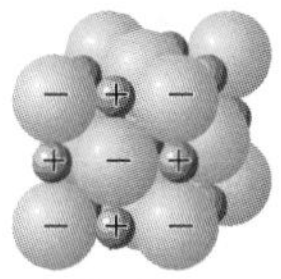

NaCl
Sodium chloride (melting point 801 °C)
Ions held strongly in ionic lattice

Ions in solid ionic compounds are held in a crystal lattice. The ionic bond is the sum of all the electrostatic attractions (and repulsions) within the lattice. A large amount of energy is required to break the lattice so ionic compounds tend to have high melting points. Many are soluble in water as the hydration energy of the ions provides the energy to overcome the lattice enthalpy. Solid ionic compounds cannot conduct electricity as the ions are held in fixed positions. When molten the ions are free to move and conduct electricity as they are chemically decomposed at the respective electrodes.

Covalent bonding

SINGLE COVALENT BONDS

Covalent bonding involves the sharing of one or more pairs of electrons so that each atom in the molecule achieves a noble gas configuration. The simplest covalent molecule is hydrogen. Each hydrogen atom has one electron in its outer shell. The two electrons are shared and attracted electrostatically by both positive nuclei resulting in a directional bond between the two atoms to form a molecule. When one pair of electrons is shared the resulting bond is known as a single covalent bond. Another example of a diatomic molecule with a single covalent bond is chlorine, Cl_2.

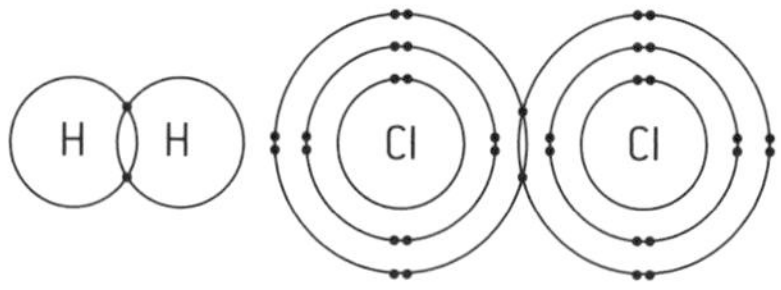

LEWIS STRUCTURES

In the Lewis structure (also known as electron dot structure) all the valence electrons are shown. There are various different methods of depicting the electrons. The simplest method involves using a line to represent one pair of electrons. It is also acceptable to represent single electrons by dots, crosses or a combination of the two. The four methods below are all correct ways of showing the Lewis structure of fluorine.

$$|\overline{F}-\overline{F}| \qquad \overset{xx\ xx}{\underset{xx\ xx}{x F x F x}} \qquad :\!\overset{..}{F}\!:\!\overset{..}{F}\!: \qquad \overset{xx}{\underset{xx}{x F x}}\overset{..}{\underset{..}{F}}:$$

Sometimes just the shared pairs of electrons are shown, e.g. F–F. This gives information about the bonding in the molecule, but it is not the Lewis structure as it does not show all the valence electrons.

SINGLE COVALENT BONDS

$$H-\overset{\overset{\displaystyle H}{|}}{\underset{\underset{\displaystyle H}{|}}{C}}-H \quad |\overline{F}-\overset{\overset{\displaystyle |\overline{F}|}{|}}{\underset{\underset{\displaystyle |\underline{F}|}{|}}{C}}-\overline{F}| \quad H-\overset{\overset{\displaystyle \overline{N}}{|}}{\underset{\underset{\displaystyle H}{|}}{N}}-H \quad H-\overset{\displaystyle \overline{O}|}{\underset{\displaystyle H}{|}} \quad H-\overline{F}|$$

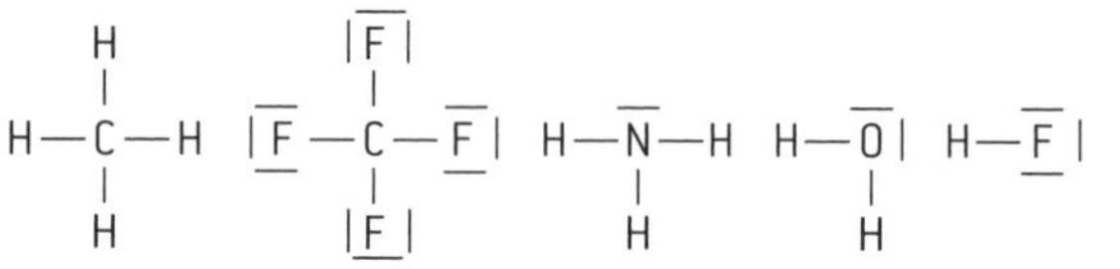

methane tetrafluoro- ammonia water hydrogen
methane fluoride

The carbon atom (electron configuration $1s^2 2s^2 2p^2$) has four electrons in its outer shell and requires a share in four more electrons. It forms four single bonds with elements that only require a share in one more electron, such as hydrogen or chlorine. Nitrogen ($1s^2 2s^2 2p^3$) forms three single bonds with hydrogen in ammonia leaving one non-bonded pair of electrons (also known as a lone pair). In water there are two non-bonded pairs and in hydrogen fluoride three non-bonded pairs.

BOND LENGTH AND BOND STRENGTH

The strength of attraction that the two nuclei have for the shared electrons affects both the length and strength of the bond. Although there is considerable variation in the bond lengths and strengths of single bonds in different compounds, double bonds are generally much stronger and shorter than single bonds. The strongest covalent bonds are shown by triple bonds.

		Length / nm	Strength / kJ mol^{-1}
Single bonds	Cl–Cl	0.199	242
	C–C	0.154	346
Double bonds	C=C	0.134	614
	O=O	0.121	498
Triple bonds	C≡C	0.120	839
	N≡N	0.110	945

e.g. ethanoic acid:

0.124 nm
O
‖
C 0.143 nm
$$H-\overset{\overset{\displaystyle H}{|}}{\underset{\underset{\displaystyle H}{|}}{C}}$$
OH

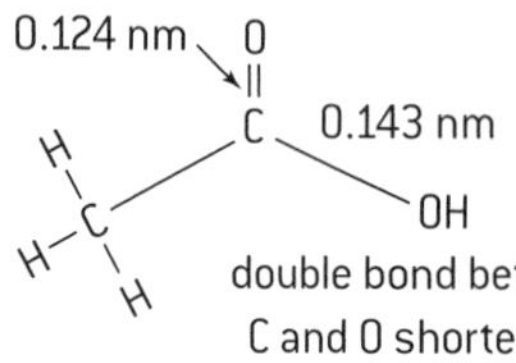

double bond between C and O shorter and stronger than single bond

MULTIPLE COVALENT BONDS

In some compounds atoms can share more than one pair of electrons to achieve an noble gas configuration.

$$\langle O=O\rangle \quad |N\equiv N| \quad \langle O=C=O\rangle \quad \overset{H}{\underset{H}{>}}C=C\overset{H}{\underset{H}{<}} \quad H-C\equiv C-H$$

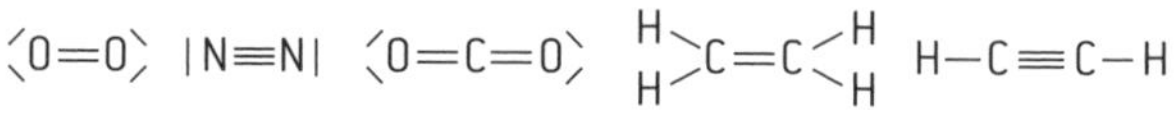

oxygen nitrogen carbon dioxide ethene ethyne

COORDINATE (DATIVE) BONDS

The electrons in the shared pair may originate from the same atom. This is known as a coordinate covalent bond.

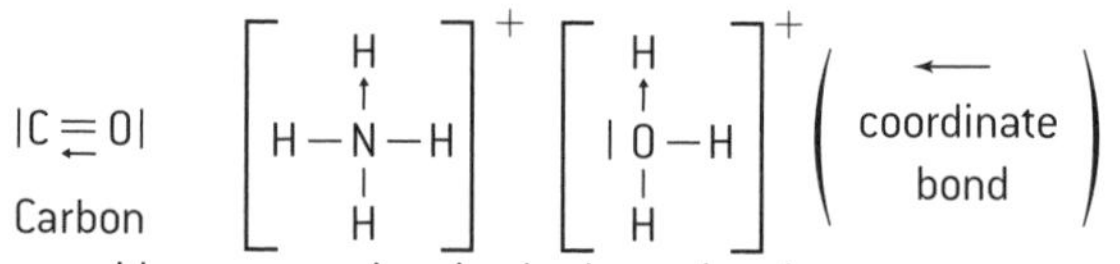

$$|C\equiv O| \qquad \left[H-\overset{\overset{\displaystyle H}{\uparrow}}{\underset{\underset{\displaystyle H}{|}}{N}}-H\right]^+ \qquad \left[|\overset{\overset{\displaystyle H}{\uparrow}}{\underset{\underset{\displaystyle H}{|}}{O}}-H\right]^+ \quad \left(\overset{\longleftarrow}{\text{coordinate bond}}\right)$$

Carbon ammonium ion hydroxonium ion
monoxide

Sulfur dioxide and sulfur trioxide are both sometimes shown as having a coordinate bond between sulfur and oxygen or they are shown as having double bonds between the sulfur and the oxygen. Both are acceptable.

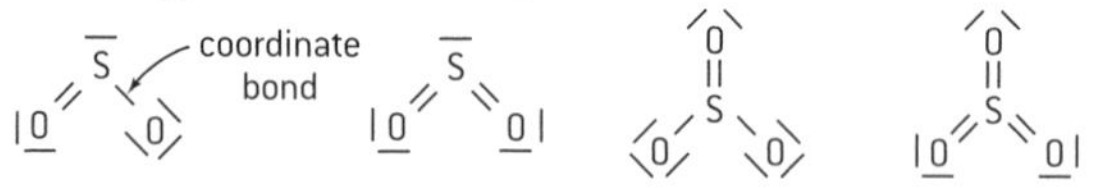

8 valence electrons around S	10 valence electrons around S	2 coordinate bonds, 8 valence electrons around S	12 valence electrons around S

BOND POLARITY

In diatomic molecules containing the same element (e.g. H_2 or Cl_2) the electron pair will be shared equally, as both atoms exert an identical attraction. However, when the atoms are different the more electronegative atom exerts a greater attraction for the electron pair. One end of the molecule will thus be more electron rich than the other end, resulting in a polar bond. This relatively small difference in charge is represented by δ+ and δ−. The bigger the difference in electronegativities the more polar the bond.

$$\overset{\delta+}{H}-\overset{\delta-}{F} \qquad \overset{\delta+}{H}-\overset{2\delta-}{\underset{\underset{\displaystyle H^{\delta+}}{|}}{O}} \qquad \overset{\delta+}{H}-\overset{3\delta-}{\underset{\underset{\displaystyle H^{\delta+}}{|}}{N}}-\overset{\delta+}{H} \qquad \overset{\delta-}{Cl}-\overset{4\delta+}{\underset{\underset{\displaystyle Cl^{\delta-}}{|}}{\overset{\overset{\displaystyle Cl^{\delta-}}{|}}{C}}}-\overset{\delta-}{Cl}$$

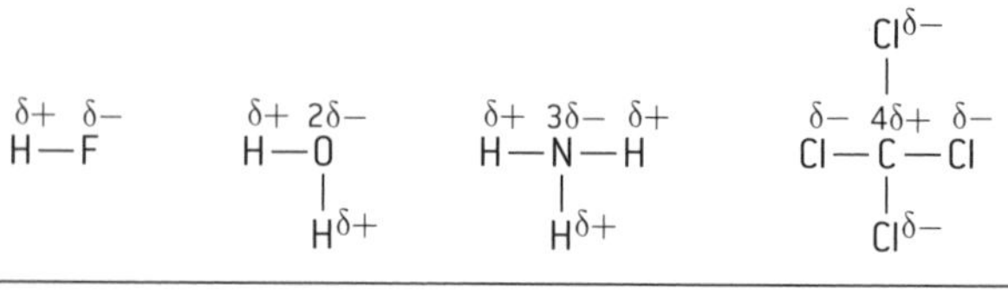

Shapes of simple molecules and ions

VSEPR THEORY

The shapes of simple molecules and ions can be determined by using the **valence shell electron pair repulsion (VSEPR)** theory. This states that pairs of electrons arrange themselves around the central atom so that they are as far apart from each other as possible. There will be greater repulsion between non-bonded pairs of electrons than between bonded pairs. Since all the electrons in a multiple bond must lie in the same direction, double and triple bonds count as one pair of electrons. Strictly speaking the theory refers to electron domains, but for most molecules this equates to pairs of electrons.

This results in five basic shapes depending on the number of pairs. For 5 and 6 electron domains the octet needs to be expanded and this can only happen if there are readily available d orbitals present that can also be utilized. For Standard Level only 2, 3 and 4 electron domains need to be considered.

No. of electron domains	*Shape*	*Name of shape*	*Bond angle(s)*
2		linear	180°
3	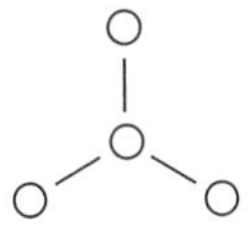	trigonal planar	120°
4	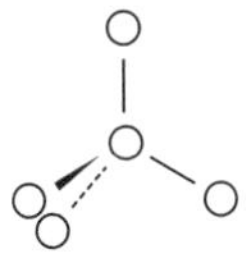	tetrahedral	109.5°
5	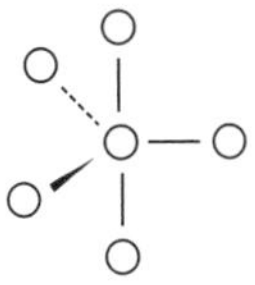	trigonal bipyramidal	90°, 120°, 180°
6	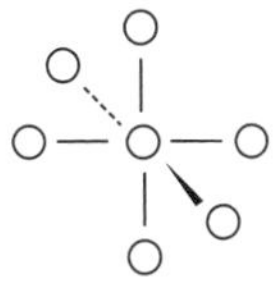	octahedral	90°, 180°

WORKING OUT THE ACTUAL SHAPE

To work out the actual shape of a molecule calculate the number of pairs of electrons around the *central* atom, then work out how many are bonding pairs and how many are non-bonding pairs. (For ions, the number of electrons that equate to the charge on the ion must also be included when calculating the total number of electrons.)

2 ELECTRON DOMAINS

Cl—Be—Cl   O=C=O   H—C≡C—H   H—C≡N

double bond counts as one pair   triple bond counts as one pair

3 ELECTRON DOMAINS

3 bonding pairs – trigonal planar

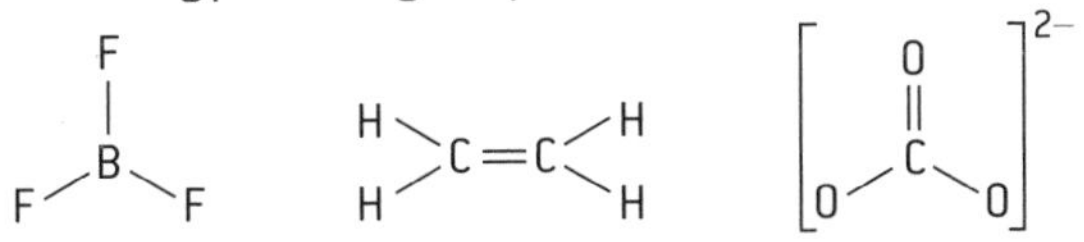

carbonate ion

2 bonding pairs, 1 non-bonded pair – bent or V-shaped

nitrite ion

5 AND 6 ELECTRON DOMAINS

5 and 6 negative charge centres

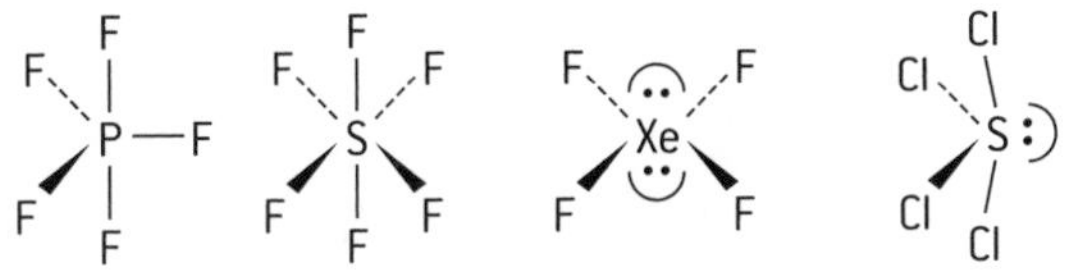

trigonal bipyramidal | octahedral | square planar non-bonding pairs as far apart as possible above and below plane | distorted tetrahedral

4 ELECTRON DOMAINS

4 bonding pairs – tetrahedral

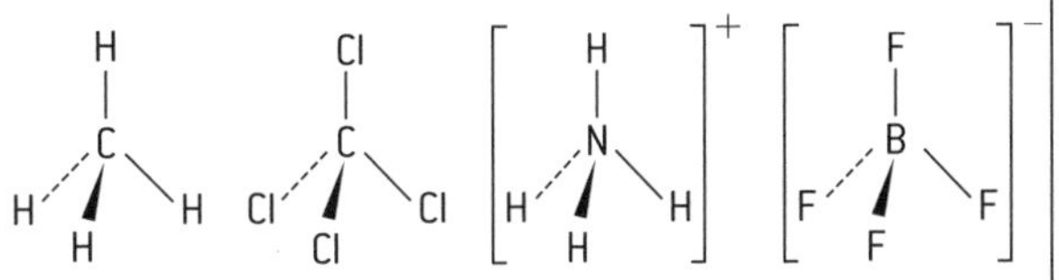

ammonium ion | tetrafluoroborate ion

3 bonding pairs, 1 non-bonding pair – trigonal pyramidal

ammonia

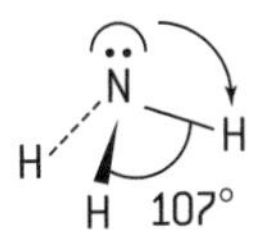

greater repulsion by non-bonding pair ∴ bond angle smaller than 109.5°

2 bonding pairs, 2 non-bonding pairs – bent or V-shaped

water

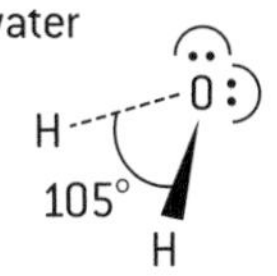

even greater repulsion by two non-bonding pairs so bond angle even smaller

Resonance hybrids and allotropes of carbon

RESONANCE STRUCTURES

When writing the Lewis structures for some molecules it is possible to write more than one correct structure. For example, ozone can be written:

These two structures are known as resonance hybrids. They are extreme forms of the true structure, which lies somewhere between the two. Evidence that this is true comes from bond lengths, as the bond lengths between the oxygen atoms in ozone are both the same and are intermediate between an O=O double bond and an O–O single bond. Resonance structures are usually shown with a double headed arrow between them. Other common compounds which can be written using resonance structures are shown here.

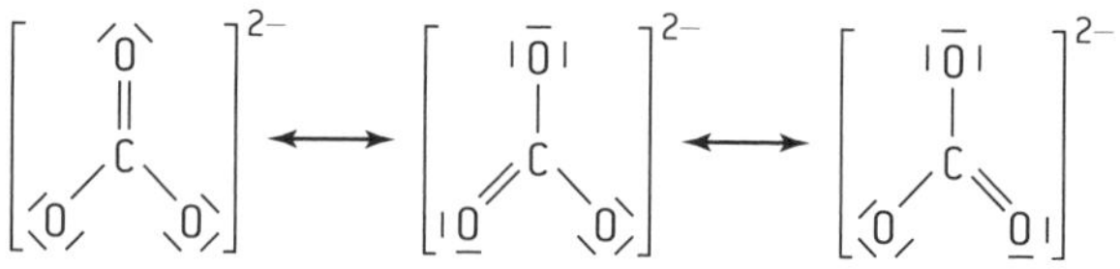

carbonate ion

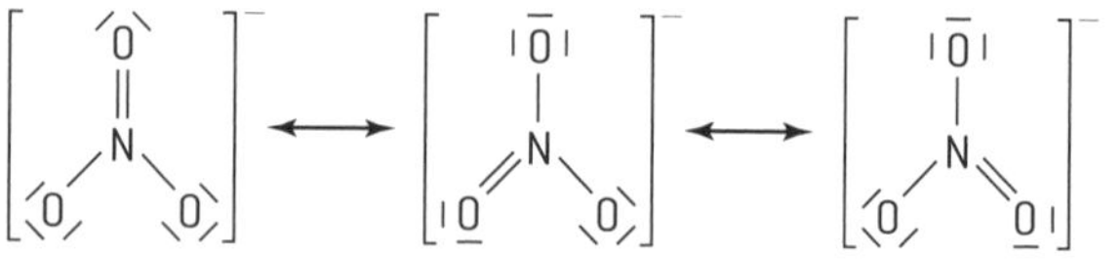

nitrate ion

ethanoate ion

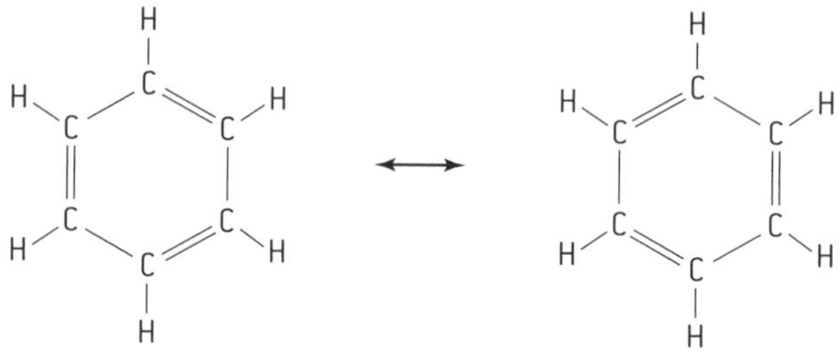

benzene

ALLOTROPES OF CARBON

Allotropes occur when an element can exist in different crystalline forms. In diamond each carbon atom is covalently bonded to four other carbon atoms to form a giant covalent structure. All the bonds are equally strong and there is no plane of weakness in the molecule so diamond is exceptionally hard, and because all the electrons are localized it does not conduct electricity. Both silicon and silicon dioxide, SiO_2, form similar giant tetrahedral structures.

diamond

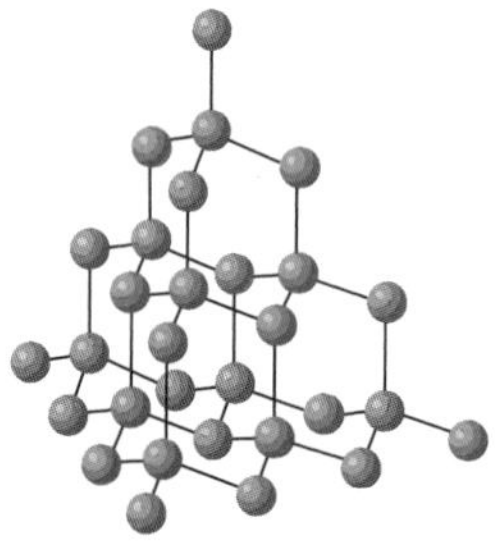

graphite

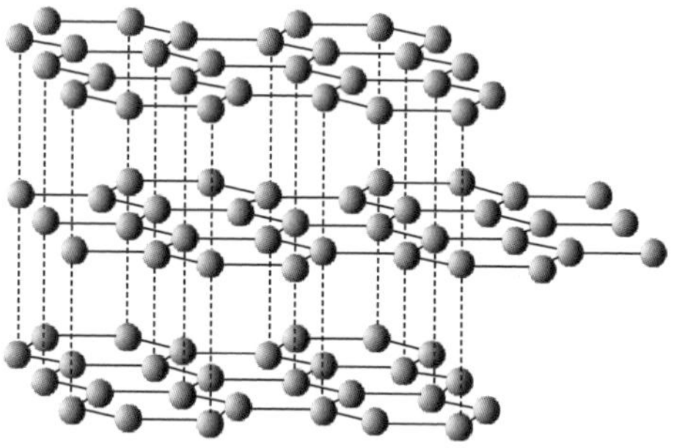

In graphite each carbon atom has very strong bonds to three other carbon atoms to give layers of hexagonal rings. There are only very weak bonds between the layers. The layers can slide over each other so graphite is an excellent lubricant and because the electrons are delocalized between the layers it is a good conductor of electricity. In 2010 the Nobel prize for Physics was awarded for the discovery of graphene. Graphene is a single layer of hexagonally arranged carbon atoms, i.e. it is essentially a form of graphite which is just one atom thick. It is extremely light, functions as a semiconductor and is 200 times stronger than steel. More recently chemists have developed a new magnetic form of graphene which is called graphone.

A third allotrope of carbon is buckminsterfullerene, C_{60}. Sixty carbon atoms are arranged in hexagons and pentagons to give a geodesic spherical structure similar to a football. Following the initial discovery of buckminsterfullerene many other similar carbon molecules have been isolated. This has led to a new branch of science called nanotechnology.

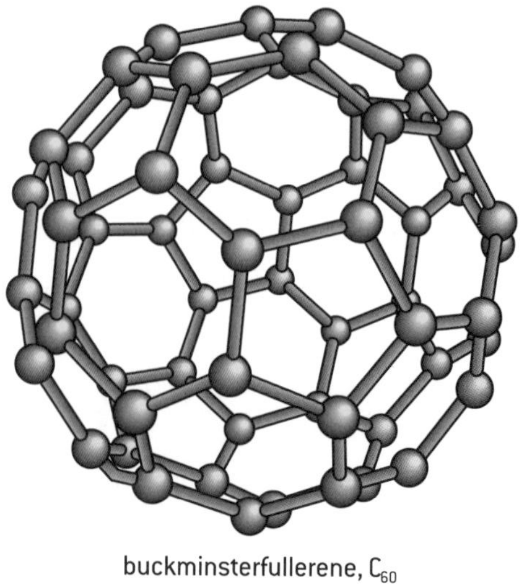

buckminsterfullerene, C_{60}

Intermolecular forces

MOLECULAR POLARITY

Whether a molecule is polar, or not, depends both on the relative electronegativities of the atoms in the molecule and on its shape. If the individual bonds are polar then it does not necessarily follow that the molecule will be polar as the resultant dipole may cancel out all the individual dipoles.

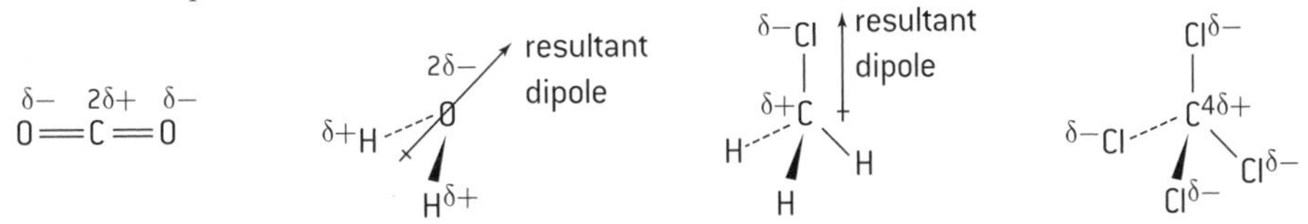

non-polar (resultant dipole zero) | polar | polar | non-polar (resultant dipole zero)

LONDON DISPERSION FORCES

Even in non-polar molecules the electrons can at any one moment be unevenly spread. This produces temporary instantaneous dipoles. An instantaneous dipole can induce another dipole in a neighbouring particle resulting in a weak attraction between the two particles. London dispersion forces increase with increasing mass.

increasing London dispersion forces →

	F_2	Cl_2	Br_2	I_2
M_r	38.0	70.9	160	254
b. pt / °C	−188	−34.0	58.0	183

increasing London dispersion forces →

	CH_4	C_2H_6	C_3H_8	C_4H_{10}
M_r	16.0	30.0	44.0	58.0
b. pt / °C	−162	−88.6	−42.2	−0.5

DIPOLE–DIPOLE FORCES

Polar molecules are attracted to each other by electrostatic forces. Although still relatively weak the attraction is stronger than London dispersion forces.

non-polar | polar

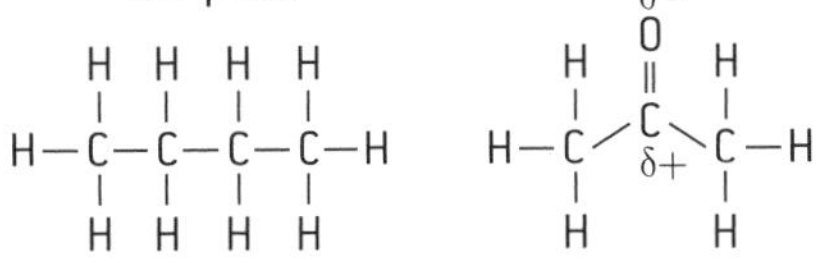

butane $M_r = 58$
b. pt −0.5 °C

propanone $M_r = 58$
b. pt 56.2 °C

identical masses
(different intermolecular forces)

INTERMOLECULAR FORCES

The covalent bonds between the atoms *within* a molecule are very strong. The forces of attraction *between* the molecules are much weaker. These intermolecular forces depend on the polarity of the molecules. Be careful with the terminology. The weakest intermolecular forces are called London dispersion forces. This description refers to instantaneous dipole-induced dipole forces that exist between any atoms or groups of atoms and should be used for non-polar entities. A more general inclusive term is van der Waals' forces which includes dipole–dipole and dipole-induced dipole as well as London dispersion forces.

HYDROGEN BONDING

Hydrogen bonding occurs when hydrogen is bonded directly to a small highly electronegative element, such as fluorine, oxygen, or nitrogen. As the electron pair is drawn away from the hydrogen atom by the electronegative element, all that remains is the proton in the nucleus as there are no inner electrons. The proton attracts a non-bonding pair of electrons from the F, N, or O resulting in a much stronger dipole–dipole attraction. Water has a much higher boiling point than the other group 16 hydrides as the hydrogen bonding between water molecules is much stronger than the dipole–dipole bonding in the remaining hydrides. A similar trend is seen in the hydrides of group 15 and group 17. Hydrogen bonds between the molecules in ice result in a very open structure. When ice melts the molecules can move closer to each other so that water has its maximum density at 4 °C.

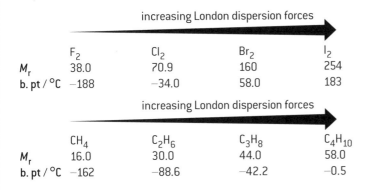

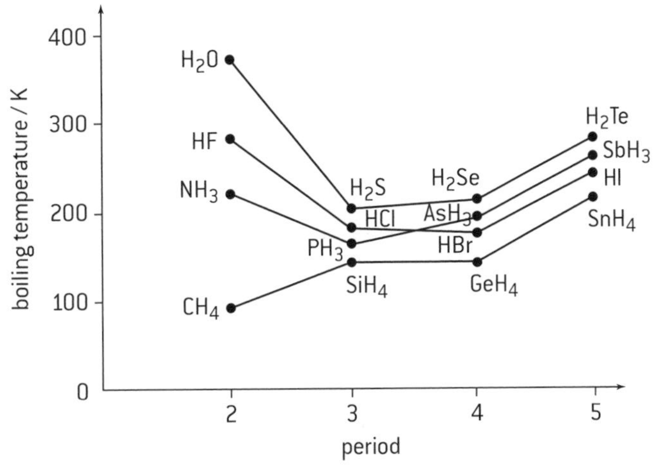

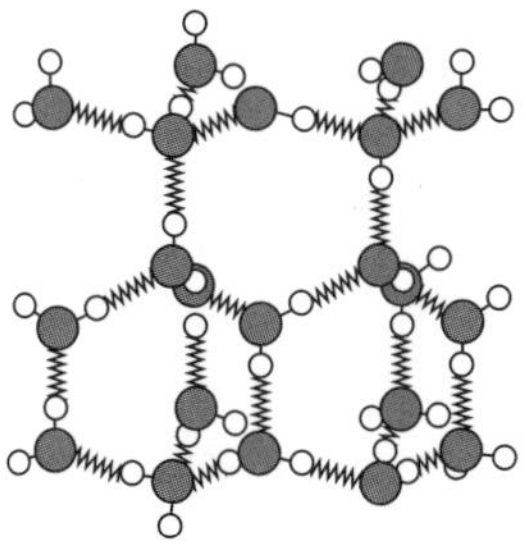

the ice lattice
⦚ = hydrogen bond

Physical properties related to bonding type

MELTING AND BOILING POINTS

When a liquid turns into a gas the attractive forces between the particles are completely broken so boiling point is a good indication of the strength of intermolecular forces. When solids melt, the crystal structure is broken down, but there are still some attractive forces between the particles. Melting points do give an indication of the strength of intermolecular forces but they are also determined by the way the particles pack in the crystal state. They are also affected by impurities. Impurities weaken the structure and result in lower melting points.

Covalent bonds are very strong so macromolecular covalent structures have extremely high melting and boiling points. For example, diamond, which has a giant tetrahedral structure, melts in the region of 4000 °C and silicon dioxide, SiO_2, which has a similar structure, melts at over 1600 °C. Graphite has very strong bonds between the carbon atoms in its hexagonal layers and has a similar melting point to diamond. Metals (see next page) and ionic compounds also tend to have relatively high melting and boiling points due to ionic attractions. Although it might be expected that ionic compounds with smaller more highly charged ions have higher melting points and boiling points the facts do not support this.

Ionic compound	Melting point / °C	Boiling point / °C
LiCl	605	1382
NaCl	801	1413
KCl	770	1420
$MgCl_2$	714	1412

The melting and boiling points of simple covalent molecules depend on the type of forces of attraction between the molecules. These follow the order:

hydrogen bonding > dipole–dipole > London dispersion forces

The weaker the attractive forces the more volatile the substance.

For example, propane, ethanal and ethanol have similar molar masses but there is a considerable difference in their melting points.

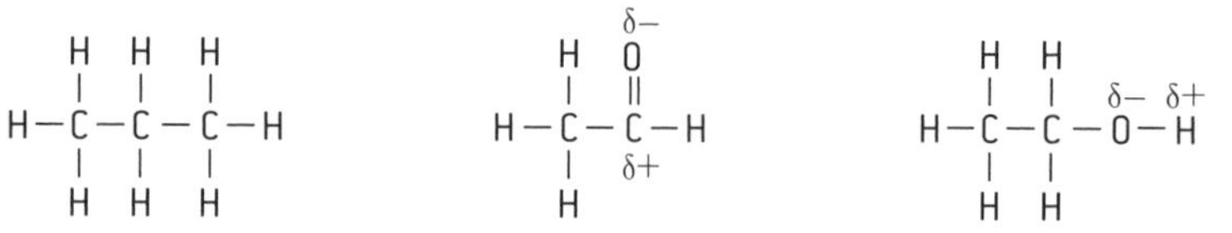

Compound	propane	ethanal	ethanol
M_r	44	44	46
M. pt / °C	−42.2	20.8	78.5
Polarity	non-polar	polar	polar
Bonding type	London dispersion	dipole–dipole	hydrogen bonding

SOLUBILITY

'Like tends to dissolve like'. Polar substances tend to dissolve in polar solvents, such as water, whereas non-polar substances tend to dissolve in non-polar solvents, such as heptane or tetrachloromethane. Organic molecules often contain a polar head and a non-polar carbon chain tail. As the non-polar carbon chain length increases in an homologous series the molecules become less soluble in water. Ethanol itself is a good solvent for other substances as it contains both polar and non-polar ends.

CH_3OH
C_2H_5OH
C_3H_7OH
C_4H_9OH

decreasing solubility in water ↓

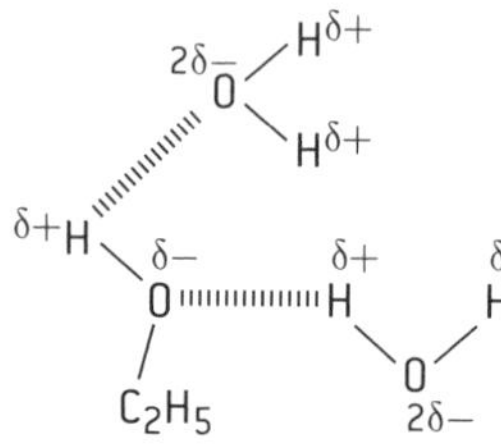

Ethanol is completely miscible with water as it can hydrogen-bond to water molecules.

CONDUCTIVITY

For conductivity to occur the substance must possess electrons or ions that are free to move. Metals (and graphite) contain delocalized electrons and are excellent conductors. Molten ionic salts also conduct electricity, but are chemically decomposed in the process. Where all the electrons are held in fixed positions, such as diamond or in simple molecules, no electrical conductivity occurs.

When a potential gradient is applied to the metal, the delocalized electrons can move towards the positive end of the gradient carrying charge.

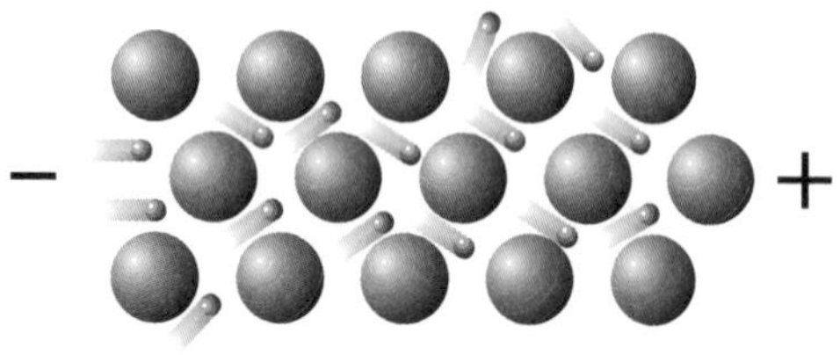

When an ionic compound melts, the ions are free to move to oppositely charged electrodes. Note: in molten ionic compounds it is the ions that carry the charge, not free electrons.

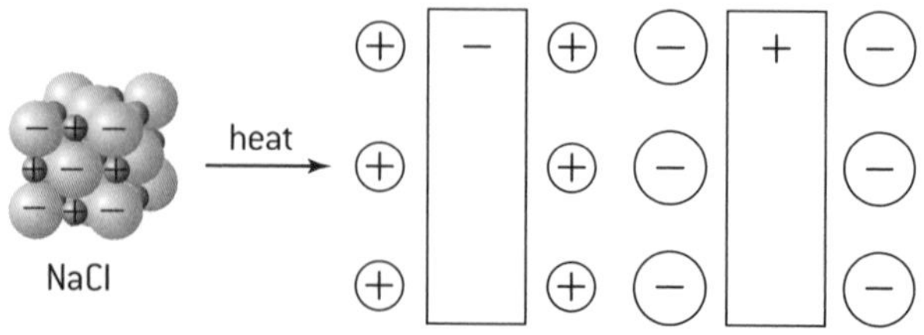

Metals and alloys

METALLIC BONDING

The valence electrons in metals become detached from the individual atoms so that metals consist of a close packed lattice of positive ions in a 'sea' of delocalized electrons. A metallic bond is the attraction that two neighbouring positive ions have for the delocalized electrons between them.

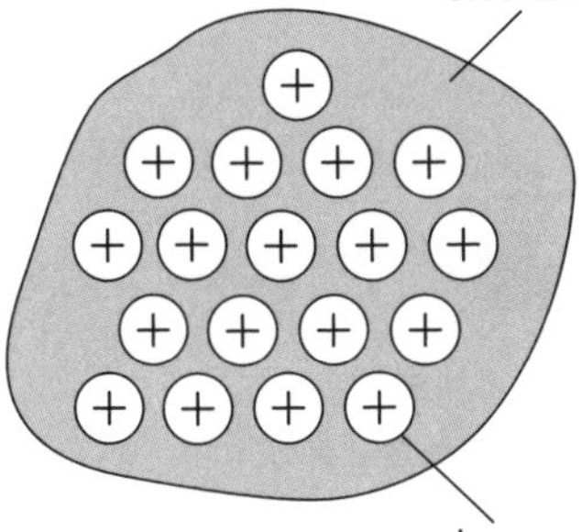

Generally the strength of a metallic bond depends on the charge of the ions and the radius of the metal ion.

Metals are malleable, that is, they can be bent and reshaped under pressure. They are also ductile, which means they can be drawn out into a wire.

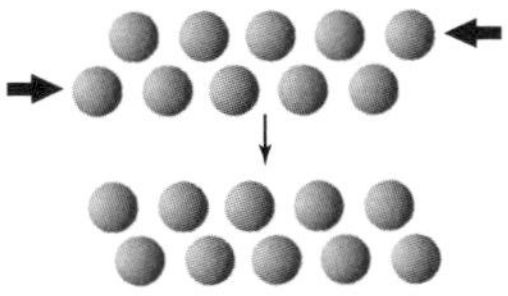

Metals are malleable and ductile because the close-packed layers of positive ions can slide over each other without breaking more bonds than are made.

MELTING POINTS OF METALS

Although most metals tend to have quite high melting points, mercury is a liquid at room temperature and the group 1 elements (alkali metals) all melt below 181 °C. The trend in group 1 clearly follows the pattern that the smaller the metal ion formed when the valence electrons delocalize the stronger the metallic bond and the higher the melting point.

	Li	Na	K	Rb	Cs
M. pt / °C	181	97.8	63.5	39.3	28.5

However this logic only just holds true across period 3 (Na to Al) even though the charge on the ion is also increasing at the same time as the size of the ion is decreasing. It breaks down in group 14 as tin, which has a smaller ionic radius than lead, has a lower boiling point.

	Na	Mg	Al	Sn	Pb
M. pt / °C	97.8	650	660	232	328

This is because the melting point does not only depend upon the size and charge of the ion formed when the valence electrons are delocalized but also on the way in which the atoms are arranged in the solid metal.

ALLOYS

Alloys are sometimes termed a metallic solid solution. They are usually made up of more than one metal although steel is an alloy of iron and carbon. Some common alloys are brass, bronze, solder, pewter and amalgams.

Alloy	**Principal metal**	**Added metal(s)**
Brass	copper	zinc
Bronze	copper	tin
Solder	lead	tin (some may have more tin than lead)
Pewter	tin	copper, antimony, bismuth or lead
Amalgams	mercury	e.g. tin, silver, gold or sodium

The addition of another metal to a metallic element alters its properties. The added metals are likely to have a different radius and may have a different charge and so distort the structure of the original metal as the bonding is less directional. One obvious example of this is that alloys may have lower melting points than their component metals. For example, before copper and plastic were used for water piping lead tended to be used (the origin of the word *plumber*). Lead melts at 328 °C and when pipes were being joined or repaired there was a danger of melting the actual pipe if too much heat was employed. Solder has a much lower melting point (typically 180–190 °C) and can be used to weld lead pipes together or to secure wires to terminals in an electrical circuit.

Generally alloys are less ductile and less malleable than pure metals as the added impurities disturb the lattice. This also tends to make alloys harder than the pure metals they are derived from. For example, aluminium is a soft, ductile and malleable metal. When it is alloyed with another soft metal such as copper the resulting aluminium alloy is much harder and stronger and yet still retains much of its low density. Small amounts of carbon added to iron produce steel with a high tensile strength. If chromium is also added it produces stainless steel, an alloy of steel with a much increased resistance to corrosion.

HL Molecular orbitals

COMBINATION OF ATOMIC ORBITALS TO FORM MOLECULAR ORBITALS

Although the Lewis representation is a useful model to represent covalent bonds it does make the false assumption that all the valence electrons are the same. A more advanced model of bonding considers the combination of atomic orbitals to form molecular orbitals.

σ bonds

A σ (sigma) bond is formed when two atomic orbitals on different atoms overlap along a line drawn through the two nuclei. This occurs when two s orbitals overlap, an s orbital overlaps with a p orbital, or when two p orbitals overlap 'head on'.

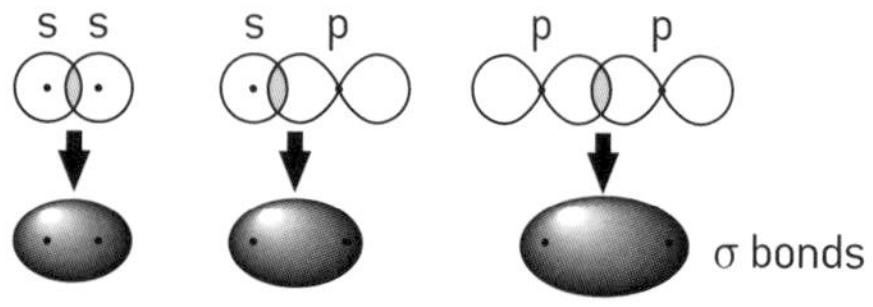

π bonds

A π (pi) bond is formed when two p orbitals overlap 'sideways on'. The overlap now occurs above and below the line drawn through the two nuclei. A π bond is made up of two regions of electron density.

overlap above and
below line of centres

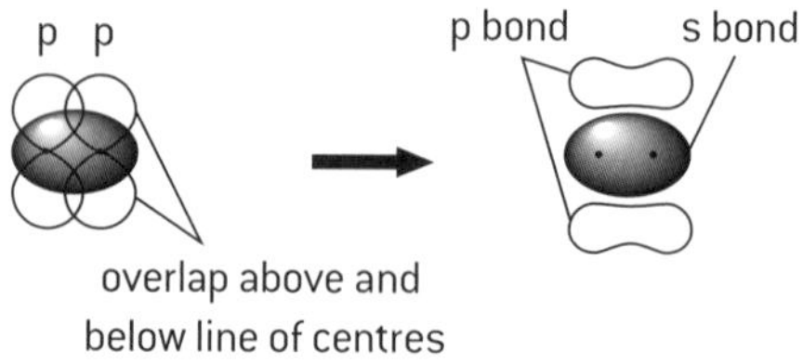

DELOCALIZATION OF ELECTRONS

Resonance involves using two or more Lewis structures to represent a particular molecule or ion where the structure cannot be described by using a single Lewis structure. They can also be explained by the delocalization of electrons. For example, in the ethanoate ion the carbon atom and the two oxygen atoms each have a p orbital containing one electron after the σ bonds have been formed. Instead of forming just one double bond between the carbon atom and one of the oxygen atoms the electrons can delocalize over all three atoms. This is energetically more favourable than forming just one double bond.

Delocalization can occur whenever alternate double and single bonds occur between carbon atoms. The delocalization energy in benzene is about 150 kJ mol^{-1}, which explains why the benzene ring is so resistant to addition reactions.

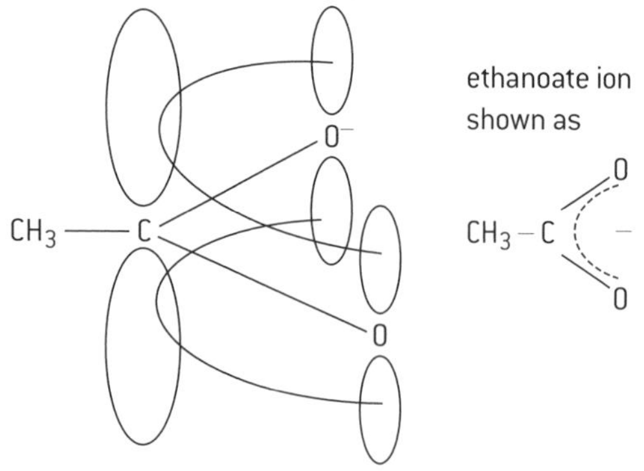

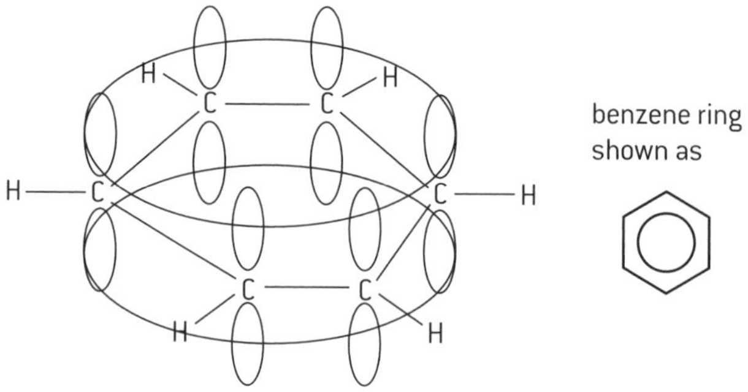

FORMAL CHARGE

Formal charge is a technique used in chemistry that is based on a false assumption but which can be useful for determining which of several potential Lewis structures is preferred when two or more are possible. It assumes that all atoms in a molecule or ion have the same electronegativity (the false assumption) and is equal to the (Number of **V**alence electrons) − (Number of **N**on-bonding electrons) − ½ (Number of **B**onding electrons). This can be described by the formula $FC = V - N - \frac{1}{2}B$ although note that this formula is not given in the IB data booklet. The preferred structure is the one where the individual atoms have the lowest possible formal charge.

For example, consider two possible structures for carbon dioxide, both of which obey the octet rule.

$$\langle O = C = O \rangle \qquad \mathord{|}O \equiv C - \overline{O}\mathord{|}$$

(0)　(0)　(0)　　　　(+1)　(0)　(−1)

Formal charges

$$C = 4 - 0 - (\tfrac{1}{2} \times 8) = 0 \qquad\qquad C = 4 - 0 - (\tfrac{1}{2} \times 8) = 0$$

$$O = 6 - 4 - (\tfrac{1}{2} \times 4) = 0 \qquad\qquad O = 6 - 2 - (\tfrac{1}{2} \times 6) = +1 \text{ (O with triple bond)}$$

$$O = 6 - 6 - (\tfrac{1}{2} \times 2) = -1 \text{ (O with single bond)}$$

Both give a total formal charge of zero but the preferred structure is the first one with the two double bonds as the individual atoms have the lowest formal charges.

Oxygen and ozone

IMPORTANCE OF THE OZONE LAYER

From the Lewis structures of both oxygen and ozone it can be seen that the double bond in oxygen is stronger than the 'one and a half' bond between the oxygen atoms in ozone.

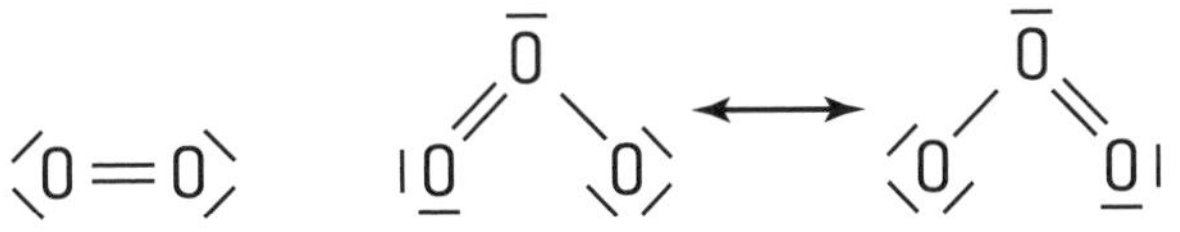

Lewis structures of oxygen and ozone

This difference in bond enthalpies helps to protect us from the Sun's harmful ultraviolet radiation. The ozone layer occurs in the stratosphere between about 12 km and 50 km above the surface of the Earth. Stratospheric ozone is in dynamic equilibrium with oxygen and is continually being formed and decomposed. The strong double bond in oxygen is broken by high energy ultraviolet light from the Sun to form atoms. These oxygen atoms are called radicals as they possess an unpaired electron and are very reactive. One oxygen radical can then react with an oxygen molecule to form ozone.

$$O{=}O\,(g) \xrightarrow{\text{UV (high energy)}} 2O^{\bullet}(g)$$

$$O^{\bullet}(g) + O_2(g) \rightarrow O_3(g)$$

The weaker bonds in ozone require ultraviolet light of less energy to break them. When they are broken the reverse process happens and the ozone breaks down back to an oxygen molecule and an oxygen radical. The radical can then react with another ozone molecule to form two oxygen molecules.

$$O_3(g) \xrightarrow{\text{UV (lower energy)}} O_2(g) + O^{\bullet}(g)$$

$$O_3(g) + O^{\bullet}(g) \rightarrow 2O_2(g)$$

Overall the rate of production of ozone is equal to the rate of ozone destruction – this process, during which a wide range of ultraviolet light is absorbed, is known as a steady state. Human-made pollutants such as CFCs and oxides of nitrogen can disrupt this process and in recent years large 'holes' in the ozone layer have appeared – particularly in winter and early spring over the South and North Poles.

WAVELENGTH OF UV LIGHT NECESSARY FOR O_2 AND O_3 DISSOCIATION

The bond enthalpy for the O=O double bond is given as 498 kJ mol^{-1} in the IB data booklet. For just one double bond this equates to 8.27×10^{-19} J. The wavelength of light that corresponds to this enthalpy value (E) can be calculated by combining the expressions $E = h\nu$ and $c = \lambda\nu$ to give $\lambda = \frac{hc}{E}$ where h is Planck's constant and c is the velocity of light.

$$\lambda = \frac{6.63 \times 10^{-34}\ (\text{J s}) \times 3.00 \times 10^{8}\ (\text{m s}^{-1})}{8.27 \times 10^{-19}\ (\text{J})} = 241\ \text{nm}$$

This is in the high energy region of the ultraviolet spectrum. Ozone is described above as two resonance hybrids. An alternative bonding model is to consider the π electrons to be delocalized over all three oxygen atoms. In both models the bond order is 1.5, i.e. with an enthalpy between an O–O single bond (144 kJ mol^{-1}) and an O=O double bond (498 kJ mol^{-1}) so ultraviolet light with a longer wavelength (lower energy) is absorbed in breaking the ozone bonds. The actual wavelength required is 330 nm. Working backwards this gives the strength of the O–O bond in ozone as 362 kJ mol^{-1}.

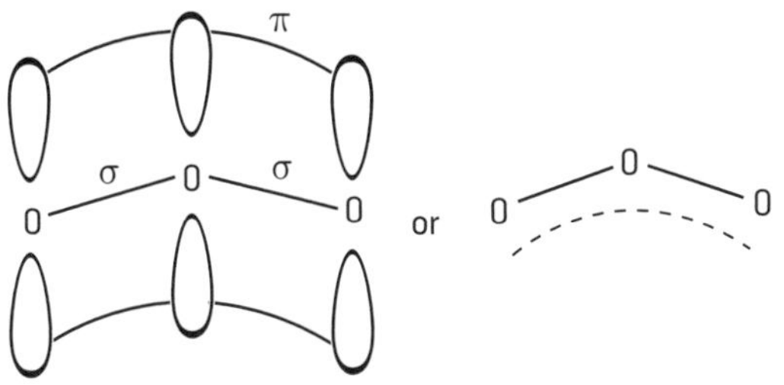

delocalized π bond in ozone

CATALYSIS AND OZONE DESTRUCTION BY CFCs AND NO_x

CFCs catalyse the destruction of ozone because the high energy ultraviolet light in the stratosphere causes the homolytic fission of the C–Cl bond to produce chlorine radicals. Note that it is the C–Cl bond that breaks, not the C–F bond, as the C–Cl bond strength is weaker. These radicals then break down ozone molecules and regenerate more radicals so that the process continues until the radicals eventually escape or terminate. It has been estimated that one molecule of a CFC can catalyse the breakdown of up to 100 000 molecules of ozone.

$CCl_2F_2(g) \longrightarrow CClF_2^{\bullet}(g) + Cl^{\bullet}(g)$ (radical initiation)

$Cl^{\bullet}(g) + O_3(g) \longrightarrow ClO^{\bullet}(g) + O_2(g)$ (propagation

$ClO^{\bullet}(g) + O^{\bullet}(g) \longrightarrow Cl^{\bullet}(g) + O_2(g)$ of radicals)

Evidence to support this mechanism is that the increase in the concentration of chlorine monoxide in the stratosphere over the Antarctic has been shown to mirror the decrease in the ozone concentration.

Nitrogen oxides also catalytically decompose ozone by a radical mechanism. The overall mechanism is complex. Essentially oxygen radicals are generated by the breakdown of NO_2 in ultraviolet light.

$$NO_2(g) \longrightarrow NO(g) + O^{\bullet}(g)$$

The oxygen radicals then react with ozone

$$O^{\bullet}(g) + O_3(g) \longrightarrow 2O_2(g)$$

The nitrogen oxide can also react with ozone to regenerate the catalyst

$$NO(g) + O_3(g) \longrightarrow NO_2(g) + O_2(g)$$

The overall reaction can be simplified as:

$$2O_3(g) \xrightarrow{NO_2(g)} 3O_2(g)$$

Hybridization (1)

sp^3 HYBRIDIZATION

Methane provides a good example of sp^3 hybridization. Methane contains four equal C–H bonds pointing towards the corners of a tetrahedron with bond angles of 109.5°. A free carbon atom has the configuration $1s^2 2s^2 2p^2$. It cannot retain this configuration in methane. Not only are there only two unpaired electrons, but the p orbitals are at 90° to each other and will not give bond angles of 109.5° when they overlap with the s orbitals on the hydrogen atoms.

When the carbon bonds in methane one of its 2s electrons is promoted to a 2p orbital and then the 2s and three 2p orbitals *hybridize* to form four new hybrid orbitals. These four new orbitals arrange themselves to be as mutually repulsive as possible, i.e. tetrahedrally. Four equal σ bonds can then be formed with the hydrogen atoms.

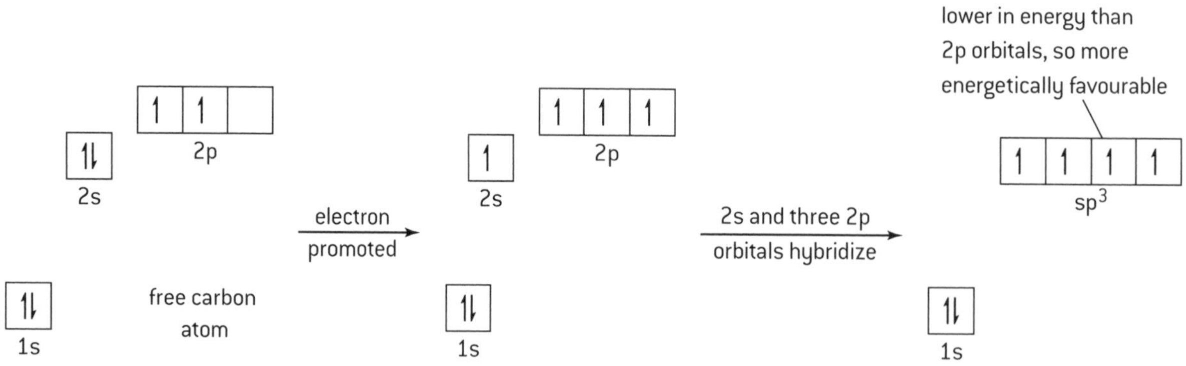

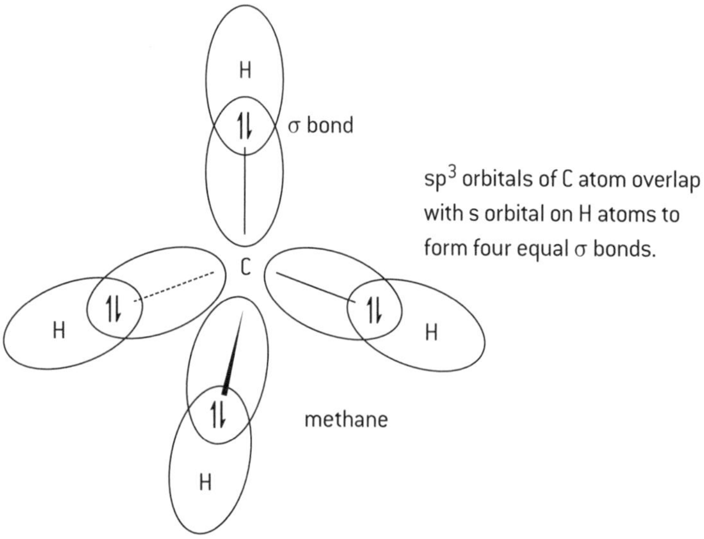

sp^2 HYBRIDIZATION

sp^2 hybridization occurs in ethene. After a 2s electron on the carbon atom is promoted the 2s orbital hybridizes with two of the 2p orbitals to form three new planar hybrid orbitals with a bond angle of 120° between them. These can form σ bonds with the hydrogen atoms and also a σ bond between the two carbon atoms. Each carbon atom now has one electron remaining in a 2p orbital. These can overlap to form a π bond. Ethene is thus a planar molecule with a region of electron density above and below the plane.

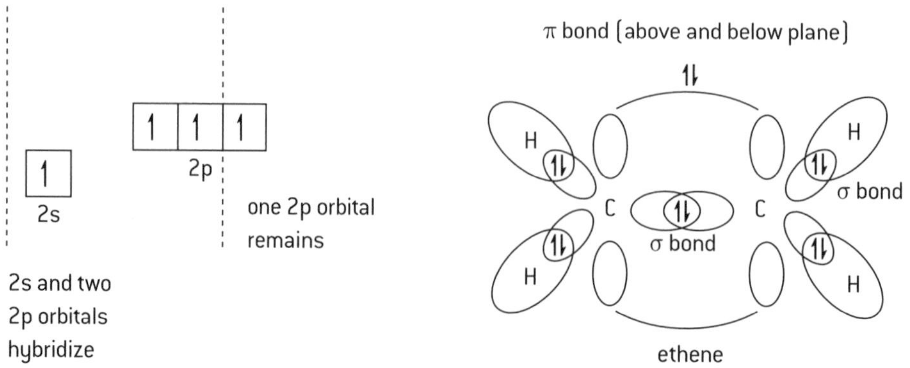

Hybridization (2)

sp HYBRIDIZATION

sp hybridization occurs when the 2s orbital hybridizes with just one of the 2p orbitals to form two new linear sp hybrid orbitals with an angle of 180° between them. The remaining two p orbitals on each carbon atom then overlap to form two π bonds. An example is ethyne.

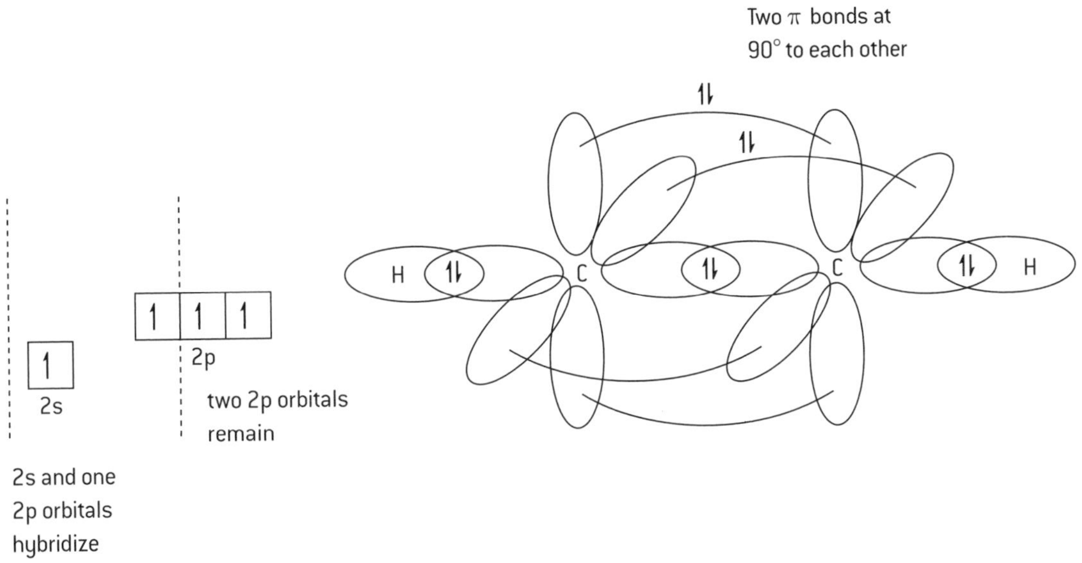

PREDICTING THE TYPE OF BONDING AND HYBRIDIZATION IN MOLECULES

In carbon compounds containing single, double or triple bonds the numbers of each different type of bond (σ or π) and the type of hybridization shown by each carbon atom can be deduced.

Vitamin A contains 5 C=C double bonds, 15 C–C bonds, 29 C–H bonds, 1 C–O bond and 1 O–H bond.

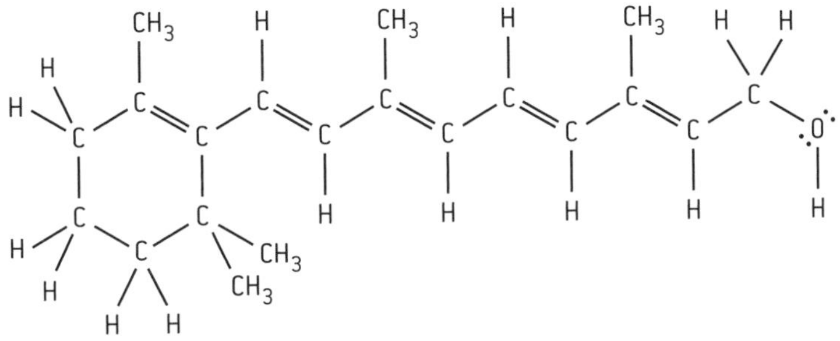

Each single bond is a σ bond and each double bond contains one σ and one π bond so there is a total of 51 σ bonds and 5 π bonds in the molecule. Each carbon atom either side of a double bond is sp^2 hybridized so 10 of the carbon atoms are sp^2 hybridized. The remaining 10 carbon atoms are sp^3 hybridized. The oxygen atom, which has four pairs of electrons around it, is also sp^3 hybridized.

RELATIONSHIP BETWEEN TYPE OF HYBRIDIZATION, LEWIS STRUCTURE, AND MOLECULAR SHAPES

Molecular shapes can be arrived at either by using the VSEPR theory or by knowing the type of hybridization. Hybridization can take place between any s and p orbital in the same energy level and is not just restricted to carbon compounds. If the shape and bond angles are known from using Lewis structures then the type of hybridization can be deduced. Similarly if the type of hybridization is known the shape and bond angles can be deduced.

Hybridization	Regular bond angle	Examples
sp^3	109.5°	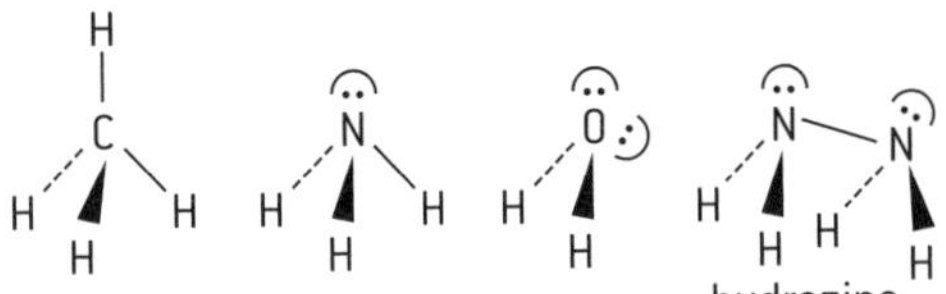
sp^2	120°	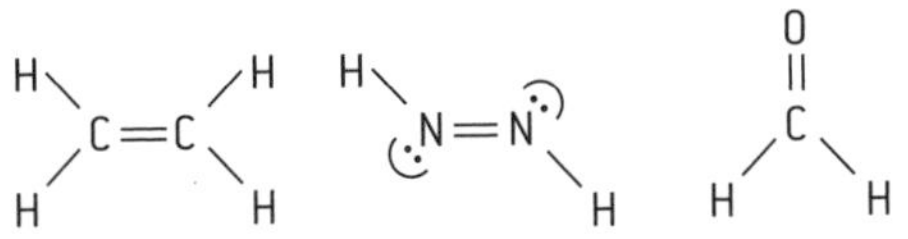
sp	180°	H—C≡C—H (:N≡N:)

MULTIPLE CHOICE QUESTIONS – CHEMICAL BONDING AND STRUCTURE

1. What are the correct formulas of the following ions?

	Nitrate	Phosphate	Carbonate	Ammonium
A.	NO_3^-	PO_4^{3-}	CO_3^-	NH_3^+
B.	NO_3^{2-}	PO_3^{2-}	CO_3^{2-}	NH_3^+
C.	NO_3^-	PO_4^{3-}	CO_3^{2-}	NH_4^+
D.	NO_3^{2-}	PO_3^{2-}	CO_3^{2-}	NH_4^+

2. Which is the correct Lewis structure for ethene?

A. 

C. 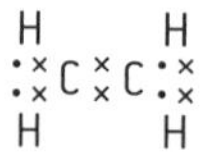

B. 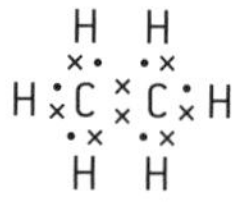

D. 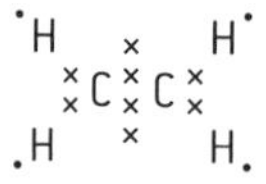

3. Given the following electronegativities, H: 2.2 N: 3.0 O: 3.5 F: 4.0, which bond would be the most polar?

A. O–H in H_2O C. N–O in NO_2

B. N–F in NF_3 D. N–H in NH_3

4. Which substance is made up of a lattice of positive ions and free moving electrons?

A. Graphite C. Sulfur

B. Sodium chloride D. Sodium

5. When CH_4, NH_3, H_2O, are arranged in order of **increasing** bond angle, what is the correct order?

A. CH_4, NH_3, H_2O C. NH_3, CH_4, H_2O

B. NH_3, H_2O, CH_4 D. H_2O, NH_3, CH_4

6. Which order is correct when the following compounds are arranged in order of **increasing** melting point?

A. $CH_4 < H_2S < H_2O$ C. $CH_4 < H_2O < H_2S$

B. $H_2S < H_2O < CH_4$ D. $H_2S < CH_4 < H_2O$

7. Which species contain a coordinate covalent bond?

I. HCHO

II. CO

III. H_3O^+

A. I and II only C. II and III only

B. I and III only D. I, II and III

8. Which is the correct order for **decreasing** H–N–H bond angles in the species NH_2^-, NH_3 and NH_4^+ (largest bond angle first)?

A. NH_3, NH_2^-, NH_4^+ C. NH_2^-, NH_4^+, NH_3

B. NH_4^+, NH_3, NH_2^- D. NH_2^-, NH_3, NH_4^+

9. Which is the correct order for **increasing** intermolecular forces of attraction (smallest force first)?

A. Covalent bonds, hydrogen bonds, dipole–dipole, London dispersion forces

B. London dispersion forces, dipole–dipole, hydrogen bonds, covalent bonds

C. London dispersion forces, hydrogen bonds, dipole–dipole, covalent bonds

D. Covalent bonds, dipole–dipole, hydrogen bonds, London dispersion forces

10. Which statement best explains why alloys tend to be less malleable than pure metals?

A. The added metal has more valence electrons so increases the amount of delocalization.

B. The added metal prevents the layers from being drawn out into a wire.

C. The added metal disturbs the lattice so the layers are less able to slide over each other.

D. The added metal acts as an impurity and so lowers the melting point.

HL

11. Which of the following species contains at least one atom that is sp^2 hybridized?

A. hydrogen cyanide, HCN

B. 2-methylpropane, $CH_3CH(CH_3)CH_3$

C. propanone, CH_3COCH_3

D. ethanol, CH_3CH_2OH

12. How many σ and π bonds are present in propanal, CH_3CH_2CHO?

A. 8σ and 2π C. 5σ and 1π

B. 8σ and 1π D. 9σ and 1π

13. Which species have delocalized electrons?

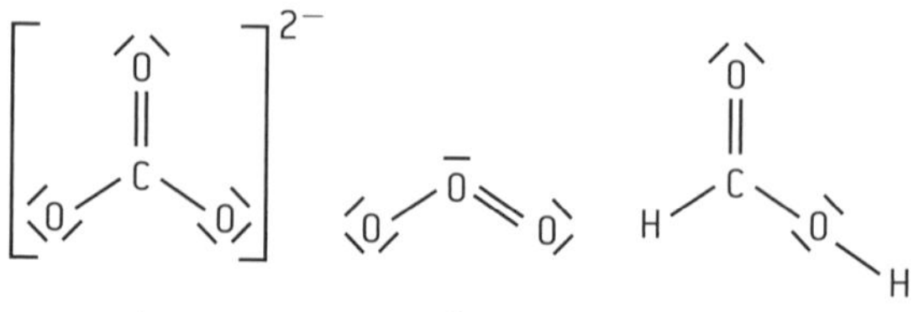

I II III

A. I and II only C. II and III only

B. I and III only D. I, II and III

14. What is the formal charge on the oxygen atom in the hydronium ion H_3O^+?

A. −2 C. 0

B. −1 D. +1

15. Which describes the shape of the SF_4 molecule?

A. Tetrahedral C. Square planar

B. Distorted tetrahedral D. Trigonal bipyramidal

16. Which statements about graphene are correct?

I. It can be considered as a single layer of graphite

II. The hybridization of the carbon atoms is sp^2

III. It is an allotrope of carbon

A. I and II only

B. I and III only

C. II and III only

D. I, II and III

SHORT ANSWER QUESTIONS – CHEMICAL BONDING AND STRUCTURE

1. PF_3, SF_2 and SiF_4 have different shapes. Draw their Lewis structures and use the VSEPR theory to predict the name of the electron domain geometry and the molecular shape of each molecule. [8]

2. a) (i) Draw the Lewis structure of NH_3, state its molecular shape and deduce and explain the H–N–H bond angle in NH_3. [4]

 (ii) The graph below shows the boiling points of the hydrides of group 15. Discuss the variation in the boiling points. [4]

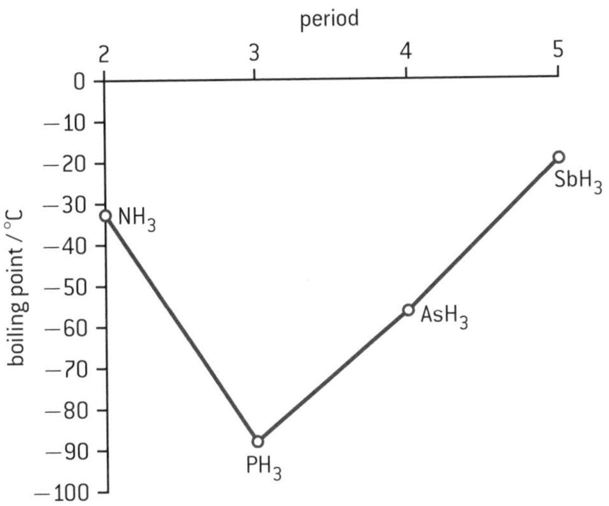

b) Explain, using diagrams, why CO and NO_2 are polar molecules but CO_2 is a non-polar molecule. [5]

3. Ethane, C_2H_6, and disilane, Si_2H_6, are both hydrides of group 14 elements with similar structures but with different chemical properties.

 a) Deduce the Lewis (electron dot) structure for Si_2H_6 showing all valence electrons. [1]

 b) State and explain the H–Si–H bond angle in Si_2H_6. [2]

 c) State which of the bonds, Si–H or C–H, is more polar. Explain your choice. [2]

 d) Predict, with an explanation, the polarity of the two molecules. [2]

 e) Explain why disilane has a higher boiling point than ethane. [2]

4. a) State and explain which of propan-1-ol, $CH_3CH_2CH_2OH$, and methoxyethane, $CH_3OCH_2CH_3$, is more volatile. [3]

 b) Propan-1-ol, $CH_3CH_2CH_2OH$, and hexan-1-ol, $CH_3(CH_2)4CH_2OH$, are both alcohols. State and explain which compound is more soluble in water. [2]

 c) Graphite is used as a lubricant and is an electrical conductor. Diamond is hard and does not conduct electricity. Explain these statements in terms of the structure and bonding of these allotropes of carbon. [6]

5. a) State the **full** and the **condensed** electron configuration for chlorine. [2]

 b) Deduce the orbital diagram for silicon using a box ☐ to represent an orbital and ↑ and ↓ to represent electrons with opposite spins. [2]

 c) Explain why chlorine forms an ionic compound with sodium but a covalent compound with silicon. [2]

 HL

6. a) Ozone and sulfur hexafluoride are greenhouse gases.

 (i) Draw the Lewis structure of sulfur hexafluoride. [1]

 (ii) Explain why sulfur can expand its octet whereas oxygen cannot. [1]

 (iii) Deduce the electron domain geometry for both ozone and sulfur hexafluoride and deduce their molecular shape. [2]

 (iv) Deduce the bond angles in ozone and sulfur hexafluoride [2]

 b) Another greenhouse gas is dichlorodifluoromethane, CCl_2F_2. This gas can also cause destruction of the ozone layer.

 (i) Determine the wavelength of light required to break the C–F and the C–Cl bonds. [4]

 (ii) Suggest why dichlorodifluoromethane is unreactive in the atmosphere near the surface of the Earth but reactive in the ozone layer. [2]

7. Two Lewis structures that obey the octet rule can be proposed for the nitronium ion, NO_2^+.

$$\left[\,|\overline{\underline{O}} - N \equiv O| \, \right]^+ \qquad \left[\langle O = N = O \rangle \right]^+$$

 I II

 Deduce the formal charge for each atom in both of the two proposed structures and determine which structure is the most likely. [4]

8. a) Describe the bonding within the carbon monoxide molecule. [2]

 b) Describe the delocalization of π (pi) electrons and explain how this can account for the structure and stability of the carbonate ion, CO_3^{2-}. [3]

 c) Explain the meaning of the term hybridization. State the type of hybridization shown by the carbon atoms in carbon dioxide, diamond, graphite and the carbonate ion. [5]

 d) (i) Explain the electrical conductivity of molten sodium oxide and liquid sulfur trioxide. [2]

 (ii) Samples of sodium oxide and solid sulfur trioxide are added to separate beakers of water. Deduce the equation for each reaction and predict the electrical conductivity of each of the solutions formed. [3]

Measuring enthalpy changes

EXOTHERMIC AND ENDOTHERMIC REACTIONS

Energy is defined as the ability to do work, that is, move a force through a distance. It is measured in joules.

Energy = force × distance
(J) (N × m)

In a chemical reaction energy is required to break the bonds in the reactants, and energy is given out when new bonds are formed in the products. The most important type of energy in chemistry is heat. If the bonds in the products are stronger than the bonds in the reactants then the reaction is said to be **exothermic**, as heat is given out to the surroundings. Examples of exothermic processes include combustion and neutralization. In **endothermic** reactions heat is absorbed from the surroundings because the bonds in the reactants are stronger than the bonds in the products.

The internal energy stored in the reactants is known as its **enthalpy**, *H*. The absolute value of the enthalpy of the reactants cannot be known, nor can the enthalpy of the products, but what can be measured is the difference between them, ΔH. By convention ΔH has a negative value for exothermic reactions and a positive value for endothermic reactions. It is normally measured under standard conditions of 100 kPa pressure at a temperature of 298 K. The **standard enthalpy change of a reaction** is denoted by $\Delta H^{\ominus}$.

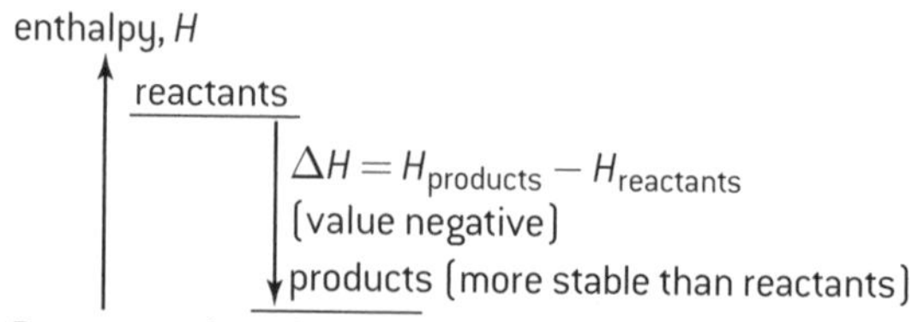

Representation of exothermic reaction using an enthalpy diagram.

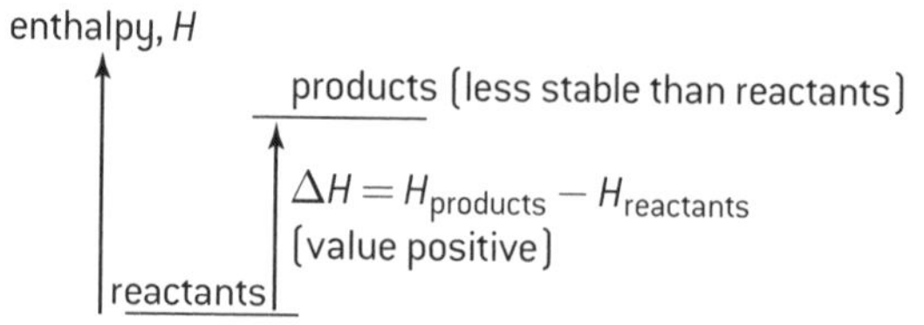

Representation of endothermic reaction using an enthalpy diagram.

TEMPERATURE AND HEAT

It is important to be able to distinguish between heat and temperature as the terms are often used loosely.

- Heat is a measure of the total energy in a given amount of substance and therefore depends on the amount of substance present.
- Temperature is a measure of the 'hotness' of a substance. It represents the average kinetic energy of the substance, but is independent of the amount of substance present.

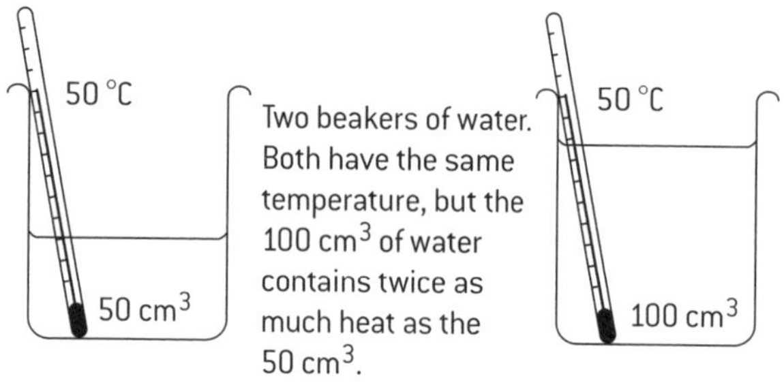

50 °C
50 cm3
Two beakers of water. Both have the same temperature, but the 100 cm3 of water contains twice as much heat as the 50 cm3.
50 °C
100 cm3

CALORIMETRY

The enthalpy change for a reaction can be measured experimentally by using a calorimeter. In a simple calorimeter all the heat evolved in an exothermic reaction is used to raise the temperature of a known mass of water. For endothermic reactions the heat transferred from the water to the reaction can be calculated by measuring the lowering of temperature of a known mass of water.

To compensate for heat lost by the water in exothermic reactions to the surroundings as the reaction proceeds a plot of temperature against time can be drawn. By extrapolating the graph, the temperature rise that would have taken place had the reaction been instantaneous can be calculated.

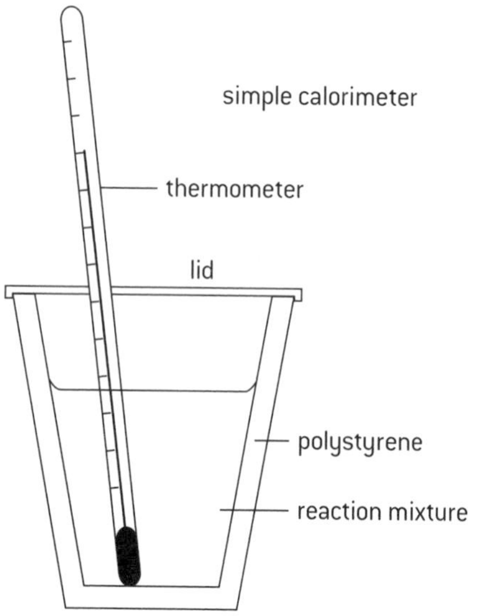

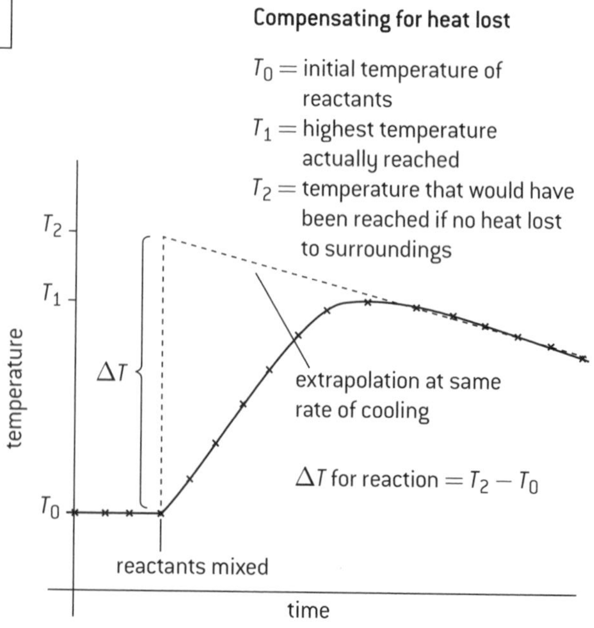

ΔH calculations

CALCULATION OF ENTHALPY CHANGES

The heat involved in changing the temperature of any substance can be calculated from the equation:

Heat energy = mass (m) × specific heat capacity (c) × temperature change (ΔT)

The specific heat capacity of water is 4.18 kJ kg^{-1} K^{-1}. That is, it requires 4.18 kilojoules of energy to raise the temperature of one kilogram of water by one kelvin.

Enthalpy changes are normally quoted in kJ mol^{-1}, for either a reactant or a product, so it is also necessary to work out the number of moles involved in the reaction which produces the heat change in the water.

WORKED EXAMPLE 1

50.0 cm^3 of 1.00 mol dm^{-3} hydrochloric acid solution was added to 50.0 cm^3 of 1.00 mol dm^{-3} sodium hydroxide solution in a polystyrene beaker. The initial temperature of both solutions was 16.7 °C. After stirring and accounting for heat loss the highest temperature reached was 23.5 °C. Calculate the enthalpy change for this reaction.

Step 1. Write equation for reaction

$HCl(aq) + NaOH(aq) \rightarrow NaCl(aq) + H_2O(l)$

Step 2. Calculate molar quantities

Amount of HCl = $\frac{50.0}{1000} \times 1.00 = 5.00 \times 10^{-2}$ mol

Amount of NaOH = $\frac{50.0}{1000} \times 1.00 = 5.00 \times 10^{-2}$ mol

Therefore the heat evolved will be for 5.00×10^{-2} mol

Step 3. Calculate heat evolved

Total volume of solution = 50.0 + 50.0 = 100 cm^3

Assume the solution has the same density and specific heat capacity as water then

mass of 'water' = 100 g = 0.100 kg

Temperature change = 23.5 − 16.7 = 6.8 °C = 6.8 K

Heat evolved in reaction = 0.100 × 4.18 × 6.8 = 2.84 kJ

= 2.84 kJ (for 5.00×10^{-2} mol)

ΔH for reaction = $-2.84 \times \frac{1}{5.00 \times 10^{-2}} = -56.8$ kJ mol^{-1}

(negative value as the reaction is exothermic)

WORKED EXAMPLE 2

A student used a simple calorimeter to determine the enthalpy change for the combustion of ethanol.

$C_2H_5OH(l) + 3O_2(g) \rightarrow 2CO_2(g) + 3H_2O(l)$

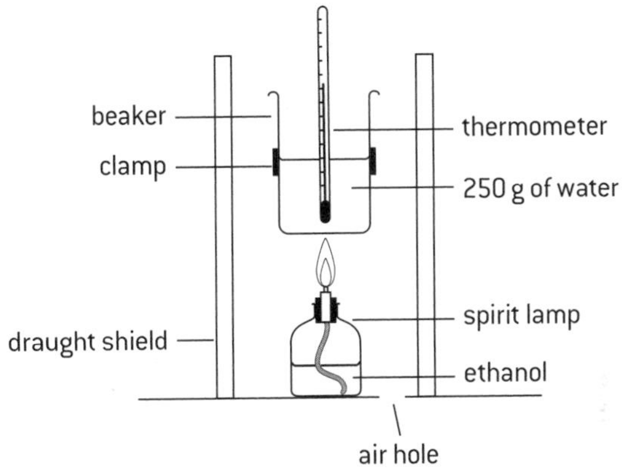

When 0.690g (0.015 mol) of ethanol was burned it produced a temperature rise of 13.2 K in 250 g of water. Calculate ΔH for the reaction.

Heat evolved by 0.015 mol = $\frac{250}{1000} \times 4.18 \times 13.2 = 13.79$ kJ

$\Delta H = -13.79 \times \frac{1}{0.015} = -920$ kJ mol^{-1}

Note: the IB data booklet value is −1371 kJ mol^{-1}. Reasons for the discrepancy include the fact that not all the heat produced is transferred to the water, the water loses some heat to the surroundings, and there is incomplete combustion of the ethanol.

WORKED EXAMPLE 3

50.0 cm^3 of 0.200 mol dm^{-3} copper(II) sulfate solution was placed in a polystyrene cup. After two minutes 1.20 g of powdered zinc was added. The temperature was taken every 30 seconds and the following graph obtained. Calculate the enthalpy change for the reaction taking place.

Step 1. Write the equation for the reaction

$Cu^{2+}(aq) + Zn(s) \rightarrow Cu(s) + Zn^{2+}(aq)$

Step 2. Determine the limiting reagent

Amount of $Cu^{+2}(aq)$ = $\frac{50.0}{1000} \times 0.200 = 0.0100$ mol

Amount of Zn(s) = $\frac{1.20}{65.37} = 0.0184$ mol

∴ $Cu^{2+}(aq)$ is the limiting reagent

Step 3. Extrapolate the graph (*already done*) to compensate for heat loss and determine ΔT

ΔT = 10.4 °C

Step 4. Calculate the heat evolved in the experiment for 0.0100 mol of reactants

Heat evolved = $\frac{50.0}{1000} \times 4.18 \times 10.4$ °C = 2.17 kJ

Step 5. Express this as the enthalpy change for the reaction

$\Delta H = -2.17 \times \frac{1}{0.0100} = -217$ kJ mol^{-1}

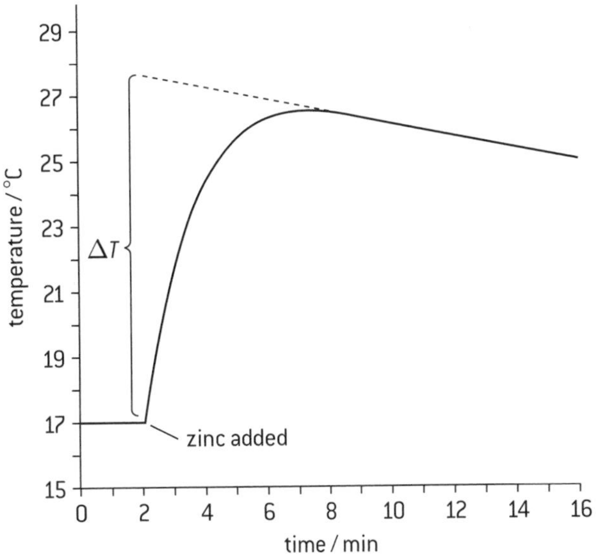

Hess' Law and standard enthalpy changes

HESS' LAW

Hess' law states that the enthalpy change for a reaction depends only on the difference between the enthalpy of the products and the enthalpy of the reactants. It is independent of the reaction pathway.

The enthalpy change going from A to B is the same whether the reaction proceeds directly to A or whether it goes via an intermediate.

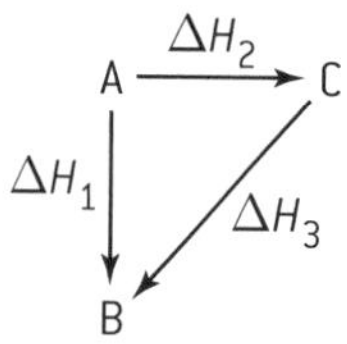

$$\Delta H_1 = \Delta H_2 + \Delta H_3$$

This law is a statement of the law of conservation of energy. It can be used to determine enthalpy changes, which cannot be measured directly. For example, the enthalpy of combustion of both carbon and carbon monoxide to form carbon dioxide can easily be measured directly, but the combustion of carbon to carbon monoxide cannot. This can be represented by an energy cycle.

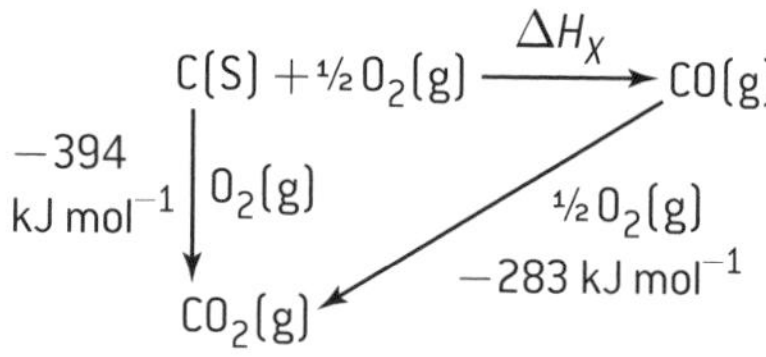

$$-394 = \Delta H_X + (-283)$$

$$\Delta H_X = -394 + 283 = -111 \text{ kJ mol}^{-1}$$

Hess' law problems can also be solved by using simultaneous equations. Write the equations that are known and then manipulate them to arrive at the required equation. For example,

I $C(s) + O_2(g) \rightarrow CO_2(g)$ $\quad \Delta H = -394 \text{ kJ mol}^{-1}$

II $CO(g) + \frac{1}{2}O_2(g) \rightarrow CO_2(g)$ $\quad \Delta H = -283 \text{ kJ mol}^{-1}$

Subtract II from I

$C(s) + \frac{1}{2}O_2(g) - CO(g) = 0$ $\quad \Delta H = -394 - (-283) \text{ kJ mol}^{-1}$

Rearrange equation

$C(s) + \frac{1}{2}O_2(g) \rightarrow CO(g)$ $\quad \Delta H = -111 \text{ kJ mol}^{-1}$

WORKED EXAMPLE

Calculate the standard enthalpy change when one mole of methane is formed from its elements in their standard states. The standard enthalpies of combustion $\Delta H_c^\ominus$ of carbon, hydrogen, and methane are −394, −286, and −890 kJ mol⁻¹ respectively.

Step 1. Write the equation for the enthalpy change with the unknown $\Delta H^\ominus$ value. Call this value $\Delta H_x^\ominus$

$$C(s) + 2H_2(g) \xrightarrow{\Delta H_x^\ominus} CH_4(g)$$

Step 2. Construct an energy cycle showing the different routes to the products (in this case the products of combustion)

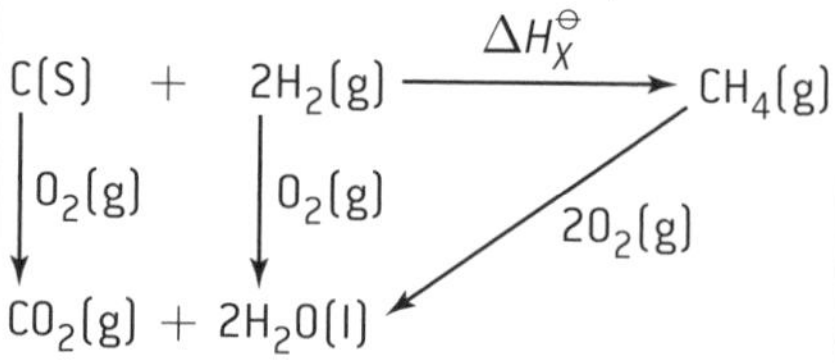

Step 3. Use Hess' law to equate the energy changes for the two different routes

$$\underbrace{\Delta H_c^\ominus(C) + 2\Delta H_c^\ominus(H_2)}_{\text{direct route}} = \underbrace{\Delta H_x^\ominus + \Delta H_c^\ominus(CH_4)}_{\text{route via methane}}$$

Step 4. Rearrange the equation and substitute the values to give the answer

$$\Delta H_x^\ominus = \Delta H_c^\ominus(C) + 2\Delta H_c^\ominus(H_2) - \Delta H_c^\ominus(CH_4)$$

$$= -394 + (2 \times -286) - (-890) \text{ kJ mol}^{-1}$$

$$= -76 \text{ kJ mol}^{-1}$$

STANDARD ENTHALPY CHANGES OF FORMATION $\Delta H_f^\ominus$ AND COMBUSTION $\Delta H_c^\ominus$

The standard enthalpy change of formation of a compound is the enthalpy change when one mole of the compound is formed from its elements in their standard states at 298 K and 100 kPa pressure. From this it follows that $\Delta H_f^\ominus$ for an element in its standard state will be zero. The standard enthalpy change of combustion, $\Delta H_c^\ominus$, is the enthalpy change when one mole of a substance is completely combusted in oxygen under standard conditions (298 K and 100 kPa pressure).

APPLICATION OF HESS' LAW

The standard enthalpy change of formation of ethanol, $C_2H_5OH(l)$ cannot be determined directly but an accurate value can be obtained indirectly by using the experimental values (in kJ mol⁻¹) for the standard enthalpy changes of combustion of carbon (−393.5), hydrogen (−285.8) and ethanol (−1371). Although this could be solved by just using simultaneous equations it is neater to show the energy cycle. Hess' law is then applied by equating the energy changes involved in combusting carbon and hydrogen directly with the energy changes involved when they are first combined to form ethanol and then combusting the ethanol.

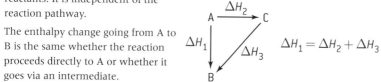

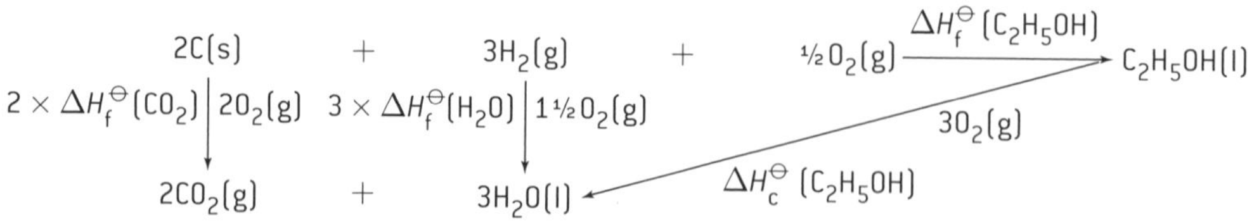

By Hess' law: $\Delta H_f^\ominus(C_2H_5OH) = 2 \times \Delta H_f^\ominus(CO_2) + 3 \times \Delta H_f^\ominus(H_2O) - \Delta H_c^\ominus(C_2H_5OH)$

Substituting the relevant values $\Delta H_f^\ominus(C_2H_5OH) = (2 \times -393.5) + (3 \times -285.8) - (-1371) = -273.4 \text{ kJ mol}^{-1}$

Bond enthalpies

BOND ENTHALPIES

Enthalpy changes can also be calculated directly from bond enthalpies. The bond enthalpy is defined as the enthalpy change for the process

$X{-}Y(g) \rightleftharpoons X(g) + Y(g)$ Note the gaseous state.

For bond formation the value is negative as energy is evolved and for bond breaking energy has to be put in so the value is positive. For simple diatomic molecules where there are only two atoms the values can be known precisely. However for bonds such as the C–H bond there are many compounds containing a C–H bond and the value can differ slightly depending upon the surrounding atoms. The term **average bond enthalpy** is used. This is defined as the energy needed to break one mole of a bond in a gaseous molecule averaged over similar compounds.

If the bond enthalpy values are known for all the bonds in the reactants and products then the overall enthalpy change can be calculated.

Some bond enthalpies and average bond enthalpies

All values in kJ mol^{-1}

H–H	436	C=C	614	C≡C	839
C–C	346	O=O	498	N≡N	945
C–H	414				
O–H	463				
N–H	391				
N–N	158				

Note that the O=O bond is stronger in O_2 than the O–O bond in ozone, O_3, (362 kJ mol^{-1}). This is important as the ozone layer protects the Earth from damaging ultraviolet radiation by absorbing both high and low energy uv light to break these bonds (see *Importance of the ozone layer* on page 33).

WORKED EXAMPLE 1

Hydrogenation of ethene

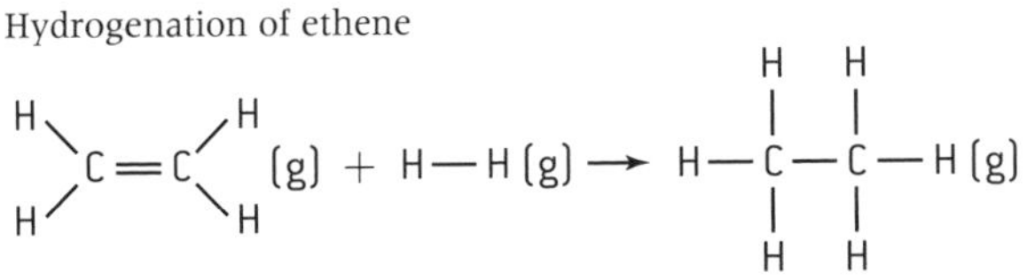

energy absorbed to break bonds:

C=C 614
4 C–H 4 × 414 } 2706 kJ
H–H 436

energy released when bonds are formed:

C–C 346
6 C–H 6 × 414 } 2830 kJ

There is more energy released than absorbed so the reaction is exothermic.

$\Delta H = -(2830 - 2706) = -124$ kJ mol^{-1}

WORKED EXAMPLE 2

Combustion of hydrazine in oxygen (this reaction has been used to power spacecraft)

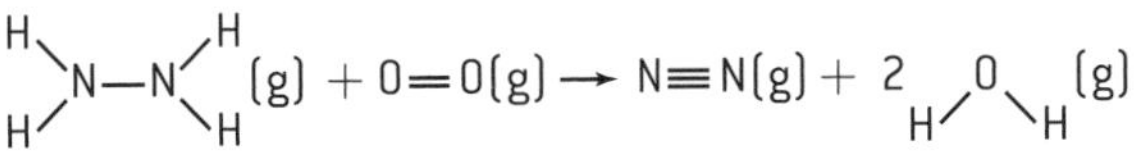

energy absorbed:

N–N 158
4N–H 4 × 391 } 2220
O=O 498

energy released:

N≡N 945
4O–H 4 × 463 } 2797

$\Delta H = -(2797 - 2220) = -577$ kJ mol^{-1}

LIMITATIONS OF USING BOND ENTHALPIES

Bond enthalpies can only be used on their own if all the reactants and products are in the gaseous state. If water were a liquid product in the above example then even more heat would be evolved since the enthalpy change of vaporization of water would also be needed to be included in the calculation.

In the above calculations some average bond enthalpies have been used. These have been obtained by considering a number of similar compounds containing the bond in question. In practice the energy of a particular bond will vary in different compounds. For this reason ΔH values obtained from using bond enthalpies will not necessarily be very accurate. Both these points are illustrated by the determination of the enthalpy change of combustion of methane by using bond enthalpies.

The equation for the reaction using bond enthalpies is:

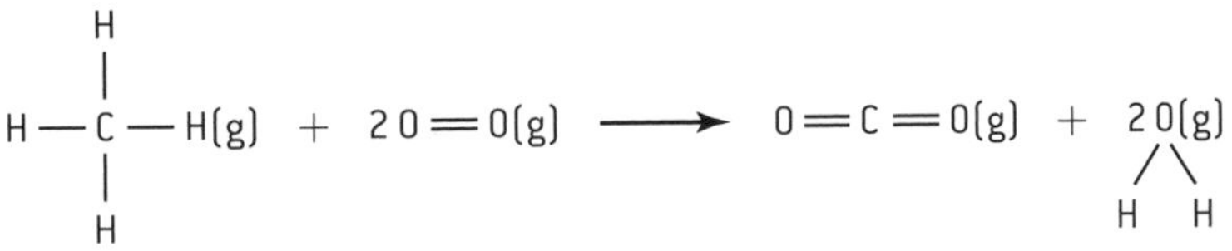

ΔH reaction = $\Sigma(\Delta H_f$ products) – $\Sigma(\Delta H_f$ reactants)

Energy taken in / kJ mol^{-1}	Energy given out / kJ mol^{-1}
4 × C–H = 4 × 414 = + 1656	2 × C=O = 2 × (–804) = –1608
2 × O=O = 2 × 498 = + 996	4 × O–H = 4 × (–463) = –1852
Total = + 2652	Total = –3468

The calculated enthalpy change for the reaction using bond enthalpies is therefore equal to –816 kJ mol^{-1}. However this is considerably different to the value of –891 kJ mol^{-1} given in Section 13 of the IB data booklet. It is to be expected that there will be a difference as the definition of enthalpy of combustion is that the reactants and products should be in their normal states under standard conditions so we need to consider the extra 2 × 44 kJ mol^{-1} of energy given out when the two moles of gaseous water product turn to liquid water. This will now bring the enthalpy of combustion value to -904 kj mol^{-1}. This is much closer to -891 kj mol^{-1} with a difference of about 1.5%. This difference is due to the fact that average bond enthalpies have been used throughout and the fact that the bond enthalpy for C=O in carbon dioxide, where there are two double bonds to oxygen on the same carbon atom, may be different to the average C=O enthalpy.

Energy cycles

BORN–HABER CYCLES

Born–Haber cycles are simply energy cycles for the formation of ionic compounds. The enthalpy change of formation of sodium chloride can be considered to occur through a number of separate steps.

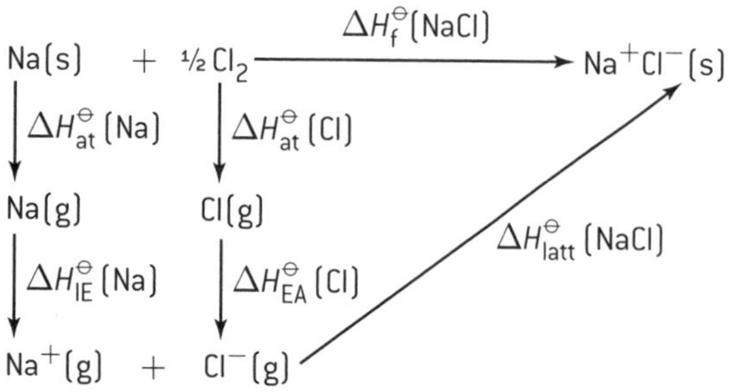

Using Hess' law:

$$\Delta H_f^\ominus(\text{NaCl}) = \Delta H_{at}^\ominus(\text{Na}) + \Delta H_{IE}^\ominus(\text{Na}) + \Delta H_{at}^\ominus(\text{Cl}) + \Delta H_{EA}^\ominus(\text{Cl}) + \Delta H_{latt}^\ominus(\text{NaCl})$$

Substituting the relevant values:

$$\Delta H_f^\ominus(\text{NaCl}) = +108 + 496 + 121 - 349 - 790 = -414 \text{ kJ mol}^{-1}$$

Note: it is the large lattice enthalpy that mainly compensates for the endothermic processes and leads to the enthalpy of formation of ionic compounds having a negative value.

Sometimes Born–Haber cycles are written as energy level diagrams with the arrows for endothermic processes in the opposite direction to the arrows for exothermic processes.

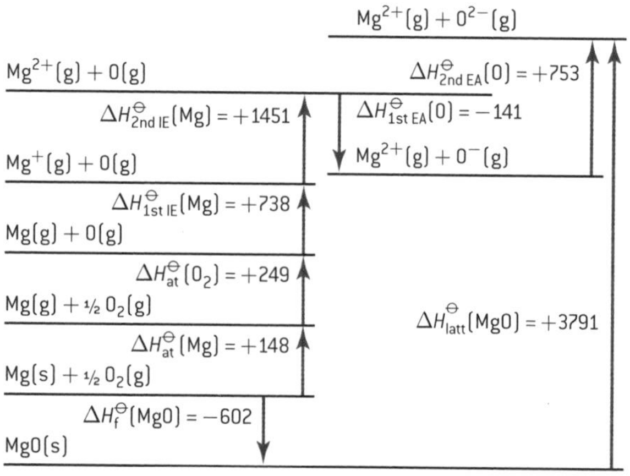

ENTHALPY OF ATOMIZATION

The standard enthalpy of atomization is the standard enthalpy change when one mole of gaseous atoms is formed from the element in its standard state under standard conditions. For diatomic molecules this is equal to half the bond dissociation enthalpy.

$$\tfrac{1}{2}Cl_2(g) \rightarrow Cl(g) \qquad \Delta H_{at}^\ominus = +121 \text{ kJ mol}^{-1}$$

LATTICE ENTHALPY

The lattice enthalpy relates either to the endothermic process of turning a crystalline solid into its gaseous ions or to the exothermic process of turning gaseous ions into a crystalline solid.

$$MX(s) \rightleftharpoons M^+(g) + X^-(g)$$

The sign of the lattice enthalpy indicates whether the lattice is being formed (−) or broken (+).

The size of the lattice enthalpy depends both on the size of the ions and on the charge carried by the ions.

The smaller the ion and the greater the charge, the higher the lattice enthalpy.

cation size increasing →

LiCl	NaCl	KCl
Lattice enthalpy / kJ mol^{-1}		
864	790	720

anion size increasing →

NaCl	NaBr	NaI
790	754	705

charge on cation increasing →

NaCl	$MgCl_2$
Lattice enthalpy / kJ mol^{-1}	
790	2540

charge on anion increasing →

$MgCl_2$	MgO
2540	3791

SOLUBILITY OF SALTS

It takes considerable energy to melt sodium chloride (melting point 801 °C) due to the strong electrostatic attractions in its lattice and yet the lattice can easily be broken down by dissolving salt in water at room temperature. An energy cycle can be drawn to explain why.

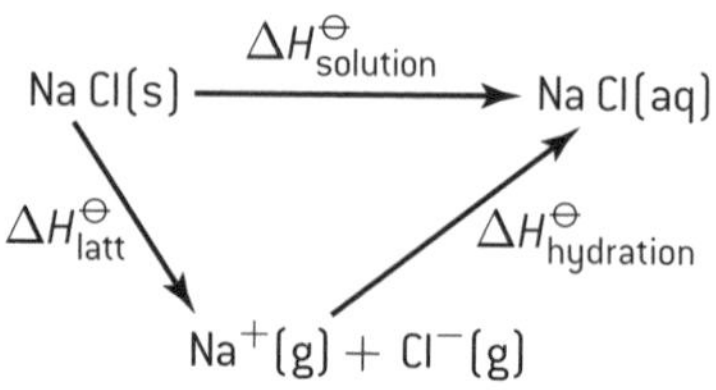

The overall step is known as the enthalpy change of solution – the enthalpy change when 1 mole of an ionic substance dissolves in water to give a solution of infinite dilution. This can be considered to proceed in two stages. The first involves the lattice enthalpy to break the lattice into gaseous ions, which will be highly endothermic and the second to hydrate the gaseous ions into aqueous ions. This second step is known as the **hydration energy** and can be defined as the enthalpy change when 1 mole of gaseous ions dissolves in sufficient water to give an infinitely dilute solution. It is a highly exothermic process. Generally the smaller and more highly charged the ion the greater the hydration energy. In the case of sodium chloride the value for the sum of the hydration energies of the Na^+ and Cl^- ions is very similar to the lattice enthalpy of NaCl and the small difference of about 7 kJ mol^{-1} can be made up by taking some heat from the water so it dissolves with a slight lowering of temperature.

$$\Delta H_{solution}^\ominus = \Delta H_{latt}^\ominus + \Delta H_{hydration}^\ominus = +790 + (-783) = +7 \text{ kJ mol}^{-1}$$

Whether or not other salts are soluble in water depends upon the relative size of the lattice enthalpy compared with the hydration energy. The highly exothermic nature of hydration energy can explain why it is inadvisable to add water to sulfur trioxide. As sulfur trioxide is not ionic there is no strong lattice enthalpy to overcome and the hydration energy released as heat is so strong that the resulting sulfuric acid can boil.

HL Entropy and spontaneity

ENTROPY

Entropy (S) refers to the distribution of available energy among the particles in a system. The more ways the energy can be distributed the higher the entropy. This is sometimes equated to a measure of the disorder of a system. In nature, systems naturally tend towards an increase in entropy. An increase in entropy (disorder) can result from:

- mixing different types of particles, e.g. the dissolving of sugar in water
- a change in state where the distance between the particles increases, e.g. liquid water → steam
- the increased movement of particles, e.g. heating a liquid or gas
- increasing the number of particles, e.g. $2H_2O_2(l) \rightarrow 2H_2O(l) + O_2(g)$.

The greatest increase in disorder is usually found where the number of particles in the gaseous state increases.

The change in the disorder of a system is known as the entropy change, ΔS. The more disordered the system becomes the more positive the value of ΔS becomes. Systems which become more ordered will have negative ΔS values.

$NH_3(g) + HCl(g)$	$\rightarrow$	$NH_4Cl(s)$	$\Delta S = -284\ J\ K^{-1}\ mol^{-1}$
(two moles of gas)		(one mole of solid)	

ABSOLUTE ENTROPY VALUES

The standard entropy of a substance is the entropy change per mole that results from heating the substance from 0 K to the standard temperature of 298 K. Unlike enthalpy, absolute values of entropy can be measured. The standard entropy change for a reaction can then be determined by calculating the difference between the entropy of the products and the reactants.

$$\Delta S^{\ominus} = S^{\ominus}\ (\text{products}) - S^{\ominus}\ (\text{reactants})$$

e.g. for the formation of ammonia

$$3H_2(g) + N_2(g) \rightleftharpoons 2NH_3(g)$$

the standard entropies of hydrogen, nitrogen and ammonia are respectively 131, 192 and 192 $J\ K^{-1}\ mol^{-1}$.

Therefore per mole of reaction

$$\Delta S^{\ominus} = 2 \times 192 - [(3 \times 131) + 192] = -201\ J\ K^{-1}\ mol^{-1}$$

(or per mole of ammonia $\Delta S^{\ominus} = \frac{-201}{2} = -101\ J\ K^{-1}\ mol^{-1}$)

SPONTANEITY

A reaction is said to be spontaneous if it causes a system to move from a less stable to a more stable state. This will depend both upon the enthalpy change and the entropy change. These two factors can be combined and expressed as the Gibbs energy change ΔG, often known as the 'free energy change'.

The standard free energy change $\Delta G^{\ominus}$ is defined as:

$$\Delta G^{\ominus} = \Delta H^{\ominus} - T\Delta S^{\ominus}$$

where all the values are measured under standard conditions. For a reaction to be spontaneous it must be able to do work, that is $\Delta G^{\ominus}$ must have a negative value.

Note: the fact that a reaction is spontaneous does not necessarily mean that it will proceed without any input of energy. For example, the combustion of coal is a spontaneous reaction and yet coal is stable in air. It will only burn on its own accord after it has received some initial energy so that some of the molecules have the necessary activation energy for the reaction to occur.

FREE ENERGY, $\Delta G^{\ominus}$, AND THE POSITION OF EQUILIBRIUM

As a reaction proceeds, the composition of the reactants and products is continually changing and the free energy will also be changing. The position of equilibrium corresponds to a maximum value of entropy and a minimum in the value of the Gibbs free energy change. The reaction will not proceed any further at this point, i.e. the rate of the forward reaction will equal the rate of the reverse reaction. The equilibrium composition of an equilibrium mixture thus depends upon the value of $\Delta G^{\ominus}$. From this it can also be deduced that the equilibrium constant for the reaction, K_c, will also depend upon the value of $\Delta G^{\ominus}$(see page 56).

ⓗ Spontaneity of a reaction

POSSIBLE COMBINATIONS FOR FREE ENERGY CHANGES

Some reactions will always be spontaneous. If ΔH^{\ominus} is negative or zero and ΔS^{\ominus} is positive then ΔG^{\ominus} must always have a negative value. Conversely if ΔH^{\ominus} is positive or zero and ΔS^{\ominus} is negative then ΔG^{\ominus} must always be positive and the reaction will never be spontaneous.

For some reactions whether or not they will be spontaneous depends upon the temperature. If ΔH^{\ominus} is positive or zero and ΔS^{\ominus} is positive, then ΔG^{\ominus} will only become negative at high temperatures when the value of $T\Delta S^{\ominus}$ exceeds the value of ΔH^{\ominus}.

Type	ΔH^{\ominus}	ΔS^{\ominus}	$T\Delta S^{\ominus}$	$\Delta H^{\ominus} - T\Delta S^{\ominus}$	ΔG^{\ominus}
1	0	+	+	(0) − (+)	−
2	0	−	−	(0) − (−)	+
3	−	+	+	(−) − (+)	−
4	+	−	−	(+) − (−)	+
5	+	+	+	(+) − (+)	− or +
6	−	−	−	(−) − (−)	+ or −

Type 1. Mixing two gases. ΔG^{\ominus} is negative so gases will mix of their own accord. Gases do not unmix of their own accord (Type 2) as ΔG^{\ominus} is positive.

Type 3. $(NH_4)_2Cr_2O_7(s) \rightarrow N_2(g) + Cr_2O_3(s) + 4H_2O(g)$

The decomposition of ammonium dichromate is spontaneous at all temperatures.

Type 4. $N_2(g) + 2H_2(g) \rightarrow N_2H_4(g)$

The formation of hydrazine from its elements will never be spontaneous.

Type 5. $CaCO_3(s) \rightarrow CaO(s) + CO_2(g)$

The decomposition of calcium carbonate is only spontaneous at high temperatures.

Type 6. $C_2H_4(g) + H_2(g) \rightarrow C_2H_6(g)$

Above a certain temperature this reaction will cease to be spontaneous.

DETERMINING THE VALUE OF ΔG^{\ominus}

The precise value of ΔG^{\ominus} for a reaction can be determined from ΔG_f^{\ominus} values using an energy cycle, e.g. to find the standard free energy of combustion of methane given the standard free energies of formation of methane, carbon dioxide, water, and oxygen.

$$CH_4(g) + 2O_2 \xrightarrow{\Delta G_x^{\ominus}} CO_2(g) + 2H_2O(l)$$

$$C(s) + 2O_2(g) + 2H_2(g)$$

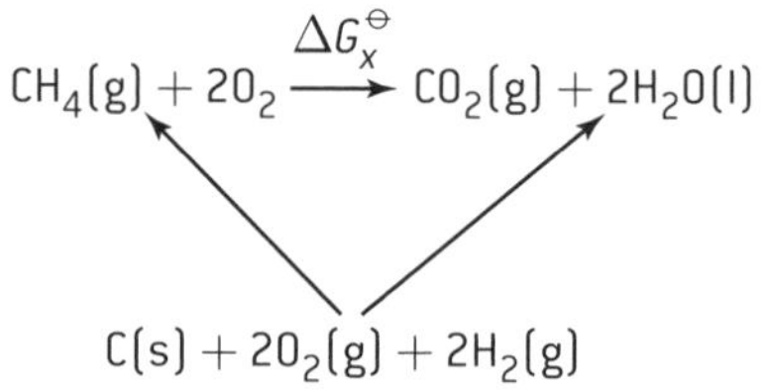

By Hess' law

$$\Delta G_x^{\ominus} = [\Delta G_f^{\ominus}(CO_2) + 2\Delta G_f^{\ominus}(H_2O)] - [\Delta G_f^{\ominus}(CH_4) + 2\Delta G_f^{\ominus}(O_2)]$$

Substituting the actual values

$$\Delta G_x^{\ominus} = [-394 + 2 \times (-237)] - [-50 + 2 \times 0]$$
$$= -818 \text{ kJ mol}^{-1}$$

ΔG^{\ominus} values can also be calculated from using the equation $\Delta G^{\ominus} = \Delta H^{\ominus} - T\Delta S^{\ominus}$. For example, in Type 5 in the adjacent list the values for ΔH^{\ominus} and ΔS^{\ominus} for the thermal decomposition of calcium carbonate are $+178$ kJ mol^{-1} and $+165.3$ J K^{-1} mol^{-1} respectively. Note that the units of ΔS^{\ominus} are different to those of ΔH^{\ominus}.

At 25 °C (298 K) the value for $\Delta G^{\ominus} = 178 - 298 \times \dfrac{165.3}{1000}$
$$= +129 \text{ kJ mol}^{-1}$$

which means that the reaction is not spontaneous.

The reaction will become spontaneous when $T\Delta S^{\ominus} > \Delta H^{\ominus}$.

$$T\Delta S^{\ominus} = \Delta H^{\ominus} \text{ when } T = \frac{\Delta H^{\ominus}}{\Delta S^{\ominus}} = \frac{178}{165.3/1000} = 1077K \ (804\,°C)$$

Therefore above 804 °C the reaction will be spontaneous.

Note: this calculation assumes that the entropy value is independent of temperature, which is not strictly true.

MULTIPLE CHOICE QUESTIONS – ENERGETICS / THERMOCHEMISTRY

1. Which statement about the reaction below is correct?

 $2NO(g) + O_2(g) \rightarrow 2NO_2(g) \ \Delta H^{\ominus} = -114$ kJ

 A. 114 kJ of energy are absorbed for every mole of NO reacted.

 B. 114 kJ of energy are released for every mole of NO reacted.

 C. 57 kJ of energy are absorbed for every mole of NO reacted.

 D. 57 kJ of energy are released for every mole of NO reacted.

2. When an aqueous solution of sulfuric acid is added to an aqueous solution of potassium hydroxide the temperature increases. Which describes the reaction taking place?

	Type	Sign of ΔH^{\ominus}
A.	Exothermic	+
B.	Exothermic	−
C.	Endothermic	−
D.	Endothermic	+

3. A student measured the temperature of a reaction mixture over time using a temperature probe. By considering the graph, which of the following deductions can be made?

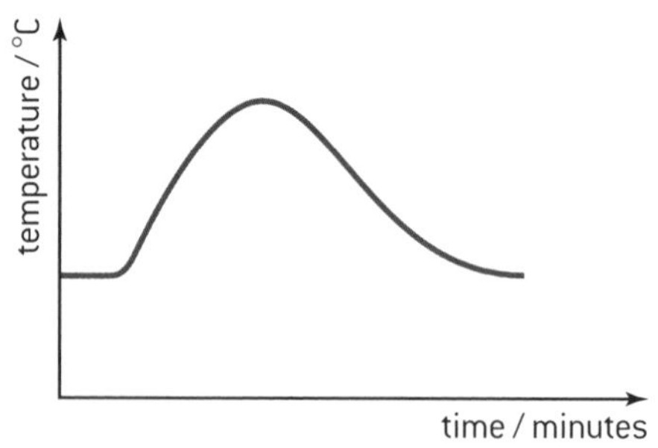

I. The reaction is exothermic.

 II. The products are more stable than the reactants.

 III. The reactant bonds are stronger than the product bonds.

 A. I and II only C. II and III only

 B. I and III only D. I, II and III

4. 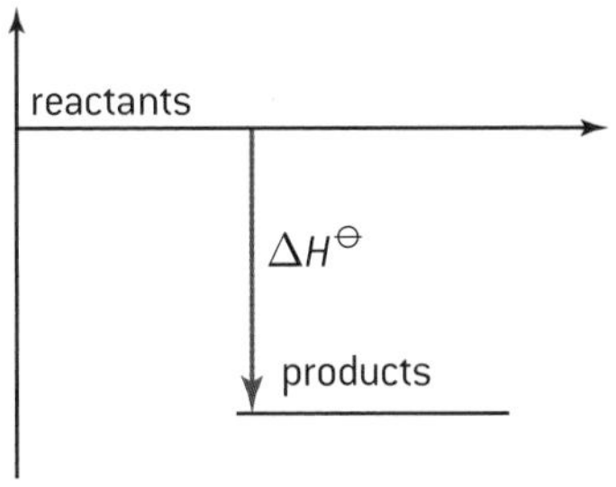

What can be deduced about the relative stability of the reactants and products and the sign of $\Delta H^\ominus$, from the enthalpy level diagram above?

	Relative stability	Sign of $\Delta H^\ominus$
A.	Products more stable	−
B.	Products more stable	+
C.	Reactants more stable	−
D.	Reactants more stable	+

5. Consider the following reactions.

$CH_4(g) + O_2(g) \rightarrow HCHO(l) + H_2O(l)$ $\quad \Delta H^\ominus = x$

$HCHO(l) + \frac{1}{2}O_2(g) \rightarrow HCOOH(l)$ $\quad \Delta H^\ominus = y$

$2HCOOH(l) + \frac{1}{2}O_2(g) \rightarrow (COOH)_2(s) + H_2O(l)$ $\quad \Delta H^\ominus = z$

What is the enthalpy change of the reaction below?

$2CH_4(g) + 3\frac{1}{2}O_2(g) \rightarrow (COOH)_2(s) + 3H_2O(l)$

A. $x + y + z$
B. $2x + y + z$
C. $2x + 2y + z$
D. $2x + 2y + 2z$

6. Which equation represents the H–F bond enthalpy?

A. $2HF(g) \rightarrow H_2(g) + F_2(g)$
B. $HF(g) \rightarrow H^+(g) + F^-(g)$
C. $HF(g) \rightarrow H^-(g) + F^+(g)$
D. $HF(g) \rightarrow H(g) + F(g)$

7. 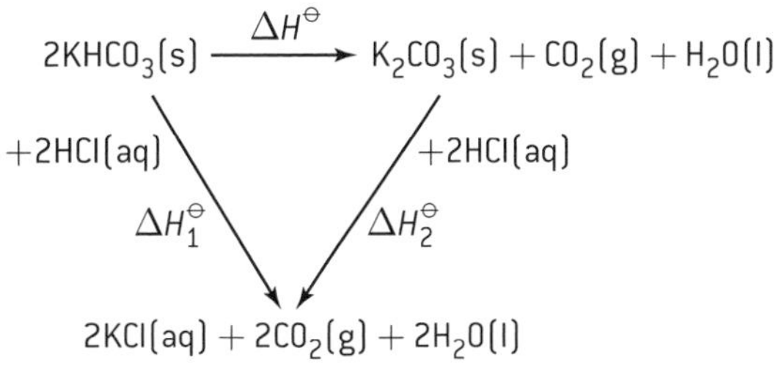

This cycle may be used to determine $\Delta H^\ominus$ for the decomposition of potassium hydrogencarbonate. Which expression can be used to calculate $\Delta H^\ominus$?

A. $\Delta H^\ominus = \Delta H_1^\ominus + \Delta H_2^\ominus$
B. $\Delta H^\ominus = \Delta H_1^\ominus - \Delta H_2^\ominus$
C. $\Delta H^\ominus = \frac{1}{2}\Delta H_1^\ominus - \Delta H_2^\ominus$
D. $\Delta H^\ominus = \Delta H_2^\ominus - \Delta H_1^\ominus$

8. Which process is endothermic?

A. $HNO_3(aq) + NaOH(aq) \rightarrow NaNO_3(aq) + H_2O(l)$
B. $Cl(g) + e^- \rightarrow Cl^-(g)$
C. $H_2O(l) \rightarrow H_2O(g)$
D. $C_2H_5OH(l) + 3O_2(g) \rightarrow 2CO_2(g) + 3H_2O(l)$

HL

9. Which is a correct definition of lattice enthalpy?

A. It is the enthalpy change that occurs when an electron is removed from 1 mol of gaseous atoms.
B. It is the enthalpy change that occurs when 1 mol of a compound is formed from its elements.
C. It is the enthalpy change that occurs when 1 mol of solid crystal changes into a liquid.
D. It is the enthalpy change that occurs when 1 mol of solid crystal is formed from its gaseous ions.

10. The diagram represents the Born–Haber cycle for the lattice enthalpy of sodium chloride.

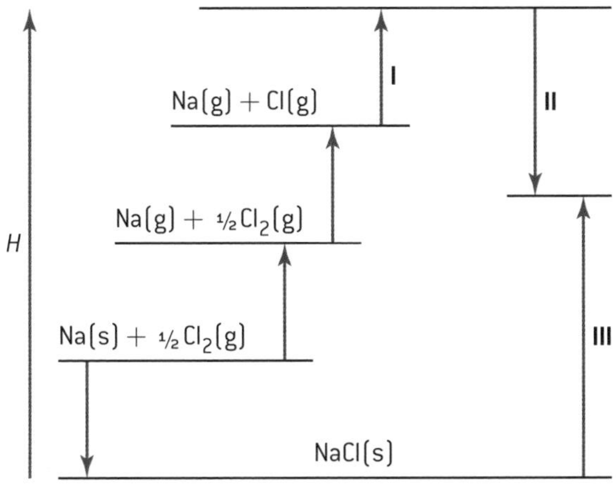

What is the name of the enthalpy changes **I**, **II** and **III**?

	I	II	III
A.	ionization energy of Na	electron affinity of Cl	lattice enthalpy of NaCl
B.	lattice enthalpy of NaCl	ionization energy of Na	electron affinity of Cl
C.	electron affinity of Cl	ionization energy of Na	lattice enthalpy of NaCl
D.	ionization energy of Na	lattice enthalpy of NaCl	electron affinity of Cl

11. Which reaction has the largest increase in entropy?

A. $H_2(g) + Cl_2(g) \rightarrow 2HCl(g)$
B. $Al(OH)_3(s) + NaOH(aq) \rightarrow Al(OH)_4^-(aq) + Na^+(aq)$
C. $Na_2CO_3(s) + 2HCl(aq) \rightarrow 2NaCl(aq) + CO_2(g) + H_2O(l)$
D. $BaCl_2(aq) + Na_2SO_4(aq) \rightarrow BaSO_4(s) + 2NaCl(aq)$

12. Which statements about entropy for the following reaction at 298 K are correct?

$2NO(g) + O_2(g) \rightarrow 2NO_2(g)$

I. $S^\ominus(O_2) = 0$
II. $\Delta S^\ominus = 2S^\ominus(NO_2) - 2S^\ominus(NO) - S^\ominus(O_2)$
III. $\Delta S^\ominus < 0$

A. I and II only
B. I and III only
C. II and III only
D. I, II and III

SHORT ANSWER QUESTIONS – ENERGETICS / THERMOCHEMISTRY

1. In an experiment to measure the enthalpy change of combustion of ethanol, a student heated a copper calorimeter containing 100 cm^3 of water with a spirit lamp and collected the following data.

   Initial temperature of water: 20.0 °C

   Final temperature of water: 55.0 °C

   Mass of ethanol burned: 1.78 g

   Density of water: 1.00 g cm^{-3}

   a) Use the data to calculate the heat evolved (in kJ) when the ethanol was combusted. [2]

   b) Calculate the enthalpy change of combustion per mole of ethanol. [2]

   c) Suggest two reasons why the result is not the same as the value in the IB data booklet. [2]

2. Ethanol is used as a component in fuel for some vehicles. One fuel mixture contains 10% by mass of ethanol in unleaded petrol (gasoline). This mixture is often referred to as Gasohol E10.

   a) Assume that the other 90% by mass of Gasohol E10 is octane. 1.00 kg of this fuel mixture was burned.

   $$CH_3CH_2OH(l) + 3O_2(g) \rightarrow 2CO_2(g) + 3H_2O(l) \qquad \Delta H^\ominus = -1367 \text{ kJ mol}^{-1}$$

   $$C_8H_{18}(l) + 12\tfrac{1}{2}\, O_2(g) \rightarrow 8CO_2(g) + 9H_2O(l) \qquad \Delta H^\ominus = -5470 \text{ kJ mol}^{-1}$$

   (i) Calculate the mass, in g, of ethanol and octane in 1.00 kg of the fuel mixture. [1]

   (ii) Calculate the amount, in mol, of ethanol and octane in 1.00 kg of the fuel mixture. [1]

   (iii) Calculate the total amount of energy, in kJ, released when 1.00 kg of the fuel mixture is completely burned. [3]

   b) If the fuel blend was vaporized before combustion, predict whether the amount of energy released would be greater, less or the same. Explain your answer. [2]

3. The data shown are from an experiment to measure the enthalpy change for the reaction of aqueous copper(II) sulfate, $CuSO_4$ (aq) and zinc, Zn (s).

   $$Cu^{2+}(aq) + Zn\ (s) \rightarrow Cu\ (s) + Zn^{2+}(aq)$$

   50.0 cm^3 of 1.00 mol dm^{-3} copper(II) sulfate solution was placed in a polystyrene cup and zinc powder was added after 100 seconds. The temperature–time data was taken from a data-logging software program. The table shows the initial 19 readings.

	A	B	C	D	E	F	G	H
1	time / s	temperature / °C						
2	0.0	24.8						
3	1.0	24.8						
4	2.0	24.8						
5	3.0	24.8						
6	4.0	24.8						
7	5.0	24.8						
8	6.0	24.8						
9	7.0	24.8						
10	8.0	24.8						
11	9.0	24.8						
12	10.0	24.8						
13	11.0	24.8						
14	12.0	24.8						
15	13.0	24.8						
16	14.0	24.8						
17	15.0	24.8						
18	16.0	24.8						
19	17.0	24.8						
20	18.0	24.8						
21								
22								
23								
24								

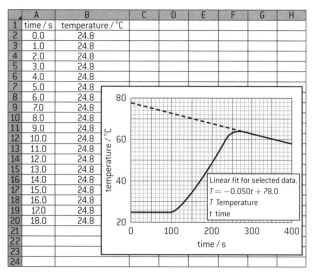

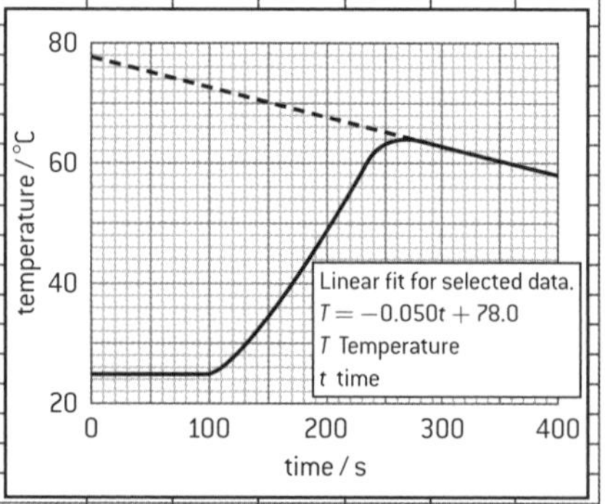

A straight line has been drawn through some of the data points. The equation for this line is given by the data-logging software as $T = -0.050t + 78.0$

The heat produced by the reaction can be calculated from the temperature change, ΔT, using the expression

Heat change = Volume of $CuSO_4$(aq) × Specific heat capacity of $H_2O \times \Delta T$

   a) Describe two assumptions made in using this expression to calculate heat changes. [2]

   b) (i) Use the data presented by the data-logging software to deduce the temperature change, ΔT, which would have occurred if the reaction had taken place instantaneously with no heat loss. [2]

   (ii) State the assumption made in part b) (i). [1]

   (iii) Calculate the heat, in kJ, produced during the reaction using the expression given in part a). [1]

   c) The colour of the solution changed from blue to colourless. Deduce the amount, in moles, of zinc which reacted in the polystyrene cup. [1]

   d) Calculate the enthalpy change, in kJ mol^{-1}, for this reaction. [1]

HL

4. a) The standard enthalpy change of three combustion reactions is given below in kJ.

   $$2C_2H_6(g) + 7O_2(g) \rightarrow 4CO_2(g) + 6H_2O(l) \qquad \Delta H^\ominus = -3120$$

   $$2H_2(g) + O_2(g) \rightarrow 2H_2O(l) \qquad \Delta H^\ominus = -572$$

   $$C_2H_4(g) + 3O_2(g) \rightarrow 2CO_2(g) + 2H_2O(l) \qquad \Delta H^\ominus = -1411$$

   Based on the above information, calculate the standard change in enthalpy, $\Delta H^\ominus$, for the following reaction.

   $$C_2H_6(g) \rightarrow C_2H_4(g) + H_2(g)$$ [4]

   b) Predict, stating a reason, whether the sign of $\Delta S^\ominus$ for the above reaction would be positive or negative. [2]

   c) Discuss why the above reaction is non-spontaneous at low temperature but becomes spontaneous at high temperatures. [2]

   d) Using bond enthalpy values, calculate $\Delta H^\ominus$ for the following reaction.

   $$C_2H_6(g) \rightarrow C_2H_4(g) + H_2(g)$$ [3]

   e) Suggest with a reason, why the values obtained in parts a) and d) are different. [1]

5. 'Synthesis gas' is produced by the following reaction.

   $$CH_4(g) + H_2O(g) \rightarrow 3H_2(g) + CO(g) \qquad \Delta H^\ominus = +210 \text{ kJ}$$

   For this reaction $\Delta S^\ominus = +216$ J K^{-1}.

   a) Explain why this reaction is not spontaneous at 298 K. [2]

   b) Calculate the temperature at which this reaction becomes spontaneous. [2]

Rates of reaction and collision theory

RATE OF REACTION

Chemical kinetics is the study of the factors affecting the rate of a chemical reaction. The rate of a chemical reaction can be defined either as the increase in the concentration of one of the products per unit time or as the decrease in the concentration of one of the reactants per unit time. It is measured in mol dm^{-3} s^{-1}.

The change in concentration can be measured by using any property that differs between the reactants and the products. Common methods include mass or volume changes when a gas is evolved, absorption using a spectrometer when there is a colour change, pH changes when there is a change in acidity, and electrical conductivity when there is a change in the ionic concentrations. Data loggers could be used for all these methods. A graph of concentration against time is then usually plotted. The rate at any stated point in time is then the gradient of the graph at that time. Rates of reaction usually decrease with time as the reactants are used up.

The reaction of hydrochloric acid with calcium carbonate can be used to illustrate three typical curves that could be obtained depending on whether the concentration of reactant, the volume of the product or the loss in mass due to the carbon dioxide escaping is followed.

$$CaCO_3(s) + 2HCl(aq) \rightarrow CaCl_2(aq) + H_2O(l) + CO_2(g)$$

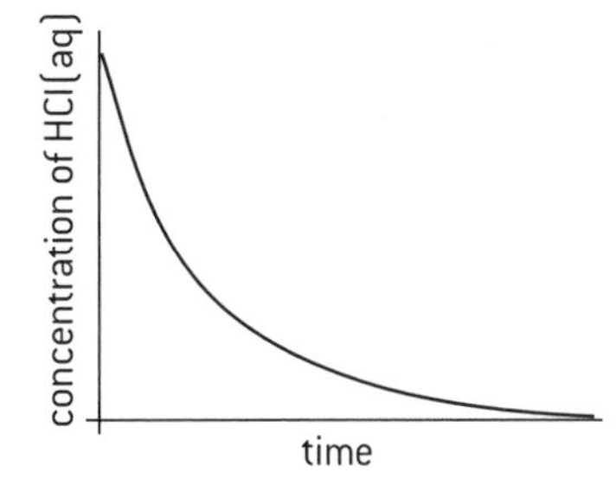

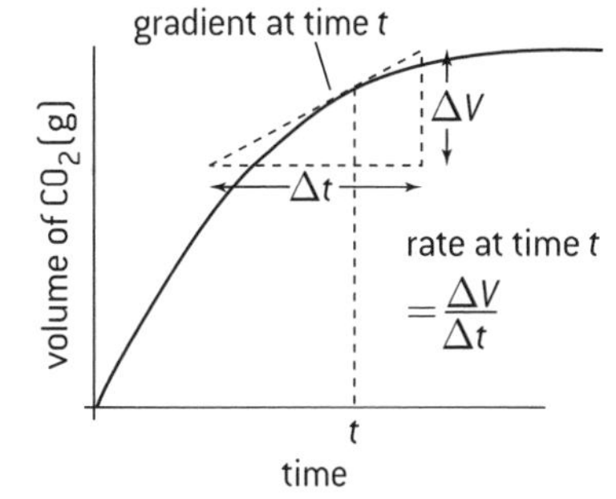

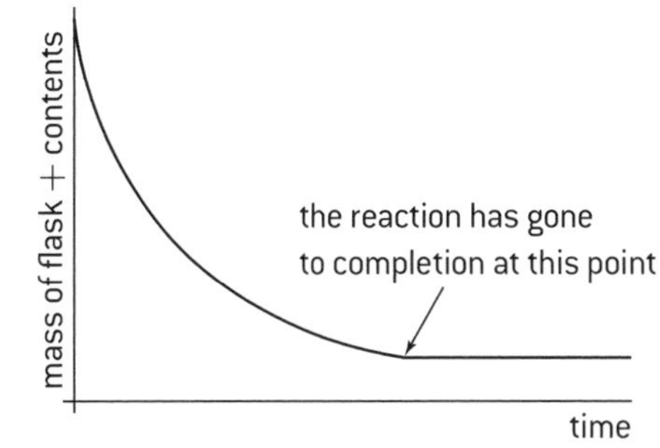

MAXWELL–BOLTZMANN DISTRIBUTION

The moving particles in a gas or liquid do not all travel with the same velocity. Some are moving very fast and others much slower. The faster they move the more kinetic energy they possess. The distribution of kinetic energies is shown by a Maxwell–Boltzmann curve.

As the temperature increases, the area under the curve does not change as the total number of particles remains constant. More particles have a very high velocity resulting in an increase in the average kinetic energy, which leads to a broadening of the curve.

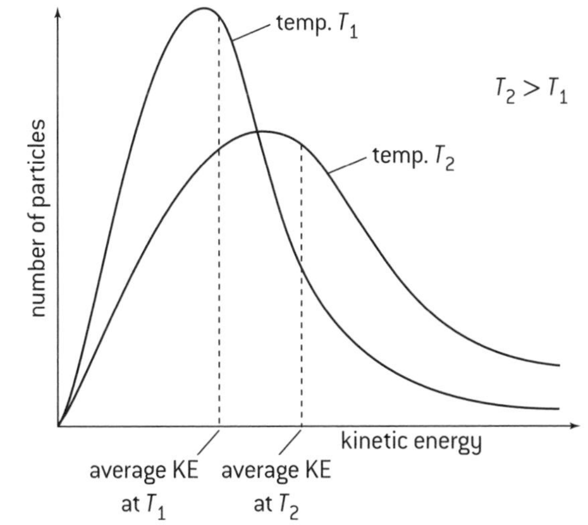

COLLISION THEORY

For a reaction between two particles to occur three conditions must be met.

- The particles must collide.
- They must collide with the appropriate geometry or orientation so that the reactive parts of the particles come into contact with each other.
- They must collide with sufficient energy to bring about the reaction.

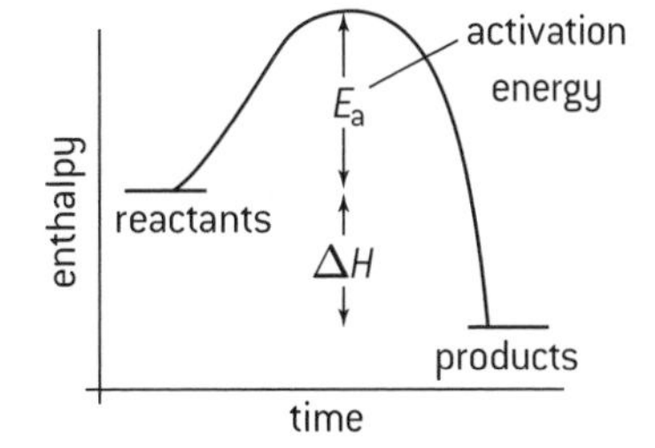

This minimum amount of energy required is known as the **activation energy**. Any factor that either increases the frequency of the collisions or increases the energy with which they collide will make the reaction go faster.

Factors affecting the rate of reaction

TEMPERATURE

As the temperature increases, the particles will move faster so there will be more collisions per second. However, the main reason why an increase in temperature increases the rate is that more of the colliding particles will possess the necessary activation energy resulting in more successful collisions. As a rough rule of thumb an increase of 10 °C doubles the rate of a chemical reaction.

$$CaCO_3(s) + 2HCl(aq) \rightarrow CaCl_2(aq) + CO_2(g) + H_2O(l)$$

volume of CO_2 (g)

lower temperature

higher temperature

time

T K

$(T + 10$ K$)$

more reactant particles possess necessary activation energy at higher temperature

number of reactant particles

energy

E_a

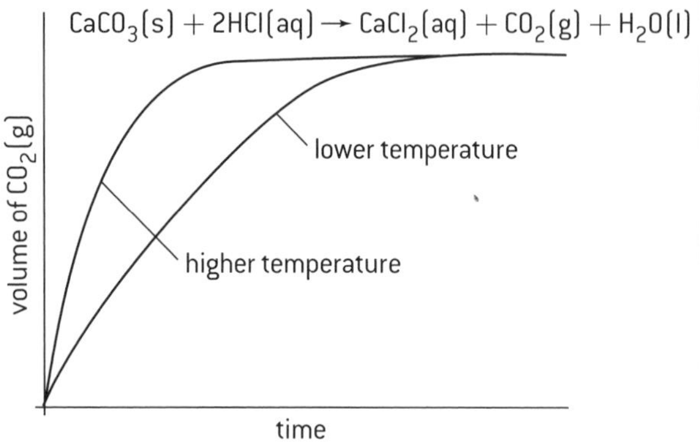

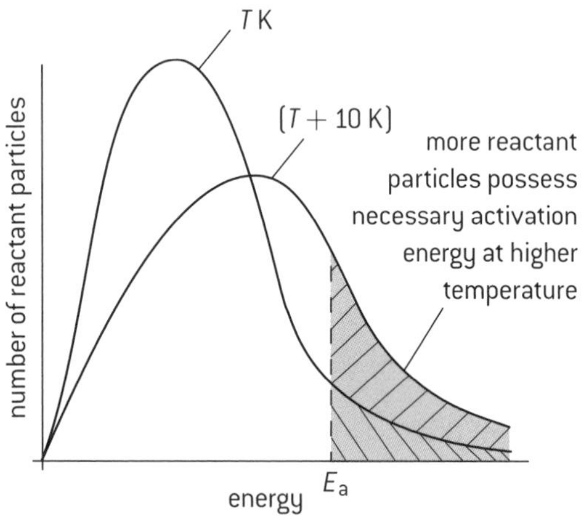

SURFACE AREA

In a solid substance only the particles on the surface can come into contact with a surrounding reactant. If the solid is in powdered form then the surface area increases dramatically and the rate increases correspondingly.

$$CaCO_3(s) + 2HCl(aq) \rightarrow CaCl_2(aq) + CO_2(g) + H_2O(l)$$

volume of CO_2 (g)

large pieces of $CaCO_3$

powdered $CaCO_3$

time

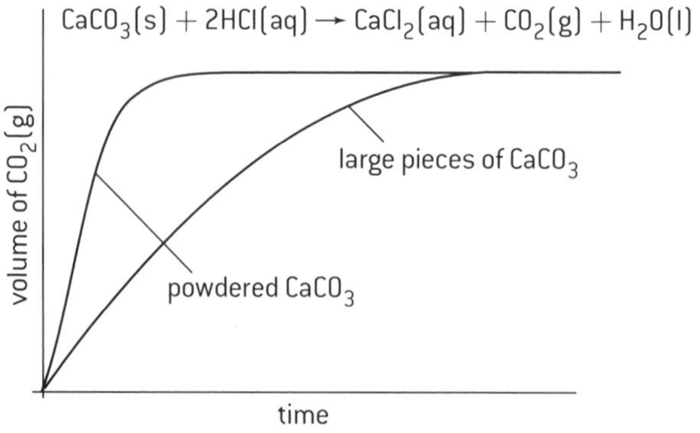

CONCENTRATION

The more concentrated the reactants the more collisions there will be per second per unit volume. As the reactants get used up their concentration decreases. This explains why the rate of most reactions gets slower as the reaction proceeds. (Some exothermic reactions do initially speed up if the heat that is given out more than compensates for the decrease in concentration.)

$$CaCO_3(s) + 2HCl(aq) \rightarrow CaCl_2(aq) + CO_2(g) + H_2O(l)$$

volume of CO_2 (g)

1 mol dm^{-3} HCl

2 mol dm^{-3} HCl

time

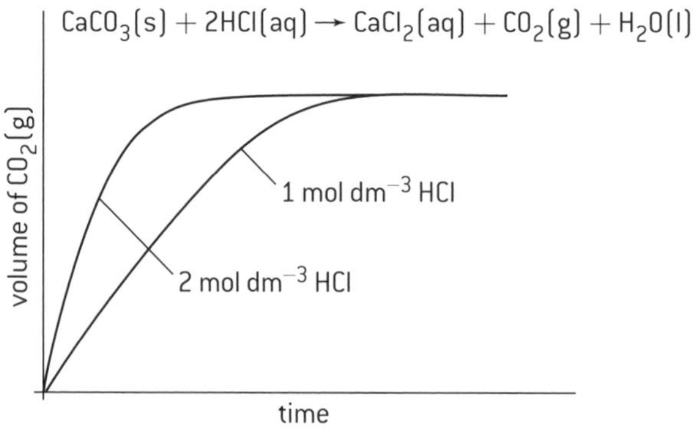

Note: this graph assumes that calcium carbonate is the limiting reagent or that equal amounts (mol) of acid have been added.

CATALYST

Catalysts increase the rate of a chemical reaction without themselves being chemically changed at the end of the reaction. They work essentially by bringing the reactive parts of the reactant particles into close contact with each other. This provides an alternative pathway for the reaction with a lower activation energy. More of the reactants will possess this lower activation energy, so the rate increases.

activated complex for uncatalysed reaction

activation energy for uncatalysed reaction

activated complex for catalysed pathway

activation energy for catalysed reaction

E_a

E_a'

enthalpy

reactants

ΔH

products

progress of reaction

fraction of particles with $E > E_a$ for the uncatalysed reaction

fraction of particles with $E > E_a$ cat

lower E_a for catalysed reaction

activation energy for uncatalysed reaction

number of reactant particles

E_acat E_a

kinetic energy

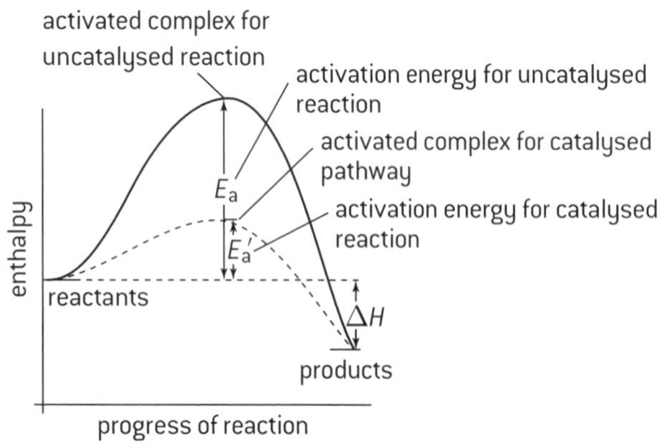

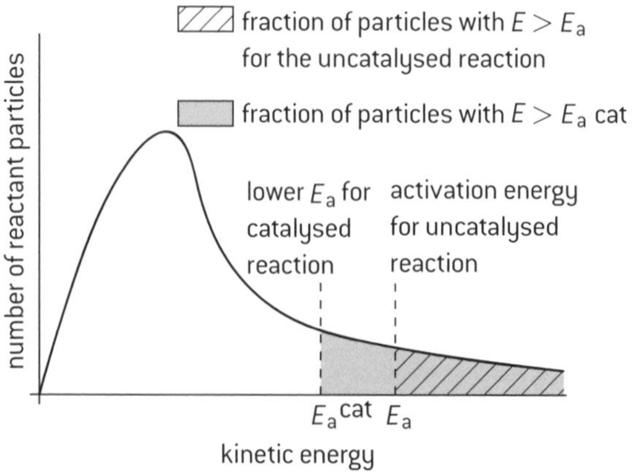

Rate expression and order of reaction

RATE EXPRESSIONS

The rate of reaction between two reactants, A and B, can be followed experimentally. The rate will be found to be proportional to the concentration of A raised to some power and also to the concentration of B raised to a power. If square brackets are used to denote concentration this can be written as rate $\propto [A]^x$ and rate $\propto [B]^y$. They can be combined to give the rate expression:

$$\text{rate} = k[A]^x[B]^y$$

where k is the constant of proportionality and is known as the **rate constant**.

x is known as the **order of the reaction** with respect to A.

y is known as the order of the reaction with respect to B.

The overall order of the reaction $= x + y$.

Note: the order of the reaction and the rate expression can only be determined experimentally. They cannot be deduced from the balanced equation for the reaction.

UNITS OF RATE CONSTANT

The units of the rate constant depend on the overall order of the reaction.

First order: rate $= k[A]$

$$k = \frac{\text{rate}}{[A]} = \frac{\text{mol dm}^{-3}\,\text{s}^{-1}}{\text{mol dm}^{-3}} = \text{s}^{-1}$$

Second order: rate $= k[A]^2$ or $k = [A][B]$

$$k = \frac{\text{rate}}{[A]^2} = \frac{\text{mol dm}^{-3}\,\text{s}^{-1}}{(\text{mol dm}^{-3})^2} = \text{dm}^3\,\text{mol}^{-1}\,\text{s}^{-1}$$

Third order: rate $= k[A]^2[B]$ or rate $= k[A][B]^2$

$$k = \frac{\text{rate}}{[A]^2[B]} = \frac{\text{mol dm}^{-3}\,\text{s}^{-1}}{(\text{mol dm}^{-3})^3} = \text{dm}^6\,\text{mol}^{-2}\,\text{s}^{-1}$$

GRAPHICAL REPRESENTATIONS OF REACTIONS

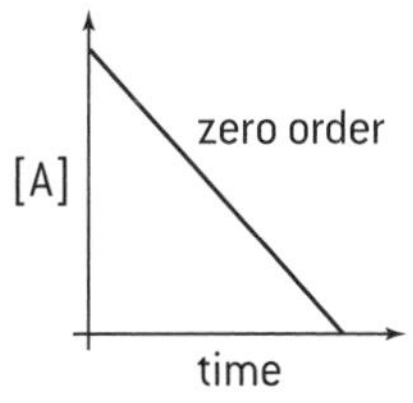

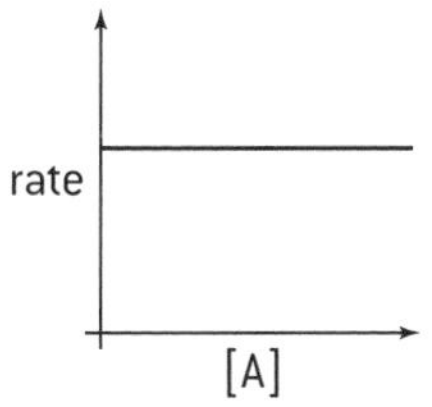

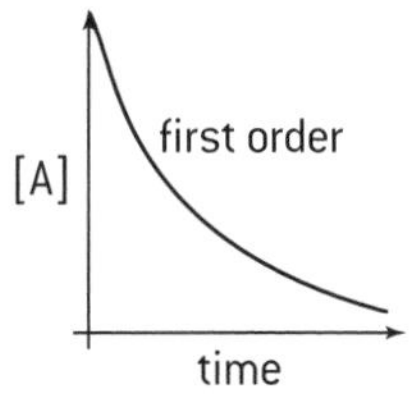

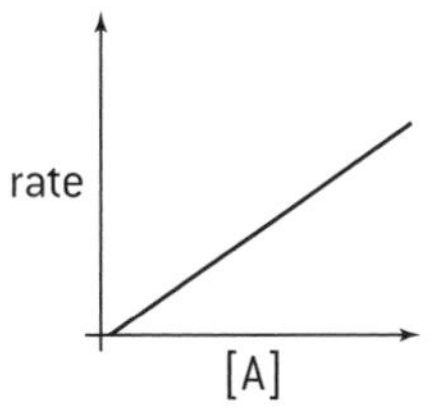

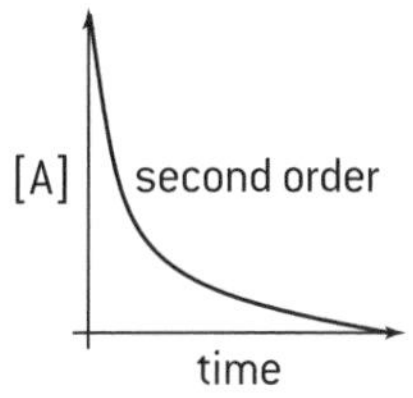

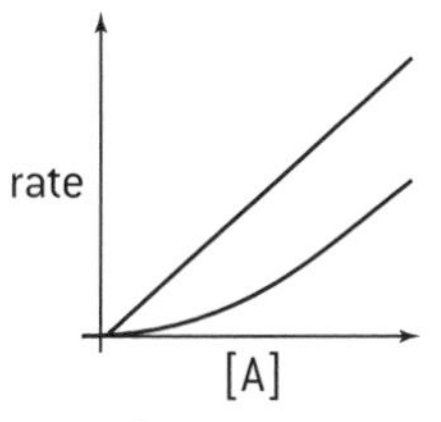

$[A]^2$ is straight line

DERIVING A RATE EXPRESSION BY INSPECTION OF DATA

Experimental data obtained from the reaction between hydrogen and nitrogen monoxide at 1073 K:

$$2H_2(g) + 2NO(g) \rightarrow 2H_2O(g) + N_2(g)$$

Experiment	Initial concentration of H_2(g) / mol dm^{-3}	Initial concentration of NO(g) / mol dm^{-3}	Initial rate of formation of N_2(g) / mol dm^{-3} s^{-1}
1	1×10^{-3}	6×10^{-3}	3×10^{-3}
2	2×10^{-3}	6×10^{-3}	6×10^{-3}
3	6×10^{-3}	1×10^{-3}	0.5×10^{-3}
4	6×10^{-3}	2×10^{-3}	2.0×10^{-3}

From experiments 1 and 2 doubling $[H_2]$ doubles the rate so rate $\propto [H_2]$.

From experiments 3 and 4 doubling [NO] quadruples the rate so rate $\propto [NO]^2$.

Rate expression given by rate $= k[H_2][NO]^2$.

The rate is first order with respect to hydrogen, second order with respect to nitrogen monoxide, and third order overall. The value of k can be found by substituting the values from any one of the four experiments:

$$k = \frac{\text{rate}}{[H_2][NO]^2} = 8.33 \times 10^4\ \text{dm}^6\,\text{mol}^{-2}\,\text{s}^{-1}$$

HALF-LIFE $t_{½}$

For a first-order reaction the rate of change of concentration of A is equal to k[A]. This can be expressed as $\frac{d[A]}{dt} = k[A]$.

If this expression is integrated then $kt = \ln [A]_0 - \ln [A]$ where $[A]_0$ is the initial concentration and [A] is the concentration at time t. This expression is known as the integrated form of the rate equation.

The half-life is defined as the time taken for the concentration of a reactant to fall to half of its initial value.

At $t_{½}$ $[A] = ½[A]_0$ the integrated rate expression then becomes $kt_{½} = \ln [A]_0 - \ln ½[A]_0 = \ln 2$ since $\ln 2 = 0.693$ this

simplifies to $t_{½} = \frac{0.693}{k}$

From this expression it can be seen that the half-life of a first-order reaction is independent of the original concentration of A, i.e. first-order reactions have a constant half-life.

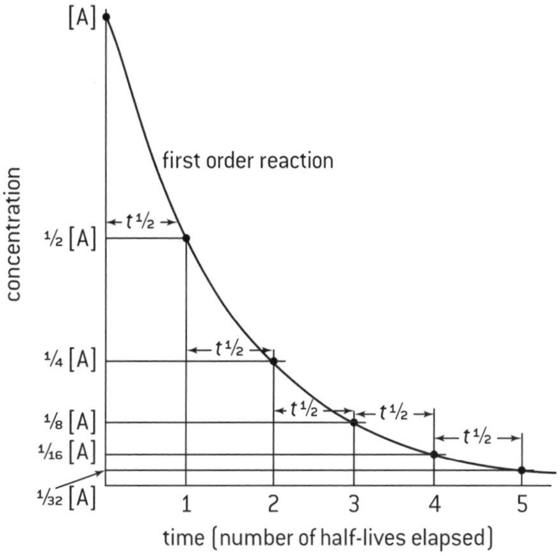

HL Reaction mechanisms and activation energy

REACTION MECHANISMS

Many reactions do not go in one step. This is particularly true when there are more than two reactant molecules as the chances of a successful collision between three or more particles is extremely small. When there is more than one step then each step will proceed at its own rate. No matter how fast the other steps are the overall rate of the reaction will depend only upon the rate of the slowest step. This slowest step is known as the **rate-determining step.**

e.g. consider the reaction between A and B to form A_2B: $2A + B \rightarrow A_2B$. A possible mechanism might be:

Step 1. $A + A \xrightarrow{\text{slow}} A\text{–}A$ rate-determining step

Step 2. $A\text{–}A + B \xrightarrow{\text{fast}} A_2B$

However fast A–A reacts with B the rate of production of A_2B will only depend on how fast A–A is formed.

When the separate steps in a chemical reaction are analysed there are essentially only two types of processes. Either a single species can break down into two or more products by what is known as a **unimolecular process**, or two species can collide and interact by a **bimolecular process**.

In a bimolecular process the species collide with the necessary activation energy to give initially an **activated complex**. An activated complex is not a chemical substance which can be isolated, but consists of an association of the reacting particles in which bonds are in the process of being broken and formed. An activated complex either breaks down to form the products or reverts back to the original reactants.

The number of species taking part in any specified step in the reaction is known as the **molecularity**. In most cases the molecularity refers to the slowest step, that is the rate-determining step.

In the reaction on the previous page, between nitrogen monoxide and hydrogen, the stoichiometry of the reaction involves two molecules of hydrogen and two molecules of nitrogen monoxide. Any proposed mechanism must be consistent with the rate expression. For third-order reactions, such as this, the rate-determining step will never be the first step. The proposed mechanism is:

$NO(g) + NO(g) \xrightarrow{\text{fast}} N_2O_2(g)$

$N_2O_2(g) + H_2(g) \xrightarrow{\text{slow}} N_2O(g) + H_2O(g)$ rate-determining step

$N_2O(g) + H_2(g) \xrightarrow{\text{fast}} N_2(g) + H_2O(g)$

Overall $2NO(g) + 2H_2(g) \longrightarrow N_2(g) + 2H_2O(g)$

If the first step was the slowest step the the rate expression would be rate $= k[NO]^2$ and the rate would be zero order with respect to hydrogen. The rate for the second step depends on $[H_2]$ and $[N_2O_2]$. However, the concentration of N_2O_2 depends on the first step. So the rate expression for the second step becomes rate $= k[H_2][NO]^2$, which is consistent with the experimentally determined rate expression. The molecularity of the reaction is two, as two reacting species are involved in the rate-determining step.

ARRHENIUS EQUATION

The rate constant for a reaction is only constant if the temperature remains constant. As the temperature increases the reactants possess more energy and the rate constant increases. The relationship between rate constant and absolute temperature is given by the Arrhenius equation:

$$k = Ae^{(-E_a/RT)}$$

where E_a is the activation energy and R is the gas constant. A is known as the frequency factor (or pre-exponential factor) and is indicative of the frequency of collisions with the correct orientation for the reaction to occur. This equation is often expressed in its logarithmic form:

$$\ln k = \frac{-E_a}{RT} + \ln A$$

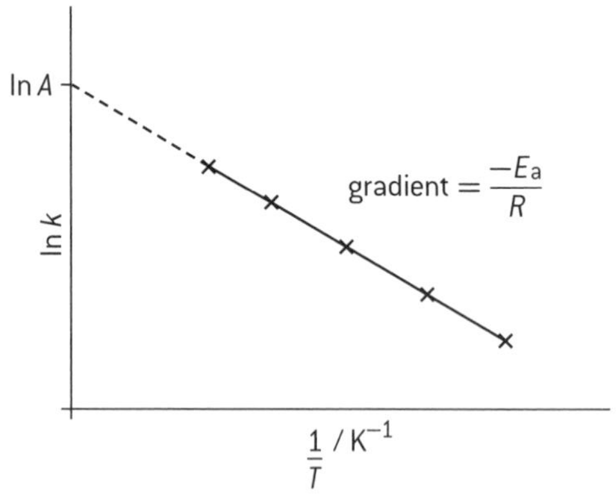

The equation can be used to determine both the frequency factor and the activation energy for the reaction. This can be done either by substitution using simultaneous equations or by plotting ln k against $\frac{1}{T}$ to give a straight line graph. The gradient of the graph will be equal to $\frac{-E_a}{R}$ from which the activation energy can be calculated. Extrapolating the graph back to the ln k axis will give an intercept with a value equal to ln A.

MULTIPLE CHOICE QUESTIONS – CHEMICAL KINETICS

1. A piece of zinc was added to aqueous nitric acid and the volume of hydrogen gas produced was measured every minute. The results are plotted on the graph below.

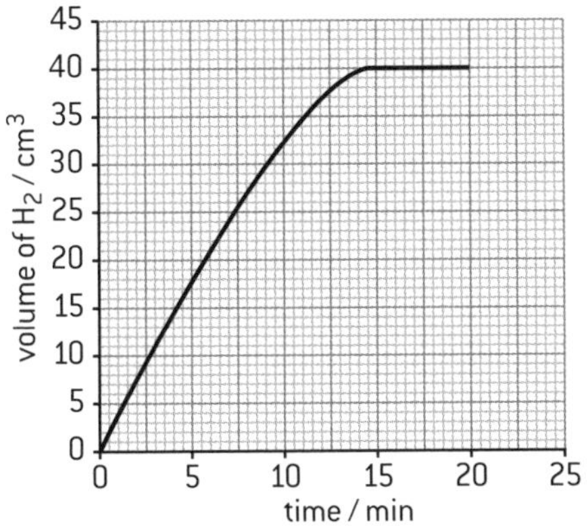

Which graph would you expect if the same mass of powdered zinc was added to nitric acid with the same concentration?

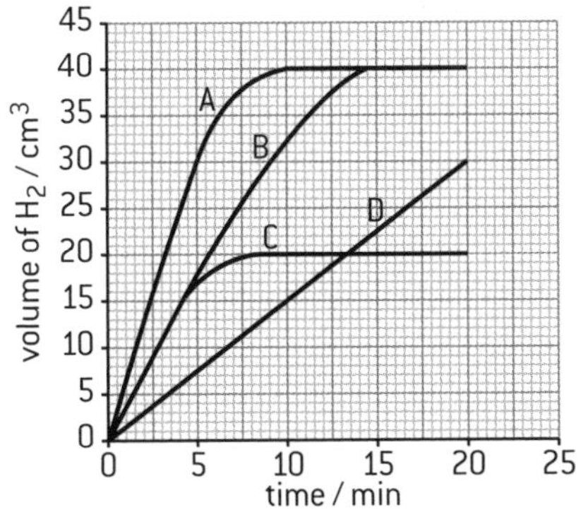

2. Which changes increase the rate of the reaction below?

$$C_4H_{10}(g) + Cl_2(g) \rightarrow C_4H_9Cl(l) + HCl(g)$$

I. Increase of pressure

II. Increase of temperature

III. Removal of HCl(g)

A. I and II only

B. I and III only

C. II and III only

D. I, II and III

3. The reaction between excess calcium carbonate and hydrochloric acid can be followed by measuring the volume of carbon dioxide produced with time. The results of one such reaction are shown below. How does the rate of this reaction change with time and what is the main reason for this change?

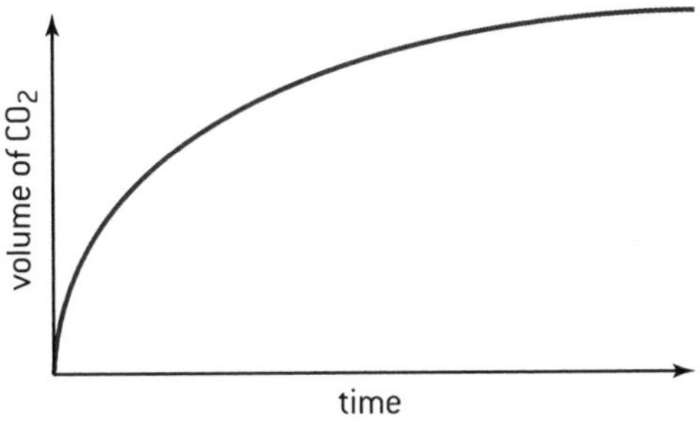

A. The rate increases with time because the calcium carbonate particles get smaller.

B. The rate increases with time because the acid becomes more dilute.

C. The rate decreases with time because the calcium carbonate particles get smaller.

D. The rate decreases with time because the acid becomes more dilute.

4. Hydrochloric acid is reacted with large pieces of calcium carbonate, the reaction is then repeated using calcium carbonate powder. How does this change affect the activation enery and the collision frequency?

	Activation energy	Collision frequency
A.	increases	increases
B.	stays constant	increases
C.	increases	stays constant
D.	stays constant	stays constant

5. Which statement is true about using sulfuric acid as a catalyst in the following reaction?

$$CH_3\text{–}CO\text{–}CH_3(aq) + I_2(aq) \xrightarrow{H^+(aq)} CH_3\text{–}CO\text{–}CH_2\text{–}I(aq) + HI(aq)$$

I. The catalyst increases the rate of reaction.

II. The catalyst lowers the activation energy for the reaction.

III. The catalyst has been consumed at the end of the chemical reaction.

A. I and II only

B. I and III only

C. II and III only

D. I, II and III

6. Which are appropriate units for the rate of a reaction?

A. $mol\ dm^{-3}\ s$

B. $mol\ dm^{-3}\ s^{-1}$

C. $mol\ dm^{-3}$

D. s

7. The following enthalpy level diagram shows the effect of the addition of a catalyst to a chemical reaction. What do ***m***, ***n*** and ***o*** represent?

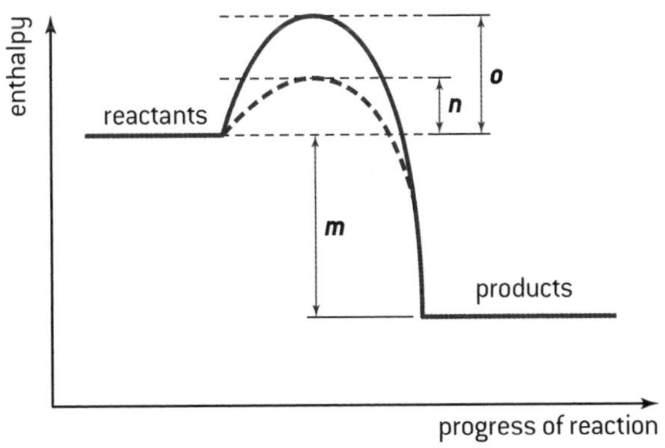

	m	*n*	*o*
A.	ΔH	E_a (without a catalyst)	E_a (with a catalyst)
B.	E_a (with a catalyst)	ΔH	E_a (without a catalyst)
C.	E_a (with a catalyst)	E_a (without a catalyst)	ΔH
D.	ΔH	E_a (with a catalyst)	E_a (without a catalyst)

MULTIPLE CHOICE QUESTIONS – CHEMICAL KINETICS

8. Consider the reaction between magnesium and hydrochloric acid. Which factors will affect the reaction rate?

   I. The collision frequency of the reactant particles

   II. The number of reactant particles with $E \geq E_a$

   III. The number of reactant particles that collide with the appropriate geometry

   A. I and II only

   B. I and III only

   C. II and III only

   D. I, II and III

9. What is the best definition of *rate of reaction*?

   A. The time it takes to use up all the reactants

   B. The rate at which all the reactants are used up

   C. The increase in concentration of a product per unit time

   D. The time it takes for one of the reactants to be used up

10. At 25 °C, 200 cm³ of 1.0 mol dm⁻³ nitric acid is added to 5.0 g of magnesium powder. If the experiment is repeated using the same mass of magnesium powder, which conditions will result in the same initial reaction rate?

	Volume of HNO_3 / cm^3	Concentration of HNO_3 / mol dm^{-3}	Temperature / °C
A.	200	2.0	25
B.	200	1.0	50
C.	100	2.0	25
D.	100	1.0	25

HL

11. The decomposition of N_2O_5 occurs according to the following equation.

    $2N_2O_5(g) \rightarrow 4NO_2(g) + O_2(g)$

    The reaction is first order with respect to N_2O_5. What combination of variables could the axes represent on the graph below?

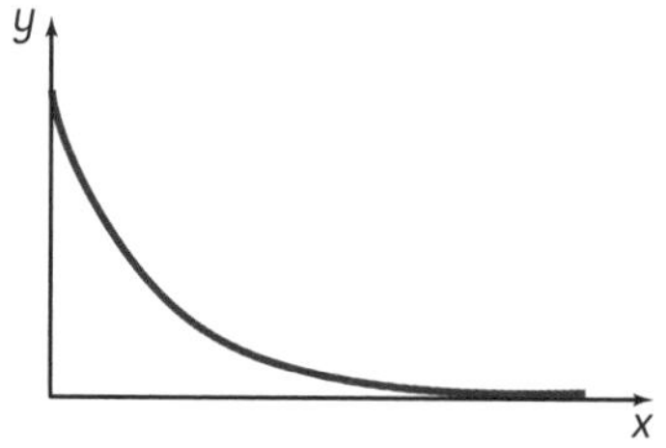

	x-axis	*y*-axis
A.	time	$[N_2O_5]$
B.	$[N_2O_5]$	time
C.	$[N_2O_5]$	rate of reaction
D.	rate of reaction	$[N_2O_5]$

12. Which graph represents a reaction that is second order with respect to X for the reaction X → products?

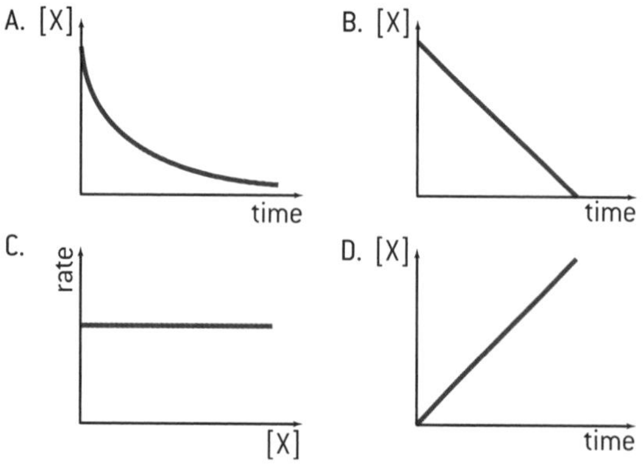

13. Consider the reaction:

    $2NO(g) + Br_2(g) \rightarrow 2NOBr(g)$

    One suggested mechanism is:

    $NO(g) + Br_2(g) \rightleftharpoons NOBr_2(g)$ *fast*

    $NOBr_2(g) + NO(g) \longrightarrow 2NOBr(g)$ *slow*

    Which statements are correct?

    I. $NOBr_2(g)$ is an intermediate.

    II. The second step is the rate-determining step.

    III. rate = $k\,[NO]^2[Br_2]$

    A. I and II only

    B. I and III only

    C. II and III only

    D. I, II and III

14. Consider the following reaction.

    $2P + Q \rightarrow R + S$

    This reaction occurs according to the following mechanism.

    $P + Q \rightarrow X$ *slow*

    $P + X \rightarrow R + S$ *fast*

    What is the rate expression?

    A. rate = $k[P]$

    B. rate = $k[P]\,[X]$

    C. rate = $k[P]\,[Q]$

    D. rate = $k[P]^2\,[Q]$

SHORT ANSWER QUESTIONS – CHEMICAL KINETICS

1. a) (i) Draw a graph that shows the distribution of molecular energies in a sample of a gas at two different temperatures, T_1 and T_2 such that T_2 is greater than T_1. [2]

   (ii) Define the term activation energy. [1]

   (iii) State and explain the effect of a catalyst on the rate of an endothermic reaction. [2]

   b) (i) Magnesium is added to a solution of hydrochloric acid. Sketch a graph of acid concentration on the y-axis against time on the x-axis to illustrate the progress of the reaction. [1]

   (ii) Describe how the slope of the line changes with time. [1]

   (iii) Use the collision theory to state and explain the effect of decreasing concentration on the rate of the reaction. [2]

2. Hydrogen peroxide, H_2O_2 (aq), releases oxygen gas, O_2 (g), as it decomposes according to the equation:

   $2H_2O_2(aq) \downarrow 2H_2O(l) + O_2(g)$

   50.0 cm^3 of hydrogen peroxide solution was placed in a boiling tube, and a drop of liquid detergent was added to create a layer of bubbles on the top of the hydrogen peroxide solution as oxygen gas was released. The tube was placed in a water bath at 75 °C and the height of the bubble layer was measured every thirty seconds. A graph was plotted of the height of the bubble layer against time.

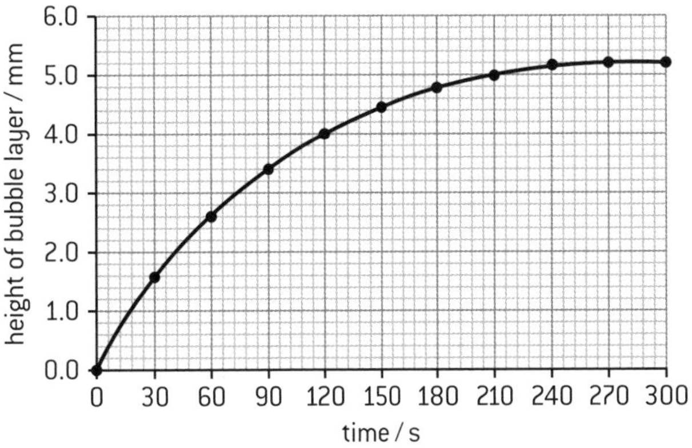

   a) Explain why the curve reaches a maximum. [1]

   b) Use the graph to calculate the rate of decomposition of hydrogen peroxide at 120 s. [3]

   c) The experiment was repeated using solid manganese(IV) oxide, MnO_2 (s), as a catalyst.

   (i) Draw a curve on the graph above to show how the height of the bubble layer changes with time when manganese(IV) oxide is present. [1]

   (ii) Explain the effect of the catalyst on the rate of decomposition of hydrogen peroxide. [2]

HL

3. Hydrogen and nitrogen(II) oxide react together exothermically as follows.

   $2H_2(g) + 2NO(g) \rightarrow 2H_2O(g) + N_2(g)$

   The rate of this reaction was investigated in a series of experiments carried out at the same temperature, the results of which are shown below.

Experiment	Initial $[H_2(g)]$ / mol dm^{-3}	Initial [NO(g)] / mol dm^{-3}	Initial rate of reaction / mol dm^{-3} s^{-1}
1	2.0×10^{-3}	4.0×10^{-3}	4.0×10^{-3}
2	4.0×10^{-3}	4.0×10^{-3}	8.0×10^{-3}
3	6.0×10^{-3}	4.0×10^{-3}	
4	2.0×10^{-3}	2.0×10^{-3}	1.0×10^{-3}
5	2.0×10^{-3}	1.0×10^{-3}	

   a) Explain how the results from Experiments 1 and 2 can be used to deduce that the order of reaction with respect to hydrogen is 1. [1]

   b) Deduce the order of reaction with respect to nitrogen(II) oxide, giving a reason for your answer. [2]

   c) Use your answers from parts a) and b) to deduce the rate expression for the reaction. [1]

   d) Calculate the rate of reaction for each of Experiments 3 and 5. [2]

   e) Use the results from Experiment 1 to determine the value of, and the units for, the rate constant, k, for the reaction. [2]

   f) Suggest a mechanism for the reaction that is consistent with the rate expression. [2]

   g) The reaction is faster in the presence of a heterogeneous catalyst. Explain the meaning of the term heterogeneous as applied to a catalyst. Draw a labelled enthalpy level diagram that shows the effect of the catalyst. [3]

4. a) The reaction between nitrogen(II) oxide and chlorine was studied at 263 K.

   $2NO(g) + Cl_2(g) \rightleftharpoons 2NOCl(g)$

   It was found that the forward reaction is first order with respect to Cl_2 and second order with respect to NO. The reverse reaction is second order with respect to NOCl.

   (i) State the rate expression for the forward reaction. [1]

   (ii) Predict the effect on the rate of the forward reaction and on the rate constant if the concentration of NO is halved. [2]

   b) Consider the following reaction.

   $NO_2(g) + CO(g) \rightarrow NO(g) + CO_2(g)$

   Possible reaction mechanisms are:

Above 775 K:	$NO_2 + CO \rightarrow NO + CO_2$	*slow*
Below 775 K:	$2NO_2 \rightarrow NO + NO_3$	*slow*
	$NO_3 + CO \rightarrow NO_2 + CO_2$	*fast*

   Based on the mechanisms, deduce the rate expressions above and below 775 K. [2]

   c) State two situations when the rate of a chemical reaction is equal to the rate constant. [2]

The equilibrium law

DYNAMIC EQUILIBRIUM

$$A + B \underset{\text{reverse reaction}}{\overset{\text{forward reaction}}{\rightleftharpoons}} C + D$$

Most chemical reactions do not go to completion. Once some products are formed the reverse reaction can take place to reform the reactants. In a closed system the concentrations of all the reactants and products will eventually become constant. Such a system is said to be in a state of **dynamic equilibrium**. The forward and reverse reactions continue to occur, but at equilibrium the rate of the forward reaction is equal to the rate of the reverse reaction.

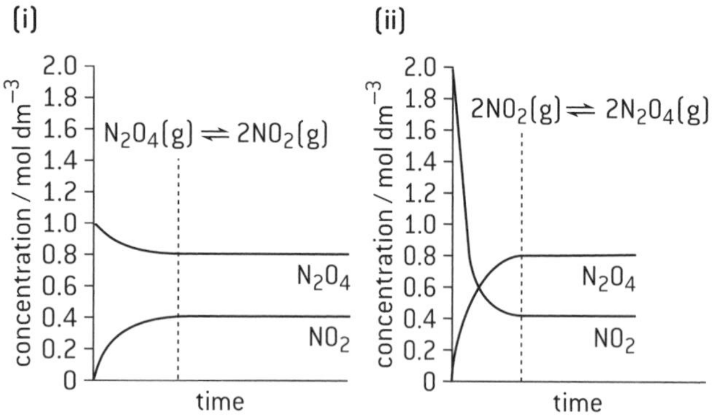

Graph (i) shows the decomposition of N_2O_4. Graph (ii) shows the reverse reaction starting with NO_2. Once equilibrium is reached (shown by the dotted line), the composition of the mixture remains constant and is independent of the starting materials.

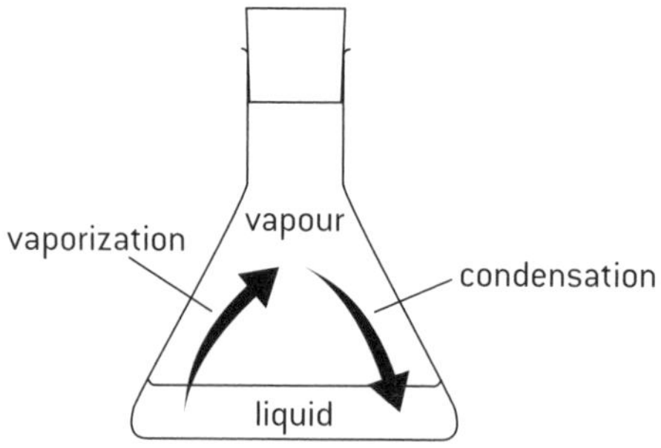

Dynamic equilibrium also occurs when physical changes take place. In a closed flask, containing some water, equilibrium will be reached between the liquid water and the water vapour. The faster moving molecules in the liquid will escape from the surface to become vapour and the slower moving molecules in the vapour will condense back into liquid. Equilibrium will be established when the rate of vaporization equals the rate of condensation.

$$H_2O(l) \rightleftharpoons H_2O(g)$$

CLOSED SYSTEM

A closed system is one in which neither matter nor energy can be lost or gained from the system, that is, the macroscopic properties remain constant. If the system is open some of the products from the reaction could escape and equilibrium would never be reached.

REACTION QUOTIENT AND EQUILIBRIUM CONSTANT

Consider the following general reversible reaction in which w moles of A react with x moles of B to produce y moles of C and z moles of D.

$$wA + xB \rightleftharpoons yC + zD$$

At any particular point in time the concentrations of A, B, C and D can be written as [A], [B], [C] and [D] respectively. The reaction quotient, Q, is defined as being

$$Q = \frac{[C]^y \times [D]^z}{[A]^w \times [B]^x}$$

As the reaction proceeds, the reaction quotient will change until the point of equilibrium is reached. At that point the concentrations of A, B, C and D remain constant and the reaction quotient is known as the equilibrium constant, K_c.

The equilibrium law states that for this reaction at a particular temperature

$$K_c = \frac{[C]^y_{eqm} \times [D]^z_{eqm}}{[A]^w_{eqm} \times [B]^x_{eqm}}$$

Examples

Formation of sulfur trioxide in the Contact process

$$2SO_2(g) + O_2(g) \rightleftharpoons 2SO_3(g)$$

$$K_c = \frac{[SO_3]^2_{eqm}}{[SO_2]^2_{eqm} \times [O_2]_{eqm}}$$

Formation of an ester from ethanol and ethanoic acid

$$C_2H_5OH(l) + CH_3COOH(l) \rightleftharpoons CH_3COOC_2H_5(l) + H_2O(l)$$

$$K_c = \frac{[CH_3COOC_2H_5]_{eqm} \times [H_2O]_{eqm}}{[C_2H_5OH]_{eqm} \times [CH_3COOH]_{eqm}}$$

In both of these examples all the reactants and products are in the same phase. In the first example they are all in the gaseous phase and in the second example they are all in the liquid phase. Such reactions are known as *homogeneous reactions*. Another example of a homogeneous system would be where all the reactants and products are in the aqueous phase.

MAGNITUDE OF THE EQUILIBRIUM CONSTANT

Since the equilibrium expression has the concentration of products on the top and the concentration of reactants on the bottom it follows that the magnitude of the equilibrium constant is related to the position of equilibrium. When the reaction goes nearly to completion $K_c \gg 1$. If the reaction hardly proceeds then $K_c \ll 1$. If the value for K_c lies between about 10^{-2} and 10^2 then both reactants and products will be present in the system in noticeable amounts. The value for K_c in the esterification reaction above is 4 at 100 °C. From this it can be inferred that the concentration of the products present in the equilibrium mixture is roughly twice that of the reactants.

Le Chatelier's principle and factors affecting the position of equilibrium

LE CHATELIER'S PRINCIPLE

Provided the temperature remains constant the value for K_c must remain constant. If the concentration of the reactants is increased, or one of the products is removed from the equilibrium mixture then more of the reactants must react in order to keep K_c constant, i.e. the position of equilibrium will shift to the right (towards more products). This is the explanation for Le Chatelier's principle, which states that if a system at equilibrium is subjected to a small change the equilibrium tends to shift so as to minimize the effect of the change.

2. equilibrium shifts to minimize effect

$$A + B \rightleftharpoons C + D$$

1. remove product

FACTORS AFFECTING THE POSITION OF EQUILIBRIUM

CHANGE IN CONCENTRATION

$C_2H_5OH(l) + CH_3COOH(l) \rightleftharpoons CH_3COOC_2H_5(l) + H_2O(l)$

If more ethanoic acid is added the concentration of ethanoic acid increases so that at the point of addition:

$$K_c \neq \frac{[\text{ester}] \times [\text{water}]}{[\text{acid}] \times [\text{alcohol}]}$$

To restore the system so that the equilibrium law is obeyed the equilibrium will move to the right, so that the concentration of ester and water increases and the concentration of the acid and alcohol decreases.

CHANGE IN PRESSURE

If there is an overall volume change in a gaseous reaction then increasing the pressure will move the equilibrium towards the side with less volume. This shift reduces the total number of molecules in the equilibrium system and so tends to minimize the pressure.

$2NO_2(g)$	$\rightleftharpoons N_2O_4(g)$
brown	colourless
(2 vols)	(1 vol)

If the pressure is increased the mixture will initially go darker as the concentration of NO_2 increases then become lighter as the position of equilibrium is re-established with a greater proportion of N_2O_4.

CHANGE IN TEMPERATURE

In exothermic reactions heat is also a product. Taking the heat away will move the equilibrium towards the right, so more products are formed. The forward reaction in exothermic reactions is therefore increased by lowering the temperature

$2NO_2(g) \rightleftharpoons N_2O_4(g)$ $\Delta H^\ominus = -24 \text{ kJ mol}^{-1}$

brown colourless

Lowering the temperature will cause the mixture to become lighter as the equilibrium shifts to the right.

For an endothermic reaction the opposite will be true.

Unlike changing the concentration or pressure, a change in temperature will also change the value of K_c. For an exothermic reaction the concentration of the products in the equilibrium mixture decreases as the temperature increases, so the value of K_c will decrease. The opposite will be true for endothermic reactions.

e.g. $H_2(g) + CO_2(g) \rightleftharpoons H_2O(g) + CO(g)$ $\Delta H^\ominus = +41 \text{ kJ mol}^{-1}$

***T* / K**	***K*$_c$**	
298	1.00×10^{-5}	increase ↓
500	7.76×10^{-3}	
700	1.23×10^{-1}	
900	6.01×10^{-1}	

ADDING A CATALYST

A catalyst will increase the rate at which equilibrium is reached, as it will speed up both the forward and reverse reactions equally, but it will have no effect on the position of equilibrium and hence on the value of K_c.

MANIPULATING EQUILIBRIUM CONSTANTS

When a reaction is reversed the equilibrium constant for the reverse reaction will be the reciprocal of the equilibrium constant for the forward reaction, K_c. For example, for the reverse reaction of the Haber process, $2NH_3(g) \rightleftharpoons N_2(g) + 3H_2(g)$

$$K_c' = \frac{[N_2] \times [H_2]^3}{[NH_3]^2} = \frac{1}{K_c}$$

If there are multiple steps in a reaction, each with its own equilibrium constant, then the equilibrium constants are multiplied to give the overall value of K_c. For example,

For the step $A + B \rightleftharpoons C$ $$K_c^1 = \frac{[C]}{[A] \times [B]}$$

For the step $C + D \rightleftharpoons X + Y$ $$K_c^2 = \frac{[X] \times [Y]}{[C] \times [D]}$$

For the overall reaction $A + B + D \rightleftharpoons X + Y$ $$K_c = \frac{[X] \times [Y]}{[A] \times [B] \times [D]} = K_c^1 \times K_c^2$$

Equilibrium calculations

EQUILIBRIUM CALCULATIONS

The equilibrium law can be used either to find the value for the equilibrium constant, or to find the value of an unknown equilibrium concentration.

a) 23.0 g (0.50 mol) of ethanol was reacted with 60.0 g (1.0 mol) of ethanoic acid and the reaction allowed to reach equilibrium at 373 K. 37.0 g (0.42 mol) of ethyl ethanoate was found to be present in the equilibrium mixture. Calculate K_c to the nearest integer at 373 K.

	$C_2H_5OH(l)$	+	$CH_3COOH(l)$	$\rightleftharpoons$	$CH_3COOC_2H_5(l)$	+	$H_2O(l)$
Initial amount / mol	0.50		1.00		–		–
Equilibrium amount / mol	(0.50 – 0.42)		(1.00 – 0.42)		0.42		0.42
Equilibrium concentration / mol dm^{-3} (where V = total volume)	$\frac{(0.50-0.42)}{V}$		$\frac{(1.00-0.42)}{V}$		$\frac{0.42}{V}$		$\frac{0.42}{V}$

$$K_c = \frac{[\text{ester}] \times [\text{water}]}{[\text{alcohol}] \times [\text{acid}]} = \frac{(0.42/V) \times (0.42/V)}{(0.08/V) \times (0.58/V)} = 4 \text{ (to the nearest integer)}$$

b) What mass of ester will be formed at equilibrium if 2.0 moles of ethanoic acid and 1.0 moles of ethanol are reacted under the same conditions?

Let x moles of ester be formed and let the total volume be V dm^3.

$$K_c = 4 = \frac{[\text{ester}] \times [\text{water}]}{[\text{alcohol}] \times [\text{acid}]} = \frac{x^2/V^2}{(1.0-x)/V \times (2.0-x)/V} = \frac{x^2}{(x^2 - 3x + 2)}$$

$\Rightarrow 3x^2 - 12x + 8 = 0$

solve by substituting into the quadratic expression $x = \frac{-b \pm \sqrt{b^2 - 4ac}}{2a} \Rightarrow x = \frac{12 \pm \sqrt{144 - 96}}{6}$

$x = 0.845$ or ~~3.15~~ (it can not be 3.15 as only 1.0 mol of ethanol was taken)

Mass of ester = 0.845 × 88.08 = 74.4 g (Note: IB Diploma Programme chemistry does not examine the use of the quadratic expression.)

c) 1.60 mol of hydrogen and 1.00 mol of iodine are allowed to reach equilibrium at a temperature of 704 K in a 4.00 dm^3 flask, the amount of hydrogen iodide formed in the equilibrium mixture is 1.80 mol. Determine the value of the equilibrium constant at this temperature.

	$H_2(g)$	$I_2(g)$	$\rightleftharpoons$	$2HI(g)$
Initial amount / mol	1.60	1.00		0
Equilibrium amount / mol	0.70	0.10		1.80
Equilibrium concentration / mol dm^{-3}	0.175	0.025		0.450

$$K_c = \frac{[HI(g)]^2}{[H_2(g)] \times [I_2(g)]} = \frac{0.450^2}{0.175 \times 0.025} = 46.3 \text{ at 704 K}$$

RELATIONSHIP BETWEEN FREE ENERGY CHANGE AND THE EQUILIBRIUM CONSTANT

The position of equilibrium corresponds to a maximum value of entropy and a minimum in the value of the Gibbs free energy change. This means that the equilibrium constant, K_c and the Gibbs free energy change, $\Delta G^\ominus$ can both be used to measure the position of equilibrium in a reaction. They are related by the equation

$\Delta G = -RT \ln K$

This can be illustrated by the dissociation of water according to the equation

$H_2O(l) \rightleftharpoons H^+(aq) + OH^-(aq) \quad \Delta H^\ominus = +55.8 \text{ kJ mol}^{-1}$

The relevant entropy values are:

	$H_2O(l)$	$H^+(aq)$	$OH^-(aq)$
$S^\ominus$ / J K^{-1} mol^{-1}	+70.0	0	– 10.9

The change in entropy, $\Delta S^\ominus = (\Sigma S^\ominus products) - (\Sigma S^\ominus reactants) = (-10.9) - (+70.0) = -80.9$ J K^{-1} mol^{-1}

$\Delta G^\ominus = \Delta H^\ominus - T\Delta S^\ominus = 55.8 \times 1000 - (298 \times -80.9) = +79908$ J mol^{-1}

Using the expression $\Delta G = -RT \ln K$

$$\ln K = \frac{-79908}{8.314 \times 298} = -32.2$$

$K = e^{-32.2} = 1.00 \times 10^{-14}$ at 298 K

This is the value for the equilibrium constant of water at 298 K, known as the ionic product constant for water (K_w), given in Section 2 of the IB data booklet.

MULTIPLE CHOICE QUESTIONS – EQUILIBRIUM

1. Which statement is true about a chemical reaction at equilibrium?

   A. The reaction has completely stopped.

   B. The concentrations of the products are equal to the concentrations of the reactants.

   C. The rate of the forward reaction is equal to the rate of the reverse reaction.

   D. The concentrations of the products and reactants are constantly changing.

2. What is the equilibrium constant expression, K_c, for the following reaction?

   $2NOBr(g) \rightleftharpoons 2NO(g) + Br_2(g)$

   A. $K_c = \frac{[NO][Br_2]}{[NOBr]}$

   B. $K_c = \frac{[NO]^2[Br_2]}{[NOBr]^2}$

   C. $K_c = \frac{2[NO] + [Br_2]}{[2NOBr]}$

   D. $K_c = \frac{[NOBr]^2}{[NO]^2[Br_2]}$

3. The following are K_c values for a reaction, with the same starting conditions, carried out at different temperatures. Which equilibrium mixture has the highest concentration of products?

   A. 1×10^{-2}

   B. 1

   C. 1×10^{1}

   D. 1×10^{2}

4. What effect will an increase in temperature have on the K_c value and the position of equilibrium in the following reaction?

   $N_2(g) + 3H_2(g) \rightleftharpoons 2NH_3(g) \quad \Delta H = -92\,kJ$

	K_c	Equilibrium position
A.	increases	shifts to the right
B.	decreases	shifts to the left
C.	increases	shifts to the left
D.	decreases	shifts to the right

5. Consider the following equilibrium reaction.

   $2SO_2(g) + O_2(g) \rightleftharpoons 2SO_3(g) \quad \Delta H^{\ominus} = -197\,kJ$

   Which change in conditions will increase the amount of SO_3 present when equilibrium is re-established?

   A. Decreasing the concentration of SO_2

   B. Increasing the volume

   C. Decreasing the temperature

   D. Adding a catalyst

6. The Haber process uses an iron catalyst to convert hydrogen gas, $H_2(g)$, and nitrogen gas, $N_2(g)$, to ammonia gas, $NH_3(g)$.

   $3H_2(g) + N_2(g) \rightleftharpoons 2NH_3(g)$

   Which statements are correct for this equilibrium system?

   I. The iron catalyst increases rates of the forward and reverse reactions equally.

   II. The iron catalyst does not affect the value of the equilibrium constant, K_c.

   III. The iron catalyst increases the yield for ammonia gas, $NH_3(g)$.

   A. I and II only

   B. I and III only

   C. II and III only

   D. I, II and III

7. The formation of nitric acid, $HNO_3(aq)$, from nitrogen dioxide, $NO_2(g)$, is exothermic and is a reversible reaction.

   $4NO_2(g) + O_2(g) + 2H_2O(l) \rightleftharpoons 4HNO_3(aq)$

   What is the effect of a catalyst on this reaction?

   A. It increases the yield of nitric acid.

   B. It increases the rate of the forward reaction only.

   C. It increases the equilibrium constant.

   D. It has no effect on the equilibrium position.

8. The value of K_c for the reaction $H_2(g) + Br_2(g) \rightleftharpoons 2HBr(g)$ is 4.0×10^{-2}. What is the value of the equilibrium constant for the reaction $2HBr(g) \rightleftharpoons H_2(g) + Br_2(g)$ at the same temperature?

   A. 4.0×10^{-2}

   B. 2.0×10^{-1}

   C. 25

   D. 400

HL

9. 0.50 mol of $I_2(g)$ and 0.50 mol of $Br_2(g)$ are placed in a closed flask. The following equilibrium is established.

   $I_2(g) + Br_2(g) \rightleftharpoons 2IBr(g)$

   The equilibrium mixture contains 0.80 mol of IBr(g). What is the value of K_c?

   A. 0.64

   B. 1.3

   C. 2.6

   D. 64

10. A 2.0 dm^3 reaction vessel initially contains 4.0 mol of P and 5.0 mol of Q. At equilibrium 3 mol of R is present. What is the value of K_c for the following reaction?

    $P(g) + Q(g) \rightleftharpoons R(g) + S(g)$

    A. $\frac{2}{9}$

    B. $\frac{9}{20}$

    C. 4.5

    D. 9

11. At 35 °C $K_c = 1.6 \times 10^{-5}\,mol\,dm^{-3}$ for the reaction

    $2NOCl(g) \rightleftharpoons 2NO(g) + Cl_2(g)$

    Which relationship must be correct at equilibrium?

    A. [NO] = [NOCl]

    B. 2[NO] = $[Cl_2]$

    C. [NOCl] < $[Cl_2]$

    D. [NO] < [NOCl]

12. Free energy change and the equilibrium constant are related by the equation $\Delta G = -RT\ln K_c$. Which combination is most likely for a reaction to go to completion at all temperatures?

	$\Delta H^{\ominus}$	$\Delta S^{\ominus}$	K_c
A	–	+	> 1
B	+	+	> 1
C	–	–	< 1
D	–	–	> 1

SHORT ANSWER QUESTIONS – EQUILIBRIUM

1. Ethanol is manufactured by the hydration of ethene according to the equation below.

   $C_2H_4(g) + H_2O(g) \rightleftharpoons C_2H_5OH(g)$

   a) State the expression for the equilibrium constant, K_c, for this reaction. [1]

   b) Under certain conditions, the value of K_c for this reaction is 3.7×10^{-3}. When the temperature is increased the value is 4.9×10^{-4}.

      (i) State what can be deduced about the position of equilibrium at the higher temperature from these values of K_c. [1]

      (ii) State what can be deduced about the sign of ΔH for the reaction, explaining your choice. [3]

   c) The process used to manufacture ethanol is carried out at high pressure. State and explain two advantages of using high pressure. [4]

2. Ammonia is produced by the Haber process according to the following reaction

   $N_2(g) + 3H_2(g) \rightleftharpoons 2NH_3(g)$ ΔH is negative

   a) State the equilibrium expression for the above reaction. [1]

   b) Predict, giving a reason, the effect on the position of equilibrium when the pressure in the reaction vessel is increased. [2]

   c) State and explain the effect on the value of K_c when the temperature is increased. [2]

   d) Explain why a catalyst has no effect on the position of equilibrium. [1]

3. Consider the following equilibrium:

   $4NH_3(g) + 5O_2(g) \rightleftharpoons 4NO(g) + 6H_2O(g)$ $\Delta H^{\ominus} = -909\,kJ$

   a) Deduce the equilibrium constant expression, K_c, for the reaction. [1]

   b) Predict the direction in which the equilibrium will shift when the following changes occur. [4]

      (i) The volume increases.

      (ii) The temperature decreases.

      (iii) $H_2O(g)$ is removed from the system.

      (iv) A catalyst is added to the reaction mixture.

   c) Define the term *activation energy*. [1]

4. An example of a homogeneous reversible reaction is the reaction between hydrogen and iodine.

   $H_2(g) + I_2(g) \rightleftharpoons 2HI(g)$

   a) Outline the characteristics of a homogeneous chemical system that is in a state of equilibrium. [2]

   b) Formulate the expression for the equilibrium constant, K_c. [1]

   c) Predict what would happen to the position of equilibrium and the value of K_c if the pressure is increased from 1 atm to 2 atm. [2]

   d) The value of K_c at 500 K is 160 and the value of K_c at 700 K is 54. Deduce what this information tells us about the enthalpy change of the forward reaction. [1]

   (e) Deduce the value of the equilibrium constant, K_c' at 500 K for the reaction below: [1]

   $2HI(g) \rightleftharpoons H_2(g) + I_2(g)$

HL

5. Consider the two equilibrium systems involving bromine gas illustrated below.

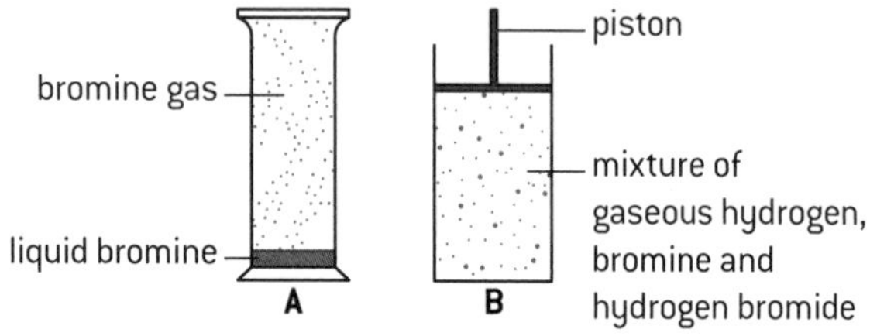

   a) Formulate equations to represent the equilibria in A and B with $Br_2(g)$ on the left-hand side in both equilibria. [2]

   b) (i) Describe what you would observe if a small amount of liquid bromine is introduced into **A**. [1]

      (ii) Predict what happens to the position of equilibrium if a small amount of hydrogen is introduced into **B**. [1]

      (iii) State and explain the effect of increasing the pressure in **B** on the position of equilibrium. [2]

   c) (i) Deduce the equilibrium constant expression, K_c, for the equilibrium in **B**. [1]

      (ii) State the effect of increasing $[H_2]$ in **B** on the value of K_c. [1]

6. Ammonia production is important in industry.

   $N_2(g) + 3H_2(g) \rightleftharpoons 2NH_3(g)$ $\Delta H = -92$ kJ

   a) The standard entropy values, S, at 298 K for $N_2(g)$, $H_2(g)$ and $NH_3(g)$ are 193, 131 and 192 J K^{-1} mol^{-1} respectively. Calculate $\Delta S^{\ominus}$ for the reaction as shown by the equation above. [2]

   b) Determine $\Delta G^{\ominus}$ for the reaction at 298 K. [2]

   c) Describe and explain the effect of increasing temperature on the spontaneity of the reaction. [2]

   d) Determine the value of the equilibrium constant at 298 K by using the value of $\Delta G^{\ominus}$ that you obtained in b). [3]

   e) 0.20 mol of N(g) and 0.20 mol of $H_2(g)$ were allowed to reach equilibrium in a 1 dm^3 closed container at a temperature T_2 which is different to 298 K. At equilibrium the concentration of $NH_3(g)$ was found to be 0.060 mol dm^{-3}. Determine the value of K_c at temperature T_2. [3]

   f) Comment on the two different values for K_c that you have obtained. [2]

   g) Describe how increasing the pressure affects the yield of ammonia. [2]

   h) In practice, typical conditions used in the Haber process are a temperature of 500°C and a pressure of 200 atmospheres. Suggest why these conditions are used rather than those that give the highest yield. [2]

   i) Iron is used as a catalyst in this manufacturing process. A catalyst has no effect on the value of K_c or on the position of equilibrium. Suggest why a catalyst is used in this process. [1]

Theories and properties of acids and bases

THE IONIC THEORY

An acid was originally distinguished by its sour taste. Later it was said to be the oxide of a non-metal combined with water although hydrochloric acid does not fit into this definition. The ionic theory which is still commonly used today states that an acid is a substance which produces hydrogen ions, $H^+(aq)$, in aqueous solution, e.g.

$$HCl(aq) \rightarrow H^+(aq) + Cl^-(aq)$$

In aqueous solution hydrogen ions are hydrated to form hydroxonium ions, $H_3O^+(aq)$. In the International Baccalaureate it is correct to write either $H^+(aq)$ or $H_3O^+(aq)$ to represent the hydrogen ions in an aqueous solution. Strictly speaking an acid gives a hydrogen ion concentration in aqueous solution greater than 1.0×10^{-7} mol dm^{-3}. A base is a substance that can neutralize an acid. An alkali is a base that is soluble in water.

BRØNSTED–LOWRY ACIDS AND BASES

A Brønsted–Lowry acid is a substance that can *donate* a proton. A Brønsted–Lowry base is a substance that can *accept* a proton.

Consider the reaction between hydrogen chloride gas and water.

$$\underset{\text{acid}}{HCl(g)} + \underset{\text{base}}{H_2O(l)} \rightleftharpoons \underset{\text{acid}}{H_3O^+(aq)} + \underset{\text{base}}{Cl^-(aq)}$$

Under this definition both HCl and H_3O^+ are acids as both can donate a proton. Similarly both H_2O and Cl^- are bases as both can accept a proton. Cl^- is said to be the **conjugate base** of HCl and H_2O is the conjugate base of H_3O^+. The conjugate base of an acid is the species remaining after the acid has lost a proton. Every base also has a conjugate acid, which is the species formed after the base has accepted a proton. In the reaction with hydrogen chloride water is behaving as a base. Water can also behave as an acid.

$$\underset{\text{base}}{NH_3(g)} + \underset{\text{acid}}{H_2O(l)} \rightleftharpoons \underset{\text{acid}}{NH_4^+(aq)} + \underset{\text{base}}{OH^-(aq)}$$

Substances such as water, which can act both as an acid and as a base, are described as **amphiprotic**.

Many acids, particularly organic acids, contain one or more non-acidic hydrogen atoms. The location of the acidic hydrogen atom(s) should be clearly identified. For example, ethanoic acid should be written as CH_3COOH rather than $C_2H_4O_2$ so that the conjugate base can be identified as the carboxylate anion CH_3COO^- rather than just $C_2H_3O_2^-$.

TYPICAL PROPERTIES OF ACIDS AND BASES

The typical reactions of acids are:

1. With indicators.

   Acid–base indicators can be used to determine whether or not a solution is acidic. Common indicators include:

Indicator	**Colour in acidic solution**	**Colour in alkaline solution**
litmus	red	blue
phenolphthalein	colourless	pink
methyl orange	red	yellow

2. Neutralization reactions with bases.
   a) With hydroxides to form a salt and water,
      e.g. $CH_3COOH(aq) + NaOH(aq) \rightarrow NaCH_3COO(aq) + H_2O(l)$
   b) With metal oxides to form a salt and water,
      e.g. $H_2SO_4(aq) + CuO(s) \rightarrow CuSO_4(aq) + H_2O(l)$
   c) With ammonia to form a salt.
      e.g. $HCl(aq) + NH_3(aq) \rightarrow NH_4Cl(aq)$
3. With reactive metals (those above copper in the activity series) to form a salt and hydrogen, e.g.
   $2HCl(aq) + Mg(s) \rightarrow MgCl_2(aq) + H_2(g)$
4. With carbonates (soluble or insoluble) to form a salt, carbon dioxide and water, e.g.
   $2HNO_3(aq) + Na_2CO_3(aq) \rightarrow 2NaNO_3(aq) + CO_2(g) + H_2O(l)$
   $2HCl(aq) + CaCO_3(s) \rightarrow CaCl_2(aq) + CO_2(g) + H_2O(l)$
5. With hydrogencarbonates to form a salt, carbon dioxide and water, e.g.
   $HCl(aq) + NaHCO_3(aq) \rightarrow NaCl(aq) + CO_2(g) + H_2O(aq)$

The pH scale

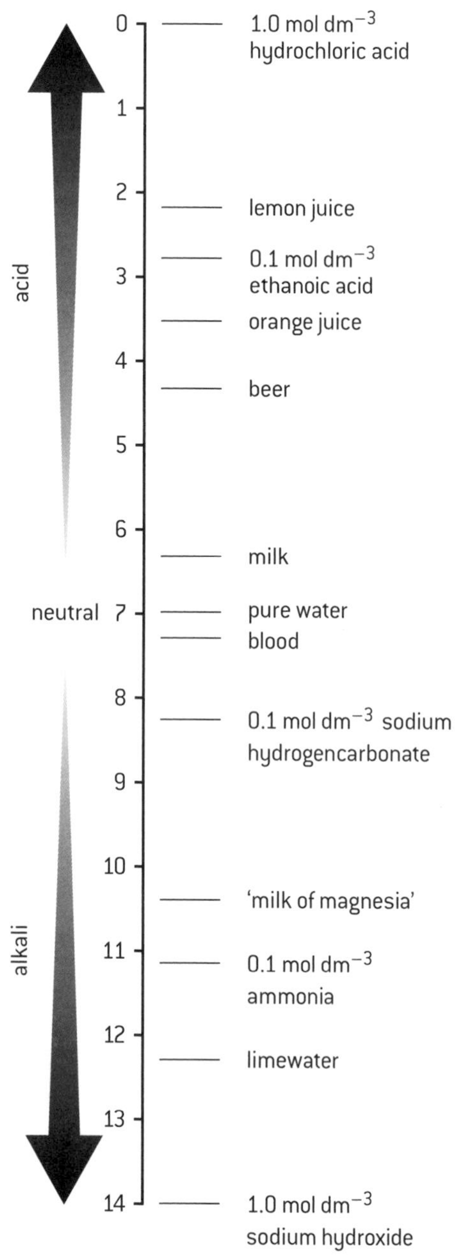

THE pH SCALE

Pure water is very slightly dissociated:

$H_2O(l) \rightleftharpoons H^+(aq) + OH^-(aq)$

At 25°C the equilibrium constant for this reaction is 1×10^{-14}.

i.e. $K_w = [H^+(aq)][OH^-(aq)] = 1 \times 10^{-14}$.

The concentration of the hydrogen ions (which is the same as the concentration of the hydroxide ions) equals 1×10^{-7} mol dm^{-3}.

pH (which stands for **p**ower of **H**ydrogen) is defined as being equal to minus the logarithm to the base ten of the hydrogen ion concentration.

i.e. $pH = -\log_{10}[H^+(aq)]$

In practice this means that it is equal to the power of ten of the hydrogen ion concentration with the sign reversed. The pH of pure water is thus 7.

Pure water is neutral, so the pH of any neutral solution is 7. If the solution is acidic the hydrogen ion concentration will be greater than 10^{-7} mol dm^{-3} and the pH will decrease. Similarly alkaline solutions will have a pH greater than 7.

The pH scale runs from 0 to 14. Because it depends on the power of ten a change in one unit in the pH corresponds to a tenfold change in the hydrogen ion concentration. A 0.1 mol dm^{-3} solution of a strong monoprotic acid will have a pH of 1, a 0.001 mol dm^{-3} solution of the same acid will have a pH of 3.

THE LOG_{10} SCALE AND p-SCALE

Normal scale – the distance between each number is equal

−5 −4 −3 −2 −1 0 1 2 3 4 5

Log_{10} scale – the distances between powers of ten are equal

0.00001 0.0001 0.001 0.01 0.1 1 10 100 1000 10 000 100 000

which can be written

10^{-5} 10^{-4} 10^{-3} 10^{-2} 10^{-1} 10^{0} 10^{1} 10^{2} 10^{3} 10^{4} 10^{5}

p-scale – sometimes used by chemists to express equilibrium constants and concentration. It is equal to minus the power of ten in the logarithmic scale so the scale becomes:

5 4 3 2 1 0 −1 −2 −3 −4 −5

DETERMINATION OF pH

The pH of a solution can be determined by using a pH meter or by using 'universal' indicator, which contains a mixture of indicators that give a range of colours at different pH values.

pH	$[H^+]$/ mol dm^{-3}	$[OH^-]$/ mol dm^{-3}	Description	Colour of universal indicator
0	1	1×10^{-14}	very acidic	red
4	1×10^{-4}	1×10^{-10}	acidic	orange
7	1×10^{-7}	1×10^{-7}	neutral	green
10	1×10^{-10}	1×10^{-4}	basic	blue
14	1×10^{-14}	1	very basic	purple

STRONG, CONCENTRATED AND CORROSIVE

In English the words strong and concentrated are often used interchangeably. In chemistry they have very precise meanings:

- **strong:** completely dissociated into ions
- **concentrated:** a high number of moles of solute per litre (dm^3) of solution
- **corrosive:** chemically reactive.

Similarly weak and dilute also have very different chemical meanings:

- **weak:** only slightly dissociated into ions
- **dilute:** a low number of moles of solute per litre of solution.

Strong and weak acids and bases and simple pH calculations

STRONG AND WEAK ACIDS AND BASES

A strong acid is completely dissociated (ionized) into its ions in aqueous solution. Similarly a strong base is completely dissociated into its ions in aqueous solution. Examples of strong acids and bases include:

Strong acids	**Strong bases**
hydrochloric acid, HCl	sodium hydroxide, $NaOH$
nitric acid, HNO_3	potassium hydroxide, KOH
sulfuric acid, H_2SO_4	barium hydroxide, $Ba(OH)_2$

Note: because one mole of HCl produces one mole of hydrogen ions it is known as a **monoprotic** acid. Sulfuric acid is known as a **diprotic** acid as one mole of sulfuric acid produces two moles of hydrogen ions.

Weak acids and bases are only slightly dissociated (ionized) into their ions in aqueous solution.

Weak acids	**Weak bases**
ethanoic acid, CH_3COOH	ammonia, NH_3
'carbonic acid' (CO_2 in water), H_2CO_3	aminoethane, $C_2H_5NH_2$

The difference can be seen in their reactions with water:

Strong acid: $HCl(g) + H_2O(l) \rightarrow H_3O^+(aq) + Cl^-(aq)$
reaction goes to completion

Weak acid:

$CH_3COOH(aq) + H_2O(l) \rightleftharpoons CH_3COO^-(aq) + H_3O^+(aq)$
equilibrium lies on the left

i.e. a solution of hydrochloric acid consists only of hydrogen ions and chloride ions in water, whereas a solution of ethanoic acid contains mainly undissociated ethanoic acid with only very few hydrogen and ethanoate ions.

Strong base: $KOH(s) \xrightarrow{H_2O(l)} K^+(aq) + OH^-(aq)$

Weak base: $NH_3(g) + H_2O(l) \rightleftharpoons NH_4^+(aq) + OH^-(aq)$
equilibrium lies on the left

EXPERIMENTS TO DISTINGUISH BETWEEN STRONG AND WEAK ACIDS AND BASES

1. pH measurement
   Because a strong acid produces a higher concentration of hydrogen ions in solution than a weak acid, with the same concentration, the pH of a strong acid will be lower than a weak acid. Similarly a strong base will have a higher pH in solution than a weak base, with the same concentration. The most accurate way to determine the pH of a solution is to use a pH meter.

$0.10\ mol\ dm^{-3}\ HCl(aq)$	$pH = 1.0$
$0.10\ mol\ dm^{-3}\ CH_3COOH$	$pH = 2.9$

2. Conductivity measurement
   Strong acids and strong bases in solution will give much higher readings on a conductivity meter than **equimolar** (equal concentration) solutions of weak acids or bases, because they contain more ions in solution.
3. Concentration measurement
   As the concentration of hydrogen ions is much greater, the rate of reaction of strong acids with metals, metal oxides, metal hydroxides, metal hydrogen carbonates and metal carbonates is greater than that of weak acids with the same concentration.

STRONG ACID AND BASE pH CALCULATIONS

For pure water the pH must be 7 at 25 °C as the concentration of $H^+(aq)$ is equal to the concentration of $OH^-(aq)$ and $[H^+(aq)] \times [OH^-(aq)] = 1 \times 10^{-14}$. Strong acids are completely dissociated so, for example, the hydrogen ion concentration of $0.100\ mol\ dm^{-3}$ hydrochloric acid, HCl(aq) will be $0.100\ mol\ dm^{-3}$ as each mole of acid produces one mole of hydrogen ions when it dissociates (ionizes). The pH of $0.100\ mol\ dm^{-3}$ HCl(aq) will therefore be equal to $-\log_{10}(0.100) = 1$.

If the acid is diluted ten times the new hydrogen ion concentration will be $0.0100\ mol\ dm^{-3}$ and the $pH = -\log_{10}(0.0100)$ or $-\log_{10}(1.00 \times 10^{-2}) = 2$.

Sulfuric acid is assumed for simplicity to be a strong diprotic acid. The hydrogen ion concentration of $0.0100\ mol\ dm^{-3}\ H_2SO_4(aq)$ will therefore be 2×0.0100 or $2.00 \times 10^{-2}\ mol\ dm^{-3}$ and the pH will equal $-\log_{10}(2.00 \times 10^{-2}) = 1.7$.

Note that pH is a measure of concentration so $10.0\ cm^3$ of $0.100\ mol\ dm^{-3}$ HCl(aq) will have the same pH as $100\ cm^3$ of $0.100\ mol\ dm^{-3}$ HCl(aq). Also note that $[H^+(aq)] = 10^{-pH}$ so if the pH of an acid is 3 then $[H^+(aq)] = 10^{-3} = 1.0 \times 10^{-3}\ mol\ dm^{-3}$.

To calculate the pH of a strong base work out the hydroxide concentration first then calculate the hydrogen ion concentration from the expression $[H^+(aq)] \times [OH^-(aq)] = 1 \times 10^{-14}$.

Example 1: To calculate the pH of $0.100\ mol\ dm^{-3}$ NaOH(aq).

$[OH^-(aq)] = 0.100\ mol\ dm^{-3}$

$[H^+(aq)] = \frac{1 \times 10^{-14}}{0.100} = 1 \times 10^{-13}\ mol\ dm^{-3}$

$pH = -\log_{10}(1.00 \times 10^{-13}) = 13.$

Example 2: To calculate the pH of $0.100\ mol\ dm^{-3}\ Ba(OH)_2(aq)$.

$[OH^-(aq)] = 0.200\ mol\ dm^{-3}$

$[H^+(aq)] = \frac{1 \times 10^{-14}}{0.200} = 5 \times 10^{-14}\ mol\ dm^{-3}$

$pH = -\log_{10}(5.00 \times 10^{-14}) = 13.3$

Acid deposition

OXIDES OF SULFUR SO_x

Sulfur dioxide occurs naturally from volcanoes. It is produced industrially from the combustion of sulfur-containing fossil fuels and the smelting of sulfide ores.

$$S(s) + O_2(g) \rightarrow SO_2(g)$$

In the presence of sunlight sulfur dioxide is oxidized to sulfur trioxide.

$$SO_2(g) + \tfrac{1}{2}O_2(g) \rightarrow SO_3(g)$$

The oxides can react with water in the air to form sulfurous acid and sulfuric acid:

$$SO_2(g) + H_2O(l) \rightarrow H_2SO_3(aq)$$

and

$$SO_3(g) + H_2O(l) \rightarrow H_2SO_4(aq)$$

OXIDES OF NITROGEN NO_x

Nitrogen oxides occur naturally from electrical storms and bacterial action. Nitrogen monoxide is produced in the internal combustion engine and in jet engines.

$$N_2(g) + O_2(g) \rightarrow 2NO(g)$$

Oxidation to nitrogen dioxide occurs in the air.

$$2NO(g) + O_2(g) \rightarrow 2NO_2(g)$$

The nitrogen dioxide then reacts with water to form nitric acid and nitrous acid:

$$2NO_2(g) + H_2O(l) \rightarrow HNO_3(aq) + HNO_2(aq)$$

or is oxidized directly to nitric acid by oxygen in the presence of water:

$$4NO_2(g) + O_2(g) + 2H_2O(l) \rightarrow 4HNO_3(aq)$$

ACID DEPOSITION

Pure rainwater is naturally acidic with a pH of 5.65 due to the presence of dissolved carbon dioxide. Carbon dioxide itself is *not* responsible for acid rain since acid rain is defined as rain with a pH less than 5.6. It is the oxides of sulfur and nitrogen present in the atmosphere which are responsible for **acid deposition** – the process by which acidic particles, gases and precipitation leave the atmosphere. Wet deposition, due to the acidic oxides dissolving and reacting with water in the air, is known as 'acid rain' and includes fog, snow and dew as well as rain. Dry deposition includes acidic gases and particles.

VEGETATION

Increased acidity in the soil leaches important nutrients, such as Ca^{2+}, Mg^{2+} and K^+. Reduction in Mg^{2+} can cause reduction in chlorophyll and consequently lowers the ability of plants to photosynthesize. Many trees have been seriously affected by acid rain. Symptoms include stunted growth, thinning of tree tops, and yellowing and loss of leaves. The main cause is the aluminium leached from rocks into the soil water. The Al^{3+} ion damages the roots and prevents the tree from taking up enough water and nutrients to survive.

LAKES AND RIVERS

Increased levels of aluminium ions in water can kill fish. Aquatic life is also highly sensitive to pH. Below pH 6 the number of sensitive fish, such as salmon and minnow, decline as do insect larvae and algae. Snails cannot survive a pH less than 5.2 and below pH 5.0 many microscopic animal species disappear. Below pH 4.0 lakes are effectively dead. The nitrates present in acid rain can also lead to eutrophication.

BUILDINGS

Stone, such as marble, that contains calcium carbonate is eroded by acid rain. With sulfuric acid the calcium carbonate reacts to form calcium sulfate, which can be washed away by rainwater thus exposing more stone to corrosion. Salts can also form within the stone that can cause the stone to crack and disintegrate.

$$CaCO_3(s) + H_2SO_4(aq) \rightarrow CaSO_4(aq) + CO_2(g) + H_2O(l)$$

HUMAN HEALTH

The acids formed when NO_x and SO_x dissolve in water irritate the mucous membranes and increase the risk of respiratory illnesses, such as asthma, bronchitis and emphysema. In acidic water there is more probability of poisonous ions, such as Cu^{2+} and Pb^{2+}, leaching from pipes and high levels of aluminium in water may be linked to Alzheimer's disease.

METHODS TO LOWER OR COUNTERACT THE EFFECTS OF ACID RAIN

1. Lower the amounts of NO_x and SO_x formed, e.g. by improved engine design, the use of catalytic converters, and removing sulfur before, during, and after combustion of sulfur-containing fuels.
2. Switch to alternative methods of energy (e.g. wind and solar power) and reduce the amount of fuel burned, e.g. by reducing private transport and increasing public transport and designing more efficient power stations.
3. Liming of lakes – adding calcium oxide or calcium hydroxide (lime) neutralizes the acidity, increases the amount of calcium ions and precipitates aluminium from solution. This has been shown to be effective in many, but not all, lakes where it has been tried.

Lewis acids and bases

LEWIS ACIDS AND BASES

Brønsted–Lowry bases must contain a non-bonding pair of electrons to accept the proton. The Lewis definition takes this further and describes bases as substances which can *donate* a pair of electrons, and acids as substances which can *accept* a pair of electrons. In the process a coordinate (both electrons provided by one species) covalent bond is formed between the base and the acid.

The Lewis theory is all-embracing, so the term Lewis acid is usually reserved for substances which are not also Brønsted–Lowry acids. Many Lewis acids do not even contain hydrogen.

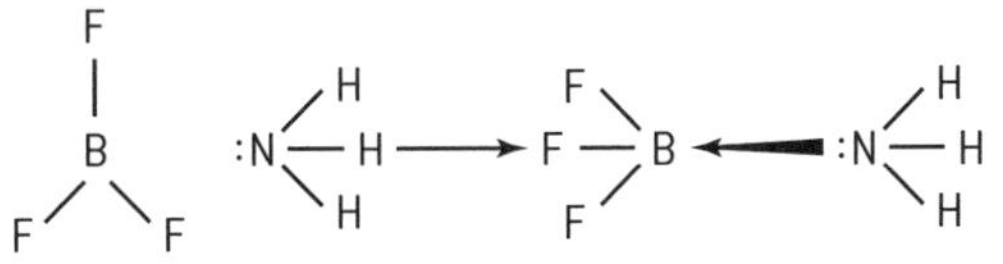

acid base

BF_3 is a good Lewis acid as there are only six electrons around the central boron atom which leaves room for two more. Other common Lewis acids are aluminium chloride, $AlCl_3$, and also transition metal ions in aqueous solution which can accept a pair of electrons from each of six surrounding water molecules, e.g. $[Fe(H_2O)_6]^{3+}$.

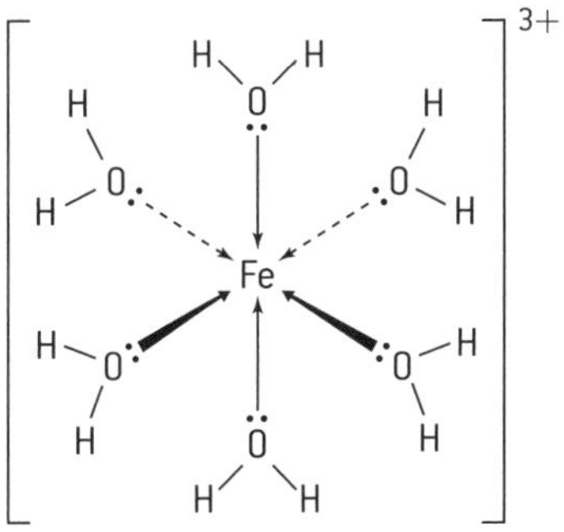

Note that hydrated transition metal ions, such as $[Fe(H_2O)_6]^{3+}$ are acidic in solution as the +3 charge is spread over a very small ion which gives the ion a high charge density. The non-bonded pair of electrons on one of the water molecules surrounding the ion will be strongly attracted to the ion and the water molecule will lose a hydrogen ion in the process. This process can continue until iron(III) hydroxide is formed. The equilibrium can be further moved to the right by adding hydroxide ions, $OH^-(aq)$ or back to the left by adding hydrogen ions, $H^+(aq)$ which exemplifies the ion's amphoteric nature.

$$[Fe(H_2O)_6]^{3+} \underset{H^+}{\overset{-H^+}{\rightleftharpoons}} [Fe(H_2O)_5OH]^{2+} \underset{H^+}{\overset{-H^+}{\rightleftharpoons}} [Fe(H_2O)_4(OH)_2]^{+} \underset{H^+}{\overset{-H^+}{\rightleftharpoons}} Fe(H_2O)_3(OH)_3$$

$$Fe(H_2O)_3(OH)_3 \underset{H^+}{\overset{OH^-}{\rightleftharpoons}} [Fe(H_2O)_2(OH)_4]^-$$

The Lewis acid and base concept is also used in organic chemistry particularly to identify reacting species and in the use of 'curly arrows' to explain the movement of pairs of electrons in organic reaction mechanisms.

For example the addition of hydrogen bromide to an alkene proceeds by an electrophilic addition mechanism (see page 90). The δ^+ hydrogen atom of the hydrogen bromide molecule acts as the electrophile and accepts a pair of electrons from the double bond of the alkene. Hence, the electrophile is acting as a Lewis acid and the alkene is acting as a Lewis base. Since a 'curly arrow' shows the movement of a pair of electrons the arrow always originates from the Lewis base and the head of the 'curly arrow' always points towards the Lewis acid. A second Lewis acid–base reaction occurs when the bromide ion (acting as the Lewis base) donates a pair of electrons to the positive carbon atom in the carbocation intermediate to form the brominated addition product.

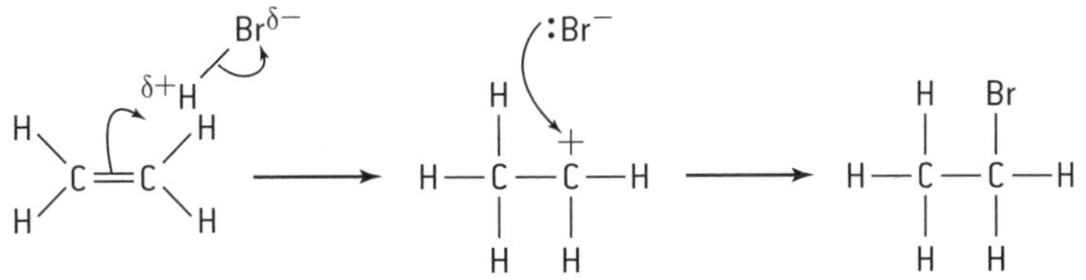

Another good example of Lewis acids in organic chemistry is the function of halogen carriers as catalysts in the electrophilic reactions of benzene (see page 88). Benzene is electron-rich due to its delocalized π bond and can react with chlorine to form chlorobenzene. A halogen carrier such as aluminium chloride is added to provide a positive chloride ion which acts as the electrophile. So in this reaction aluminium chloride (Lewis acid) and chlorine (Lewis base) undergo a Lewis acid–base reaction and then a second Lewis acid–base reaction occurs between the positive chloride ion (Lewis acid) and the benzene molecule (Lewis base). A third Lewis acid–base reaction occurs when the hydrogen atom (Lewis acid) is removed from the intermediate to form hydrogen chloride and regenerate the aluminium chloride catalyst.

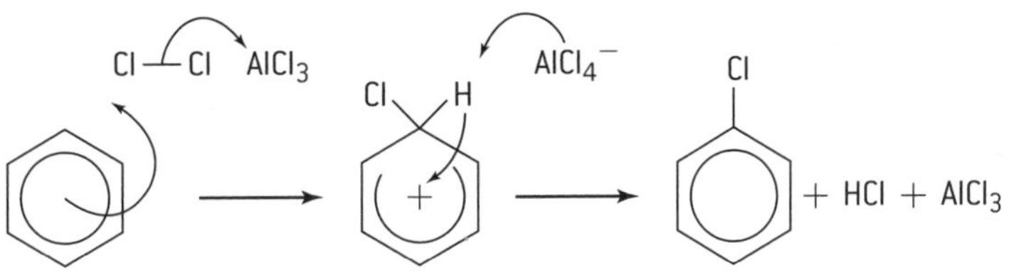

Calculations involving pH, pOH and pK_w

THE IONIC PRODUCT OF WATER

Pure water is very slightly ionized:

$H_2O(l) \rightleftharpoons H^+(aq) + OH^-(aq) \qquad \Delta H^\ominus = +57.3 \text{ kJ mol}^{-1}$

$$K_c = \frac{[H^+(aq)] \times [OH^-(aq)]}{[H_2O(l)]}$$

Since the equilibrium lies far to the left the concentration of water can be regarded as constant so

$K_w = [H^+(aq)] \times [OH^-(aq)] = 1.00 \times 10^{-14}$ at 298 K, where K_w is known as the ionic product of water.

The dissociation of water into its ions is an endothermic process, so the value of K_w will increase as the temperature is increased.

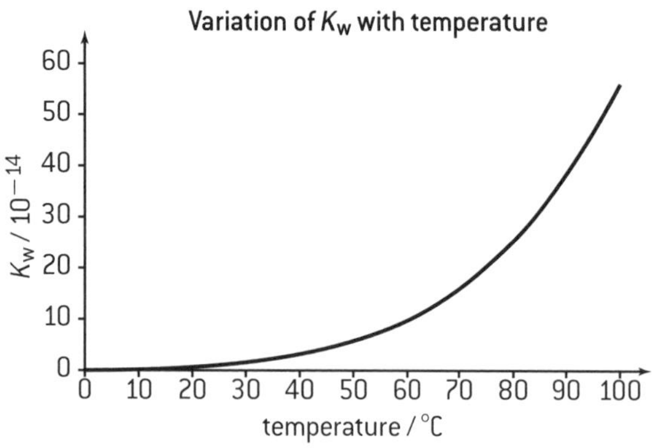

For pure water $[H^+(aq)] = [OH^-(aq)]$
$= 1.00 \times 10^{-7}$ mol dm^{-3} at 298 K

From the graph the value for $K_w = 1.00 \times 10^{-13}$ at 334 K (61°C)

At this temperature $[H^+(aq)] = \sqrt{1.00 \times 10^{-13}}$
$= 3.16 \times 10^{-7}$ mol dm^{-3}

pH, pOH AND pK_w FOR STRONG ACIDS AND BASES

As stated earlier in this chapter the pH of a solution depends only upon the hydrogen ion concentration and is independent of the volume of the solution.

$pH = -\log_{10} [H^+(aq)]$

For strong monoprotic acids the hydrogen ion concentration will be equal to the concentration of the acid and will be twice the value of the acid concentration for strong diprotic acids.

The use of the logarithmic scale can be extended to other values, e.g. pOH and pK_w.

$pOH = -\log_{10} [OH^-(aq)]$ and $pK_w = -\log_{10} K_w$

$K_w = [H^+(aq)] \times [OH^-(aq)]$

If logarithms to the base ten are taken then

$\log_{10}K_w = \log_{10} [H^+(aq)] + \log_{10} [OH^-(aq)]$ which can also be written as

$-\log_{10}K_w = -\log_{10} [H^+(aq)] - \log_{10} [OH^-(aq)]$

This leads to the useful expression

$pK_w = pH + pOH$

At 25 °C $K_w = 10^{-14}$ and pH + pOH = 14

This expression gives another way of calculating the pH of a strong base since the pOH can be determined directly from the hydroxide ion concentration then subtracted from 14. For example, to determine the pH of 4.00×10^{-3} mol dm^{-3} $Ba(OH)_2$

$[OH^-(aq)] = 2 \times 4.00 \times 10^{-3} = 8.00 \times 10^{-3}$ mol dm^{-3}

$pOH = -\log_{10} 8.00 \times 10^{-3} = 2.10$

$pH = 14 - 2.10 = 11.9$

HL Calculations with weak acids and bases

WEAK ACIDS

The dissociation of a weak acid HA in water can be written:

$$HA(aq) \rightleftharpoons H^+(aq) + A^-(aq)$$

The equilibrium expression for this reaction is:

$$K_a = \frac{[H^+] \times [A^-]}{[HA]}$$ where K_a is known as the acid dissociation constant

For example, to calculate the pH of 0.10 mol dm^{-3} CH_3COOH given that $K_a = 1.8 \times 10^{-5}$ mol dm^{-3} at 298 K:

$$CH_3COOH(aq) \rightleftharpoons CH_3COO^-(aq) + H^+(aq)$$

Initial concentration / mol dm^{-3}		
0.10	–	–
Equilibrium concentration / mol dm^{-3}		
$(0.10 - x)$	x	x

$$K_a = \frac{[CH_3COO^-] \times [H^+]}{[CH_3COOH]} = \frac{x^2}{(0.10 - x)}$$

$= 1.8 \times 10^{-5}$ mol dm^{-3}

$\Rightarrow x^2 + (1.8 \times 10^{-5}x) - 1.8 \times 10^{-6} = 0$

by solving the quadratic equation
$x = 1.33 \times 10^{-3}$ mol dm^{-3}

$pH = -\log_{10} 1.33 \times 10^{-3} = 2.88$

If the acids are quite weak the equilibrium concentration of the acid can be assumed to be the same as its initial concentration. Provided the assumption is stated it is usual to simplify the expression in calculations to avoid a quadratic equation. In the above example:

$$K_a = \frac{[CH_3COO^-] \times [H^+]}{[CH_3COOH]} \approx \frac{[H^+]^2}{0.10}$$

$= 1.8 \times 10^{-5}$ mol dm^{-3}

$\Rightarrow [H^+] = \sqrt{1.8 \times 10^{-6}} = 1.34 \times 10^{-3}$ mol dm^{-3}

pH = 2.87

Examples of other weak acid calculations

1. The pH of a 0.020 mol dm^{-3} solution of a weak acid is 3.9. Find the K_a of the acid.

   $$K_a = \frac{[H^+]^2}{(0.020 - [H^+])} \approx \frac{10^{-3.9} \times 10^{-3.9}}{0.020}$$

   $= 7.92 \times 10^{-7}$ mol dm^{-3}

2. An acid whose K_a is 4.1×10^{-6} mol dm^{-3} has a pH of 4.5. Find the concentration of the acid.

   $$[HA] = \frac{[H^+]^2}{K_a} = \frac{10^{-4.5} \times 10^{-4.5}}{4.1 \times 10^{-6}}$$

   $= 2.44 \times 10^{-4}$ mol dm^{-3}

Note that the weaker the weak acid the **smaller** the value of $\mathbf{K_a}$ and the **larger** the value of $\mathbf{pK_a}$. Thus ethanoic acid ($pK_a = 4.76$) is a weaker acid than methanoic acid ($pK_a = 3.75$).

WEAK BASES

The reaction of a weak base can be written:

$$B(aq) + H_2O(l) \rightleftharpoons BH^+(aq) + OH^-(aq)$$

Since the concentration of water is constant:

$$K_b = \frac{[BH^+] \times [OH^-]}{[B]}$$

where K_b is the base dissociation constant

If one considers the reverse reaction of BH^+ acting as an acid to give B and H^+ then:

$$K_a = \frac{[B] \times [H^+]}{[BH^+]}$$

then

$$K_a \times K_b = \frac{[B] \times [H^+]}{[BH^+]} \times \frac{[BH^+] \times [OH^-]}{[B]}$$

$$= [H^+] \times [OH^-] = K_w$$

since $pK_a = -\log_{10} K_a$; $pK_b = -\log_{10} K_b$ and $pK_w = -\log_{10} K_w = 14$ this can also be expressed as:

$$pK_a + pK_b = 14$$

Examples of calculations

1. The K_b value for ammonia is 1.8×10^{-5} mol dm^{-3}. Find the pH of a 1.00×10^{-2} mol dm^{-3} solution.

   Since $[NH_4^+] = [OH^-]$ then

   $$K_b = \frac{[OH^-]^2}{[NH_3]} \approx \frac{[OH^-]^2}{1.00 \times 10^{-2}} = 1.8 \times 10^{-5}$$

   $[OH^-] = \sqrt{1.8 \times 10^{-7}} = 4.24 \times 10^{-4}$ mol dm^{-3}

   $\Rightarrow pOH = -\log_{10} 4.24 \times 10^{-4} = 3.37$

   $\Rightarrow pH = 14 - 3.37 = 10.6$

2. The pH of a 3.00×10^{-2} mol dm^{-3} solution of weak base is 10.0. Calculate the pK_b value of the base.

   pH = 10.0 so pOH = 4.0

   $$K_b = \frac{10^{-4} \times 10^{-4}}{3.00 \times 10^{-2}} = 3.33 \times 10^{-7} \text{ mol dm}^{-3}$$

   $\Rightarrow pK_b = 6.48$

3. The value for the pK_a of methylamine (aminomethane) is 10.66. Calculate the concentration of an aqueous solution of methylamine with a pH of 10.8.

   $pK_b = 14 - 10.66 = 3.34$; $pOH = 14 - 10.8 = 3.2$

   $$[CH_3NH_2] = \frac{[OH^-]^2}{K_b} = \frac{10^{-3.2} \times 10^{-3.2}}{10^{-3.34}}$$

   $= 8.71 \times 10^{-4}$ mol dm^{-3}

Note that the weaker the weak base the **smaller** the value of $\mathbf{K_b}$ and the **larger** the value of $\mathbf{pK_b}$. Thus ammonia ($pK_b = 4.75$) is a weaker base than methylamine ($pK_b = 3.34$).

Salt hydrolysis and buffer solutions

HL

SALT HYDROLYSIS

Sodium chloride is neutral in aqueous solution. It is the salt of a strong acid and a strong base so its ions remain completely dissociated in solution. Salts made from a weak acid and a strong base, such as sodium ethanoate, are alkaline in solution. This is because the ethanoate ions will combine with hydrogen ions from water to form mainly undissociated ethanoic acid, leaving excess hydroxide ions in solution.

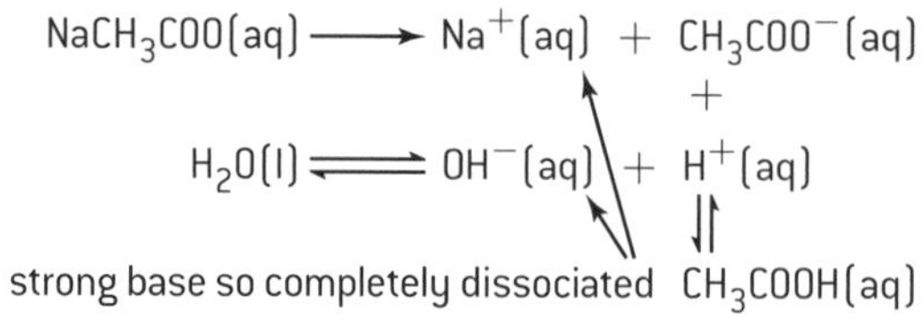

$NaCH_3COO(aq) \longrightarrow Na^+(aq) + CH_3COO^-(aq)$
+
$H_2O(l) \rightleftharpoons OH^-(aq) + H^+(aq)$
$CH_3COOH(aq)$
strong base so completely dissociated

Similarly salts derived from a strong acid and a weak base will be acidic in solution.

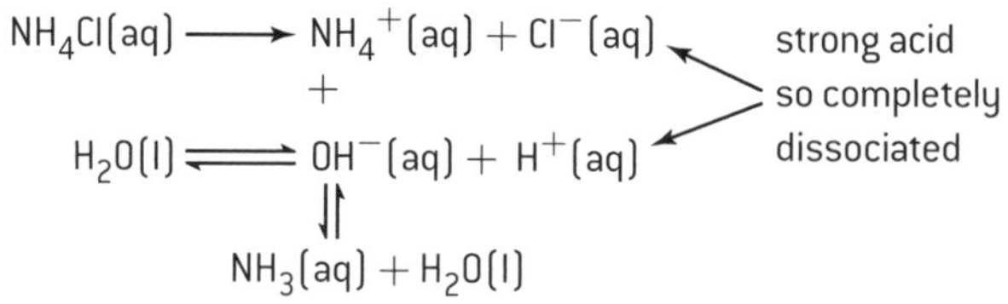

$NH_4Cl(aq) \longrightarrow NH_4^+(aq) + Cl^-(aq)$
+
$H_2O(l) \rightleftharpoons OH^-(aq) + H^+(aq)$
$NH_3(aq) + H_2O(l)$
strong acid so completely dissociated

BUFFER SOLUTIONS

A buffer solution resists changes in pH when small amounts of acid or alkali are added to it.

An acidic buffer solution can be made by mixing a weak acid together with the salt of that acid and a strong base. An example is a solution of ethanoic acid and sodium ethanoate. The weak acid is only slightly dissociated in solution, but the salt is fully dissociated into its ions, so the concentration of ethanoate ions is high.

$NaCH_3COO(aq) \rightarrow Na^+(aq) + CH_3COO^-(aq)$

$CH_3COOH(aq) \rightleftharpoons CH_3COO^-(aq) + H^+(aq)$

If an acid is added the extra H^+ ions coming from the acid are removed as they combine with ethanoate ions to form undissociated ethanoic acid, so the concentration of H^+ ions remains unaltered.

$CH_3COO^-(aq) + H^+(aq) \rightleftharpoons CH_3COOH(aq)$

If an alkali is added the hydroxide ions from the alkali are removed by their reaction with the undissociated acid to form water, so again the H^+ ion concentration stays constant.

$CH_3COOH(aq) + OH^-(aq) \rightarrow CH_3COO^-(aq) + H_2O(l)$

In practice acidic buffers are often made by taking a solution of a strong base and adding excess weak acid to it, so that the solution contains the salt and the unreacted weak acid.

$NaOH(aq) + CH_3COOH(aq) \rightarrow NaCH_3COO(aq) + H_2O(l) + CH_3COOH(aq)$

limiting reagent — salt — excess weak acid

(salt + excess weak acid) = buffer solution

An alkali buffer with a fixed pH greater than 7 can be made from a weak base together with the salt of that base with a strong acid. An example is ammonia with ammonium chloride.

$NH_4Cl(aq) \rightarrow NH_4^+(aq) + Cl^-(aq)$

$NH_3(aq) + H_2O(l) \rightleftharpoons NH_4^+(aq) + OH^-(aq)$

If H^+ ions are added they will combine with OH^- ions to form water and more of the ammonia will dissociate to replace them. If more OH^- ions are added they will combine with ammonium ions to form undissociated ammonia. In both cases the hydroxide ion concentration and the hydrogen ion concentration remain constant.

HL Titration curves and indicators

STRONG ACID – STRONG BASE TITRATION

The change in pH during an acid–base titration can be followed using a pH meter. Consider starting with 50 cm^3 of 1.0 mol dm^{-3} hydrochloric acid. Since $[H^+(aq)] = 1.0$ mol dm^{-3} the initial pH will be 0. After 49 cm^3 of 1.0 mol dm^{-3} NaOH have been added there will be 1.0 cm^3 of the original 1.0 mol dm^{-3} hydrochloric acid left in 99 cm^3 of solution. At this point $[H^+(aq)] \approx 1.0 \times 10^{-2}$ mol dm^{-3} so the pH = 2.

When 50 cm^3 of the NaOH solution has been added the solution will be neutral and the pH will be 7. This is indicated by the point of inflexion, which is known as the equivalence point. It can be seen that there is a very large change in pH around the equivalence point. Almost all of the common acid–base indicators change colour (reach their end point) within this pH region. This means that it does not matter which indicator is used.

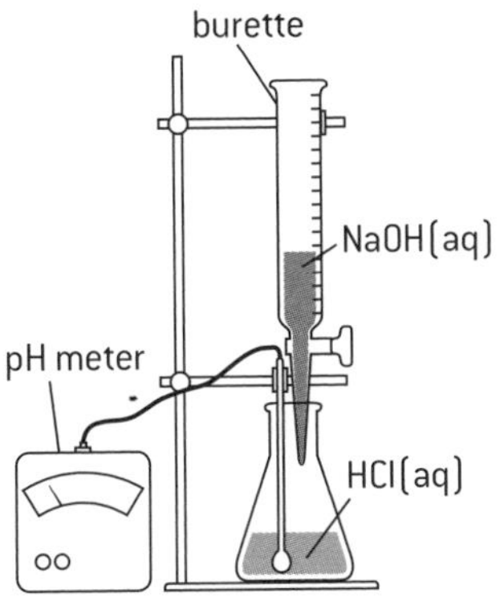

This curve shows what happens when 1.0 mol dm^{-3} sodium hydroxide is added to 50cm^3 of 1.0 mol dm^{-3} hydrochloric acid

$$NaOH(aq) + HCl(aq) \rightarrow NaCl(aq) + H_2O(l)$$

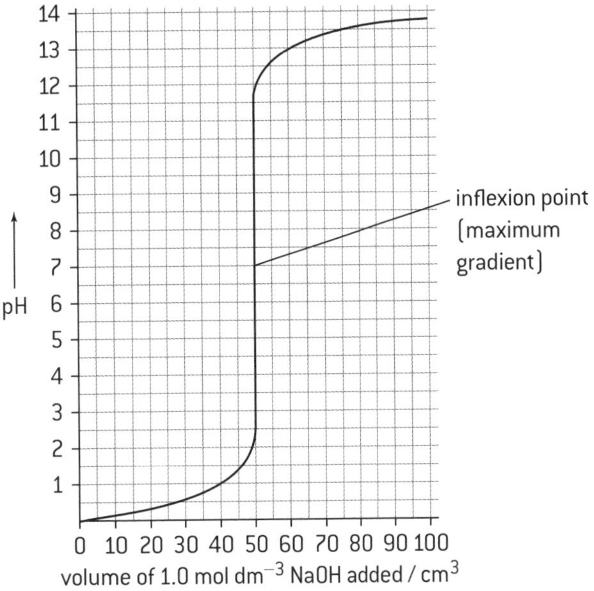

WEAK ACID – STRONG BASE TITRATION

Consider titrating 50.0 cm^3 of 1.0 mol dm^{-3} CH_3COOH with 1.0 mol dm^{-3} NaOH.

$K_a = 1.8 \times 10^{-5}$. Making the usual assumptions the initial $[H^+] = \sqrt{K_a \times [CH_3COOH]}$ and pH = 2.37.

When 49.0 cm^3 of the 1.0 mol dm^{-3} NaOH has been added $[CH_3COO^-] \approx 0.05$ mol dm^{-3} and $[CH_3COOH] \approx 1.0 \times 10^{-2}$ mol dm^{-3}.

$$[H^+] = \frac{K_a \times [CH_3COOH]}{[CH_3COO^-]} \approx \frac{1.8 \times 10^{-5} \times 1 \times 10^{-3}}{0.05}$$

$$= 3.6 \times 10^{-7} \text{ mol dm}^{-3} \text{ and pH} = 6.44$$

After the equivalence point the graph will follow the same pattern as the strong acid–strong base curve as more sodium hydroxide is simply being added to the solution.

$$CH_3COOH(aq) + NaOH(aq) \rightarrow NaCH_3COO(aq) + H_2O(l)$$

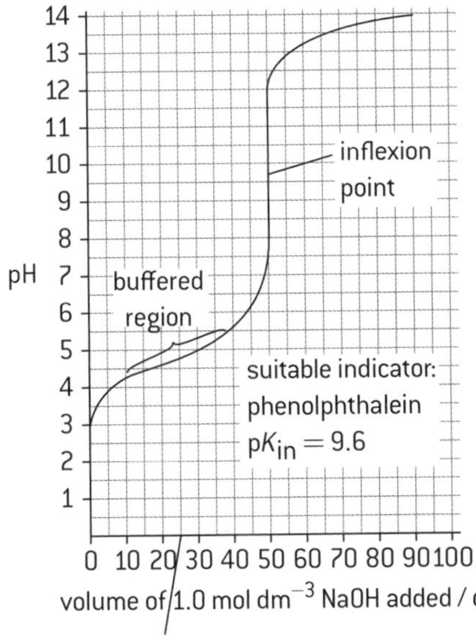

When 25cm^3 of alkali have been added, half the acid has been turned into its salt, so pK_a = pH.

INDICATORS

An indicator is a weak acid (or base) in which the dissociated form is a different colour to the undissociated form.

$$HIn(aq) \rightleftharpoons H^+(aq) + In^-(aq)$$

colour A (colour in acid solution) — colour B (colour in alkali solution)

$K_{in} = [H^+] \times \frac{[In^-]}{[HIn]}$ Assuming the colour changes when $[In^-] \approx [HIn]$ then the end point of the indicator will be when $[H^+] \approx K_{in}$, i.e. when pH $\approx pK_{in}$. Different indicators have different K_{in} values and so change colour within different pH ranges.

Indicator	pK_{in}	pH range	Use
methyl orange	3.7	3.1–4.4	titrations with strong acids
phenolphthalein	9.6	8.3–10.0	titrations with strong bases

Similar arguments can be used to explain the shapes of pH curves for strong acid–weak base, and weak acid–weak base titrations. Since there is no sharp inflexion point titrations involving weak acids with weak bases should not be used in analytical chemistry.

Strong acid – weak base e.g. $HCl(aq)$ and $NH_3(aq)$

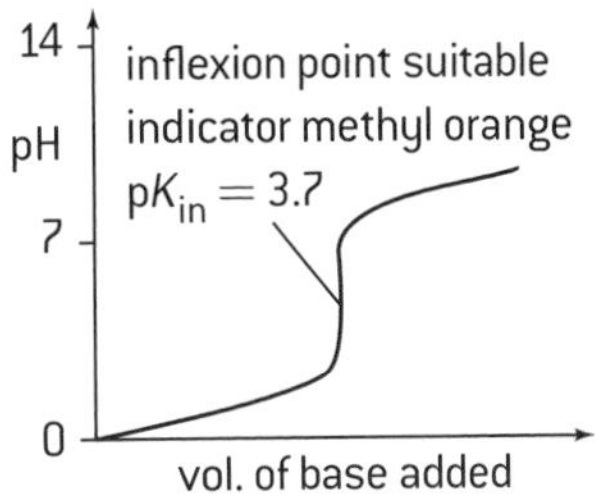

Weak acid – weak base e.g. $CH_3COOH(aq)$ and $NH_3(aq)$

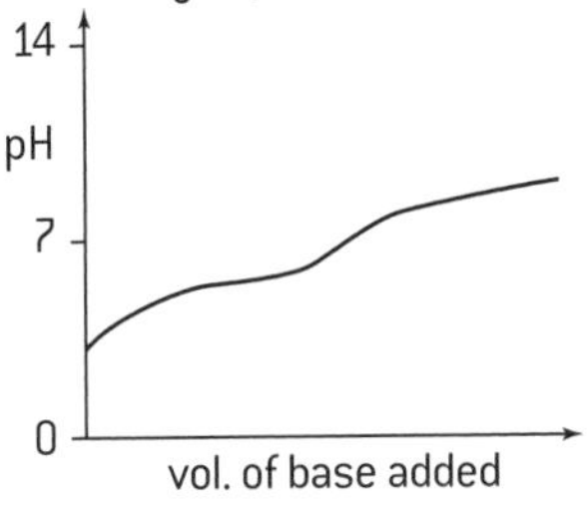

MULTIPLE CHOICE QUESTIONS – ACIDS AND BASES

1. Which statement about hydrochloric acid is false?
   A. It can react with copper to give hydrogen
   B. It can react with sodium carbonate to give carbon dioxide
   C. It can react with ammonia to give ammonium chloride
   D. It can react with copper oxide to give water

2. 1.00 cm^3 of a solution has a pH of 3. 100 cm^3 of the same solution will have pH of:
   A. 1
   B. 3
   C. 5
   D. Impossible to calculate from the data given.

3. Which statement(s) is/are true about separate solutions of a strong acid and a weak acid both with the same concentration?
   I. They both have the same pH.
   II. They both have the same electrical conductivity.
   A. I and II
   B. I only
   C. II only
   D. Neither I nor II

4. Identify the correct statement about 25 cm^3 of a solution of 0.1 mol dm^{-3} ethanoic acid CH_3COOH.
   A. It will contain more hydrogen ions than 25 cm^3 of 0.1 mol dm^{-3} hydrochloric acid.
   B. It will have a pH greater than 7.
   C. It will react exactly with 25 cm^3 of 0.1 mol dm^{-3} sodium hydroxide.
   D. It is completely dissociated into ethanoate and hydrogen ions in solution.

5. What is the pH of 1.0×10^{-4} mol dm^{-3} sulfuric acid, $H_2SO_4(aq)$?
   A. −4
   B. between 3 and 4
   C. 4
   D. between 4 and 5

6. $NH_3(aq)$, $HCl(aq)$, $NaOH(aq)$, $CH_3COOH(aq)$
   When 1.0 mol dm^{-3} solutions of the substances above are arranged in order of **decreasing** pH the order is:
   A. $NaOH(aq)$, $NH_3(aq)$, $CH_3COOH(aq)$, $HCl(aq)$
   B. $NH_3(aq)$, $NaOH(aq)$, $HCl(aq)$, $CH_3COOH(aq)$
   C. $CH_3COOH(aq)$, $HCl(aq)$, $NaOH(aq)$, $NH_3(aq)$
   D. $HCl(aq)$, $CH_3COOH(aq)$, $NH_3(aq)$, $NaOH(aq)$

7. A solution with a pH of 8.5 would be described as:
   A. very basic
   B. slightly basic
   C. slightly acidic
   D. very acidic

8. Which statement is true about two solutions, one with a pH of 3 and the other with a pH of 6?
   A. The solution with a pH of 3 is twice as acidic as the solution with a pH of 6
   B. The solution with a pH of 6 is twice as acidic as the solution with a pH of 3
   C. The hydrogen ion concentration in the solution with a pH of 6 is one thousand times greater than that in the solution with a pH of 3
   D. The hydrogen ion concentration in the solution with a pH of 3 is one thousand times greater than that in the solution with a pH of 6

9. Which of the following is not a conjugate acid–base pair?
   A. HNO_3/NO_3^-
   B. H_2SO_4/HSO_4^-
   C. NH_3/NH_2^-
   D. H_3O^+/OH^-

10. Which gas cannot lead to acid deposition?
    A. CO_2
    B. SO_2
    C. NO
    D. NO_2

HL

11. During the titration of a known volume of a strong acid with a strong base:
    A. there is a steady increase in pH
    B. there is a sharp increase in pH around the end point
    C. there is a steady decrease in pH
    D. there is a sharp decrease in pH around the end point.

12. Three acids, HA, HB, and HC have the following K_a values
    $K_a(HA) = 1 \times 10^{-5}$ $K_a(HB) = 2 \times 10^{-5}$
    $K_a(HC) = 1 \times 10^{-6}$
    What is the correct order of increasing acid strength (weakest first)?
    A. HA, HB, HC
    B. HC, HB, HA
    C. HC, HA, HB
    D. HB, HA, HC

13. Which of the following reagents could not be added together to make a buffer solution?
    A. $NaOH(aq)$ and $CH_3COOH(aq)$
    B. $NaCH_3COO(aq)$ and $CH_3COOH(aq)$
    C. $NaOH(aq)$ and $NaCH_3COO(aq)$
    D. $NH_4Cl(aq)$ and $NH_3(aq)$

14. When 1.0 cm^3 of a weak acid solution is added to 100 cm^3 of a buffer solution:
    A. the volume of the resulting mixture will be 100 cm^3
    B. there will be almost no change in the pH of the solution
    C. the pH of the solution will increase noticeably
    D. the pH of the solution will decrease noticeably.

15. Which species cannot act as a Lewis acid?
    A. NH_3
    B. BF_3
    C. Fe^{2+}
    D. $AlCl_3$

16. Which salt does not form an acidic solution in water?
    A. $MgCl_2$
    B. Na_2CO_3
    C. $FeCl_3$
    D. NH_4NO_3

17. An indicator changes colour in the pH range 8.3–10.0. This indicator should be used when titrating a known volume of:
    A. a strong acid with a weak base
    B. a weak acid with a weak base
    C. a weak base with a strong acid
    D. a weak acid with a strong base.

SHORT ANSWER QUESTIONS – ACIDS AND BASES

1. a) (i) Define a Brønsted–Lowry base. [1]

(ii) Deduce the two acids and their conjugate bases in the following reaction:

$$H_2O(l) + HCl(aq) \rightleftharpoons H_3O^+(aq) + Cl^-(aq)$$ [2]

b) Ethanoic acid, CH_3COOH, is a weak acid.

(i) Explain the difference between a strong acid and a weak acid. [2]

(ii) State the equation for the reaction of ethanoic acid with aqueous ammonia. [1]

(iii) Compare and contrast the reactions of 1.00 mol dm^{-3} hydrochloric acid and 1.00 mol dm^{-3} ethanoic acid with excess magnesium metal. [4]

2. a) 10 cm^3 of 5.00×10^{-3} mol dm^{-3} sulfuric acid, $H_2SO_4(aq)$, is added to an empty volumetric flask.

(i) Calculate the pH of the sulfuric acid solution (assume it is a completely strong diprotic acid). [2]

(ii) Determine the pH of the diluted solution if the total volume is made up to 100 cm^3 with distilled water. [2]

b) State the equation for the reaction of sulfuric acid with sodium hydroxide solution. [1]

c) The diluted solution in (a)(ii) is used to titrate 25.0 cm^3 of 1.00×10^{-4} mol dm^{-3} sodium hydroxide solution.

(i) Describe what will be observed when the end point is reached if phenolphthalein is used as the indicator for this titration. [2]

(ii) Determine the volume of acid needed to reach the equivalence point of this titration. [2]

3. When hydrochloric acid is added to a solution of sodium hydrogencarbonate, $NaHCO_3(aq)$ carbon dioxide is evolved.

a) State the equation for this reaction. [2]

b) The hydrogencarbonate ion can act either as an acid or a base according to Brønsted–Lowry theory.

(i) Deduce the formula of the conjugate base if it is behaving as an acid. [1]

(ii) Deduce the formula of the acid if it is behaving as a conjugate base. [1]

c) State the equations for the reaction of hydrochloric acid with (i) copper(II) oxide CuO and (ii) sodium carbonate, Na_2CO_3. [2]

4. a) Explain why rain water with a pH of 6 is not classified as 'acid rain' even though its pH is less than 7? [2]

b) State the equation for the formation of nitrogen(II) oxide, NO(g), in an internal combustion engine and describe with equations how it is converted into nitric acid in the atmosphere. [4]

c) Explain why marble statues become corroded by acid rain. [2]

d) Outline why adding calcium hydroxide (lime) to lakes can reduce the effects of acid deposition. [2]

HL

5. The graph below shows how the ionic product of water, K_w varies with temperature.

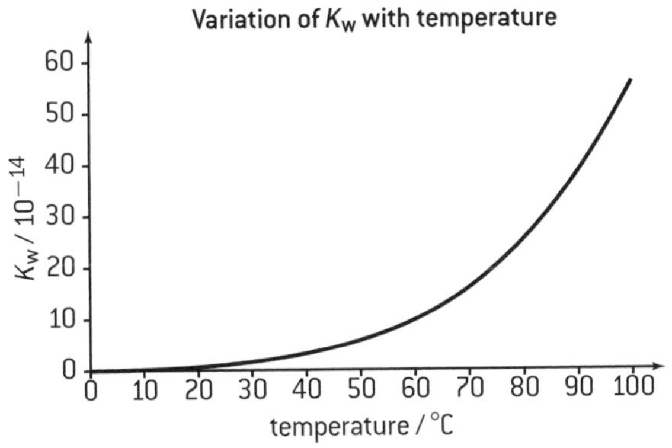

a) State the equation for the dissociation of water and deduce from the graph whether the reaction is exothermic or endothermic. [2]

b) Determine the hydrogen ion concentration and the hydroxide ion concentration in pure water at 90 °C and hence deduce the pH of pure water at this temperature. [3]

6. a) Explain why a nucleophile can also be described as a Lewis base. [1]

b) The diagram below shows the mechanism for the nucleophilic addition reaction between cyanide ions and ethanal to form 2-hydroxypropanenitrile.

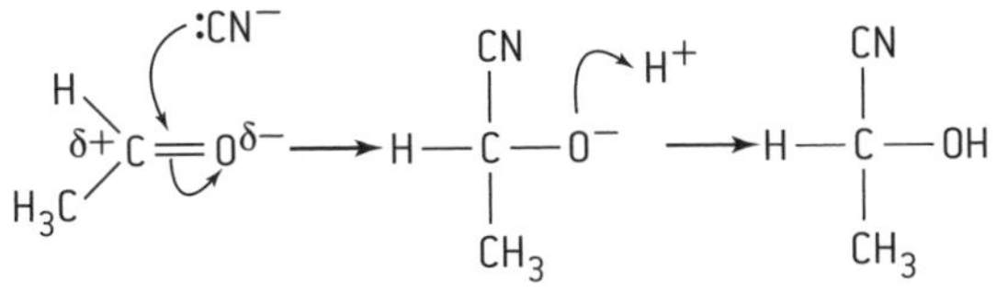

(i) Identify the nucleophile in this reaction. [1]

(ii) Explain the mechanism in terms of Lewis acid–base theory. [3]

c) Explain why the reaction between a transition metal ion and six monodentate ligands to form an octahedral complex ion is an example of a Lewis acid–base reaction. [3]

7. a) (i) State the equation for the reaction between propanoic acid, $C_2H_5COOH(aq)$ and water and deduce the equilibrium expression. [2]

(ii) Calculate the pH of a 2.00×10^{-3} mol dm^{-3} solution of propanoic acid ($pK_a = 4.87$). [3]

(iii) State any assumptions you have made in arriving at your answer to (a)(ii). [2]

b) 25.0 cm^3 of 2.00×10^{-3} mol dm^{-3} sodium hydroxide solution, NaOH(aq) was added to 50 cm^3 of 2.00×10^{-3} mol dm^{-3} propanoic acid.

(i) Identify all the chemical species present in the resulting solution. [2]

(ii) Explain how the resulting solution can function as a buffer solution is a small amount of alkali is added. [2]

8. A particular indicator is a weak acid and can be represented as HIn. The K_a for HIn is 3×10^{-10}. HIn(aq) is colourless and $In^-(aq)$ is pink in aqueous solution.

a) Identify the colour this indicator will show in strongly alkaline solution. [1]

b) Explain whether or not this indicator would be suitable to use when titrating hydrochloric acid with ammonia solution. [2]

c) Explain why no acid–base indicator is suitable to use when titrating ethanoic acid with ammonia solution. [1]

Redox reactions (1)

DEFINITIONS OF OXIDATION AND REDUCTION

Oxidation used to be narrowly defined as the addition of oxygen to a substance. For example, when magnesium is burned in air the magnesium is oxidized to magnesium oxide.

$2Mg(s) + O_2(g) \rightarrow 2MgO(s)$

The electronic configuration of magnesium is $[Ne]3s^2$. During the oxidation process it loses two electrons to form the Mg^{2+} ion with the electronic configuration of [Ne]. **Oxidation** is now defined as the *loss of one or more electrons from a substance*. This is a much broader definition, as it does not necessarily involve oxygen. Bromide ions, for example, are oxidized by chlorine to form bromine.

$2Br^-(aq) + Cl_2(aq) \rightarrow Br_2(aq) + 2Cl^-(aq)$

If a substance loses electrons then something else must be gaining electrons. *The gain of one or more electrons* is called **reduction**. In the first example oxygen is reduced as it is gaining two electrons from magnesium to form the oxide ion O^{2-}. Similarly, in the second example chlorine is reduced as each chlorine atom gains one electron from a bromide ion to form a chloride ion.

Since the processes involve the transfer of electrons oxidation and reduction must occur simultaneously. Such reactions are known as **redox reactions**. In order to distinguish between the two processes half-equations are often used:

$2Mg(s) \rightarrow 2Mg^{2+}(s) + 4e^-$	— OXIDATION —	$2Br^-(aq) \rightarrow Br_2(aq) + 2e^-$
$O_2(g) + 4e^- \rightarrow 2O^{2-}(s)$	— REDUCTION —	$Cl_2(aq) + 2e^- \rightarrow 2Cl^-(aq)$
$2Mg(s) + O_2(g) \rightarrow 2MgO(s)$	OVERALL REDOX EQUATION	$2Br^-(aq) + Cl_2(aq) \rightarrow Br_2(aq) + 2Cl^-(aq)$

Understanding that magnesium must lose electrons and oxygen must gain electrons when magnesium oxide MgO is formed from its elements is a good way to remember the definitions of oxidation and reduction. Some students prefer to use the mnemonic OILRIG: **O**xidation **I**s the **L**oss of electrons, **R**eduction **I**s the **G**ain of electrons.

RULES FOR DETERMINING OXIDATION STATES

It is not always easy to see how electrons have been transferred in redox processes. Oxidation states can be a useful tool to identify which species have been oxidized and which reduced. Oxidation states are assigned according to a set of rules:

1. In an ionic compound between two elements the oxidation state of each element is equal to the charge carried by the ion, e.g.

   Na^+Cl^- (Na = +1; Cl = −1) $Ca^{2+}Cl^-_2$ (Ca = +2; Cl = −1)

2. For covalent compounds assume the compound is ionic with the more electronegative element forming the negative ion, e.g.

   CCl_4 (C = +4; Cl = −1) NH_3 (N = −3; H= +1)

3. The algebraic sum of all the oxidation states in a compound = zero, e.g.

   CCl_4 [(+4) + 4 × (−1) = 0];
   H_2SO_4 [2 × (+1) + (+6) + 4 × (−2) = 0]

4. The algebraic sum of all the oxidation states in an ion = the charge on the ion, e.g.

   SO_4^{2-} [(+6) + 4 × (−2) = −2];
   MnO_4^- [(+7) + 4 × (−2) = −1]; NH_4^+ [(−3) + 4 × (+1) = +1]

5. Elements not combined with other elements have an oxidation state of zero, e.g. O_2; P_4; S_8.
6. Oxygen when combined always has an oxidation state of −2 except in peroxides (e.g. H_2O_2) when it is −1.
7. Hydrogen when combined always has an oxidation state of +1 except in certain metal hydrides (e.g. NaH) when it is −1.

Many elements can show different oxidation states in different compounds, e.g. nitrogen in:

NH_3	N_2H_4	N_2	N_2O	NO	NO_2	NO_3^-
(−3)	(−2)	(0)	(+1)	(+2)	(+4)	(+5)

When elements show more than one oxidation state the oxidation number is represented by using Roman numerals when naming the compound,

e.g. $FeCl_2$ iron(II) chloride; $FeCl_3$ iron(III) chloride
$K_2Cr_2O_7$ potassium dichromate(VI); $KMnO_4$ potassium manganate(VII)
Cu_2O copper(I) oxide; CuO copper(II) oxide.

OXIDIZING AND REDUCING AGENTS

A substance that readily oxidizes other substances is known as an **oxidizing agent**. Oxidizing agents are thus substances that readily accept electrons. Usually they contain elements that are in their highest oxidation state,

e.g. O_2, Cl_2, F_2, SO_3 (SO_4^{2-} in solution), MnO_4^-, and $Cr_2O_7^{2-}$.

Reducing agents readily donate electrons and include H_2, Na, C, CO, and SO_2 (SO_3^{2-} in solution),

e.g. $Cr_2O_7^{2-}(aq) + 3SO_3^{2-}(aq) + 8H^+(aq) \rightarrow 2Cr^{3+}(aq) + 3SO_4^{2-}(aq) + 4H_2O(l)$

$Cr_2O_7^{2-}$: (+6) (orange) (oxidizing agent); SO_3^{2-}: (+4) (reducing agent); Cr^{3+}: (+3) (green); SO_4^{2-}: (+6)

Redox reactions (2)

OXIDATION AND REDUCTION IN TERMS OF OXIDATION STATES

When an element is oxidized its oxidation state *increases*,

e.g. $Mg(s) \rightarrow Mg^{2+}(aq) + 2e^-$

(0) (+2)

When an element is reduced its oxidation state *decreases*,

e.g. $SO_4^{2-}(aq) + 2H^+(aq) + 2e^- \rightarrow SO_3^{2-}(aq) + H_2O(l)$

(+6) (+4)

The change in the oxidation state will be equal to the number of electrons involved in the half-equation.

Using oxidation states makes it easy to identify whether or not a reaction is a redox reaction.

Redox reactions (change in oxidation states)

$CuO(s) + H_2(g) \rightarrow Cu(s) + H_2O(l)$

(+2) (0) (0) (+1)

$5Fe^{2+}(aq) + MnO_4^-(aq) + 8H^+(aq) \rightarrow 5Fe^{3+}(aq) + Mn^{2+}(aq) + 4H_2O(l)$

(+2) (+7) (+3) (+2)

Not redox reactions (no change in oxidation states)

precipitation $Ag^+(aq) + Cl^-(aq) \rightarrow AgCl(s)$

(+1) (−1) (+1) (−1)

neutralization $HCl(aq) + NaOH(aq) \rightarrow NaCl(aq) + H_2O(l)$

(+1)(−1) (+1)(−2)(+1) (+1)(−1) (+1)(−2)

Note: reactions where an element is uncombined on one side of the equation and combined on the other side *must* be redox reactions since there must be a change in oxidation state,

e.g. $Mg(s) + 2HCl(aq) \rightarrow MgCl_2(aq) + H_2(g)$

BALANCING REDOX EQUATIONS

To obtain the overall redox equation the number of electrons in the oxidation half-equation must balance the number of electrons in the reduction half-equation. For many redox equations this is straightforward. For example, the oxidation of magnesium by silver(I) ions. Write the two half-equations, then double the silver half-equation so that both half-equations involve two electrons and then simply add them together:

$2Ag^+(aq) + 2e^- \rightarrow 2Ag(s)$

$Mg(s) \rightarrow Mg^{2+}(aq) + 2e^-$

Overall $2Ag^+(aq) + Mg(s) \rightarrow 2Ag(s) + Mg^{2+}(aq)$

It is less straightforward when there is a change in the number of oxygen atoms in a compound or ion. **The rule is that if oxygen atoms need to be accounted for then water is used and if hydrogen atoms need to be accounted for then hydrogen ions, $H^+(aq)$, are used.** Consider the oxidation of sulfite ions, $SO_3^{2-}(aq)$ to sulfate ions, $SO_4^{2-}(aq,)$ by purple permanganate(VII) ions, $MnO_4^-(aq)$, which are reduced to very pale pink (virtually colourless) Mn^{2+} (aq) ions in the process.

$SO_3^{2-}(aq) \rightarrow SO_4^{2-}(aq)$. Add the extra O atom by adding water and ensure the charges on both sides are equal so the balanced half-equation becomes: $SO_3^{2-}(aq) + H_2O(l) \rightarrow SO_4^{2-}(aq) + 2H^+(aq) + 2e^-$

$MnO_4^-(aq) \rightarrow Mn^{2+}(aq)$. Add 8 H^+ ions to remove the oxygen atoms as water and ensure the charges on both sides are equal so the balanced half-equation becomes: $MnO_4^-(aq) + 8H^+(aq) + 5e^- \rightarrow Mn^{2+}(aq) + 4H_2O(l)$.

Note that the number of electrons can also be obtained from the change in oxidation states. S goes from +4 to +6 so two electrons are involved and Mn goes from +7 to +2 so five electrons are involved.

To ensure both half-equations have the same number of electrons multiply the sulfite equation by 5 and the permanganate(VII) equation by 2 then obtain the overall redox equation by adding them together to cancel out the ten electrons and simplify the water and hydrogen ions.

$2MnO_4^-(aq) + 16H^+(aq) + 10e^- + 5SO_3^{2-}(aq) + 5H_2O(l) \rightarrow 2Mn^{2+}(aq) + 8H_2O(l) + 5SO_4^{2-}(aq) + 10H^+(aq) + 10e^-$

Which simplifies to:

$2MnO_4^-(aq) + 6H^+(aq) + 5SO_3^{2-}(aq) \rightarrow 2Mn^{2+}(aq) + 3H_2O(l) + 5SO_4^{2-}(aq)$

This means that acid (H^+ ions) needs to be present for the reaction to take place.

Activity series and Winkler method

ACTIVITY SERIES

Lithium, sodium and potassium all react with cold water to give similar products but the reactivity increases down the group.

$$2M(s) + 2H_2O(l) \rightarrow 2M^+(aq) + 2OH^-(aq) + H_2(g) \quad (M = \text{Li, Na or K})$$

Slightly less reactive metals react with steam and will give hydrogen with dilute acids, e.g.

$$Mg(s) + 2H_2O(g) \rightarrow Mg(OH)_2 + H_2(g)$$

$$Mg(s) + 2HCl(aq) \rightarrow Mg^{2+}(aq) + 2Cl^-(aq) + H_2(g)$$

In all of these reactions the metal is losing electrons – that is, it is being oxidized and in the process it is acting as a reducing agent. An activity series of reducing agents can be deduced by considering the reactivity of metals with water and acids, and the reactions of metals with the ions of other metals.

The series used by the IB can be found in Section 25 of the IB Chemistry data booklet. A simplified series follows the order:

Li > K > Ca > Na > Mg > Al > Zn > Fe > Sn > Pb > H > Cu > Ag > Au

Generally the more readily the metal ion loses its outer electron(s) the more reactive it is although note that lithium is actually higher than potassium in the activity series as the redox reactions also involve the hydration of the ions formed. Metals higher in the series can displace metal ions lower in the series from solution, e.g. zinc can react with copper ions to form zinc ions and precipitate copper metal.

$$Zn(s) + Cu^{2+}(aq) \rightarrow Zn^{2+}(aq) + Cu(s)$$

$Zn(s) \rightarrow Zn^{2+}(aq) + 2e^-$ Zn loses electrons in preference to Cu

$Cu^{2+}(aq) + 2e^- \rightarrow Cu(s)$ Cu^{2+} gains electrons in preference to Zn^{2+}

This also explains why only metals above hydrogen can react with acids (displace hydrogen ions) to produce hydrogen gas, e.g.

$$Zn(s) + 2H^+(aq) \rightarrow Zn^{2+}(aq) + H_2(g)$$

The series can be extended for oxidizing agents. The most reactive oxidizing agent will be the species that gains electrons the most readily. For example, in group 17

$I^-(aq) = e^- + \frac{1}{2}I_2(aq)$

$Br^-(aq) = e^- + \frac{1}{2}Br_2(aq)$

$Cl^-(aq) = e^- + \frac{1}{2}Cl_2(aq)$

$F^-(aq) = e^- + \frac{1}{2}F_2(aq)$

(↓ increasing oxidizing ability)

Oxidizing agents lower in the series gain electrons from species higher in the series, e.g.

$$Cl_2(aq) + 2Br^-(aq) \rightarrow 2Cl^-(aq) + Br_2(aq)$$

WINKLER METHOD

One application of a redox process is the Winkler method which is used to measure Biological Oxygen Demand (BOD). BOD is a measure of the dissolved oxygen (in ppm) required to decompose the organic matter in water biologically over a set time period, which is usually five days. Polluted water with a high BOD without the means of replenishing oxygen will not sustain aquatic life.

The sample of the water is saturated with oxygen so the initial concentration of dissolved oxygen is known. A measured volume of the sample is then incubated at a fixed temperature for five days while microorganisms in the water oxidize the organic material. An excess of a manganese(II) salt is then added to the sample.

Under alkaline conditions manganese(II) ions are oxidized to manganese(IV) oxide by the remaining oxygen.

$$2Mn^{2+}(aq) + 4OH^-(aq) + O_2(aq) \rightarrow 2MnO_2(s) + 2H_2O(l)$$

Potassium iodide is then added which is oxidized by the manganese(IV) oxide in acidic solution to form iodine.

$$MnO_2(s) + 2I^-(aq) + 4H^+(aq) \rightarrow Mn^{2+}(aq) + I_2(aq) + 2H_2O(l)$$

The iodine released is then titrated with standard sodium thiosulfate solution.

$$I_2(aq) + 2S_2O_3^{2-}(aq) \rightarrow S_4O_6^{2-}(aq) + 2I^-(aq)$$

By knowing the amount (in mol) of iodine produced the amount of oxygen present in the sample of water can be calculated and hence its concentration.

Worked example

100 cm^3 of a sample of water was treated with an excess of alkaline manganese(II) sulphate, $MnSO_4(aq)$. After all the dissolved oxygen had reacted the solution was then acidified and excess potassium iodide. KI(aq) added. The iodine released was titrated with sodium thiosulfate solution, $Na_2S_2O_3(aq)$, using starch as an indicator. It was found that 6.00 cm^3 of 1.00×10^{-2} mol dm^{-3} $Na_2S_2O_3(aq)$ was required to react with all the iodine. Determine the concentration of dissolved oxygen in parts per million (ppm).

Step 1. Calculate the amount of sodium thiosulfate used in the titration.

$$\text{Amount of } S_2O_3^{2-}(aq) = \frac{6.00}{1000} \times 1.00 \times 10^{-2} = 6.00 \times 10^{-5} \text{ mol}$$

Step 2. Calculate the amount of iodine that reacted with the sodium thiosulfate.

$$I_2(aq) + 2S_2O_3^{2-}(aq) \rightarrow S_4O_6^{2-}(aq) + 2I^-(aq)$$

$$\text{Amount of } I_2(aq) = \tfrac{1}{2} \times 6.00 \times 10^{-5} = 3 \times 10^{-5} \text{ mol}$$

Step 3. Calculate the amount of MnO_2 produced by the oxidation of the Mn^{2+} ions.

$$MnO_2(s) + 2I^-(aq) + 4H^+(aq) \rightarrow Mn^{2+} + I_2(aq) + 2H_2O(l)$$

$$\text{Amount of } MnO_2 = 3 \times 10^{-5} \text{ mol}$$

Step 4. Calculate the mass of oxygen dissolved in the 100 cm^3 sample of water.

$$2Mn^{2+}(aq) + 4OH^-(aq) + O_2(aq) \rightarrow 2MnO_2(s) + 2H_2O(l)$$

$$\text{Amount of } O_2 = \tfrac{1}{2} \times 3 \times 10^{-5} = 1.5 \times 10^{-5} \text{ mol}$$

$$\text{Mass of } O_2 = 32.00 \times 1.5 \times 10^{-5} = 4.8 \times 10^{-4} \text{ g}$$

Step 5. Calculate the concentration of dissolved oxygen in parts per million (ppm). (This is the same as the mass in mg dissolved in 1 dm^3 of water.)

100 cm^3 contains 4.8×10^{-4} g of O_2 so concentration $= 4.8 \times 10^{-3}$ g dm^{-3} = 4.8 ppm.

Electrochemical cells

SIMPLE VOLTAIC CELLS

A half-cell is simply a metal in contact with an aqueous solution of its own ions. A voltaic cell consists of two different half-cells, connected together to enable the electrons transferred during the redox reaction to produce energy in the form of electricity. The cells are connected by an external wire and by a salt bridge, which allows the free movement of ions.

A good example of a voltaic cell is a zinc half-cell connected to a copper half-cell. Because zinc is higher in the activity series the electrons will flow from the zinc half-cell towards the copper half-cell. To complete the circuit and to keep the half-cells electrically neutral, ions will flow through the salt bridge. The voltage produced by a voltaic cell depends on the relative difference between the two metals in the activity series. Thus the voltage from a $Mg(s)/Mg^{2+}(aq)$ half-cell connected to a $Cu(s)/Cu^{2+}(aq)$ half-cell will be greater than that obtained from a $Zn(s)/Zn^{2+}(aq)$ half-cell connected to a $Fe(s)/Fe^{2+}(aq)$ half-cell.

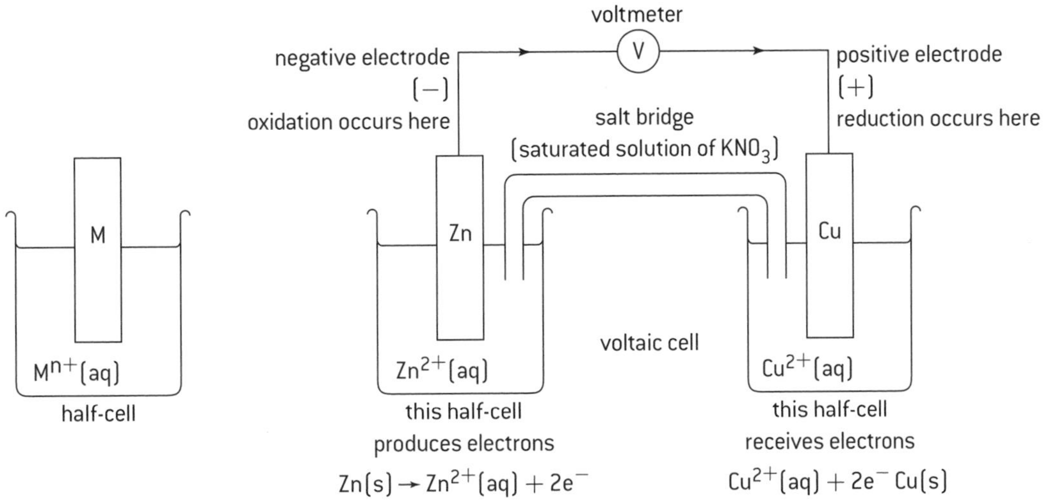

CONVENTION FOR WRITING CELLS

By convention in a cell diagram the half-cell undergoing oxidation is placed on the left of the diagram and the half-cell undergoing reduction on the right of the diagram. The two aqueous solutions are then placed either side of the salt bridge e.g.

$Zn(s)/Zn^{2+}(aq) \,||\, Cu^{2+}(aq)/Cu(s)$

The words cathode and anode can also be used to describe the electrodes. The anode is where oxidation occurs so for a voltaic cell it is the negative electrode but for an electrolytic cell (see below), where the electricity is provided from an external source, the anode is the positive electrode, which can cause some confusion.

ELECTROLYTIC CELLS

In a voltaic cell electricity is produced by the spontaneous redox reaction taking place. Electrolytic cells are used to make non-spontaneous redox reactions occur by providing energy in the form of electricity from an external source. In an electrolytic cell electricity is passed through an **electrolyte** and electrical energy is converted into chemical energy. An electrolyte is a substance which does not conduct electricity when solid, but does conduct electricity when molten or in aqueous solution and is chemically decomposed in the process. A simple example of an electrolytic cell is the electrolysis of molten sodium chloride.

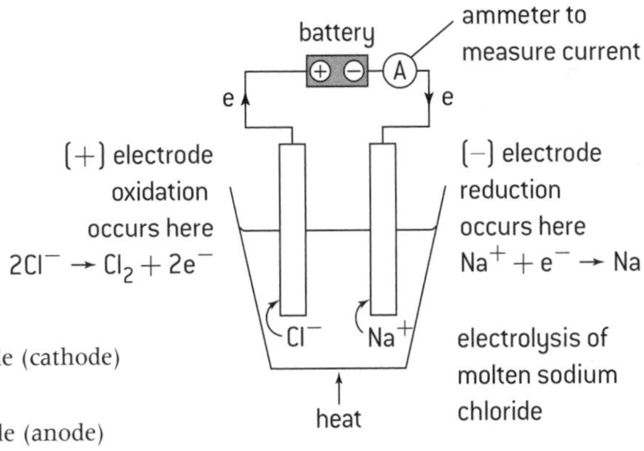

During the electrolysis:

- sodium is formed at the negative electrode (cathode)
  $Na^+(l) + e^- \rightarrow Na(l)$ reduction
- chlorine is formed at the positive electrode (anode)
  $2Cl^-(l) \rightarrow Cl_2(g) + 2e^-$ oxidation
- The current is due to the movement of electrons in the external circuit and the movement of ions in the electrolyte.

Electrolysis is an important industrial process used to obtain reactive metals, such as sodium, from their common ores.

Electrolysis

FACTORS AFFECTING THE DISCHARGE OF IONS DURING ELECTROLYSIS

During the electrolysis of molten salts there are only usually two ions present, so the cation will be discharged at the negative electrode (cathode) and the anion at the positive electrode (anode). However, for aqueous electrolytes there will also be hydrogen ions and hydroxide ions from the water present. There are three main factors that influence which ions will be discharged at their respective electrodes.

1. **Position in the electrochemical series**

   The lower the metal ion is in the electrochemical series the more readily it will gain electrons (be reduced) to form the metal at the cathode. Thus in the electrolysis of a solution of sodium hydroxide, hydrogen will be evolved at the negative electrode in preference to sodium, whereas in a solution of copper(II) sulfate, copper will be deposited at the negative electrode in preference to hydrogen.

   For negative ions the order of discharge follows $OH^- > Cl^- > SO_4^{2-}$.

2. **Concentration**

   If one of the ions is much more concentrated than another ion then it may be preferentially discharged. For example, when electricity is passed through an aqueous solution of sodium chloride both oxygen and chlorine are evolved at the positive electrode. For dilute solutions mainly oxygen is evolved, but for concentrated solutions of sodium chloride more chlorine than oxygen is evolved.

3. **The nature of the electrode**

   It is normally safe to assume that the electrode is inert, i.e. does not play any part in the reaction. However, if copper electrodes are used during the electrolysis of a solution of copper sulfate then the positive electrode is itself oxidized to release electrons and form copper(II) ions. Since copper is simultaneously deposited at the negative electrode the concentration of the solution will remain constant throughout the electrolysis.

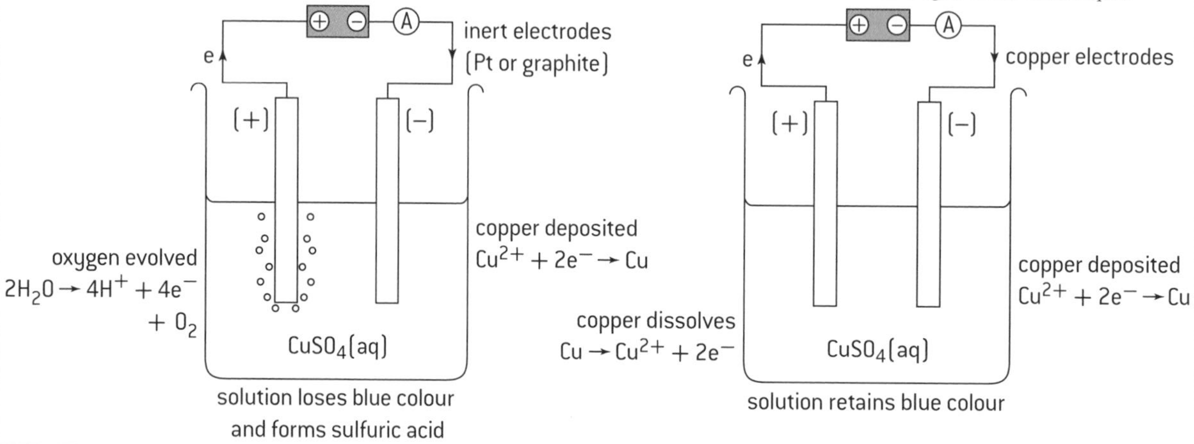

FACTORS AFFECTING THE QUANTITY OF PRODUCTS DISCHARGED DURING ELECTROLYSIS

The amount of substance deposited will depend on:

1. The number of electrons flowing through the system, i.e. the amount of charge passed. This in turn depends on the current and the time for which it flows. If the current is doubled then twice as many electrons pass through the system and twice as much product will be formed. Similarly if the time is doubled twice as many electrons will pass through the system and twice as much product will be formed.

   charge = current × time
   (1 coulomb = 1 ampere × 1 second)

2. The charge on the ion. To form one mole of sodium in the electrolysis of molten sodium chloride requires one mole of electrons to flow through the cell. However, the formation of one mole of lead during the electrolysis of molten lead(II) bromide requires two moles of electrons.

   $Na^+(l) + e^- \rightarrow Na(l)$
   $Pb^{2+}(l) + 2e^- \rightarrow Pb(l)$

If cells are connected in series then the same amount of electricity will pass through both cells and the relative amounts of products obtained can be determined.

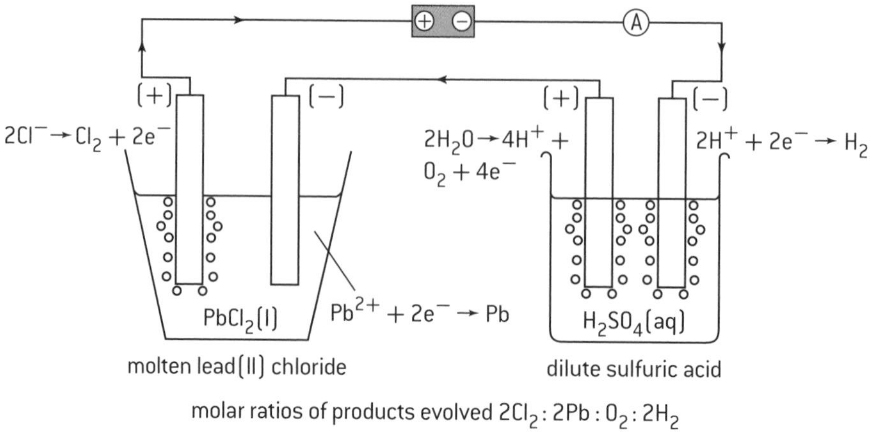

Electroplating and standard electrode potentials

ELECTROPLATING

Electrolysis can also be used in industry to coat one metal with a thin layer of another metal. This process is known as **electroplating**.

For example, in copper plating the negative electrode (cathode) is made from the metal to be copper plated. It is placed into a solution of copper(II) sulfate together with a positive electrode (anode) made from a piece of copper. As electricity is passed through the solution the copper anode dissolves in the solution to form $Cu^{2+}(aq)$ ions and the $Cu^{2+}(aq)$ ions in solution are deposited onto the cathode. By making the anode of impure copper and the cathode from a small piece of pure copper this process can also be used to purify impure copper. This is an important industrial process as one of the main uses of copper is for electrical wiring, where purity is important since impure copper has a much higher electrical resistance.

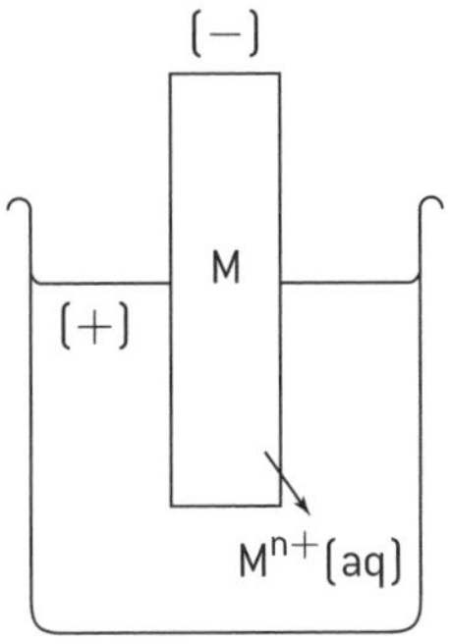

STANDARD ELECTRODE POTENTIALS

There are two opposing tendencies in a half-cell. The metal may dissolve in the solution of its own ions to leave the metal with a negative potential compared with the solution, or the metal ions may deposit on the metal, which will give the metal a positive potential compared with the solution. It is impossible to measure this potential, as any attempt to do so interferes with the system being investigated. However, the electrode potential of one half-cell can be compared against another half-cell. The hydrogen half-cell is normally used as the standard. Under standard conditions of 100 kPa pressure, 298 K, and 1.0 mol dm^{-3} hydrogen ion concentration the standard electrode potential of the hydrogen electrode is assigned a value of zero volts.

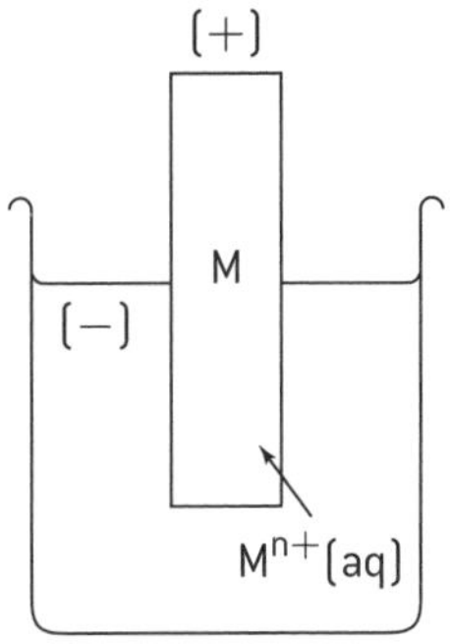

When the half-cell contains a metal above hydrogen in the reactivity series electrons flow from the half-cell to the hydrogen electrode, and the electrode potential is given a negative value. If the half-cell contains a metal below hydrogen in the reactivity series electrons flow from the hydrogen electrode to the half-cell, and the electrode potential has a positive value. The standard electrode potentials are arranged in increasing order to form the electrochemical series.

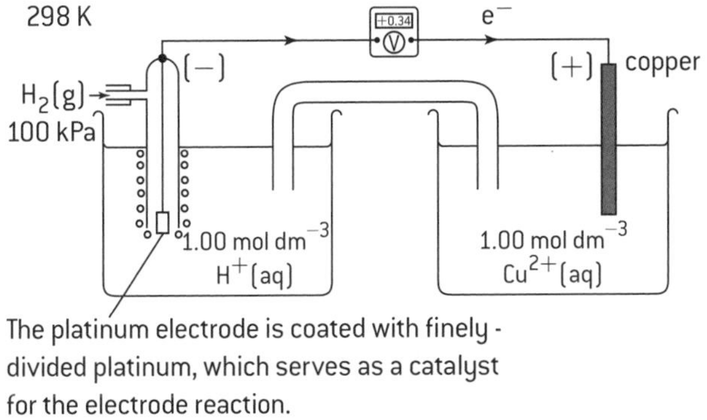

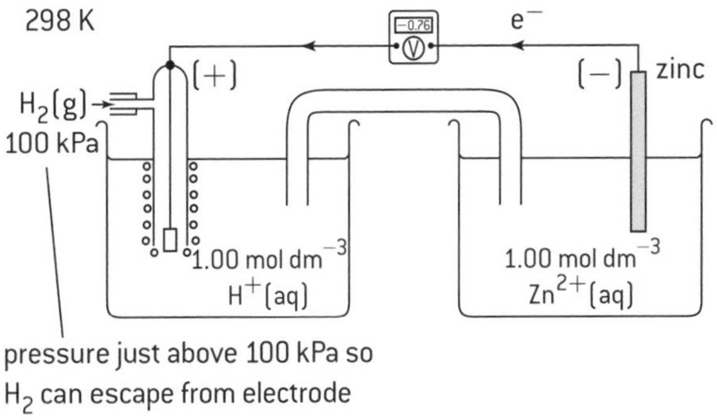

ELECTROCHEMICAL SERIES

(A more complete series can be found in the IB data booklet)

Couple	$E^\ominus$ / V
$K(s)/K^+(aq)$	−2.93
$Ca(s)/Ca^{2+}(aq)$	−2.87
$Na(s)/Na^+(aq)$	−2.71
$Mg(s)/Mg^{2+}(aq)$	−2.37
$Al(s)/Al^{3+}(aq)$	−1.66
$Zn(s)/Zn^{2+}(aq)$	−0.76
$Fe(s)/Fe^{2+}(aq)$	−0.45
$\frac{1}{2}H_2(g)/H^+(aq)$	0.00
$Cu(s)/Cu^{2+}(aq)$	+0.34
$I^-(aq)/\frac{1}{2}I_2(aq)$	+0.54
$Ag(s)/Ag^+(aq)$	+0.80
$Br^-(aq)/\frac{1}{2}Br_2(aq)$	+1.09
$Cl^-(aq)/\frac{1}{2}Cl_2(aq)$	+1.36
$F^-(aq)/\frac{1}{2}F_2(aq)$	+2.87

Spontaneity of electrochemical reactions

SHORTHAND NOTATION FOR A CELL

To save drawing out the whole cell a shorthand notation has been adopted. A half-cell is denoted by a / between the metal and its ions, and two vertical lines are used to denote the salt bridge between the two half cells,

e.g. $Cu(s)/Cu^{2+}(aq) \,||\, H^+(aq)/H_2(g)$ and
$Zn(s)/Zn^{2+}(aq) \,||\, H^+(aq)/H_2(g)$

The standard electronmotive force (emf) of any cell $E^\ominus_{cell}$ is simply the difference between the standard electrode potentials of the two half-cells,

e.g. $Zn(s)/Zn^{2+}(aq) \,||\, Cu^{2+}(aq)/Cu(s)$

$E^\ominus$ −0.76 V   +0.34 V

$E^\ominus_{cell} = 1.10$ V

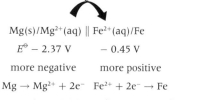

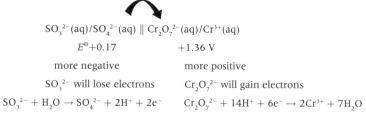

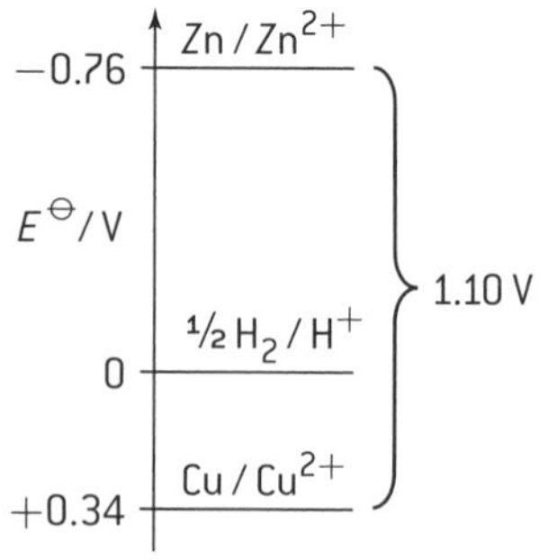

ELECTRON FLOW, SPONTANEOUS REACTIONS AND FREE ENERGY

By using standard electrode potentials it is easy to see what will happen when two half-cells are connected together. The electrons will always flow **from** the more negative half-cell **to** the more positive half-cell, e.g. consider an iron half-cell connected to a magnesium half-cell:

e^-

$Mg(s)/Mg^{2+}(aq) \,||\, Fe^{2+}(aq)/Fe$

$E^\ominus$ − 2.37 V   − 0.45 V

more negative   more positive

$Mg \rightarrow Mg^{2+} + 2e^-$ $Fe^{2+} + 2e^- \rightarrow Fe$

Spontaneous reaction: $Mg(s) + Fe^{2+}(aq) \rightarrow Mg^{2+}(aq) + Fe(s)$ $E^\ominus_{cell} = 1.92$ V

From the expression $\Delta G^\ominus = -nFE^\ominus$ it can be seen that positive $E^\ominus_{cell}$ values give negative $\Delta G^\ominus$ values, which is obvious as the cell is producing electrical energy that can do work as the reaction is spontaneous. The actual amount of energy can be calculated as F is one Faraday and has a value equal to 9.65×10^4 C mol^{-1} and 1 Joule = 1 Volt × 1 Coulomb. The transfer of two mol of electrons is involved. $\Delta G^\ominus = -2$ (mol) $\times 9.65 \times 10^4$ (C mol^{-1}) × 1.92 (V) $= -3.71 \times 10^5$ J = −371 kJ.

The reverse reaction ($Fe(s) + Mg^{2+}(aq) \rightarrow Fe^{2+}(aq) + Mg(s)$) has a negative $E^\ominus_{cell}$ value which gives a positive value of + 371 kJ for $\Delta G^\ominus$ and the reaction will not be spontaneous. This reaction could only proceed if an external source of energy is provided greater than 371 kJ, i.e. a voltage greater than 1.92 V is supplied to force the reaction in the opposite direction. It is worth noting that just because a reaction is thermodynamically spontaneous (i.e. has a negative $\Delta G^\ominus$ value) it does not mean that it will necessarily proceed, as the reaction may have a large activation energy which first needs to be overcome.

REDOX EQUATIONS

Standard electrode potentials can be extended to cover any half-equation. The values, which can be found in the Electrochemical Series in the IB data booklet, can then be used to determine whether a particular reaction is spontaneous. For example, $Cr_2O_7^{2-}(aq)/Cr^{3+}(aq)$, $E^\ominus = +1.36$ V, but this only takes place in acid solution, so hydrogen ions and water are required to balance the half-equation. Consider the reaction between acidified dichromate(VI) ions, $Cr_2O_7^{2-}(aq)$, and sulfite ions, $SO_3^{2-}(aq)$. Using standard electrode potentials the cell becomes:

e^-

$SO_3^{2-}(aq)/SO_4^{2-}(aq) \,||\, Cr_2O_7^{2-}(aq)/Cr^{3+}(aq)$

$E^\ominus$+0.17   +1.36 V

more negative   more positive

SO_3^{2-} will lose electrons $Cr_2O_7^{2-}$ will gain electrons

$SO_3^{2-} + H_2O \rightarrow SO_4^{2-} + 2H^+ + 2e^-$ $Cr_2O_7^{2-} + 14H^+ + 6e^- \rightarrow 2Cr^{3+} + 7H_2O$

As previously explained, to obtain the overall equation the number of electrons in both half-equations must be equal. This is achieved by multiplying the sulfite equation by three. The two equations are then added together and the water and hydrogen ions, which appear on both sides, are simplified to give:

$Cr_2O_7^{2-}(aq) + 8H^+(aq) + 3SO_3^{2-}(aq) \rightarrow 2Cr^{3+}(aq) + 3SO_4^{2-}(aq) + 4H_2O(l)$ $E^\ominus_{cell} = 1.19$ V

The acid that is normally added to supply the $H^+(aq)$ ions is sulfuric acid, since acidified dichromate ions cannot oxidize sulfate ions. The standard electrode potential for the ½ $Cl_2(aq)/Cl^-(aq)$ half-cell is +1.36 V. This is exactly the same as the value for the $Cr_2O_7^{2-}(aq)/Cr^{3+}(aq)$ half-cell. Under standard conditions dichromate ions will not oxidize chloride ions from hydrochloric acid to chlorine but the conditions only need to be changed slightly and this may no longer be true so hydrochloric acid should never be used to acidify the solution of dichromate ions.

MULTIPLE CHOICE QUESTIONS – REDOX PROCESSES

1. $5Fe^{2+}(aq) + MnO_4^-(aq) + 8H^+(aq) \rightarrow 5Fe^{3+}(aq) + Mn^{2+}(aq) + 4H_2O(l)$

   In the equation above:

   A. $Fe^{2+}(aq)$ is the oxidizing agent

   B. H^+ (aq) ions are reduced

   C. $Fe^{2+}(aq)$ ions are oxidized

   D. $MnO_4^-(aq)$ is the reducing agent

2. The oxidation states of nitrogen in NH_3, HNO_3 and NO_2 are, respectively

   A. −3, −5, +4 **C.** −3, +5, −4

   B. +3, +5, +4 **D.** −3, +5, +4

3. Which one of the following reactions is **not** a redox reaction?

   A. $Ag^+(aq) + Cl^-(aq) \rightarrow AgCl(s)$

   B. $2Na(s) + Cl_2(g) \rightarrow 2NaCl(s)$

   C. $Mg(s) + 2HCl(aq) \rightarrow MgCl_2(aq) + H_2(g)$

   D. $Cu^{2+}(aq) + Zn(s) \rightarrow Cu(s) + Zn^{2+}(aq)$

4. Which substance does not have the correct formula?

   A. iron(III) sulfate $Fe_2(SO_4)_3$

   B. iron(II) oxide Fe_2O

   C. copper(I) sulfate Cu_2SO_4

   D. copper(II) nitrate $Cu(NO_3)_2$

5. For which conversion is an oxidizing agent required?

   A. $2H^+(aq) \rightarrow H_2(g)$ **C.** $SO_3(g) \rightarrow SO_4^{2-}(aq)$

   B. $2Br^-(aq) \rightarrow Br_2(aq)$ **D.** $MnO_2(s) \rightarrow Mn^{2+}(aq)$

6. Ethanol can be oxidized to ethanal by an acidic solution of dichromate(VI) ions.

   $\_C_2H_5OH(aq) + \_H^+(aq) + \_Cr_2O_7^{2-}(aq) \rightarrow$
   $\_CH3CHO(aq) + \_Cr^{3+}(aq) + \_H_2O(l)$

   The sum of all the coefficients in the balanced equation is:

   A. 24 **B.** 26 **C.** 28 **D.** 30

7. When an $Fe(s)/Fe^{2+}(aq)$ half-cell is connected to a $Cu(s)/Cu^{2+}(aq)$ half-cell by a salt bridge and a current allowed to flow between them

   A. the electrons will flow from the copper to the iron.

   B. the salt bridge allows the flow of ions to complete the circuit.

   C. the salt bridge allows the flow of electrons to complete the circuit.

   D. the salt bridge can be made of copper or iron.

8. During the electrolysis of molten sodium chloride using platinum electrodes

   A. sodium is formed at the negative electrode.

   B. chlorine is formed at the negative electrode.

   C. sodium is formed at the positive electrode.

   D. oxygen is formed at the positive electrode.

9. Which statement is true?

   A. Lead chloride is ionic so solid lead chloride will conduct electricity.

   B. When a molten ionic compound conducts electricity free electrons pass through the liquid.

   C. When liquid mercury conducts electricity mercury ions move towards the negative electrode.

   D. During the electrolysis of a molten salt reduction will always occur at the negative electrode.

10. The following information is given about reactions involving the metals X, Y and Z and solutions of their sulfates.

    $X(s) / YSO_4(aq) \rightarrow$ no reaction

    $Z(s) + YSO_4(aq) \rightarrow Y(s) + ZSO_4(aq)$

    When the metals are listed in decreasing order of reactivity (most reactive first), what is the correct order?

    A. Z > Y > X **C.** Y > X > Z

    B. X > Y > Z **D.** Y > Z > X

HL

11. Which are correct for a spontaneous reaction occurring in a voltaic cell?

    I. $E^{\ominus}$ for the cell has a positive value

    II. $\Delta G^{\ominus}$ for the reaction has a negative value

    III. The reaction must occur under standard conditions

    A. I and II only **C.** II and III only

    B. I and III only **D.** I, II and III

12. When the same quantity of electricity was passed through a dilute solution of sodium hydroxide and through a molten solution of lead bromide, 0.100 mol of lead was produced. What amount (in mol) of hydrogen gas was evolved?

    A. 0.014 **C.** 0.100

    B. 0.050 **D.** 0.200

Use the following information to answer questions 13 and 14.

$Sn^{2+}(aq) + 2e^- \rightleftharpoons Sn(s)$	$E^{\ominus} = -0.14V$
$Sn^{4+}(aq) + 2e^- \rightleftharpoons Sn^{2+}(aq)$	$E^{\ominus} = +0.15V$
$Fe^{2+}(aq) + 2e^- \rightleftharpoons Fe(s)$	$E^{\ominus} = -0.44V$
$Fe^{3+}(aq) + e^- \rightleftharpoons Fe^{2+}(aq)$	$E^{\ominus} = +0.77V$

13. Under standard conditions which statement is correct?

    A. $Sn^{2+}(aq)$ can reduce $Fe^{3+}(aq)$.

    B. $Fe(s)$ can oxidize $Sn^{2+}(aq)$.

    C. $Sn(s)$ can reduce $Fe(s)$.

    D. $Fe^{3+}(aq)$ can reduce $Sn^{4+}(aq)$.

14. When a half-cell of $Fe^{2+}(aq)/Fe^{3+}(aq)$ is connected by a salt bridge to a half-cell of $Sn^{2+}(aq)/Sn^{4+}(aq)$ under standard conditions and a current allowed to flow in an external circuit the total emf of the spontaneous reaction will be:

    A. +0.92V **B.** −0.92V **C.** +0.62V **D.** −0.62V

SHORT ANSWER QUESTIONS – REDOX PROCESSES

1. The data below is from an experiment used to determine the percentage of iron present in a sample of iron ore. This sample was dissolved in acid and all of the iron was converted to Fe^{2+}. The resulting solution was titrated with a standard solution of potassium manganate(VII), $KMnO_4$. This procedure was carried out three times. In acidic solution, MnO_4^- reacts with Fe^{2+} ions to form Mn^{2+} and Fe^{3+} and the end point is indicated by a slight pink colour.

Titre	**1**	**2**	**3**
Initial burette reading / cm^3	1.00	23.60	10.00
Final burette reading / cm^3	24.60	46.10	32.50

Mass of iron ore / g	3.682×10^{-1}
Concentration of $KMnO_4$ solution / mol dm^{-3}	2.152×10^{-2}

a) Deduce the balanced redox equation for this reaction in **acidic** solution. [2]

b) Identify the reducing agent in the reaction. [1]

c) Calculate the amount, in moles, of MnO_4^- used in the titration. [2]

d) Calculate the amount, in moles, of Fe present in the 3.682×10^{-1} g sample of iron ore. [2]

e) Determine the percentage by mass of Fe present in the 3.682×10^{-1} g sample of iron ore. [2]

2. Chemical energy can be converted to electrical energy in the voltaic cell below.

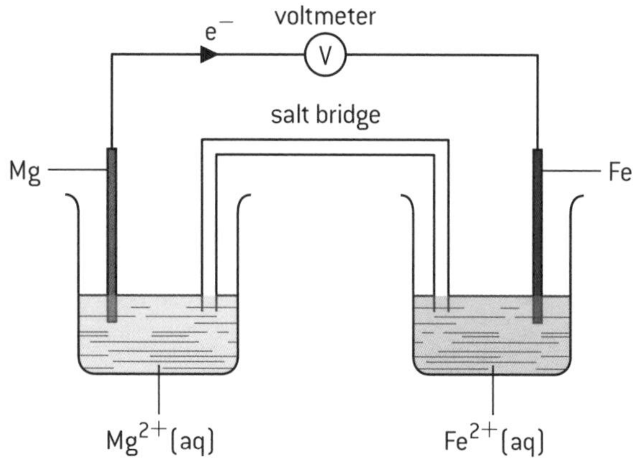

a) Explain how the diagram confirms that magnesium is above iron in the activity series. [2]

b) Identify the positive electrode (anode) of the cell. [1]

c) (i) State the half-equation for the reaction occurring at the iron electrode. [1]

(ii) State the overall equation for the reaction when the cell is producing electricity. [2]

d) Deduce whether the voltage produced by the cell would be greater or less if the iron half-cell is replaced by a copper half-cell. [2]

3. a) Molten sodium chloride can be electrolysed using graphite electrodes.

(i) Draw the essential components of this electrolytic cell and identify the products that form at each electrode. [2]

(ii) State the half-equations for the oxidation and reduction processes and deduce the overall cell reaction, including state symbols. [2]

b) Explain why solid sodium chloride does not conduct electricity. [1]

c) Using another electrolysis reaction, aluminium can be extracted from its ore, bauxite, which contains Al_2O_3. State **one** reason why aluminium is often used instead of iron in many engineering applications. [1]

4. Iodine reacts with thiosulfate ions to form the tetrathionate ion according to the equation

$$I_2(aq) + 2S_2O_3^{2-}(aq) \rightarrow S_4O_6^{2-}(aq) + 2I^-(aq)$$

a) Show that this reaction is a redox reaction. [2]

b) The thiosulfate ion has a structure similar to the sulfate ion, SO_4^{2-}, except that one of the outer oxygen atoms has been replaced by a sulfur atom.

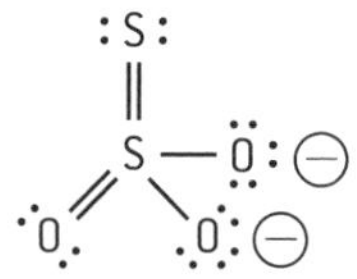

thiosulfate ion

(i) Comment on the difference in the oxidation state of sulfur in the thiosulfate ion compared with the sulfate ion. [2]

(ii) Explore the concept of oxidation state using the tetrathionate ion, $S_4O_6^{2-}$, as an example. [3]

c) Discuss whether the complete combustion of carbon in oxygen to form carbon dioxide can be described as the oxidation of carbon according to all the different definitions of oxidation. [4]

HL

5. Consider the following two half-equations:

$$Ag^+(aq) + e^- \rightleftharpoons Ag(s) \quad E^\ominus = +0.80 \text{ V}$$

$$\tfrac{1}{2} Cl_2(g) + e^- \rightleftharpoons Cl^-(aq) \quad E^\ominus = +1.36 \text{ V}$$

a) Deduce the overall equation for the spontaneous reaction that will occur when a silver half-cell is connected to a chlorine half-cell. [2]

b) Identify the direction of electron flow in the external circuit when the cell is operating. [1]

c) Determine the cell potential, $E^\ominus_{cell}$. [1]

d) Calculate the free energy produced by the cell when it is operating under standard conditions. [2]

6. a) Identify the products that will be obtained at the positive (anode) and negative (cathode) electrodes when (i) a dilute solution and (ii) a concentrated solution of sodium chloride undergoes electrolysis. [3]

b) Explain why the electrolysis of a dilute solution of sulfuric acid or a dilute solution of sodium hydroxide are both sometimes described as the electrolysis of water. [3]

c) Describe how you could use electrolysis to coat a spoon made of steel with a thin layer of silver. [2]

Fundamentals of organic chemistry

NAMING ORGANIC COMPOUNDS

Organic chemistry is concerned with the compounds of carbon. Since there are more compounds of carbon known than all the other elements put together, it is helpful to have a systematic way of naming them.

1. Identify the longest carbon chain.

   1 carbon = **meth-**

   2 carbons = **eth-**

   3 carbons = **prop-**

   4 carbons = **but-**

   5 carbons = **pent-**

   6 carbons = **hex-**

   7 carbons = **hept-**

   8 carbons = **oct-**

2. Identify the type of bonding in the chain or ring

   All single bonds in the carbon chain = **-an-**

   One double bond in the carbon chain = **-en-**

   One triple bond in the carbon chain = **-yn-**

3. Identify the functional group joined to the chain or ring.

This may come at the beginning or at the end of the name, e.g.

alkane: only hydrogen (–H) joined to chain = **-e**

hydroxyl: –OH = **-ol**

amino: $-NH_2$ = **amino-**

halo: –X: **chloro-, bromo-** or **iodo-**

aldehyde: –C(=O)–H (on the end of the chain) = **-al**

ketone: – C(=O) –(not on the end of the chain) = **-one**

carboxyl: – C(=O)–OH = **-oic acid**

ester: –C(=O)–OR: = **-oate**

4. Numbers are used to give the position of groups or bonds along the chain.

HOMOLOGOUS SERIES

The alkanes form a series of compounds all with the general formula C_nH_{2n+2}, e.g.

methane CH_4

ethane C_2H_6

propane C_3H_8

butane C_4H_{10}

If one of the hydrogen atoms is removed what is left is known as an alkyl radical R – (e.g methyl CH_3–; ethyl C_2H_5–). When other atoms or groups are attached to an alkyl radical they can form a different series of compounds. These atoms or groups attached are known as functional groups and the series formed are all homologous series.

Homologous series have the same general formula with the neighbouring members of the series differing by $-CH_2-$; for example the general formula of alcohols is $C_nH_{2n+1}OH$. The chemical properties of the individual members of a homologous series are similar and they show a gradual change in physical properties.

CLASS OF ORGANIC COMPOUND

Different compounds that all contain the same functional group are divided into classes. Sometimes the name of the class is the same as the functional group, but sometimes it is different. For example, the name of the class is the same for all **esters** that contain the **ester functional group**, **(–COOR)**, but it is different for all **alcohols** that contain the **hydroxyl**, **(–OH)**, **functional group**. The carbonyl functional group is –CO– where the carbon atom is joined by a double bond to the oxygen atom. Several classes of compounds contain this functional group as it depends upon what else is bonded to the carbon atom. For example, if H is bonded to the carbon atom then they are known as aldehydes (–COH), whereas if an alkyl group is bonded to the carbon atom then they are known as ketones (–COR). The –COOH functional group is known as the carboxyl group and the class of compounds containing this group is called carboxylic acids. Compounds containing an amino group are known as amines and compounds containing a halo group are known as halogenoalkanes (or more specifically chloroalkanes, bromoalkanes or iodoalkanes).

CLASSIFICATION OF ALCOHOLS, HALOGENOALKANES AND AMINES

Alcohols and halogenoalkanes may be classified according to how many –R groups are bonded to the carbon atom containing the functional group. Similar logic applies to amines but now it is the number of –R groups attached to the nitrogen atom of the amino functional group.

primary (one –R group bonded to the C or N atom)

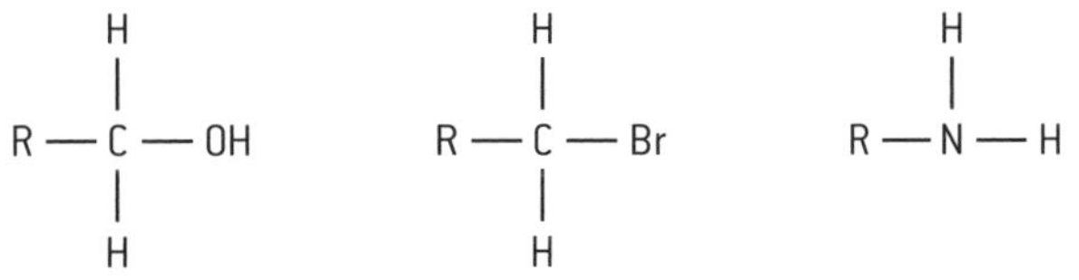

secondary (two –R groups bonded to the C or N atom)

R′ may be the same as R or different

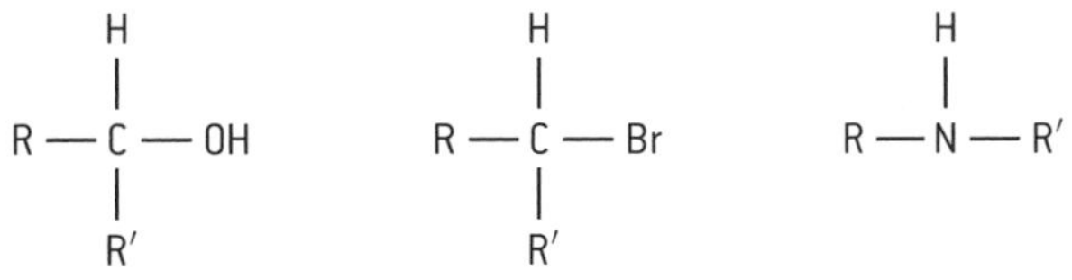

tertiary (three –R groups bonded to the C or N atom)

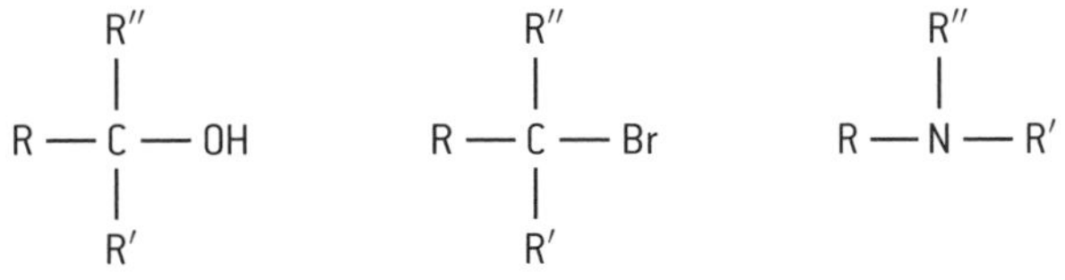

R″ may be the same as R′ or R or may be different.

Common classes of organic compounds

SOME COMMON CLASSES OF ORGANIC COMPOUNDS

Formula | **Name** | **Examples**

R—H alkane

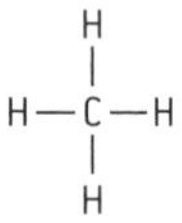
methane

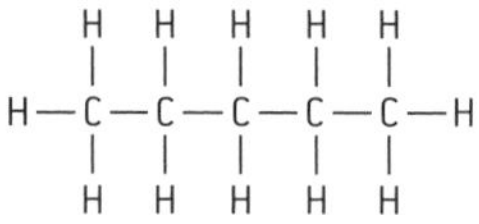
butane

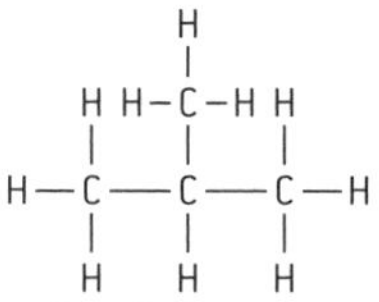
2-methylpropane

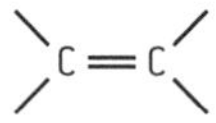
alkene
functional group: alkenyl

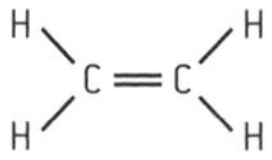
ethene

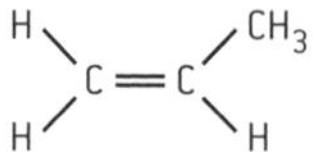
propene

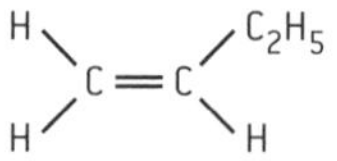
but-1-ene

—C≡C— alkyne
functional group: alkynyl

H—C≡C—H
ethyne

H—C≡C—CH_3
propyne

R—OH alcohol
functional group: hydroxy

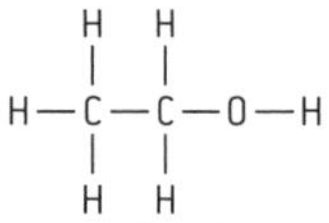
ethanol

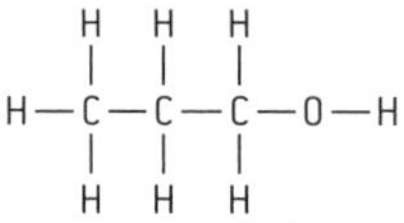
propan-1-ol

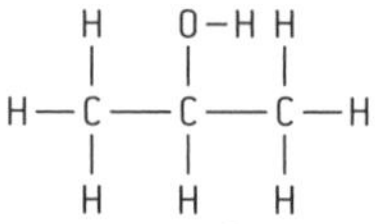
propan-2-ol

R—NH_2 amine

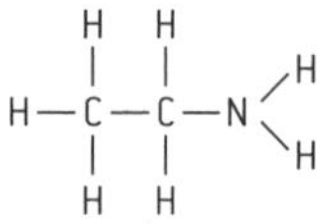
ethylamine (aminoethane)

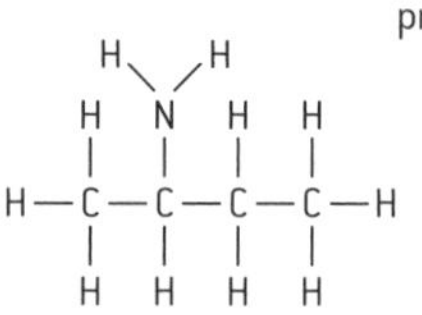
2-aminobutane

R—X halogenoalkane
(X = F, Cl, Br, or I)

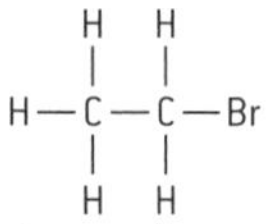
bromoethane

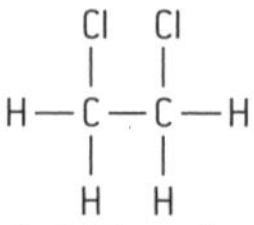
1, 2-dichloroethane

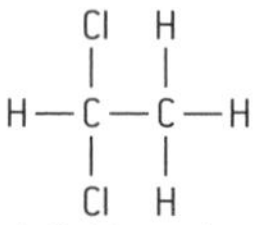
1, 1-dichloroethane

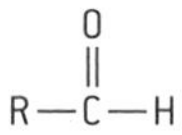
aldehyde

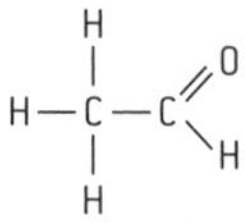
ethanal

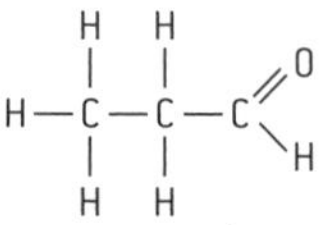
propanal

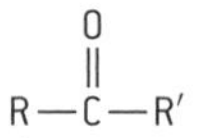
ketone
(R′ may be the same as or different to R)

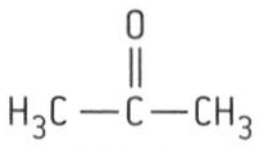
propanone

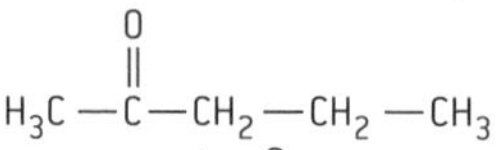
pentan-2-one

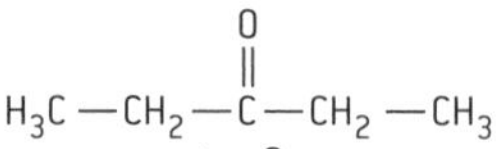
pentan-3-one

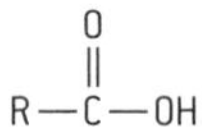
carboxylic acid

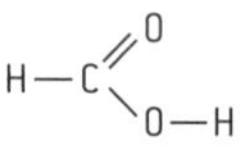
methanoic acid

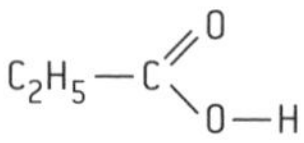
propanoic acid

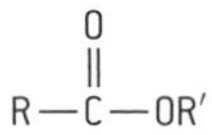
ester

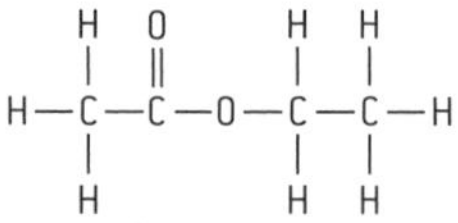
ethyl ethanoate

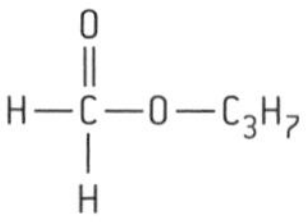
propyl methanoate

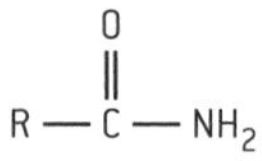
carboxamide

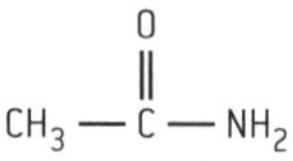
ethanamide

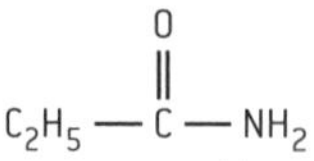
propanamide

R—C≡N nitrile

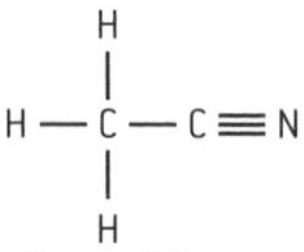
ethanenitrile

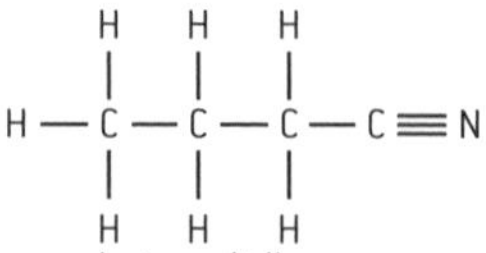
butanenitrile

R—O—R′ ether

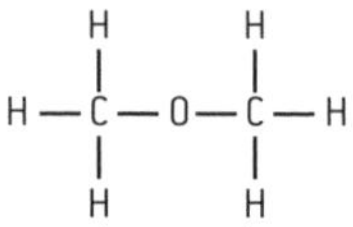
methoxymethane

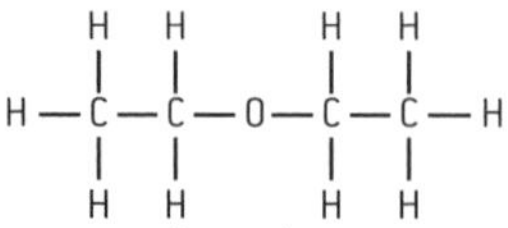
ethoxyethane

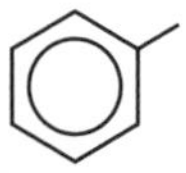
(C_6H_5–) benzene ring
functional group: phenyl

benzene

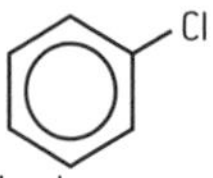
chlorobenzene

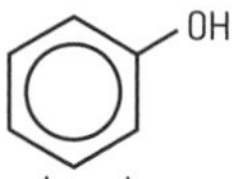
phenol

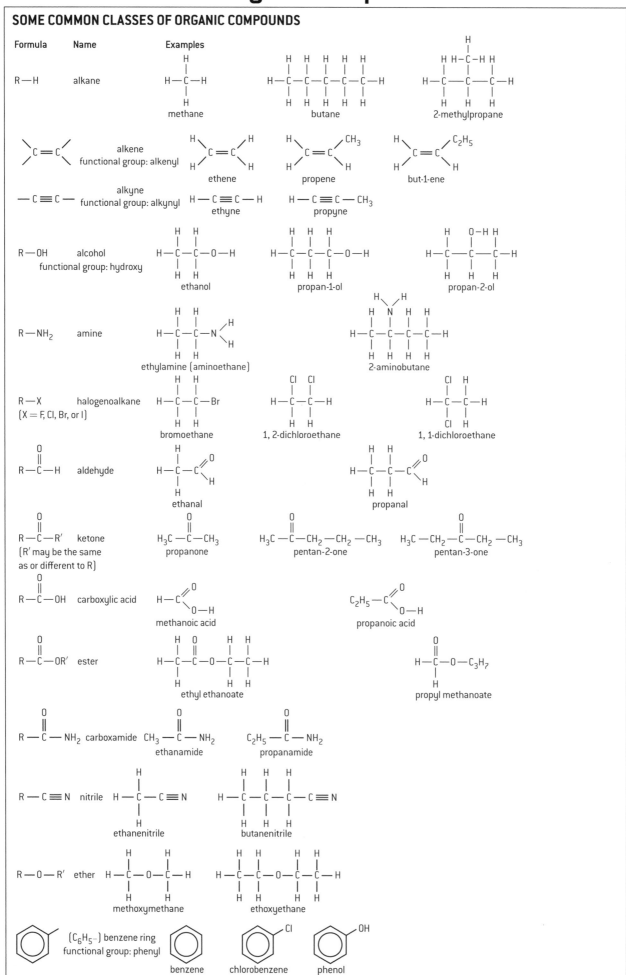

Structural formulas

STRUCTURAL FORMULAS

The difference between the empirical, molecular and structural formulas of a compound has already been covered in Topic 1 – *Stoichiometric relationships*. Because the physical and chemical properties of a compound are determined by the functional group and the arrangement of carbon and other atoms within the molecule, the structural formulas for organic compounds are often used.

These may be shown in a variety of different ways but all ways should show unambiguously how the atoms are bonded together. When drawing full structural formulas with lines representing bonds and the symbol of the element representing atoms, all the hydrogen atoms must also be included in the diagram. This is because the skeletal formula does not include any symbols for the elements and the end of a line represents a methyl group unless another atom is shown attached. Although you should understand skeletal formulas they are not normally used for simple formulas except where benzene is involved. Note that unless specifically asked for, Lewis structures, showing all the valence electrons, are not necessary. The bonding must be clearly indicated. Structures may be shown using lines as bonds or in their shortened form, e.g. $CH_3CH_2CH_2CH_2CH_3$ or $CH_3–(CH_2)_3–CH_3$ for pentane but the molecular formula, C_5H_{12} will not suffice.

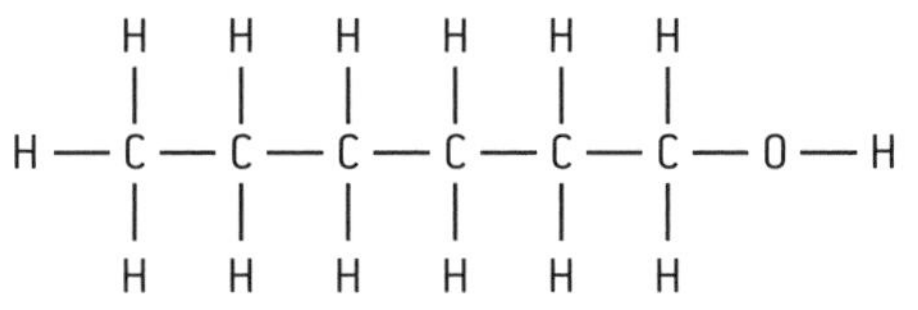

Structural formula of hexan-1-ol
also acceptable are
$CH_3CH_2CH_2CH_2CH_2CH_2OH$
and $CH_3(CH_2)_4CH_2OH$

skeletal formula of hexan-1-ol

STRUCTURAL FORMULA OF BENZENE

The simplest aromatic compound (arene) is benzene, C_6H_6. The Kekulé structure of benzene (cyclohexa-1,3,5-triene) consists of three double bonds.

There is both physical and chemical evidence to support the fact that benzene does not contain three separate double bonds but exists as a resonance hybrid structure with delocalized electrons. The two resonance hybrid forms are shown (right).

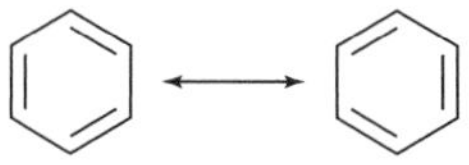

1. The C–C bond lengths are all the same and have a value of 0.140 nm which lies between the values for C–C (0.154 nm) and C=C (0.134 nm).
2. The enthalpy of hydrogenation of cyclohexene is –120 kJ mol⁻¹.

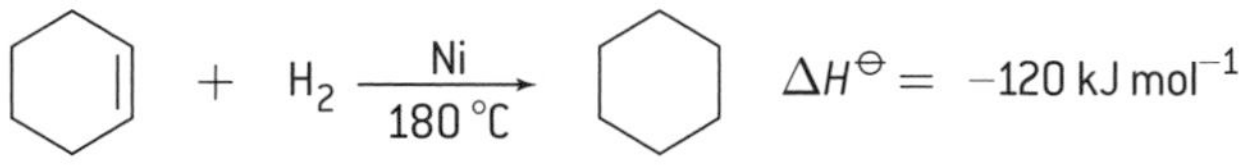

If benzene simply had the cyclohexa-1,3,5-triene structure with three double bonds the enthalpy change of hydrogenation of benzene would be expected to be equal to 3 times the enthalpy change of hydrogenation of cyclohexene, i.e. –360 kJ mol^{-1}.

However the experimentally determined value for benzene is –210 kJ mol^{-1}.

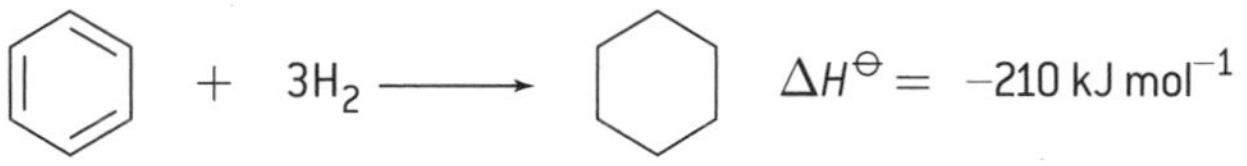

The difference of 150 kJ mol^{-1} is the extra energy associated with the delocalization.

3. Only one isomer exists for 1,2- disubstituted benzene compounds. If there were simply alternate double bonds then two isomers would exist.

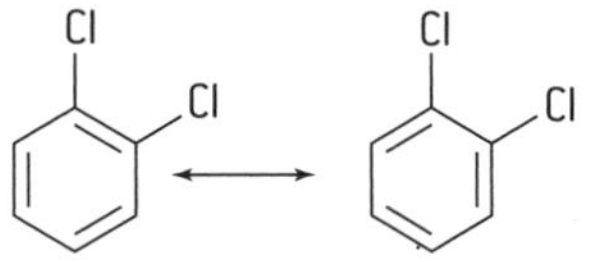

The two isomers of 1, 2- dichlorobenzene which would be expected if benzene has the cyclohexa-1, 3, 5-triene structure

4. If benzene had three normal double bonds it would be expected to readily undergo addition reactions. In fact it will only undergo addition reactions with difficulty and more commonly undergoes substitution reactions. For example, with bromine it forms bromobenzene and hydrogen bromide rather than 1,2-dibromobenzene.

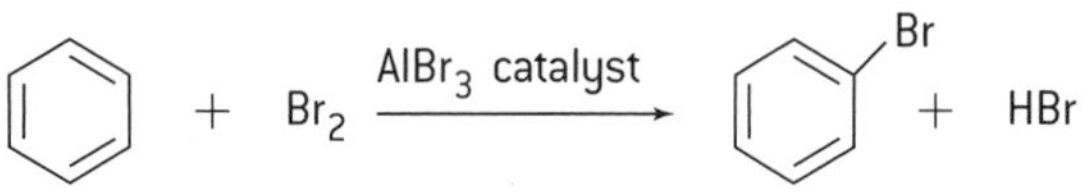

The actual bonding in benzene is best described by the delocalization of electrons. For this reason benzene is often represented by a hexagonal ring with a circle in the middle of it.

Structural isomers

STRUCTURAL FORMULAS OF HYDROCARBONS

Isomers of alkanes

Each carbon atom contains four single bonds to give a saturated compound. There is only one possible structure for each of methane, ethane and propane however two structures of butane are possible.

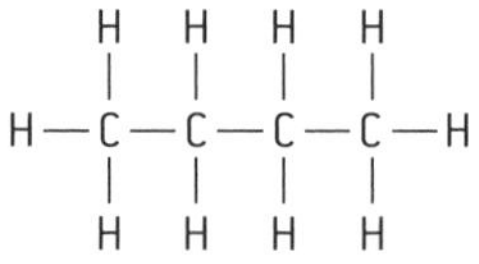

butane

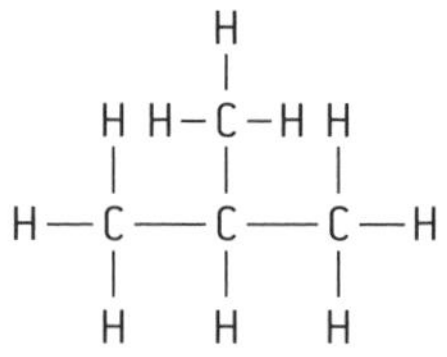

2-methylpropane

These are examples of structural isomers. **Structural isomers** have the same molecular formula but a different structural formula. They normally have similar chemical properties but their physical properties may be slightly different. There are three structural isomers of pentane.

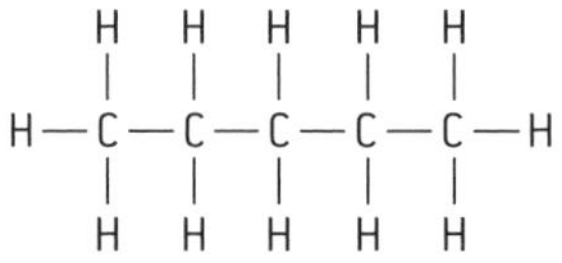

pentane

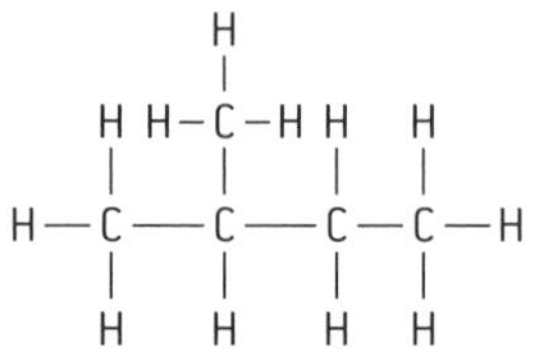

2-methylbutane

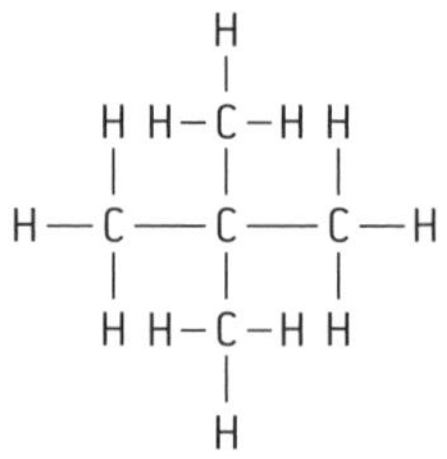

2, 2-dimethylpropane

Isomers of alkenes and alkynes

Alkenes and alkynes are unsaturated. The alkenes ethene and propene each have only one possible structure but butene has three structural isomers. Similarly the alkynes ethyne and propyne each have only one possible structure but butyne has two.

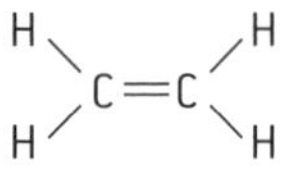

ethene

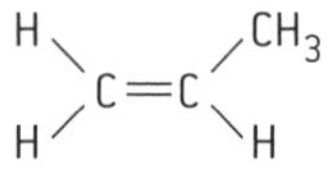

propene

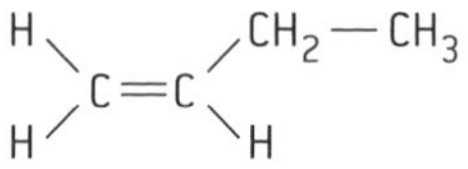

but-1-ene

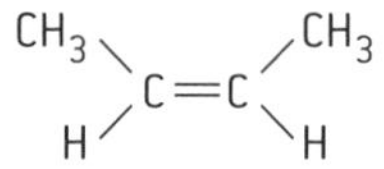

but-2-ene

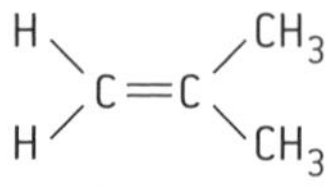

2-methylpropene

$H-C\equiv C-H$

ethyne

$H-C\equiv C-CH_3$

propyne

$H-C\equiv C-CH_2-CH_3$

but-1-yne

$CH_3-C\equiv C-CH_3$

but-2-yne

NAMING STRUCTURAL ISOMERS

The naming system explained on page 79 is known as the IUPAC (International Union of Pure and Applied Chemistry) system. The IUPAC names to distinguish between structural isomers of alkanes, alkenes, alkynes, alcohols, ethers, halogenoalkanes, aldehydes, ketones, esters and carboxylic acids, each containing up to six carbon atoms, are required.

For example, four different structural isomers with the molecular formula $C_6H_{12}O$ are shown.

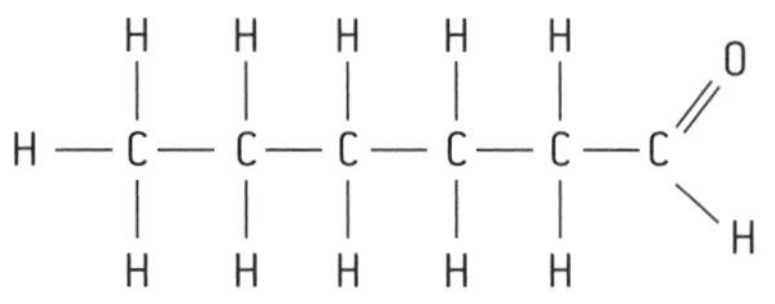

or $CH_3CH_2CH_2CH_2CH_2CHO$

hexanal

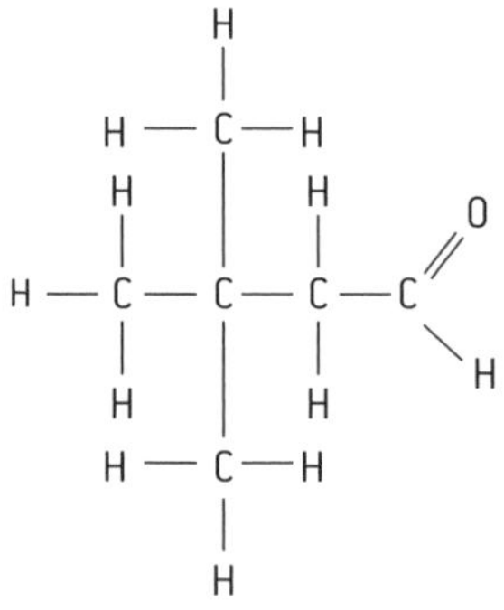

$CH_3C(CH_3)_2CH_2CHO$

3, 3-dimethylbutanal

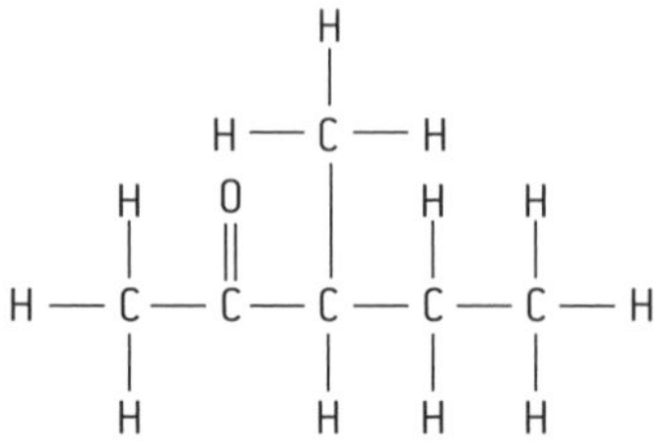

or $CH_3COCH(CH_3)CH_2CH_3$

3-methylpentan-2-one

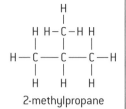

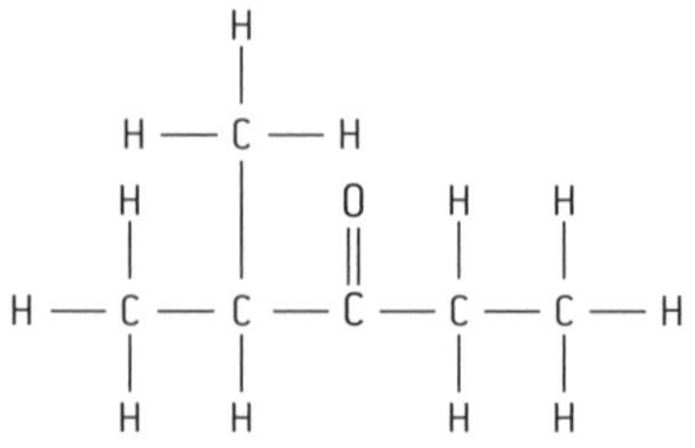

or $CH_3CH(CH_3)COCH_2CH_3$

2-methylpentan-3-one

3-D models of structural formulas

3-D REPRESENTATIONS

To help distinguish between isomers and because the chemical and physical properties of molecules depend upon their shape and the types of bonds and functional groups they contain it is helpful to be able to visualize the structures in three dimensions. Two-dimensional display diagrams of alkanes, for examples, wrongly suggest that the H–C–H bond angles are 90° instead of 109.5°. The traditional way in which chemists have approached this is to use a full 'wedge' to show an atom coming out of the two dimensional page and a dotted line to show it going behind the page.

The very best way to 'see' molecular structures in three dimensions is to use a molecular modelling kit and build the models for yourself. This can be particularly helpful when distinguishing between isomers and understanding bond angles. Two common ways to represent molecules in 3-D are 'ball and stick' models or space-fill models. You can also see virtual 3-D representations using apps on your laptop, smart phone or tablet, which are also able to show the molecules rotating. However it is easy to forget that these are still being displayed on a 2-D screen so are not truly three dimensional.

Some models of the structure of methane

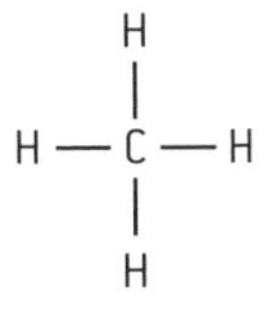

traditional 2-D 'display'

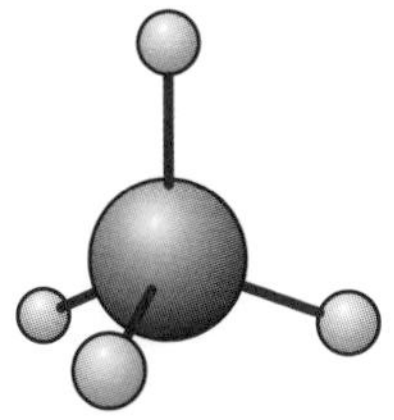

'ball and stick' model

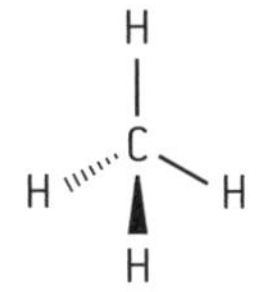

traditional 3-D representation

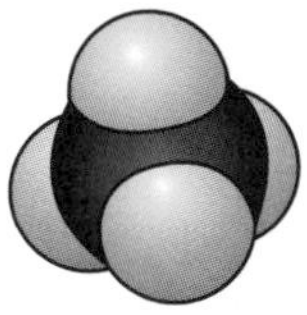

'space-fill' model

IDENTIFYING FUNCTIONAL GROUPS IN MOLECULES FROM VIRTUAL 3-D MODELS

As well as recognizing different functional groups within more complex molecules from their structural formulas you should also be able to recognize them from virtual 3-D representations.

The following three molecules, aspirin, paracetamol (acetaminophen) and ibuprofen, are all mild pain killers. As you can see they all contain a phenyl group, although this is a phenol in paracetamol as the phenyl group is directly bonded to an hydroxyl group. Aspirin and ibuprofen contain a carboxyl group (carboxylic acid), aspirin also contains an ester group and paracetamol contains a carboxamide group. If you look at past papers or in the IB data booklet you can find other molecules such as morphine and heroin where you can do a similar exercise.

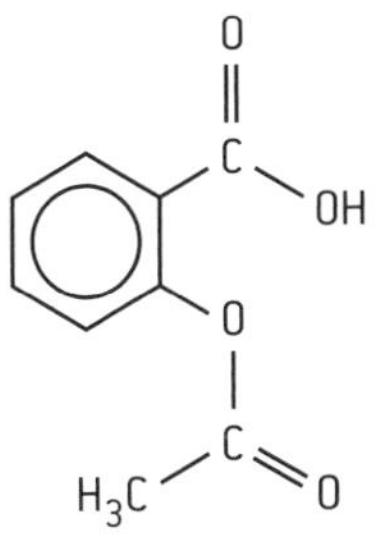

aspirin

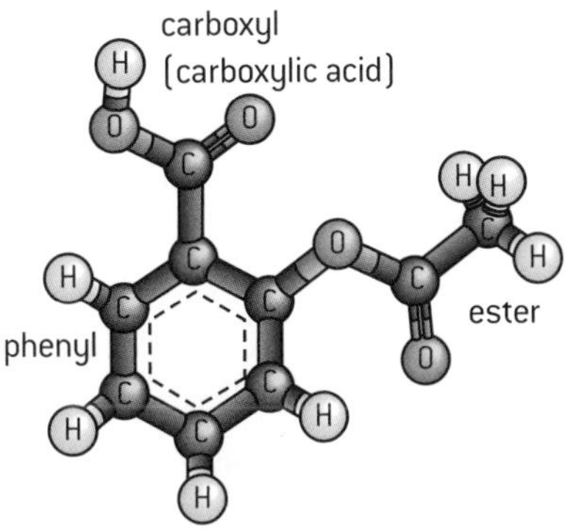

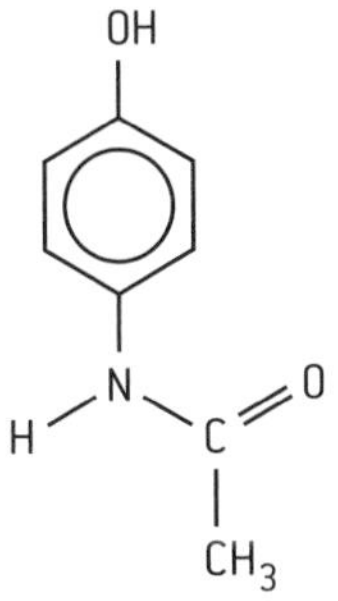

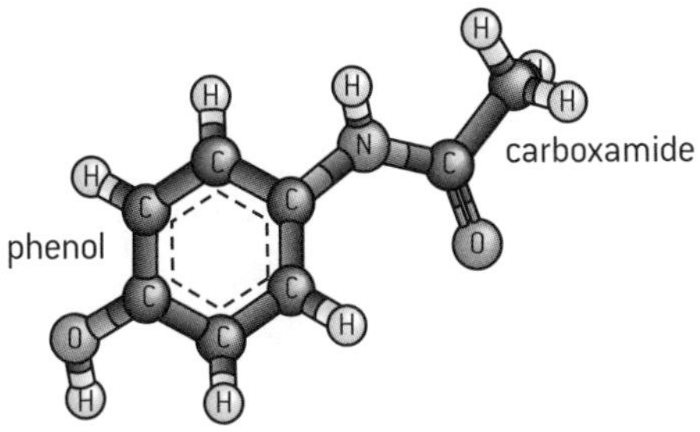

paracetamol (acetaminophen)

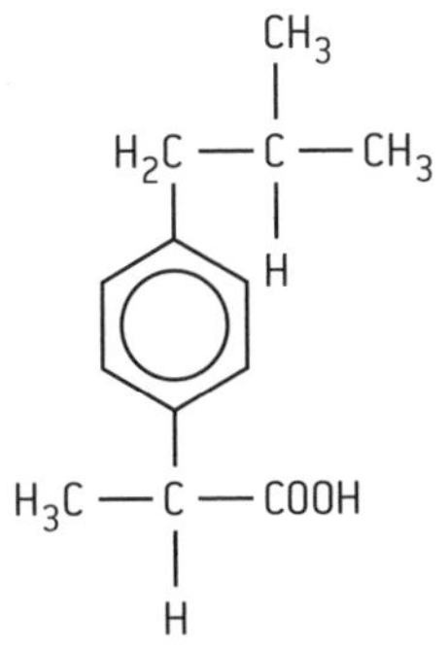

Ibuprofen

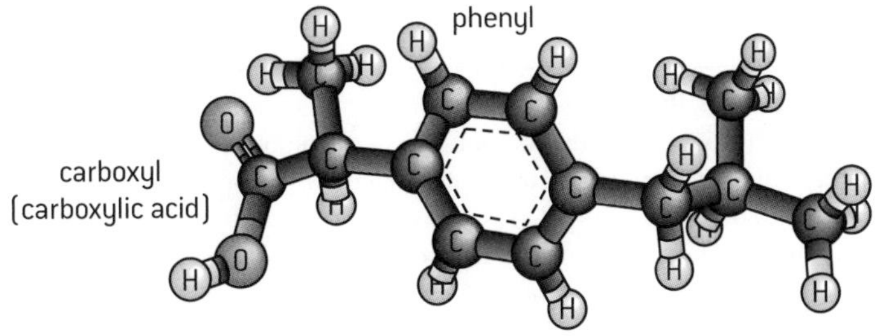

Properties of different homologous series

BOILING POINTS

As the carbon chain in the homologous series of alkanes increases the London dispersion forces of attraction increase and hence the boiling point also increases. A plot of boiling points against number of carbon atoms shows a sharp increase at first, as the percentage increase in mass is high, but as successive $-CH_2-$ groups are added the rate of increase in boiling point decreases.

When branching occurs the molecules become more spherical in shape, which reduces the contact surface area between them and lowers the boiling point.

Boiling points of the alkanes

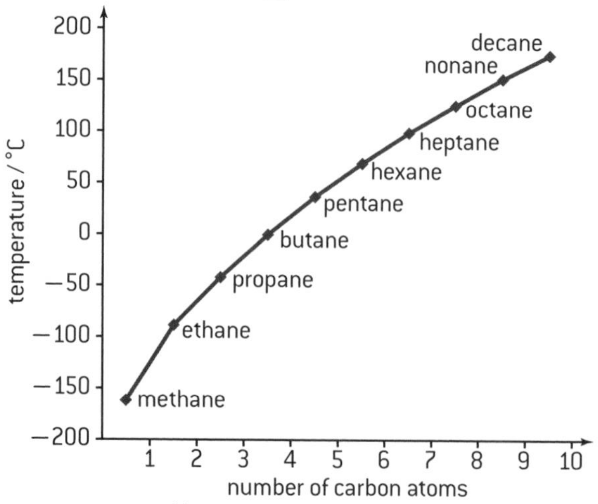

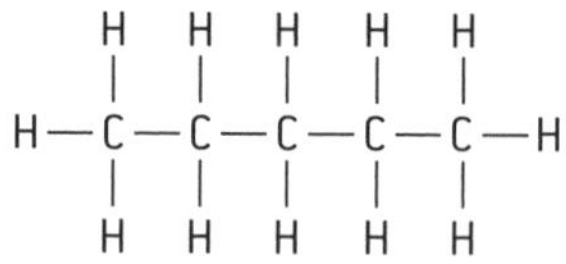

pentane
(b. pt 36.3 °C)

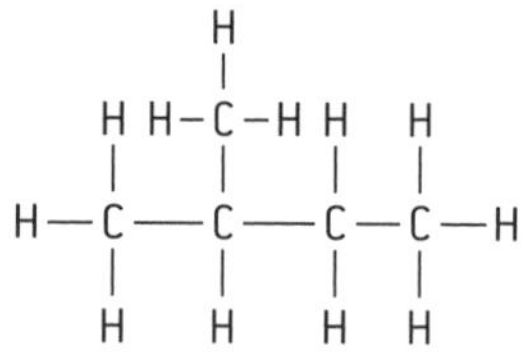

2-methylbutane
(b. pt 27.9 °C)

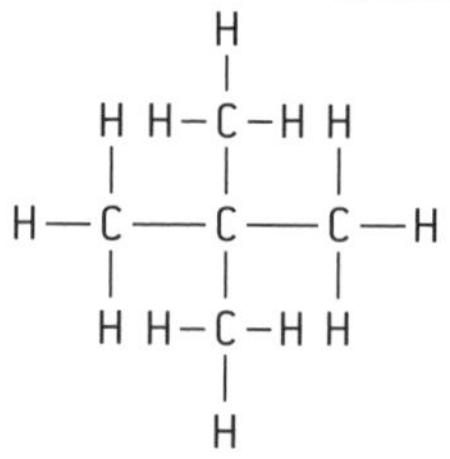

2, 2-dimethylpropane
(b. pt 9.5 °C)

Other homologous series show similar trends but the actual temperatures at which the compounds boil will depend on the types of attractive forces between the molecules. The volatility of the compounds also follows the same pattern. The lower members of the alkanes are all gases as the attractive forces are weak and the next few members are volatile liquids. Methanol, the first member of the alcohols is a liquid at room temperature, due to the presence of hydrogen bonding. Methanol is classed as volatile as its boiling point is 64.5 °C but when there are four or more carbon atoms in the chain the boiling points exceed 100 °C and the higher alcohols have low volatility.

Compound	Formula	Mr	Class of compound	Strongest type of attraction	B. pt / °C
butane	C_4H_{10}	58	alkane	London dispersion forces	−0.5
but-1-ene	C_4H_8	56	alkene	London dispersion forces	−6.5
but-1-yne	C_4H_6	54	alkyne	London dispersion forces	8.1
methyl methanoate	$HCOOCH_3$	60	ester	dipole–dipole	31.5
propanal	CH_3CH_2CHO	58	aldehyde	dipole–dipole	48.8
propanone	CH_3COCH_3	58	ketone	dipole–dipole	56.2
aminopropane	$CH_3CH_2CH_2NH_2$	59	amine	hydrogen bonding	48.6
propan-1-ol	$CH_3CH_2CH_2OH$	60	alcohol	hydrogen bonding	97.2
ethanoic acid	CH_3COOH	60	carboxylic acid	hydrogen bonding	118

SOLUBILITY IN WATER

Whether or not an organic compound will be soluble in water depends on the polarity of the functional group and on the chain length. The lower members of alcohols, amines, aldehydes, ketones and carboxylic acids are all water soluble. However, as the length of the non-polar hydrocarbon chain increases the solubility in water decreases. For example, ethanol and water mix in all proportions, but hexan-1-ol is only slightly soluble in water. Compounds with non-polar functional groups, such as alkanes and alkenes, do not dissolve in water but are soluble in other non-polar solvents. Propan-1-ol is a good solvent because it contains both polar and non-polar groups and can to some extent dissolve both polar and non-polar substances.

Alkanes

LOW REACTIVITY OF ALKANES

Because of the relatively strong C–C and C–H bonds and because they have low polarity, alkanes tend to be quite unreactive. They only readily undergo combustion reactions with oxygen and substitution reactions with halogens in ultraviolet light.

COMBUSTION

Alkanes are hydrocarbons – compounds that contain carbon and hydrogen only. All hydrocarbons burn in a plentiful supply of oxygen to give carbon dioxide and water. The general equation for the combustion of any hydrocarbon is:

$$C_xH_y + (x + \frac{y}{4})O_2 \rightarrow xCO_2 + \frac{y}{2}H_2O$$

Although the C–C and C–H bonds are strong the C=O and O–H bonds in the products are even stronger so the reaction is very exothermic and much use is made of the alkanes as fuels.

e.g. natural gas (methane)

$$CH_4(g) + 2O_2(g) \rightarrow CO_2(g) + 2H_2O(l) \quad \Delta H^{\ominus} = -890.4\ \text{kJ mol}^{-1}$$

gasoline (petrol)

$$C_8H_{18}(l) + 12\tfrac{1}{2}O_2(g) \rightarrow 8CO_2(g) + 9H_2O(l)$$

$$\Delta H^{\ominus} = -5512\ \text{kJ mol}^{-1}$$

If there is an insufficient supply of oxygen then incomplete combustion occurs and carbon monoxide and carbon are also produced as products.

SUBSTITUTION REACTIONS

Alkanes can react with chlorine (or other halogens) in the presence of ultraviolet light to form hydrogen chloride and a substituted alkane, e.g. methane can react with chlorine to form chloromethane and ethane can react with bromine to form bromoethane.

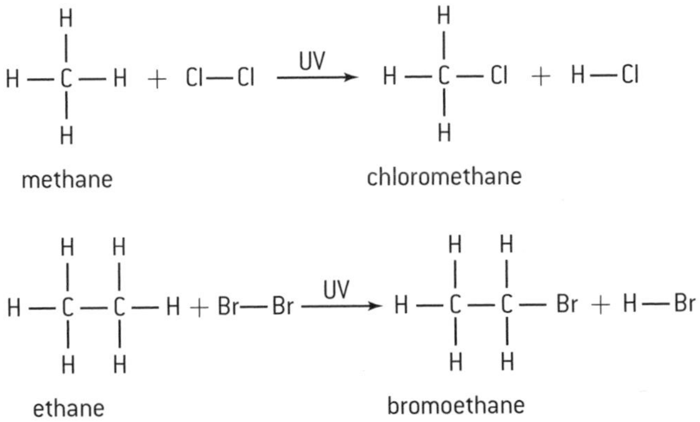

MECHANISM OF CHLORINATION OF METHANE

The mechanism of an organic reaction describes the individual steps. When chemical bonds break they may break **heterolytically** or **homolytically**. In heterolytic fission both of the shared electrons go to one of the atoms resulting in a negative and a positive ion. In homolytic fission each of the two atoms forming the bond retains one of the shared electrons resulting in the formation of two **free radicals**. The bond between two halogen atoms is weaker than the C–H or C–C bond in methane and can break homolytically in the presence of ultraviolet light.

$$Cl_2 \rightarrow Cl^\bullet + Cl^\bullet$$

This stage of the mechanism is called **initiation**.

Free radicals contain an unpaired electron and are highly reactive. When the chlorine free radicals come into contact with a methane molecule they combine with a hydrogen atom to produce hydrogen chloride and a methyl radical.

$$H_3C\text{–}H + Cl^\bullet \rightarrow H_3C^\bullet + HCl$$

Since a new radical is produced this stage of the mechanism is called **propagation**. The methyl free radical is also extremely reactive and reacts with a chlorine molecule to form the product and regenerate another chlorine radical. This is a further propagation step and enables a chain reaction to occur as the process can repeat itself.

$$CH_3^\bullet + Cl_2 \rightarrow CH_3\text{–}Cl + Cl^\bullet$$

In theory a single chlorine radical may cause up to 10 000 molecules of chloromethane to be formed. **Termination** occurs when two radicals react together.

$$\left.\begin{array}{l} Cl^\bullet + Cl^\bullet \rightarrow Cl_2 \\ CH_3^\bullet + Cl^\bullet \rightarrow CH_3Cl \\ CH_3^\bullet + CH_3^\bullet \rightarrow C_2H_6 \end{array}\right\} \text{termination}$$

Further substitution can occur when chlorine radicals react with the substituted products. For example:

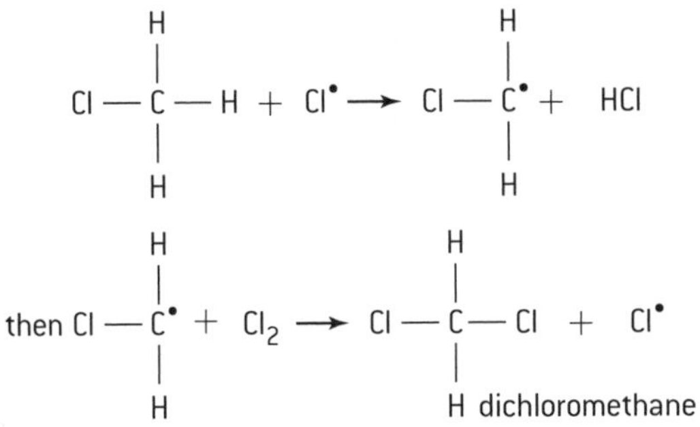

The substitution can continue even further to produce trichloromethane and then tetrachloromethane.

The overall mechanism is called **free radical substitution**. [Note that in this mechanism hydrogen radicals $H^\bullet$ are not formed.]

Alkenes

ADDITION REACTIONS

The bond enthalpy of the C=C double bond in alkenes has a value of 612 kJ mol^{-1}. This is less than twice the average value of 348kJ mol^{-1} for the C–C single bond and accounts for the relative reactivity of alkenes compared to alkanes. The most important reactions of alkenes are addition reactions. Reactive molecules are able to add across the double bond. The double bond is said to be **unsaturated** and the product, in which each carbon atom is bonded by four single bonds, is said to be **saturated**.

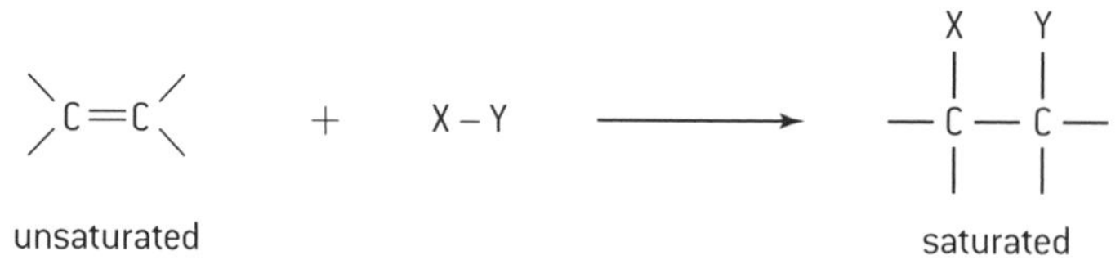

$$\text{C}=\text{C} \quad + \quad X-Y \quad \longrightarrow \quad -\overset{X}{\underset{|}{C}}-\overset{Y}{\underset{|}{C}}-$$

unsaturated saturated

Addition reactions include the addition of hydrogen, bromine, hydrogen halides and water.

(alkane)

H$_2$

bromoethane
(halogenoalkane)

◄── HBr

Br$_2$ ──►

1,2-dibromoethane
(dihalogenoalkane)

H$_2$O
(H$_2$SO$_4$ catalyst)

(alcohol)

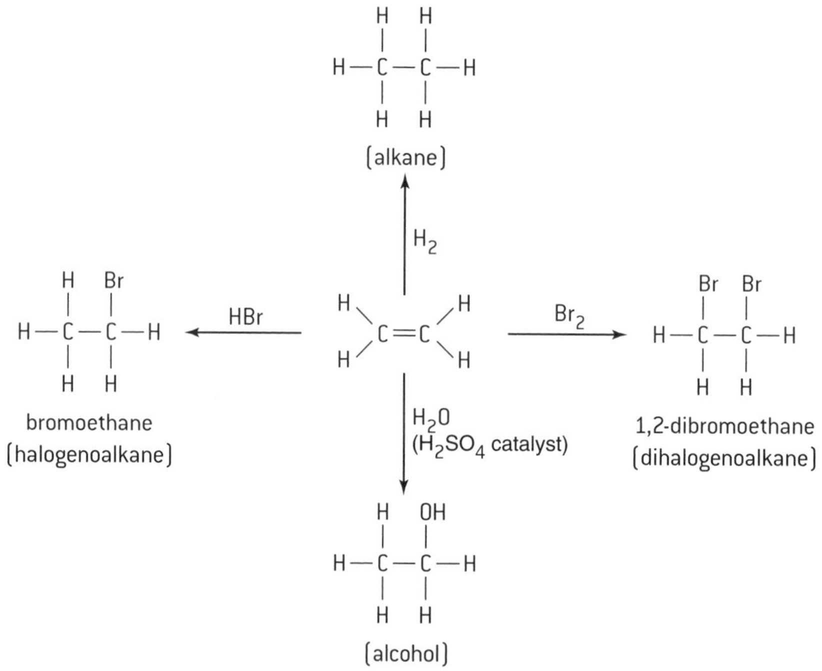

USES OF ADDITION REACTIONS

1. **Bromination**

Pure bromine is a red liquid but it has a distinctive yellow/orange colour in solution. When a solution of bromine is added to an alkene the product is colourless. This decolourization of bromine solution provides a useful test to indicate the presence of an alkene group.

2. **Hydration**

Ethene is an important product formed during the cracking of oil. Although ethanol can be made from the fermentation of starch and sugars, much industrial ethanol is formed from the addition of steam to ethene.

3. **Hydrogenation**

The addition of hydrogen to unsaturated vegetable oils is used industrially to make margarine. Hydrogenation reduces the number of double bonds in the polyunsaturated vegetable oils present in the margarine, which causes it to become a solid at room temperature.

ADDITION POLYMERIZATION

Under certain conditions ethene can also undergo addition reactions with itself to form a long chain polymer containing many thousands (typically 40 000 to 800 000) of carbon atoms.

ethene ──► poly(ethene)
(also known as polythene)

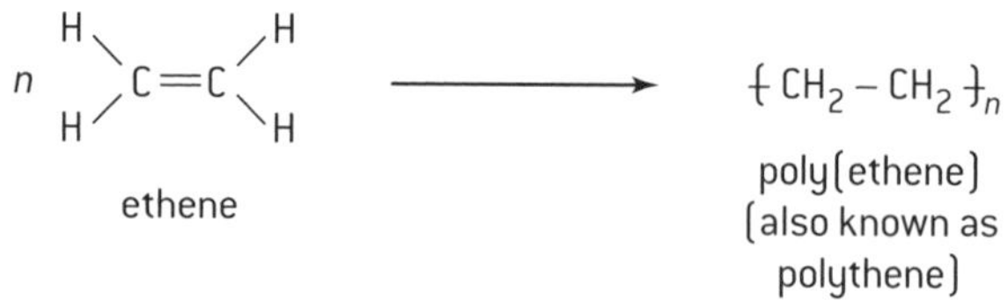

These addition reactions can be extended to other substituted alkenes to give a wide variety of different addition polymers.

e.g.

chloroethene ──► $+\text{CH}_2-\text{CHCl}\,+_n$
poly(chloroethene)
(also known as polyvinylchloride, PVC)

tetrafluoroethene ──► $+\text{CF}_2-\text{CF}_2\,+_n$
poly(tetrafluoroethene), PTFE
(also known as Teflon or 'non-stick')

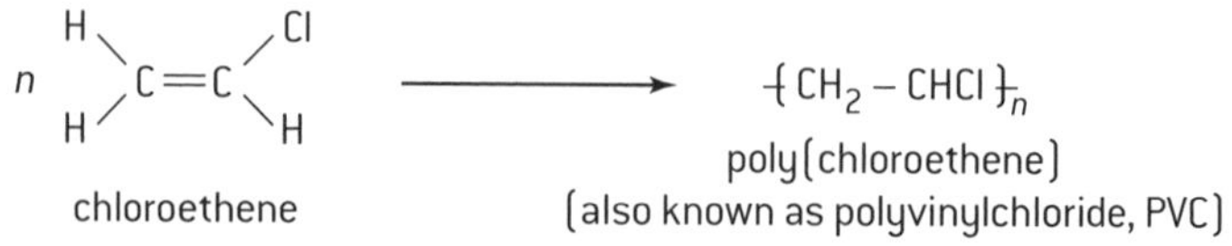

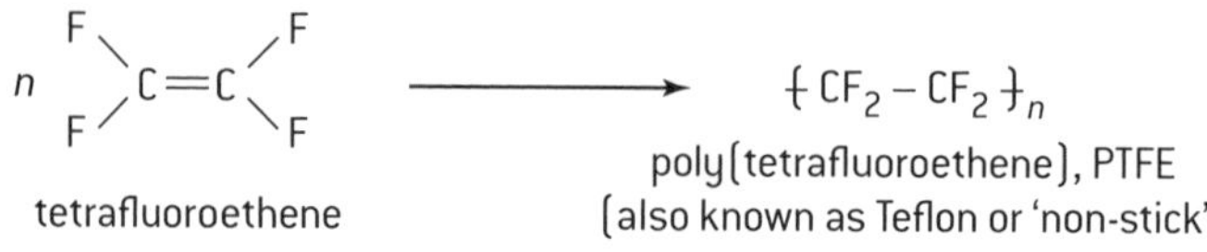

Alcohols

COMBUSTION

Ethanol is used both as a solvent and as a fuel. It combusts completely in a plentiful supply of oxygen to give carbon dioxide and water.

$C_2H_5OH(l) + 3O_2(g) \rightarrow 2CO_2(g) + 3H_2O(l) \quad \Delta H^{\ominus} = -1371 \text{ kJ mol}^{-1}$

Ethanol is already partially oxidized so it releases less energy than burning an alkane of comparable mass. However, it can be obtained by the fermentation of biomass so in some countries it is mixed with petrol to produce 'gasohol' which decreases the dependence on crude oil.

The general equation for an alcohol combusting completely in oxygen is:

$C_nH_{(2n+1)}OH + (2n-1)O_2 \rightarrow nCO_2 + (n+1)H_2O$

OXIDATION OF ETHANOL

Ethanol can be readily oxidized by warming with an acidified solution of potassium dichromate(VI). During the process the orange dichromate(VI) ion $Cr_2O_7^{2-}$ is reduced from an oxidation state of +6 to the green Cr^{3+} ion. Use is made of this in simple breathalyser tests, where a motorist who is suspected of having exceeded the alcohol limit blows into a bag containing crystals of potassium dichromate(VI).

Ethanol is initially oxidized to ethanal.

$3CH_3CH_2OH(aq) + Cr_2O_7^{2-}(aq) + 8H^+(aq) \rightarrow 3CH_3CHO(aq) + 2Cr^{3+}(aq) + 7H_2O(l)$

The ethanal is then oxidized further to ethanoic acid.

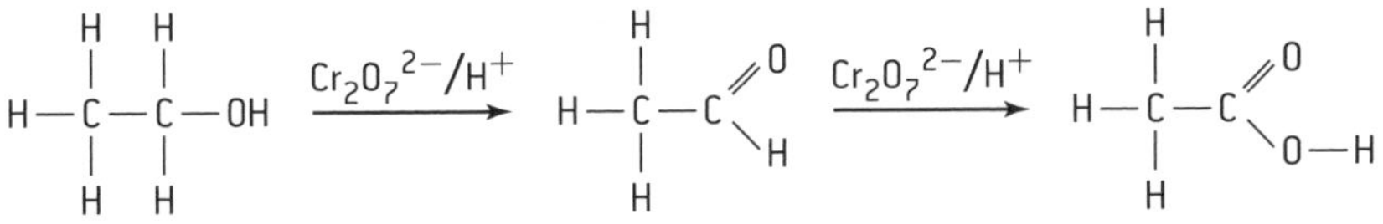

ethanol ('wine') | ethanal | ethanoic acid ('vinegar')

$3CH_3CHO(aq) + Cr_2O_7^{2-}(aq) + 8H^+(aq) \rightarrow 3CH_3COOH(aq) + 2Cr^{3+}(aq) + 4H_2O(l)$

Unlike ethanol (b. pt 78.5 °C) and ethanoic acid (b. pt 118 °C) ethanal (b. pt 20.8 °C) does not have hydrogen bonding between its molecules, and so has a lower boiling point. To stop the reaction at the aldehyde stage the ethanal can be distilled from the reaction mixture as soon as it is formed. If the complete oxidation to ethanoic acid is required, then the mixture can be heated under reflux so that none of the ethanal can escape.

OXIDATION OF ALCOHOLS

Ethanol is a primary alcohol, that is the carbon atom bonded to the –OH group is bonded to two hydrogen atoms and one alkyl group. The oxidation reactions of alcohols can be used to distinguish between primary, secondary and tertiary alcohols.

All **primary alcohols** are oxidized by acidified potassium dichromate(VI), first to aldehydes then to carboxylic acids.

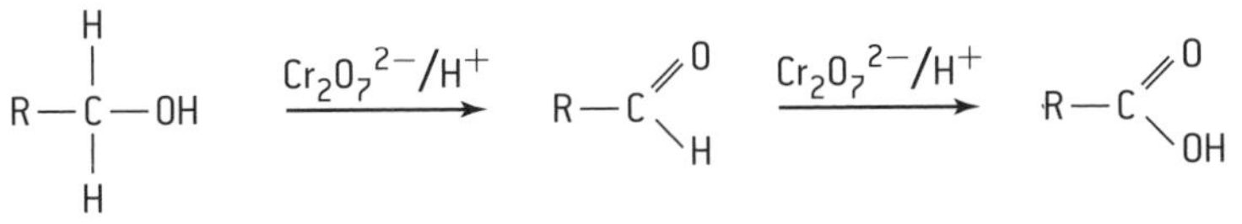

primary alcohol | aldehyde | carboxylic acid

Secondary alcohols are oxidized to ketones, which cannot undergo further oxidation.

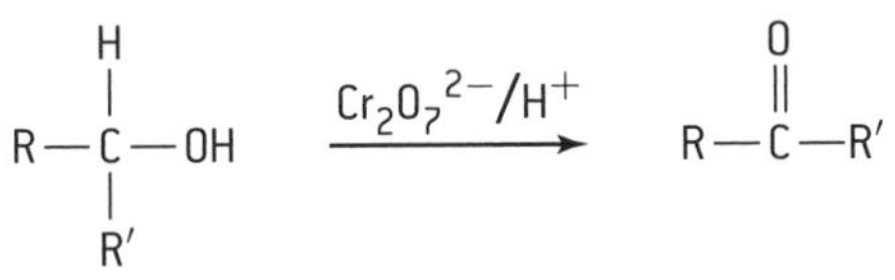

secondary alcohol | ketone

Tertiary alcohols cannot be oxidized by acidified dichromate(VI) ions as they have no hydrogen atoms attached directly to the carbon atom containing the –OH group. It is not true to say that tertiary alcohols can never be oxidized, as they burn readily, but when this happens the carbon chain is destroyed.

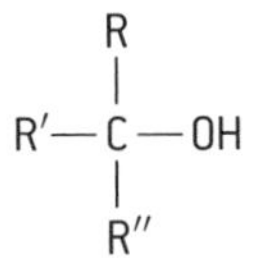

tertiary alcohol

Substitution and condensation reactions

SUBSTITUTION REACTIONS OF HALOGENOALKANES

Because of the greater electronegativity of the halogen atom compared with the carbon atom halogenoalkanes have a polar bond. Reagents that have a non-bonding pair of electrons are attracted to the electron-deficient carbon atom in halogenoalkanes and a substitution reaction occurs. Such reagents are called nucleophiles.

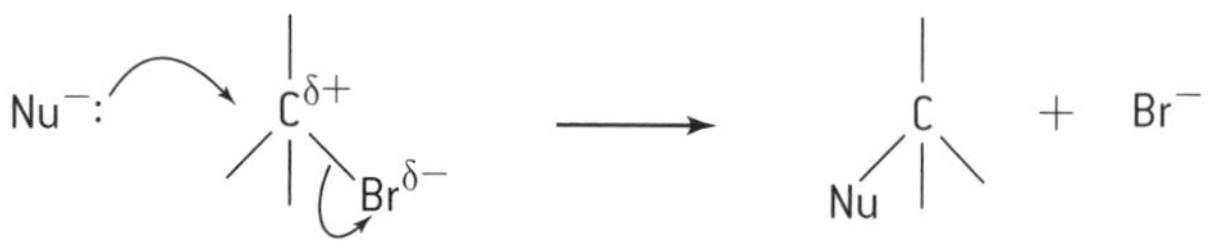

A double-headed curly arrow represents the movement of a pair of electrons. It shows where they come from and where they move to.

These reactions are useful in organic synthesis as a wide variety of different compounds can be made by varying the nucleophile. For example, with warm dilute sodium hydroxide solution the product is an alcohol, with ammonia the product is an amine and with cyanide ions, CN^-, the product is a nitrile. This last reaction is particularly useful as it provides a way to increase the number of atoms in the carbon chain.

Reaction of bromoethane with warm dilute sodium hydroxide solution:

$$C_2H_5Br(l) + NaOH(aq) \rightarrow C_2H_5OH(aq) + NaBr(aq)$$

SUBSTITUTION REACTIONS OF BENZENE

The extra stability provided by the delocalization of the electrons in the benzene ring means that benzene, unlike simple alkenes, does not readily undergo addition reactions. However benzene does undergo substitution reactions. Benzene has a high electron density so reacts with electrophiles. Electrophiles are electron-deficient species often formed *in situ* that can accept electron pairs. Two examples of electrophilic substitution reactions of benzene are the reaction with chlorine in the presence of aluminium chloride (the electrophile is Cl^+) to form chlorobenzene and the reaction with nitric acid in the presence of sulfuric acid (the electrophile is NO_2^+) to form nitrobenzene.

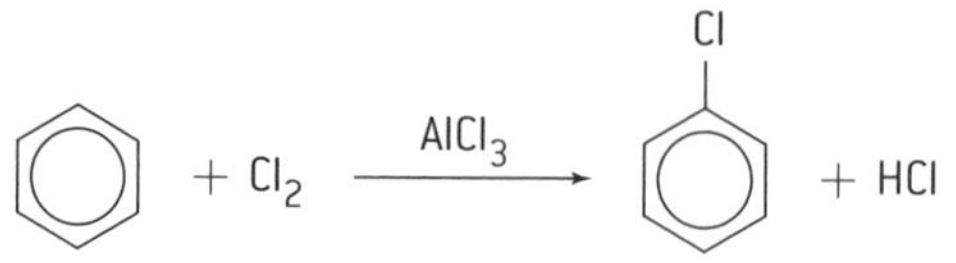

chlorobenzene

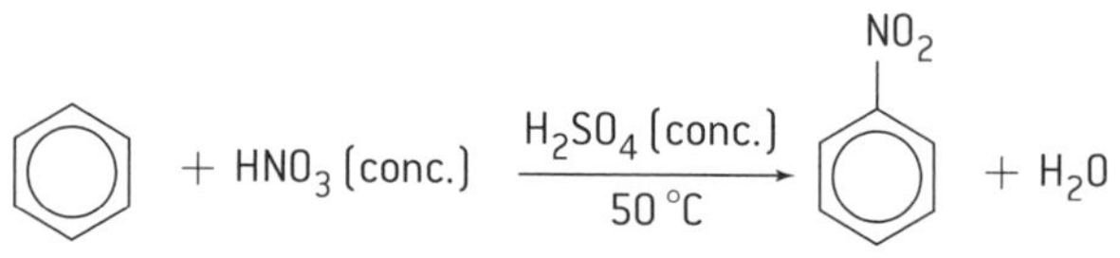

nitrobenzene

CONDENSATION REACTION BETWEEN AN ALCOHOL AND A CARBOXYLIC ACID

Alcohols can undergo a nucleophilic substitution reaction with carboxylic acids but this is more normally called esterification and is an example of a condensation reaction. A condensation reaction involves the reaction between two molecules to produce a larger molecule with the elimination of a small molecule such as water or hydrogen chloride.

Alcohols react with carboxylic acids in the presence of a small amount of concentrated sulfuric acid, which acts as a catalyst, to form an ester.

e.g.

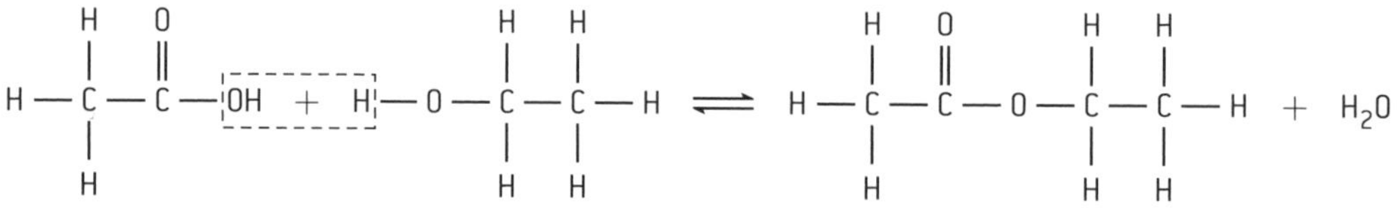

ethanoic acid　ethanol　ethyl ethanoate

Most esters have a distinctive, pleasant fruity smell and are used both as natural and artificial flavouring agents in food. For example, ethyl methanoate $HCOOCH_2CH_3$ is added to chocolate to give it the characteristic flavour of 'rum truffle'. Esters are also used as solvents in perfumes and as plasticizers (substances used to modify the properties of polymers by making them more flexible.)

Many drugs contain one or more ester groups e.g. aspirin and heroin.

Fats and oils are natural triesters formed from the reaction between glycerol (propane-1,2-3-triol) and three fatty acids. In a process known as transesterification they can react with alcohols in the presence of sodium hydroxide (which acts as a catalyst) to form alkyl esters, which can be used as biofuel.

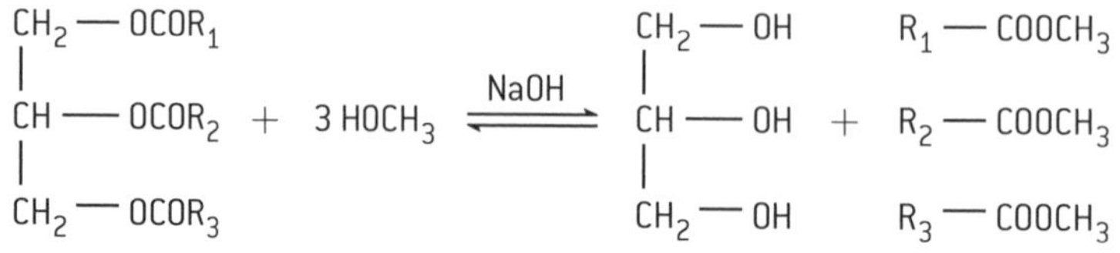

triglyceride (fat or oil)　methanol　glycerol　methyl esters (biofuel)

Nucleophilic substitution

MECHANISM OF NUCLEOPHILIC SUBSTITUTION

Primary halogenoalkanes (one alkyl group attached to the carbon atom bonded to the halogen)

For example, the reaction between bromoethane and warm dilute sodium hydroxide solution.

$C_2H_5Br + OH^- \rightarrow C_2H_5OH + Br^-$

The experimentally determined rate expression is:

rate $= k[C_2H_5Br][OH^-]$

The proposed mechanism involves the formation of a transition state which involves both of the reactants.

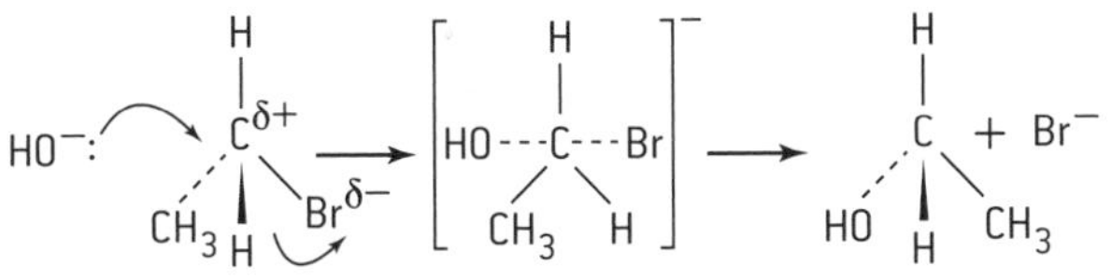

Because the molecularity of this single-step mechanism is two it is known as an S_N2 mechanism (bimolecular nucleophilic substitution).

Note that the S_N2 mechanism is stereospecific with an inversion of configuration at the central carbon atom.

Tertiary halogenoalkanes (three alkyl groups attached to the carbon atom bonded to the halogen)

For example, the reaction between 2-bromo-2-methylpropane and warm dilute sodium hydroxide solution.

$$C(CH_3)_3Br + OH^- \longrightarrow C(CH_3)_3OH + Br^-$$

The experimentally determined rate expression for this reaction is: rate $= k[C(CH_3)_3Br]$

A two-step mechanism is proposed that is consistent with this rate expression.

$$C(CH_3)_3Br \xrightarrow{\text{slow}} C(CH_3)_3^+ + Br^-$$

$$C(CH_3)_3^+ + OH^- \xrightarrow{\text{fast}} C(CH_3)_3OH$$

In this reaction it is the first step, the heterolytic fission of the C–Br bond, that is the rate-determining step. The molecularity of this step is one and the mechanism is known as S_N1 (unimolecular nucleophilic substitution).

The mechanism for the hydrolysis of secondary halogenoalkanes (e.g 2-bromopropane $CH_3CHBrCH_3$) is more complicated as they can proceed by either S_N1 or S_N2 pathways or a combination of both.

CHOICE OF SOLVENT

Whether or not the reaction proceeds by an S_N1 or S_N2 mechanism also depends upon the solvent. Protic solvents which are polar, such as water or ethanol, favour the S_N1 mechanism as they support the breakdown of halogenoalkanes into carbocations and halide ions whereas aprotic solvents which are less polar such as ethoxyethane, $C_2H_5OC_2H_5$, favour an S_N2 mechanism involving a transition state.

FACTORS AFFECTING THE RATE OF NUCLEOPHILIC SUBSTITUTION

The nature of the nucleophile

The effectiveness of a nucleophile depends on its electron density. Anions tend to be more reactive than the corresponding neutral species. This explains why the hydroxide ion is a much better nucleophile than water.

The nature of the halogen

For both S_N1 and S_N2 reactions iodoalkanes react faster than bromoalkanes, which in turn react faster than chloroalkanes. This is due to the relative bond enthalpies as the C–I bond is much weaker than the C–Cl bond and therefore breaks more readily.

	Bond enthalpy / kJ mol^{-1}
C–I	228
C–Br	285
C–Cl	324

The nature of the halogenoalakane

Tertiary halogenoalkanes react faster than secondary halogenoalkanes which in turn react faster than primary halogenoalkanes. The S_N1 route which involves the formation of an intermediate carbocation is faster than the S_N2 route which involves a transition state with a relatively high activation energy.

Electrophilic addition reactions (1)

ELECTROPHILIC ADDITION TO SYMMETRIC ALKENES

Ethene readily undergoes addition reactions. With hydrogen bromide it forms bromoethane.

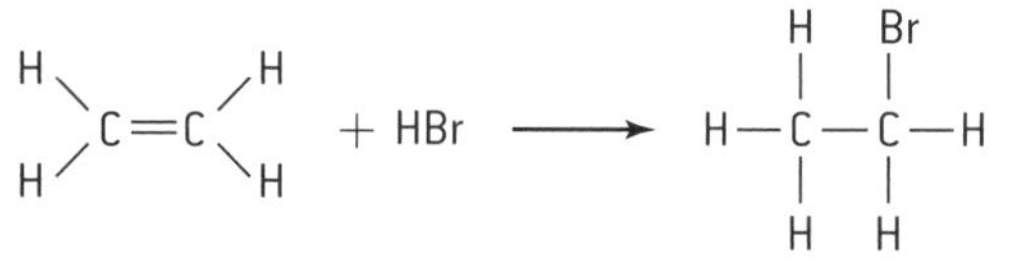

The reaction can occur in the dark which suggests that a free radical mechanism is not involved. The double bond in the ethene molecule has a region of high electron density above and below the plane of the molecule. Hydrogen bromide is a polar molecule due to the greater electronegativity of bromine compared with hydrogen. The hydrogen atom (which contains a charge of δ+) from the H–Br is attracted to the double bond and the H–Br bond breaks, forming a bromide ion. At the same time the hydrogen atom adds to one of the ethene carbon atoms leaving the other carbon atom with a positive charge. A carbon atom with a positive charge is known as a **carbocation**. The carbocation then combines with the bromide ion to form bromoethane. Because the hydrogen bromide molecule is attracted to a region of electron density it is described as an **electrophile** and the mechanism is described as **electrophilic addition**.

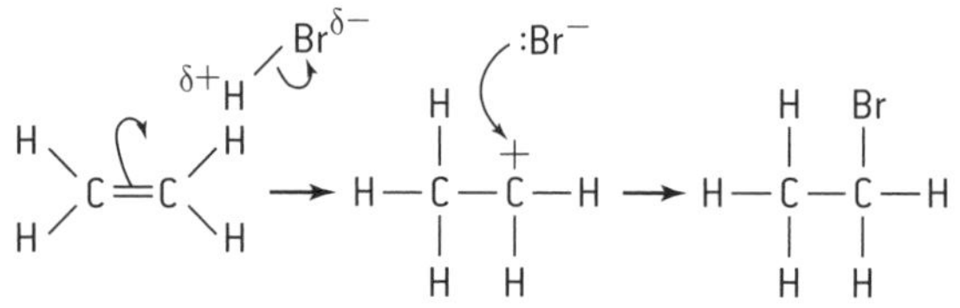

Electrophilic addition also takes place when bromine adds to ethene in a non-polar solvent to give 1,2-dibromoethane. Bromine itself is non-polar but as it approaches the double bond of the ethene an induced dipole is formed by the electron cloud.

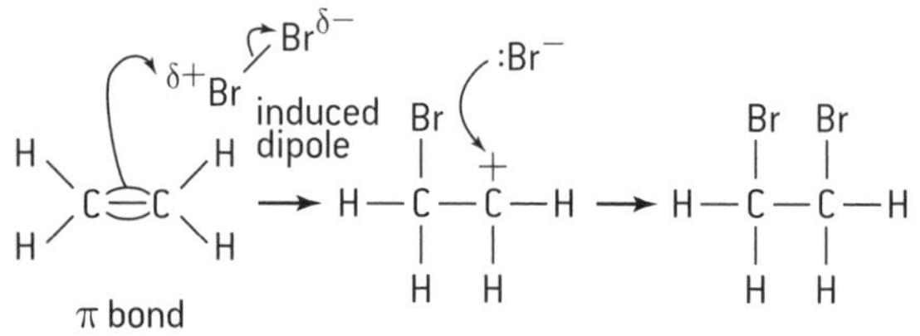

Evidence for this mechanism is that when bromine water is reacted with ethene the main product is 2-bromoethanol not 1,2-dibromoethane. This suggests that hydroxide ions from the water add to the carbocation in preference to bromide ions.

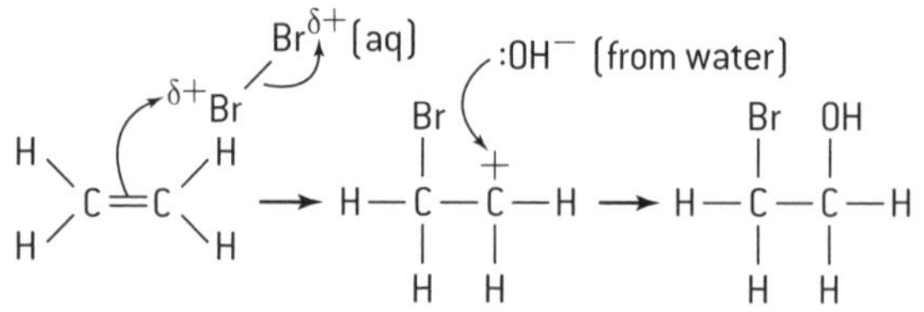

SYMMETRIC AND ASYMMETRIC ALKENES

Asymmetric alkenes contain different groups attached to the carbon atoms of the C=C bond.

Symmetric alkenes	Asymmetric alkenes
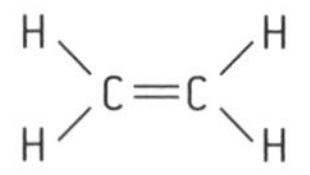 ethene	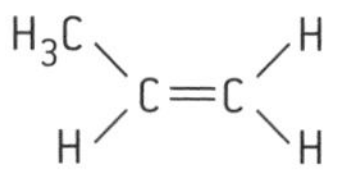 propene
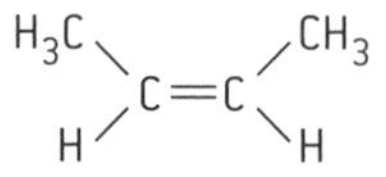 but-2-ene	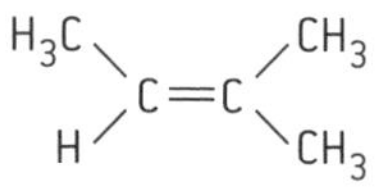 2-methylbut-2-ene
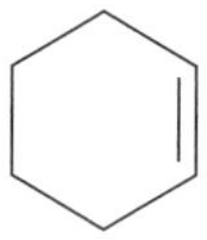 cyclohexene	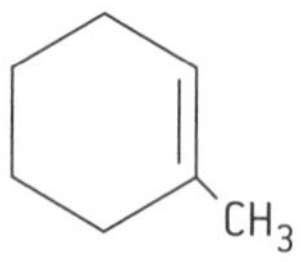 1-methylcyclohexene

CURLY ARROWS AND 'FISH HOOKS'

Curly arrows are used to show the movement of a pair of electrons as for example in the electrophilic addition reaction of hydrogen bromide with ethene shown in the opposite box. The 'tail' of the arrows shows the origin of the electron pair and the 'head' where the electron pair ends up. In S_N2 reactions such as the nucleophilic substitution reaction of bromoethane by hydroxide ions it is best to use three-dimensional representations to clarify the stereochemistry as the arrow head should go to the opposite side of the δ^+ carbon atom formed due to the polar C–Br bond (see page 89).

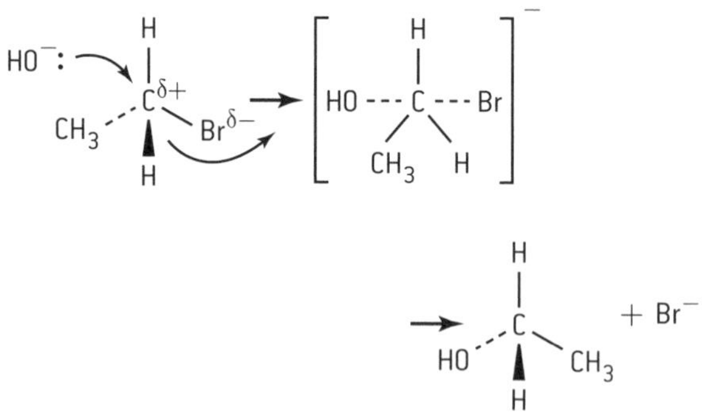

When single electrons are transferred as in the free radical substitution reactions of alkanes with halogens in ultraviolet light as described on page 85 then single-headed arrows known as 'fish hooks' are used. Thus the initiation step which involves the homolytic fission of the chlorine to chlorine bond by ultraviolet light can be represented as:

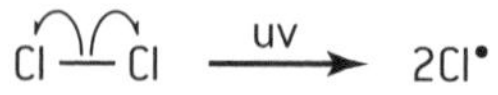

Electrophilic addition reactions (2)

MARKOVNIKOV'S RULE

When hydrogen halides add to asymmetric alkenes two products are possible depending upon which carbon atom the hydrogen atom bonds to. For example, the addition of hydrogen bromide to propene could produce 1-bromopropane or 2-bromopropane.

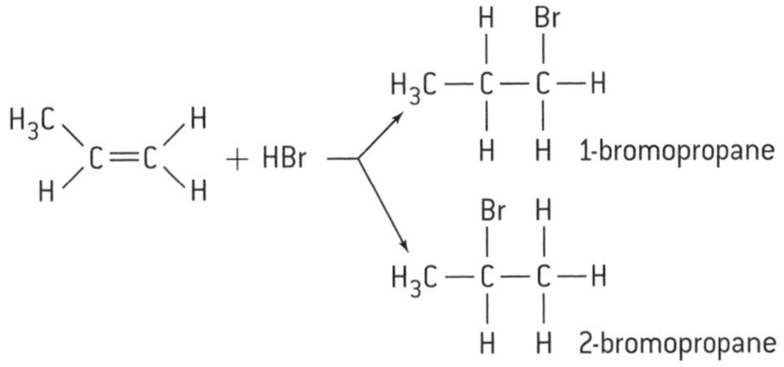

Markovnikov's rule enables you to predict which isomer will be the major product. It states that the hydrogen atom will add to the carbon atom that already contains the most hydrogen atoms bonded to it. Thus in the above example 2-bromopropane will be the major product.

EXPLANATION OF MARKOVNIKOV'S RULE

Markovnikov's rule enables the product to be predicted but it does not explain why. It can be explained by considering the nature of the possible intermediate carbocations formed during the reaction.

When hydrogen ions react with propene two different carbocation intermediates can be formed.

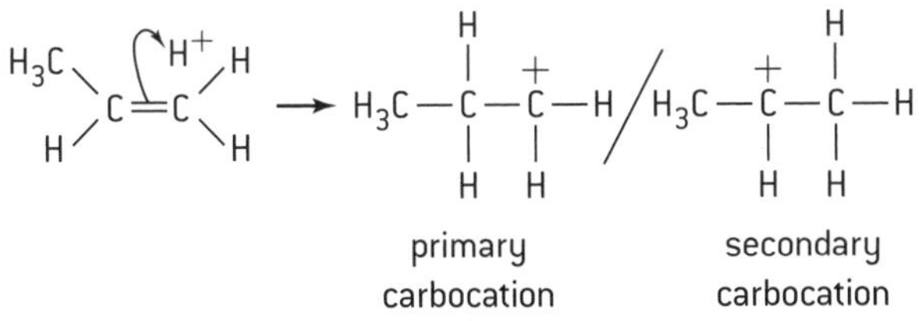

The first one has the general formula RCH_2^+ and is known as a **primary carbocation**. The second one has two R– groups attached to the positive carbon ion R_2CH^+ and is known as a **secondary carbocation**. A **tertiary carbocation** has the general formula R_3C^+. The R-groups (alkyl groups) tend to push electrons towards the carbon atom they are attached to which tends to stabilize the positive charge on the carbocation. This is known as a **positive inductive effect**. This effect will be greatest with tertiary carbocations and smallest with primary carbocations.

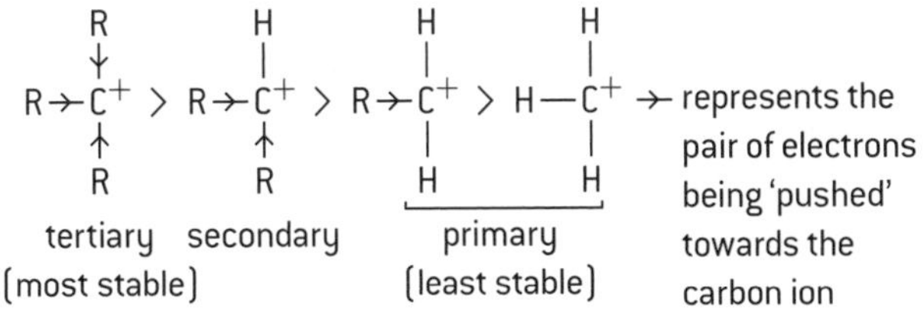

Thus in the above reaction the secondary carbocation will be preferred as it is more stable than the primary carbocation. This secondary carbocation intermediate leads to the major product, 2-bromopropane.

Understanding this mechanism enables you to predict what will happen when an interhalogen adds to an asymmetric alkene even though no hydrogen atoms are involved. Consider the reaction of iodine chloride ICl with but-1-ene. Since iodine is less electronegative than chlorine the iodine atom will act as the electrophile and add first to the alkene.

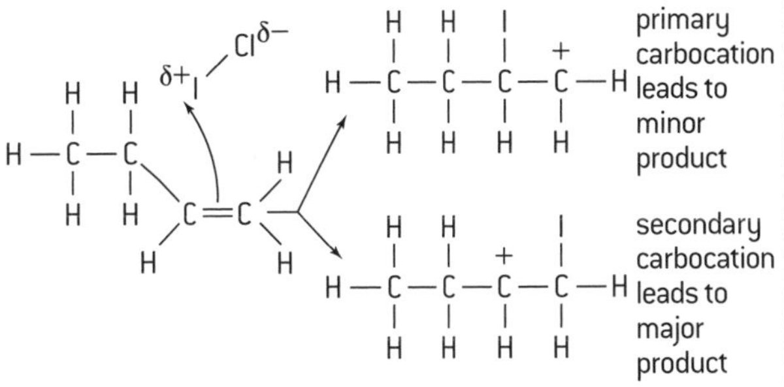

The major product will thus be 2-chloro-1-iodobutane.

HL Electrophilic substitution reactions and reduction reactions

NITRATION OF BENZENE

Benzene reacts with a mixture of concentrated nitric acid and concentrated sulfuric acid when warmed at 50 °C to give nitrobenzene and water. Note that the temperature should not be raised above 50 °C otherwise further nitration to dinitrobenzene will occur.

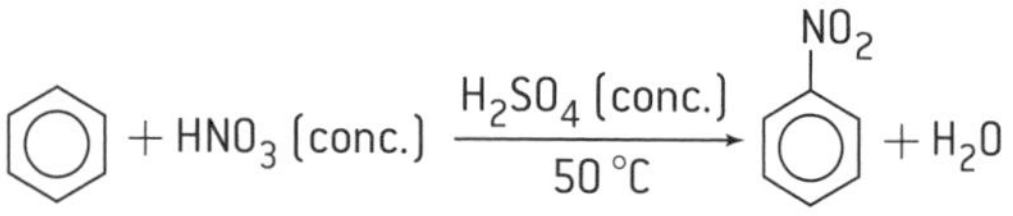

nitrobenzene

The electrophile is the **nitryl cation NO_2^+** (also called the nitronium ion). The concentrated sulfuric acid acts as a catalyst. Its function is to protonate the nitric acid which then loses water to form the electrophile. In this reaction nitric acid is acting as a base in the presence of the more acidic sulfuric acid.

$$H_2SO_4 + HNO_3 \rightleftharpoons H_2NO_3^+ + HSO_4^-$$

$$H_2NO_3^+ \rightarrow H_2O + NO_2^+$$

The NO_2^+ is attracted to the delocalized π bond and attaches to one of the carbon atoms. This requires considerable activation energy as the delocalized π bond is partially broken. The positive charge is distributed over the remains of the π bond in the intermediate. The intermediate then loses a proton and energy is evolved as the delocalized π bond is reformed. The proton can recombine with the hydrogensulfate ion to regenerate the catalyst.

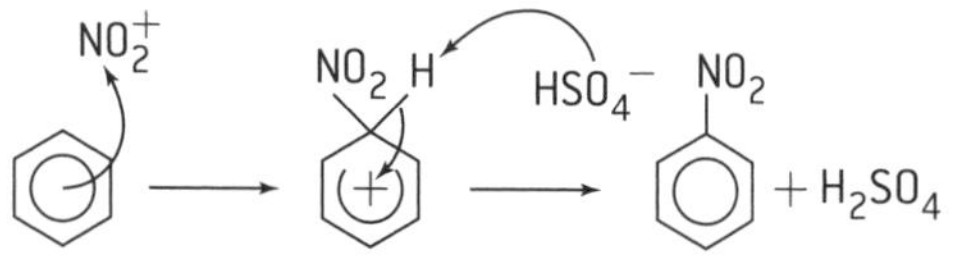

Although it is more correct to draw the intermediate as a partially delocalized π bond it can sometimes be convenient to show benzene as if it does contain alternate single and double carbon to carbon bonds. In this model the positive charge is located on a particular carbon atom.

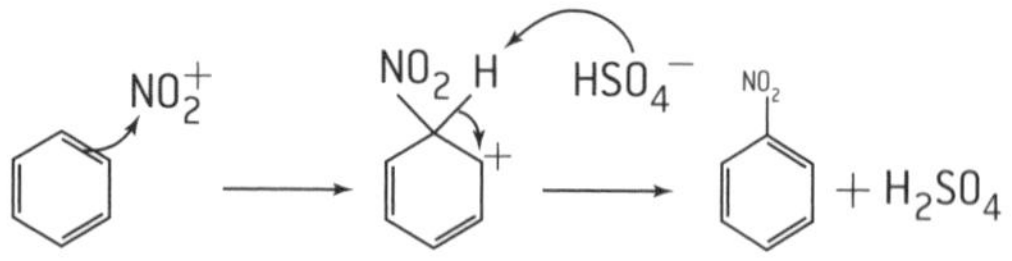

REDUCTION REACTIONS

a) Reduction of carbonyl compounds

There are several reducing agents that can be used to reduce carbonyl compounds. Typical among these are lithium tetrahydridoaluminate (also known as lithium aluminium hydride), $LiAlH_4$, and sodium tetrahydridoborate (also known as sodium borohydride), $NaBH_4$. Both effectively provide a source of H^- ions, which act as the reducing agent undergoing a nucleophilic addition reaction with the electron-deficient carbon atom of the carbonyl group. Sodium tetrahydridoborate can be used in the presence of protic solvents such as water or ethanol but is ineffectual at reducing carboxylic acids. The stronger reducing agent lithium aluminium hydride must initially be used in aprotic solvents such as ether as it reacts with water, then the reaction is acidified to obtain the product. Another reducing agent that can be used is hydrogen itself in the presence of a nickel, platinum or palladium catalyst.

Aldehydes are reduced to primary alcohols.

e.g.

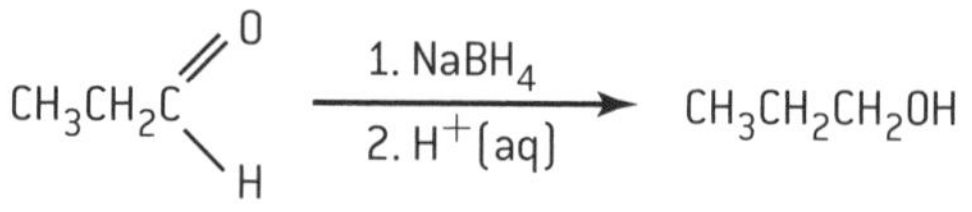

propanal propan-1-ol

Ketones are reduced to secondary alcohols.

e.g.

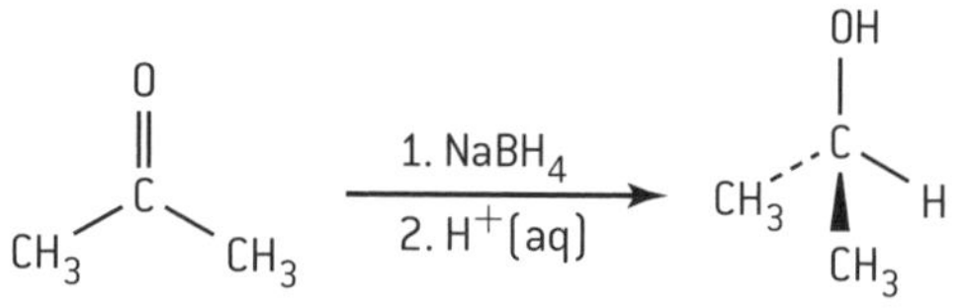

propanone propan-2-ol

Carboxylic acids are reduced to primary alcohols according to the general equation:

$$RCOOH \xrightarrow[\text{2. } H^+(aq)]{\text{1. } LiAlH_4 \text{ in ether}} RCH_2OH$$

b) Reduction of nitrobenzene

The reduction of nitrobenzene to phenylamine is usually carried out in two stages.

Stage 1. Nitrobenzene is refluxed with a mixture of tin and concentrated hydrochloric acid. The tin provides the electrons by acting as the reducing agent and the product is the phenylammonium ion.

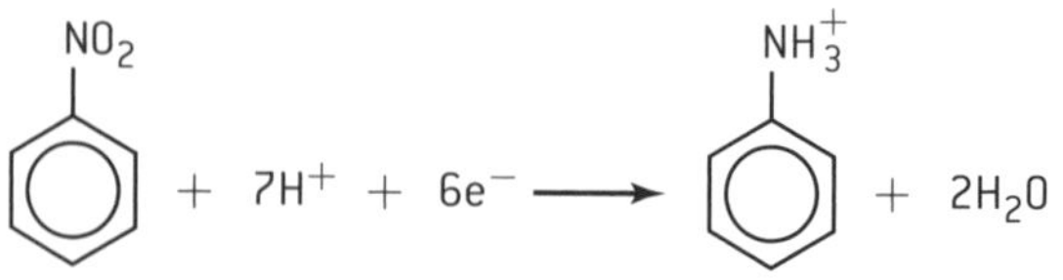

Stage 2. The addition of sodium hydroxide solution releases the free amine.

Synthetic routes

SYNTHESIS AND RETRO-SYNTHESIS

The raw materials for many organic compounds originate from coal, crude oil or natural gas. The challenge for synthetic organic chemists is to devise reaction pathways to make new compounds from simple starting materials using as few steps as possible, each with the highest yield possible. The more steps there are in an organic synthesis, then the lower the final yield is likely to be, because some material will be lost during each step. For example, but-2-ene can be obtained by cracking some of the higher boiling point fractions of crude oil. One way of obtaining butanone from but-2-ene in a two-step process would be, firstly, to hydrate the but-2-ene by heating with steam in the presence of a sulfuric acid catalyst to produce butan-2-ol. This secondary alcohol could then be oxidized with a warm acidified solution of potassium dichromate(VI).

This sub-topic requires organic synthesis using up to four different steps involving reactions that have already been covered. Rather than looking at the specific compounds it is often helpful to look at the functional groups involved and consider the reactions of these groups. For some syntheses it is helpful to work forward from the reactants but it can be equally profitable to work backwards from the product to find the simplest and cheapest starting materials – a process known as retro-synthesis. In the above example butanone is a ketone and ketones are formed by oxidizing secondary alcohols. It can then be seen that alcohols are made by hydrating alkenes. The following is a summary of the reactions covered in the Core and AHL organic chemistry.

- Combustion of hydrocarbons and alcohols
- Substitution of alkanes with halogens
- Addition reactions of alkenes with H_2, H_2O, HX, X_2, interhalogens and polymerization
- Substitution of halogenoalkanes with sodium hydroxide
- Substitution of benzene to form nitrobenzene
- Reduction of nitrobenzene to form phenylamine
- Oxidation of alcohols and aldehydes
- Reduction of aldehydes, ketones and carboxylic acids
- Esterification

When devising syntheses, the reagents and all necessary experimental conditions should be given. Often more than one solution will be possible as the same product may be obtained by different routes. For example, bromoethane can be prepared from ethene either by adding hydrogen bromide directly to ethene or by first converting the ethene into ethane then reacting it with bromine in ultraviolet light.

WORKED EXAMPLES OF SYNTHESES

1. Starting with benzene, suggest a two-step synthesis of phenylamine.

   Working backwards, phenylamine can be obtained by reducing nitrobenzene, which can be obtained from benzene by electrophilic substitution.

   Step 1. React benzene with a mixture of concentrated nitric acid and concentrated sulfuric acid at 50 °C to form nitrobenzene.

$$C_6H_6 + HNO_3(\text{conc}) \xrightarrow[50\ ^\circ C]{H_2SO_4\ (\text{conc})} C_6H_5NO_2 + H_2O$$

   Step 2. Reflux nitrobenzene with tin and concentrated hydrochloric acid, then react with sodium hydroxide solution.

$$C_6H_5NO_2 \xrightarrow[OH^-\ (aq)]{Sn/HCl,\ \text{reflux}} C_6H_5NH_2$$

2. Starting with propane, suggest a four-step synthesis of propyl propanoate.

   Propane reacts with halogens to form halogenoalkanes, which can easily be converted into alcohols which are required to make an ester. A primary alcohol can also be oxidized to a carboxylic acid which will also be required to make the ester.

   Step 1. React propane with bromine in ultraviolet light to give 1-bromopropane (other halogenated products will need to be discarded).

$$CH_3CH_2CH_3 + Br_2 \xrightarrow{UV} CH_3CH_2CH_2Br + HBr$$

   Step 2. React 1-bromopropane with warm dilute aqueous sodium hydroxide solution to give propan-1-ol.

$$CH_3CH_2CH_2Br + NaOH \xrightarrow{\text{warm}} CH_3CH_2CH_2OH + NaBr$$

   Step 3. Keep half of the propan-1-ol and reflux the remainder with warm excess acidified potassium dichromate(VI) solution to convert it into propanoic acid.

$$CH_3CH_2CH_2OH + \xrightarrow{Cr_2O_7^{2-}/H^+,\ \text{reflux}} CH_3CH_2CHOOH$$

   Step 4. Warm propan-1-ol and propanoic acid in the presence of a few drops of concentrated sulfuric acid as a catalyst to make the ester.

$$CH_3CH_2CH_2OH + CH_3CH_2COOH \xrightarrow{H^+} CH_3CH_2COOCH_2CH_2CH_3 + H_2O$$

Stereoisomerism (1)

TYPES OF ISOMERISM

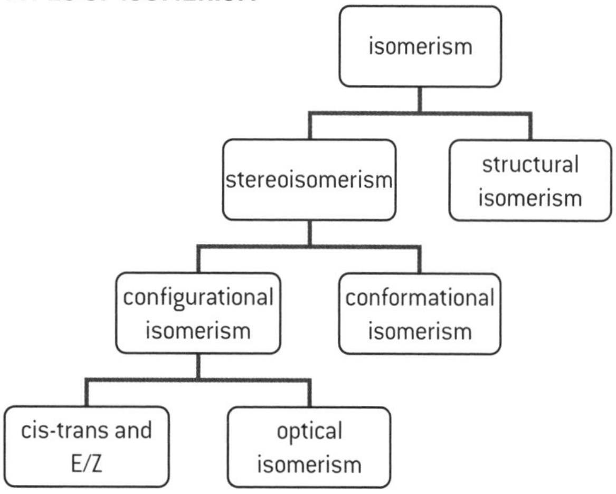

Isomers are compounds that are composed of the same elements in the same proportions but differ in properties because of differences in the arrangement of atoms. Structural isomers share the same molecular formula but have different structural formulas. That is, their atoms are bonded in different ways. Stereoisomers have the same structural formula but differ in their spatial arrangement. Stereoisomers can be sub-divided into two classes – conformational isomers, which interconvert by rotation about a σ bond, and configurational isomers that interconvert only by breaking and reforming a bond. The IB syllabus on stereoisomerism is concerned with configurational isomers which can be further sub-divided into cis–trans and *E*/*Z* isomers (which together used to be known by the now obsolete term of geometrical isomers) and optical isomers.

CIS–TRANS ISOMERISM

Both cis–trans and *E*/*Z* isomerism occur when rotation about a bond is restricted or prevented. The classic examples of cis–trans isomerism occur with asymmetric non-cyclic alkenes of the type $R_1R_2C{=}CR_1R_2$. A cis-isomer is one in which the substituents are on the same side of the double bond. In a trans-isomer the substituents are on opposite sides of the double bond. For example, consider cis-but-2-ene and trans-but-2-ene.

cis- *trans-*

When there is a single bond between two carbon atoms free rotation about the bond is possible. However, the double bond in an alkene is made up of a σ and a π bond. The π bond is formed from the combination of two p orbitals, one from each of the carbon atoms. These two p orbitals must be in the same plane to combine. Rotating the bond would cause the π bond to break so no rotation is possible.

Cis-trans isomerism will always occur in alkenes when the two groups X and Y attached to each of the two carbon atoms are different.

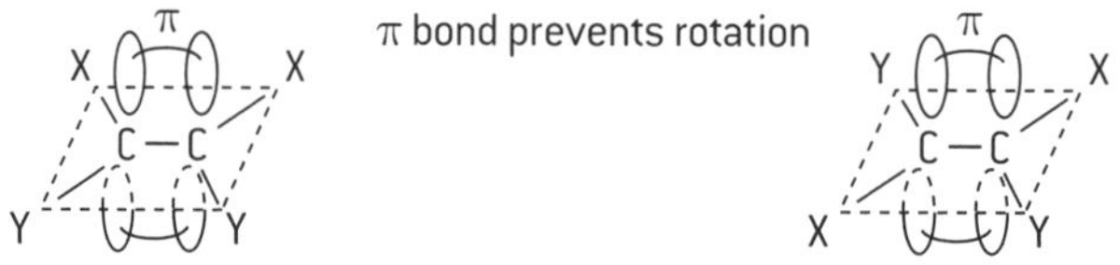

cis- *trans-*

Cis–trans isomerism can also occur in disubstituted cycloalkanes. The rotation is restricted because the C–C single bond is part of a ring system. Examples include 1,2-dichlorocyclopropane and 1,3-dichlorocyclobutane.

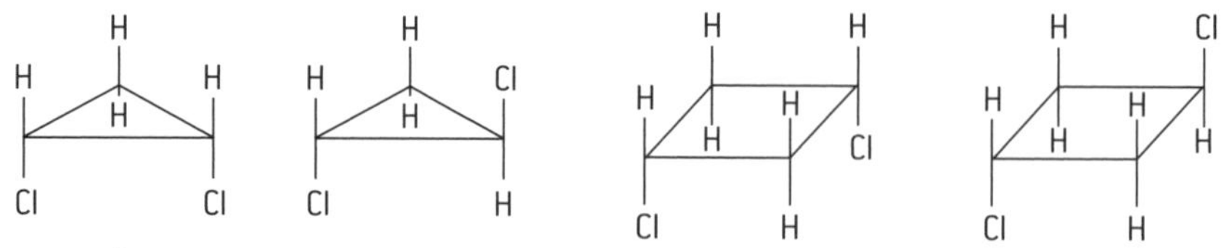

cis- *trans-*
1,2-dichlorocyclopropane

cis- *trans-*
1,3-dichlorocyclobutane

Stereoisomerism (2)

E/Z ISOMERISM

Cis–trans isomerism is a restricted form of *E*/*Z* isomerism as it only occurs when the two substituents R_1 and R_2 (or X and Y) occur on either side of the carbon to carbon double bond. *E*/*Z* isomerism covers every case where free rotation around a C=C double bond is not possible, i.e. for $R_1R_2C{=}CR_3R_4$ where ($R_1 \neq R_2$, $R_3 \neq R_4$) and where neither R_1 nor R_2 need be different from R_3 or R_4.

E/*Z* terminology is quite easy to apply and depends on what are known as the Cahn–Ingold–Prelog (CIP) rules for determining the priority of the atoms or groups attached to the two carbon atoms of the double bond. Each of these two carbon atoms on either side of the double bond is considered separately. In simple terms the higher the atomic number of the attached atoms to each carbon atom the higher the priority. Consider 1,2-dichloroethene. Chlorine has a higher atomic number than hydrogen so has a higher priority. If the two atoms/groups lie on the same side the isomer is *Z* and if they lie on opposite sides they are *E*. In this simple case *Z* is the cis-form and *E* is the trans- form.

However *Z* does not always equate to cis. If you consider 2-chlorobut-2-ene then the carbon atom of the methyl group has priority over the hydrogen atom on one side of the double bond but because chlorine has a higher atomic number than carbon the chlorine atom has priority over the methyl group on the other side of the double bond. Now the cis-isomer is the E isomer and the trans-isomer is the *Z* isomer.

E/*Z* and not cis–trans must be used when the groups are all different. Consider 2-bromo-1-chloro-2-iodoethene. It is not obvious which would be the cis- and trans- forms. However using the Cahn–Ingold–Prelog rules the *E* and *Z* forms can easily be determined.

Z and *E* both come from German words. *Z* is from ***zusammen*** (together) and *E* is from ***entgegen*** (opposite). One easy way to remember the correct application is that *E* could stand for **e**nemy, and enemies are on opposite sides.

Cl has a higher priority than H on the left-hand side and on the right-hand side.

Both highest priorities lie on the same side so *Z* isomer.

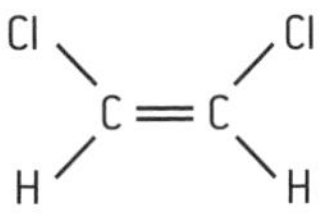

cis-1, 2-dichloroethene
(*Z*)-1, 2-dichloroethene

Highest priorities lie on opposite side so *E* isomer.

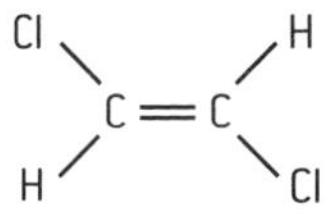

trans-1, 2-dichloroethene
(*E*)-1, 2-dichloroethene

CH_3– has a higher priority than H on the left-hand side and Cl has a higher priority than CH_3– on the right-hand side.

Both highest priorities lie on the same side so *Z* isomer.

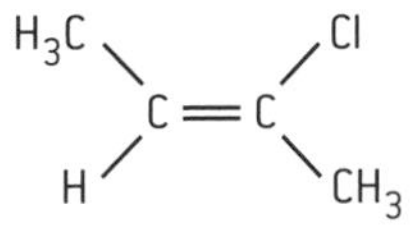

trans-2-chlorobut-2-ene
(*Z*)-2-chlorobut-2-ene

Highest priorities lie on opposite side so *E* isomer.

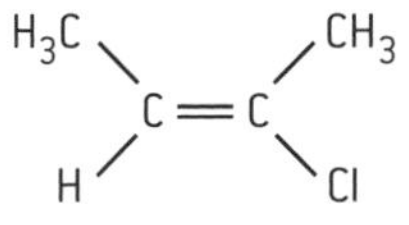

cis-2-chlorobut-2-ene
(*E*)-2-chlorobut-2-ene

Cl has a higher priority than H on the left-hand side and I has a higher priority than Br on the right-hand side.

Both highest priorities lie on the same side so *Z* isomer.

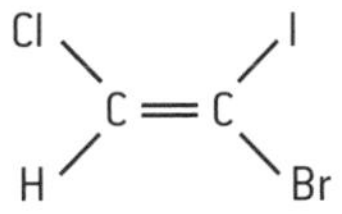

(*Z*)-2-bromo-1-chloro-2-iodoethene

Highest priorities lie on opposite side so *E* isomer.

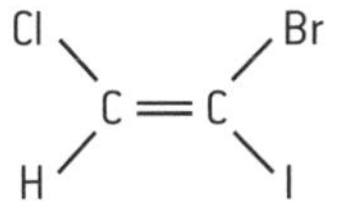

(*E*)-2-bromo-1-chloro-2-iodoethene

PHYSICAL AND CHEMICAL PROPERTIES OF E/Z AND CIS–TRANS *ISOMERS*

The chemical properties of E/Z and cis–trans isomers tend to be similar but their physical properties are different. For example, the boiling point of cis-1,2-dichloroethene is 60.3 °C whereas trans-1,2-dichloroethene boils at the lower temperature of 47.5 °C. Sometimes there can be a marked difference in both chemical and physical properties. This tends to occur when there is some sort of chemical interaction between the substituents. cis-but-2-ene-1,4-dioic acid melts with decomposition at 130–131 °C. However, trans-but-2-ene,1,4-dioic acid does not melt until 286 °C. In the cis-isomer the two carboxylic acid groups are closer together so that intramolecular hydrogen bonding is possible between them. In the trans- isomer they are too far apart to attract each other so there are stronger intermolecular forces of attraction between different molecules, resulting in a higher melting point.

The cis- isomer reacts when heated to lose water and form a cyclic acid anhydride. The trans- isomer cannot undergo this reaction.

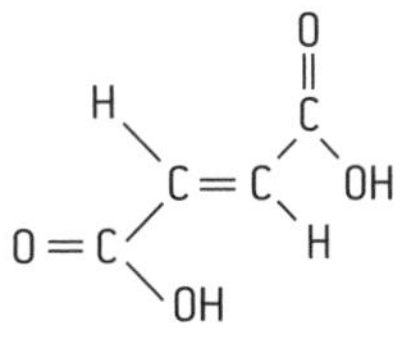

trans-but-2-ene-1,4-dioic acid

↓ heat

cannot form cyclic acid anhydride when heated

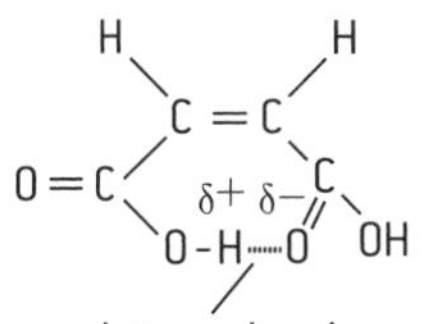

intramolecular hydrogen bonding

cis-but-2-ene-1,4-dioic acid

↓ heat

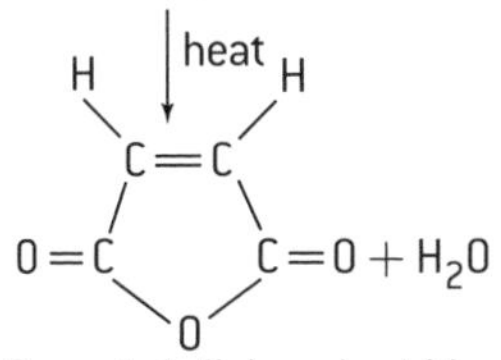

cis-but-2-ene-1,4-dioic anhydride

Stereoisomerism (3)

OPTICAL ISOMERISM

Optical isomerism is shown by all compounds that contain at least one asymmetric or chiral carbon atom within the molecule, that is, one that contains four different atoms or groups bonded to it, also known as a stereocentre. The two isomers are known as enantiomers and are mirror images of each other. Examples include butan-2-ol, $CH_3CH(OH)C_2H_5$, 2-hydroxypropanoic acid (lactic acid), $CH_3CH(OH)COOH$ and all amino acids (except glycine, NH_2CH_2COOH).

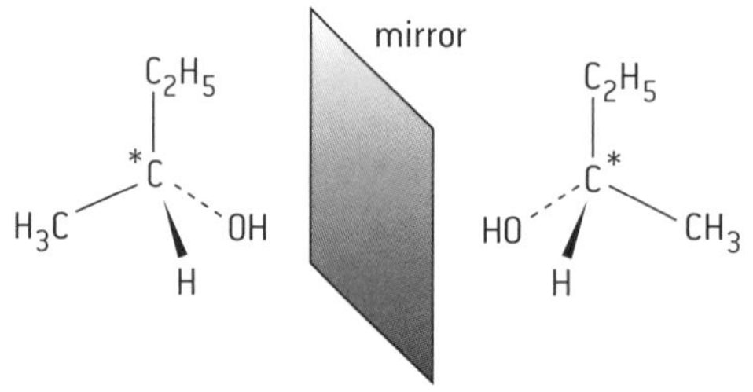

enantiomers of butan-2-ol
(* asymmetric carbon/chiral carbon/stereocentre)

The two different isomers are optically active with plane-polarized light. Normal light consists of electromagnetic radiation which vibrates in all planes. When it is passed through a polarizing filter the waves only vibrate in one plane and the light is said to be plane-polarized.

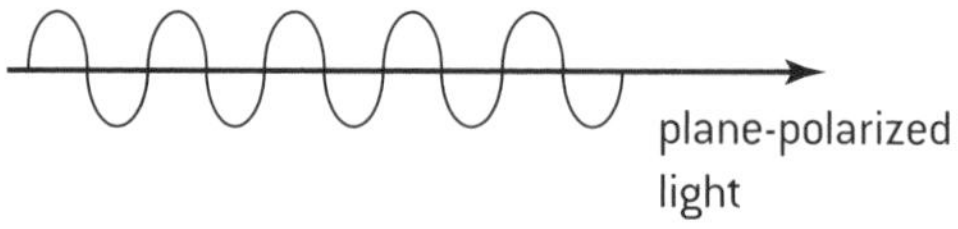

The two enantiomers both rotate the plane of plane-polarized light. One of the enantiomers rotates it to the left and the other rotates it by the same amount to the right. Apart from their behaviour towards plane-polarized light enantiomers have identical physical properties. Their chemical properties are identical too except when they interact with other optically active substances. This is often the case in the body where the different enantiomers can have completely different physiological effects. For example one of the enantiomers of the amino acid asparagine $H_2NCH(CH_2CONH_2)COOH$ tastes bitter whereas the other enantiomer tastes sweet.

If a molecule contains two or more stereocentres then several different stereoisomers are possible. They are known as enantiomers if they are mirror images and as diastereomers if they are not mirror images of each other. Diastereomerism occurs when two or more stereoisomers of a compound have different configurations at one or more (but not all) of the equivalent stereocentres. This is particularly important with many sugars and some amino acids. Diastereomers have different physical properties and different chemical reactivity. e.g.

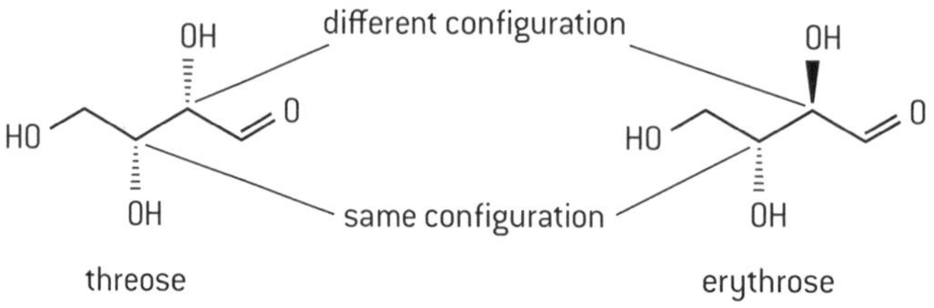

diastereomers of 2, 3, 4-trihydroxybutanal

POLARIMETRY

The optical activity of enantiomers can be detected and measured by an instrument called a polarimeter. It consists of a light source, two polarizing lenses, and between the lenses a tube to hold the sample of the enantiomer dissolved in a suitable solvent.

When light passes through the first polarizing lens (polarizer) it becomes plane-polarized. That is, it is vibrating in a single plane. With no sample present the observer will see the maximum intensity of light when the second polarizing lens (analyser) is in the same plane. Rotating the analyser by 90° will cut out all the light. When the sample is placed between the lenses the analyser must be rotated by θ degrees, either clockwise (dextrorotatory) or anticlockwise (laevorotatory) to give light of maximum intensity. The two enantiomers rotate the plane of plane-polarized light by the same amount but in opposite directions. If both enantiomers are present in equal amounts the two rotations cancel each other out and the mixture appears to be optically inactive. Such a mixture is known as a **racemic mixture** or **racemate**.

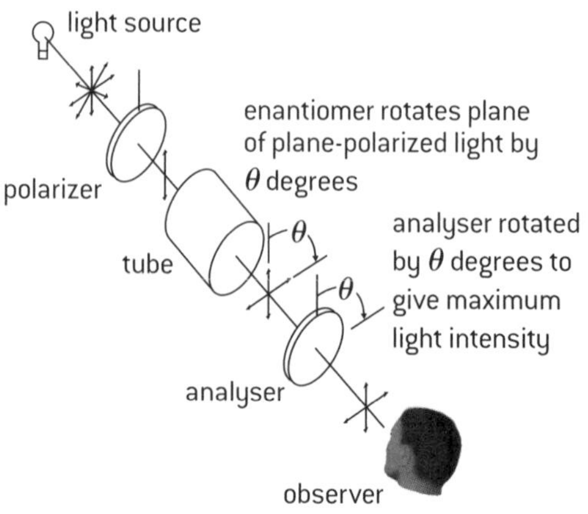

MULTIPLE CHOICE QUESTIONS – ORGANIC CHEMISTRY

1. Which of the following two compounds both belong to the same homologous series?

   A. CH_3COOH and $HCOOCH_3$

   C. C_2H_4 and C_2H_6

   B. CH_3OH and C_2H_5OH

   D. C_2H_5Cl and $C_2H_4Cl_2$

2. How many different isomers of C_5H_{12} exist?

   A. 1 | B. 2
   C. 3 | D. 4

3. Applying IUPAC rules what is the name of 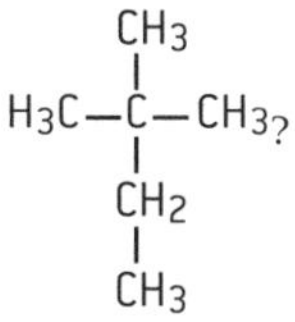

   A. 2-methyl-2-ethylpropane | C. 2,2-dimethylbutane
   B. hexane | D. 2-methylpentane

4. Which statement is correct about the reaction between methane and chlorine?

   A. It involves heterolytic fission and Cl^- ions.

   B. It involves heterolytic fission and $Cl^\bullet$ radicals.

   C. It involves homolytic fission and Cl^- ions.

   D. It involves homolytic fission and $Cl^\bullet$ radicals.

5. Which compound is an ester?

   A. CH_3COOH | C. C_2H_5CHO
   B. $CH_3OC_2H_5$ | D. $HCOOCH_3$

6. When ethanol is partially oxidized by an acidified solution of potassium dichromate(VI), the product that can be obtained by distillation as soon as it is formed is:

   A. ethanal | C. ethanoic acid
   B. ethene | D. ethane-1,2-diol

7. Which formula is that of a secondary halogenoalkane?

   A. $CH_3CH_2CH_2CH_2Br$ | C. $(CH_3)_2CHCH_2Br$
   B. $CH_3CHBrCH_2CH_3$ | D. $(CH_3)_3CBr$

8. The reaction between bromine and ethene in the dark is an example of:

   A. free radical substitution

   B. esterification

   C. nucleophilic substitution

   D. addition

9. Which alcohol cannot be oxidized by an acidified solution of potassium dichromate(VI)?

   A. $(CH_3)_3COH$ | C. $(CH_3)_2CHCH_2OH$
   B. $CH_3CH_2CH_2CH_2OH$ | D. $CH_3CHOHCH_2CH_3$

HL

10. Which statement(s) is/are true about the reactions of halogenoalkanes with warm dilute sodium hydroxide solution?

    I. CH_3I reacts faster than CH_3F

    II. $(CH_3)_3CBr$ reacts faster than CH_3Br

    III. $(CH_3)_3CBr$ and $(CH_3)_3CCl$ both react by S_N1 mechanisms.

    A. I and II only | C. II and III only
    B. I and III only | D. I, II and III

11. The product from the reaction of iodine monochloride, ICl, with pent-1-ene is:

    A. $CH_3CH_2CHICHClCH_3$ | C. $CH_3CH_2CH_2CHClCH_2I$
    B. $CH_3CH_2CH_2CHICH_2Cl$ | D. $CH_3CH_2CHClCHICH_3$

12. Ketones are reduced by sodium borohydride, $NaBH_4$, to give:

    A. a primary alcohol | C. a tertiary alcohol
    B. a secondary alcohol | D. a carboxylic acid

13. During the conversion of nitrobenzene to phenylamine (aniline) by tin and concentrated hydrochloric acid the organic intermediate is:

    A. the phenylammonium ion | C. phenol
    B. dinitrobenzene | D. chlorobenzene

14. Which can exist as enantiomers and diastereomers?

    I. 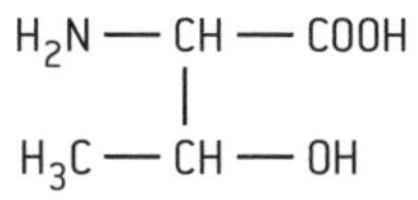

    II. 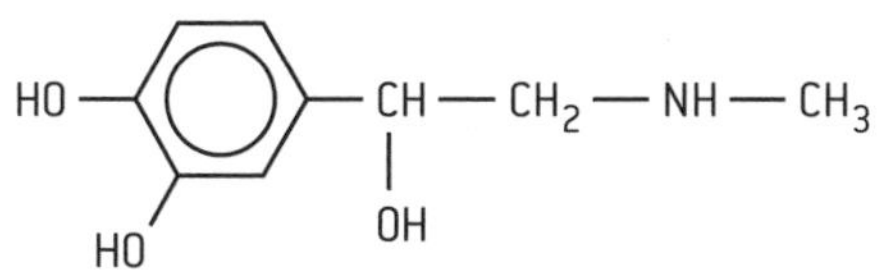

    III. $CH_3CH(OH)CH_2CH(OH)CHO$

    A. I and II only | C. II and III only
    B. I and III only | D. I, II and III

15. How many four-membered ring isomers are there of dichlorocyclobutane, $C_4H_6Cl_2$?

    A. 3 | C. 5
    B. 4 | D. 6

16. Which is an *E* isomer?

    A. 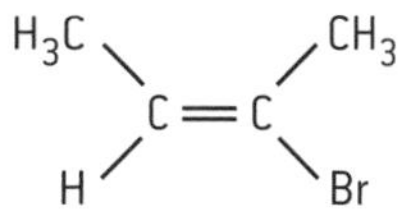

    C. 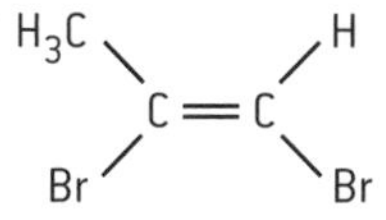

    B. 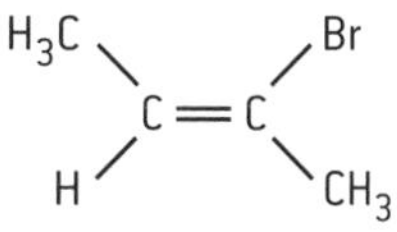

    D. 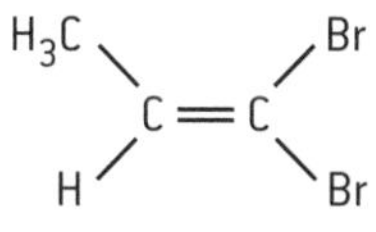

17. An organic compound is prepared from ethanol by a four-step synthesis. Each step gives a yield of 70%. What will be the yield of the organic compound based on the initial amount of ethanol?

    A. 24% | C. 34%
    B. 28% | D. 70%

SHORT ANSWER QUESTIONS – ORGANIC CHEMISTRY

1. Alkenes are important starting materials for a variety of products.

   a) State and explain the trend of the boiling points of the first five members of the alkene homologous series. [3]

   b) Describe two features of a homologous series. [2]

   c) Below is a schematic diagram representing some reactions of ethene. The letters **A–D** represent the organic compounds formed from the reactants and catalysts shown.

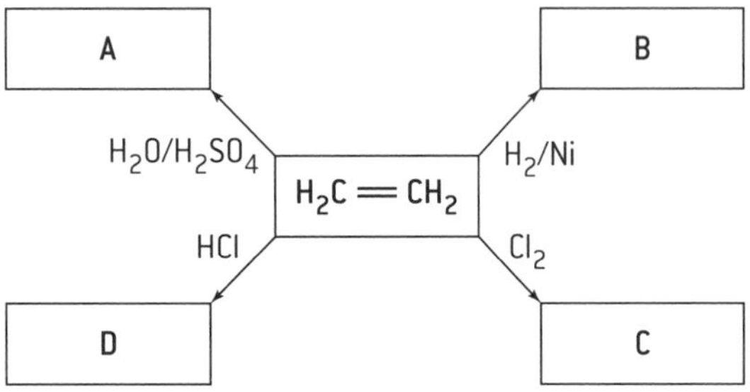

   Deduce the structural formulas of compounds **A**, **B**, **C** and **D** and state the IUPAC name of compound **C**. [5]

   d) Describe a chemical test that could be used to distinguish between pent-1-ene and pentane. [2]

   e) State and explain whether the following molecules are primary, secondary or tertiary halogenoalkanes. [4]

E:

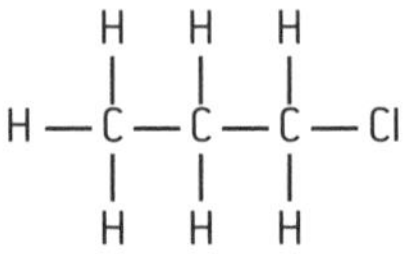

F:

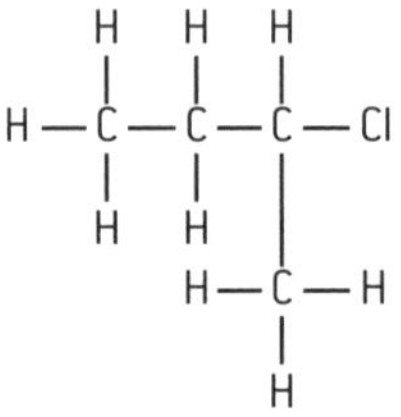

   f) Explain, using equations, the initiation, propagation and termination steps in the free-radical mechanism of the reaction of methane with chlorine. [4]

2. The following is a three-dimensional computer-generated representation of aspirin. Each carbon, oxygen and hydrogen atom has been given a unique number.

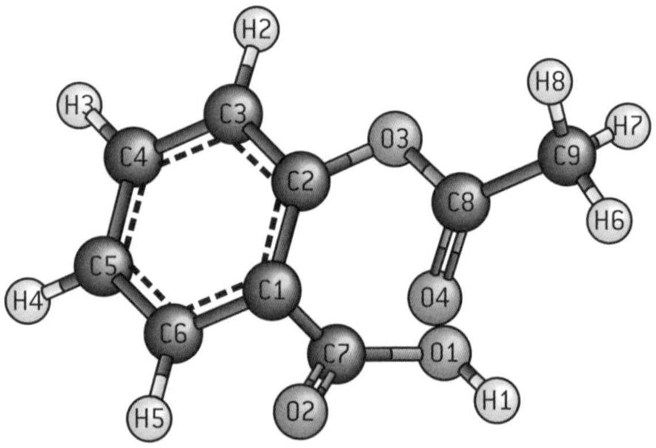

   a) Identify **one carbon** atom that is part of

      (i) the phenyl functional group.

      (ii) the carboxylic acid group.

      (iii) the ester group. [3]

   b) Explain why the bond between C_8 and O_3 is longer than the bond between C_8 and O_4. [2]

   c) Compare the length of the bond between C_1 and C_6 with the bond between C_1 and C_7. [2]

   d) Explain how a primary alcohol can be converted into

      (i) a carboxylic acid

      (ii) an ester. [4]

HL

3. Consider alkenes with the molecular formula C_5H_9Br. Give structural formulas for one isomer which shows

   a) no stereoisomerism [1]

   b) optical but no cis–trans or *E*/*Z* isomerism [1]

   c) optical, trans- and *E* isomerism [1]

   d) optical, cis- and *Z* isomerism. [1]

   e) trans- and *Z* isomerism but no optical isomerism [1]

   f) cis- and *E* isomerism but no optical isomerism. [1]

4. State the equations for each step and any necessary inorganic reagents and conditions for:

   a) The preparation of propyl propanoate by a **two**-step synthesis using propan-1-ol as the only organic reagent. [4]

   b) The preparation of propanone by a **three**-step synthesis using propene as the only organic reagent. [6]

   c) The preparation of propanal by a **four**-step synthesis using propene as the only organic reagent. [8]

5. 1-bromopropane undergoes a substitution reaction with warm aqueous sodium hydroxide solution.

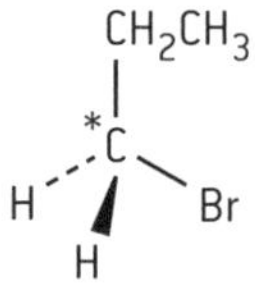

   a) Explain why the substitution occurs on the carbon atom that is marked as *C. [1]

   b) State the rate equation for this reaction and identify the name of the reaction mechanism. [2]

   c) Explain the mechanism of the reaction using curly arrows to represent the movement of electron pairs during the substitution. [4]

   d) Explain how changing the halogenoalkane to 1-chloropropane would affect the rate of the substitution reaction. [2]

6. Benzene can be nitrated to form nitrobenzene by warming with a mixture of concentrated nitric acid and concentrated sulfuric acid. Explain, with any necessary equations, the role of the concentrated sulfuric acid in this reaction. [6]

Uncertainty and error in measurement

RANDOM UNCERTAINTIES AND SYSTEMATIC ERRORS

Quantitative chemistry involves measurement. A measurement is a method by which some quantity or property of a substance is compared with a known standard. If the instrument used to take the measurements has been calibrated wrongly or if the person using it consistently misreads it then the measurements will always differ by the same amount. Such an error is known as a **systematic error**. An example might be always reading a pipette from the sides of the meniscus rather than from the middle of the meniscus.

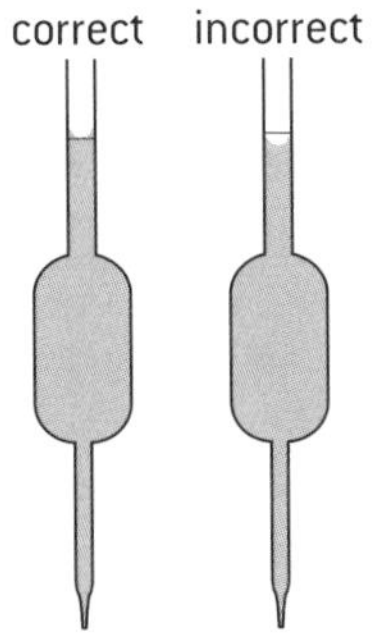

a systematic error will be introduced if a pipette is read incorrectly

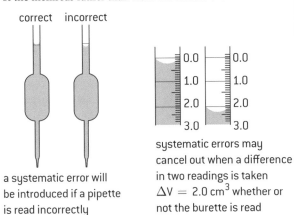

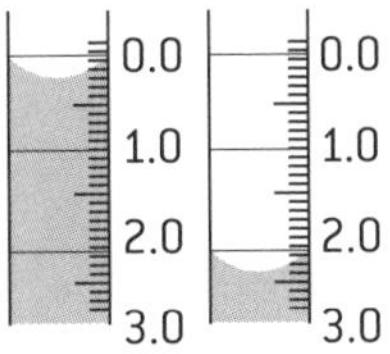

systematic errors may cancel out when a difference in two readings is taken $\Delta V = 2.0\ cm^3$ whether or not the burette is read

Random uncertainties occur if there is an equal probability of the reading being too high or too low from one measurement to the next. These might include variations in the volume of glassware due to temperature fluctuations or the decision on exactly when an indicator changes colour during an acid–base titration.

PRECISION AND ACCURACY

Precision refers to how close several experimental measurements of the same quantity are to each other. **Accuracy** refers to how close the readings are to the true value. This may be the standard value, or the literature or accepted value. A measuring cylinder used to measure exactly 25 cm^3 is likely to be much less accurate than a pipette that has been carefully calibrated to deliver exactly that volume. It is possible to have very precise readings which are inaccurate due to a systematic error. For example all the students in the class may obtain the same or very close results in a titration but if the standard solution used in all the titrations had been prepared wrongly beforehand the results would be inaccurate due to the systematic error. Because they are always either too high or too low systematic errors cannot be reduced by repeated readings. However random errors can be reduced by repeated readings because there is an equal probability of them being high or low each time the reading is taken. When taking a measurement it is usual practice to report the reading from a scale as the smallest division or the last digit capable of precise measurement even though it is understood that the last digit has been rounded up or down so that there is a random error or uncertainty of ± 0.5 of the last unit.

SIGNIFICANT FIGURES

Whenever a measurement of a physical quantity is taken there will be a random uncertainty in the reading. The measurement quoted should include the first figure that is uncertain. This should include zero if necessary. Thus a reading of 25.30 °C indicates that the temperature was measured with a thermometer that is accurate to + 0.01 °C. If a thermometer accurate to only ± 0.1 °C was used the temperature should be recorded as 25.3 °C.

Zeros can cause problems when determining the number of significant figures. Essentially zero only becomes significant when it comes *after* a non-zero digit (1,2,3,4,5,6,7,8,9).

000123.4 0.0001234	1.0234 1.2340
zero not a significant figure values quoted to 4 s.f.	zero is a significant figure values quoted to 5 s.f.

Zeros after a non-zero digit but before the decimal point may or may not be significant depending on how the measurement was made. For example 123 000 might mean exactly one hundred and twenty three thousand or one hundred and twenty three thousand to the nearest thousand. This problem can be neatly overcome by using scientific notation.

1.23000×10^6 quoted to six significant figures

1.23×10^6 quoted to three significant figures.

Calculations

1. When adding or subtracting it is the number of decimal places that is important. Thus when using a balance which measures to ± 0.01 g the answer can also be quoted to two decimal places which may increase or decrease the number of significant figures.

e.g.	7.10 g	+	3.10 g	=	10.20 g
	3 s.f.		3 s.f.		4 s.f.
	22.36 g	–	15.16 g	=	7.20 g
	4 s.f.		4 s.f.		3 s.f.

2. When multiplying or dividing it is the number of significant figures that is important. The number with the least number of significant figures used in the calculation determines how many significant figures should be used when quoting the answer.

   e.g. When the temperature of 0.125 kg of water is increased by 7.2 °C

   the heat required
   $= 0.125\ kg \times 7.2\ °C \times 4.18\ kJ\ kg^{-1}\ °C^{-1}$
   $= 3.762\ kJ$

   Since the temperature was only recorded to two significant figures the answer should strictly be given as 3.8 kJ.

Uncertainty in calculated results and graphical techniques

ABSOLUTE AND PERCENTAGE UNCERTAINTIES

When making a single measurement with a piece of apparatus the absolute uncertainty and the percentage uncertainty can both be stated relatively easily. For example consider measuring 25.0 cm^3 with a 25 cm^3 pipette which measures to $\pm$ 0.1 cm^3. The absolute uncertainty is 0.1 cm^3 and the percentage uncertainty is equal to:

$$\frac{0.1}{25.0} \times 100 = 0.4\%$$

If two volumes or two masses are simply added or subtracted then the absolute uncertainties are added. For example suppose two volumes of 25.0 cm^3 $\pm$ 0.1 cm^3 are added. In one extreme case the first volume could be 24.9 cm^3 and the second volume 24.9 cm^3 which would give a total volume of 49.8 cm^3. Alternatively the first volume might have been 25.1 cm^3 which when added to a second volume of 25.1 cm^3 gives a total volume of 50.2 cm^3. The final answer therefore can be quoted between 49.8 cm^3 and 50.2 cm^3, that is, 50.0 cm^3 $\pm$ 0.2 cm^3.

When using multiplication, division or powers then percentage uncertainties should be used during the calculation and then converted back into an absolute uncertainty when the final result is presented. For example, during a titration there are generally four separate pieces of apparatus, each of which contributes to the uncertainty.

e.g. when using a balance that weighs to $\pm$ 0.001 g the uncertainty in weighing 2.500 g will equal

$$\frac{0.001}{2.500} \times 100 = 0.04\%$$

Similarly a pipette measures 25.00 cm^3 $\pm$ 0.04 cm^3.

The uncertainty due to the pipette is thus

$$\frac{0.04}{25.00} \times 100 = 0.16\%$$

Assuming the uncertainty due to the burette and the volumetric flask is 0.50% and 0.10% respectively the overall uncertainty is obtained by summing all the individual uncertainties:

$$\text{Overall uncertainty} = 0.04 + 0.16 + 0.50 + 0.10 = 0.80\% \simeq 1.0\%$$

Hence if the answer is 1.87 mol dm^{-3} the uncertainty is 1.0% or 0.0187 mol dm^{-3}.

The answer should be given as 1.87 $\pm$ 0.02 mol dm^{-3}.

If the generally accepted 'correct' value (obtained from the data book or other literature) is known then the total error in the result is the difference between the literature value and the experimental value divided by the literature value expressed as a percentage. For example, if the 'correct' concentration for the concentration determined above is 1.90 mol dm^{-3} then:

$$\text{the total error} = \frac{(1.90 - 1.87)}{1.90} \times 100 = 1.6\%$$

GRAPHICAL TECHNIQUES

By plotting a suitable graph to give a straight line or some other definite relationship between the variables, graphs can be used to predict unknown values. There are various methods to achieve this. They include measuring the intercept, measuring the gradient, extrapolation and interpolation. **Interpolation** involves determining an unknown value within the limits of the values already measured. **Extrapolation** (see example on page 38) requires extending the graph to determine an unknown value which lies outside the range of the values measured. If possible manipulate the data to produce a straight line graph. For example, when investigating the relationship between pressure and volume for a fixed mass of gas a plot of P against V gives a curve whereas a plot of P against 1/V will give a straight line. Once the graph is in the form of $y = mx + c$ then the values for both the gradient (m) and the intercept (c) can be determined.

sketch graph of pressure against volume for a fixed mass of gas at a constant temperature

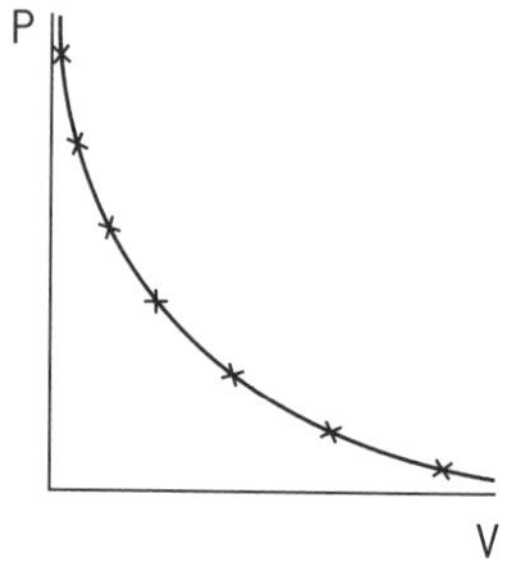

sketch graph of pressure against the reciprocal of volume for a fixed mass of gas at a constant temperature

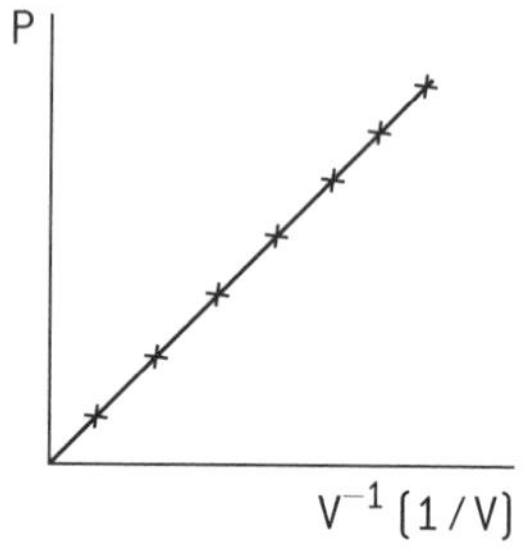

Note that a sketched graph has labelled but unscaled axes and is used to show qualitative trends. Drawn graphs have labelled and scaled axes and are used in quantitative measurements.

The following points should be observed when drawing a graph.

- Plot the independent variable on the horizontal axis and the dependent variable on the vertical axis.
- Choose appropriate scales for the axes.
- Use Standard International (SI) units wherever possible.
- Label each axis and include the units.
- Draw the line of best fit.
- Give the graph a title.

Measuring a gradient from a graph

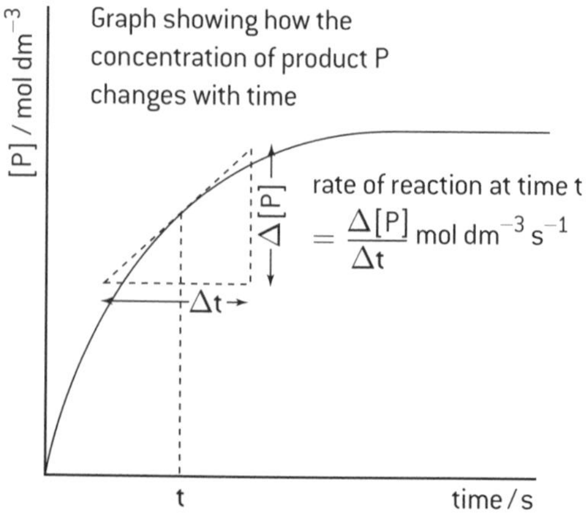

Analytical techniques

INFORMATION FROM DIFFERENT ANALYTICAL TECHNIQUES

The classic way to determine the structure of an organic compound was to determine both its empirical formula and relative molar mass experimentally, then deduce the nature of the functional groups from its chemical reactivity. However, modern well-equipped laboratories now employ a variety of instrumental techniques, which if used in combination are able to unambiguously determine the exact structural formula. They can also be used to determine the composition of the components in a mixture and to determine purity. These techniques are becoming ever more refined and some of them (e.g. mass spectrometry) can be used on extremely small samples. Before analysis can usually take place it is important to separate any mixture into its individual components – hence the need for chromatography. Often information is not obtained from a single technique but from a combination of several of them. Some examples are:

- Infrared spectroscopy – organic structural determination, information on the strength of bonds, information about the secondary structure of proteins, measuring the degree of unsaturation of oils and fats, and determining the level of alcohol in the breath.
- Mass spectrometry – organic structural determination, isotopic dating (e.g. ^{14}C dating).
- Proton nuclear magnetic resonance spectroscopy (^{1}H NMR) – organic structural determination, body scanning (known as MRI).
- Chromatography – drug testing in the blood and urine, food testing, and forensic science.

INDEX OF HYDROGEN DEFICIENCY (IHD)

Once the molecular formula of a compound is known the index of hydrogen deficiency can be calculated. This used to be known as the degree of unsaturation. A non-cyclic hydrocarbon that contains only single bonds contains the maximum number of hydrogen atoms and has the general formula $C_nH_{(2n+2)}$. If any of the bonds are replaced by double or triple bonds or the compound is cyclic then the number of hydrogen atoms falls (i.e. there is a deficiency in the number of hydrogen atoms). The amount of the deficiency therefore gives useful information about the molecule and will assist in determining the number and types of different possible isomers.

The IHD for a hydrocarbon with x carbon atoms and y hydrogen atoms, C_xH_y, is given by :

$$\text{IHD} = \frac{(2x + 2 - y)}{2}$$

where each double bond and each ring counts as one IHD and each triple bond counts as two IHD.

For example, hexane C_6H_{14} has an IHD equal to $\frac{(12 + 2 - 14)}{2} = 0$. This would be expected as hexane is a saturated hydrocarbon. If however a compound has a molecular formula of C_6H_{12} its IHD is $\frac{(12 + 2 - 12)}{2} = 1$. This means that the compound could contain one double bond and be an isomer of hexene or it could be a cyclic compound.

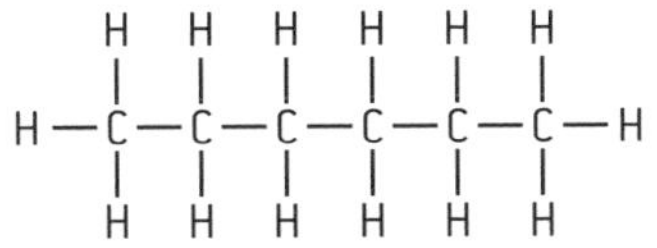

hexane

C_6H_{14} IHD = zero

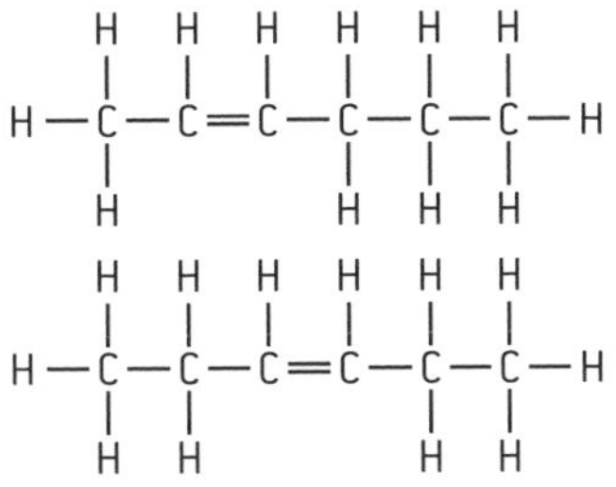

hex-2-ene and hex-3-ene

C_6H_{12} IHD = 1

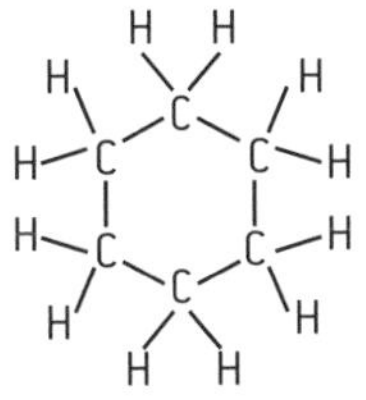

cyclohexane

C_6H_{12} IHD = 1

Determining the IHD becomes a little more complicated if elements other than carbon and hydrogen are present in the compound. In this case:

Oxygen and sulfur atoms (O and S) do not affect the IHD.

Halogens (F, Cl, Br, I) are treated like H atoms (i.e. $CHCl_3$ has the same IHD as CH_4).

For each N, add one to the number of C and one to the number of H (i.e. CH_5N has the same IHD as C_2H_6).

THE ELECTROMAGNETIC SPECTRUM

Modern spectroscopic techniques utilize different parts of the electromagnetic spectrum. The electromagnetic spectrum has already been described in Topic 2 – *Atomic structure*. You should be familiar with the relationship $c = \lambda\nu$ and know the different regions of the spectrum.

The electromagnetic spectrum

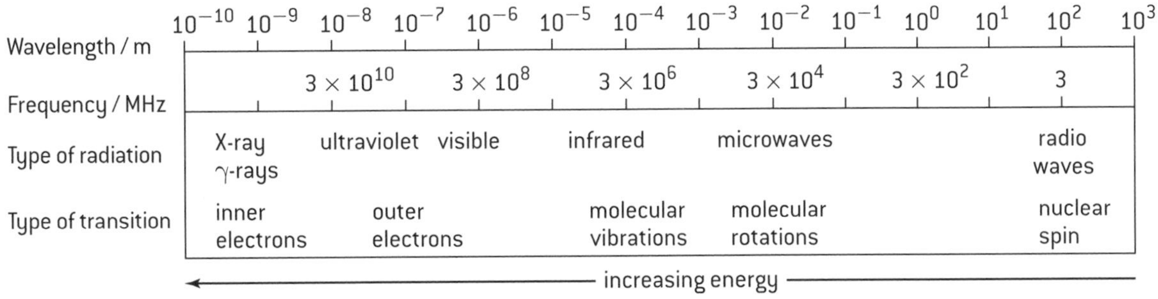

Spectroscopic identification of organic compounds – MS

MASS SPECTROMETRY

The principles of mass spectroscopy and its use to determine relative atomic masses have already been explained in Topic 2 – *Atomic structure*. It can be used in a similar way with organic compounds. However in addition to giving the precise molecular mass of the substance, mass spectroscopy gives considerable information about the actual structure of the compound from the fragmentation patterns.

When a sample is introduced into the machine the vaporized sample becomes ionized to form the molecular ion $M^+(g)$. Inside the mass spectrometer some of the molecular ions break down to give fragments, which are also deflected by the external magnetic field and which then show up as peaks on the detector. By looking at the difference in mass from the parent peak it is often possible to identify particular fragments,

e.g. $(M_r - 15)^+$ = loss of CH_3

$(M_r - 17)^+$ = loss of OH

$(M_r - 18)^+$ = loss of H_2O

$(M_r - 28)^+$ = loss of C_2H_4 or CO

$(M_r - 29)^+$ = loss of C_2H_5 or CHO

$(M_r - 31)^+$ = loss of CH_3O

$(M_r - 45)^+$ = loss of COOH

This can provide a useful way of distinguishing between structural isomers.

The mass spectra of propan-1-ol and propan-2-ol both show a peak at 60 due to the molecular ion $C_3H_8O^+$. However, the mass spectrum of propan-1-ol shows a strong peak at 31 due to the loss of $-C_2H_5$, which is absent in the mass spectrum of propan-2-ol. There is a strong peak at 45 in the spectrum of propan-2-ol as it contains two different methyl groups, which can fragment.

Mass spectrum of propan-1-ol $CH_3 - CH_2 - CH_2 - OH$

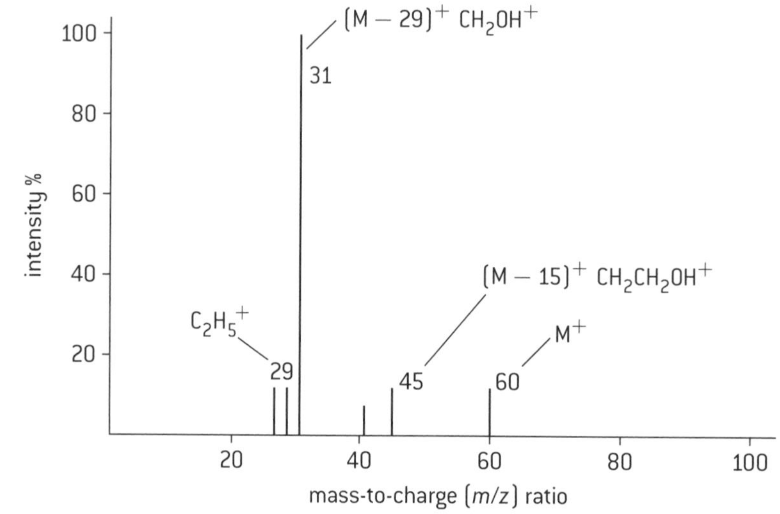

Mass spectrum of propan-2-ol $CH_3 - CH(OH) - CH_3$

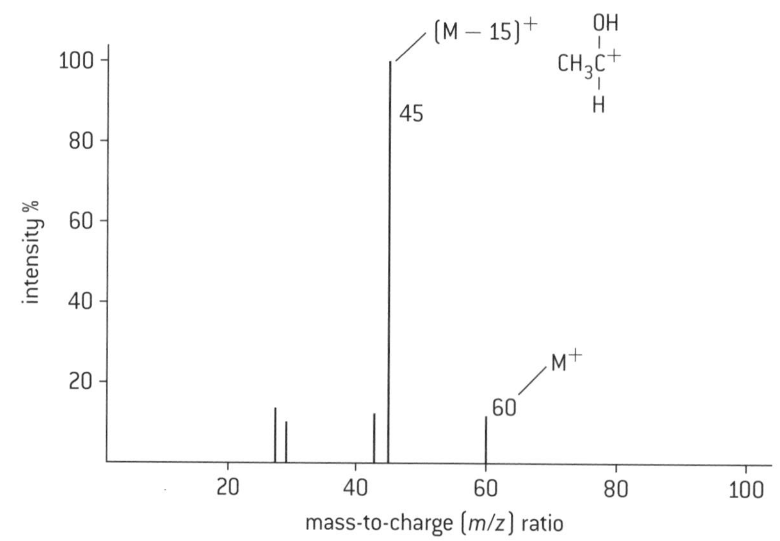

Spectroscopic identification of organic compounds – IR and ^{1}H NMR

INFRARED SPECTROSCOPY

When molecules absorb energy in the infrared region of the spectrum they vibrate (i.e. the bonds stretch and bend). The precise value of the energy they absorb depends on the particular bond and to a lesser extent on the other groups attached to the two atoms forming the bond. When infrared radiation is passed through a sample the spectrum shows the characteristic absorptions, which confirm that a particular bond is present in the molecule. Some absorptions, e.g. those due to C-H, are not particularly useful, as they are shown by most organic compounds. However others give a very clear indication of a particular functional group.

In addition to these particular absorptions, infrared spectra also possess a 'fingerprint' region. This is a characteristic pattern between about 1400–400 cm^{-1}, which is specific to a particular compound. It is often possible to identify an unknown sample by comparing the fingerprint region with a library of spectra of known compounds.

A simplified correlation chart

Bond	Wavenumber / cm^{-1}
C-Cl	600–800
C-O	1050–1410
C=C	1620–1680
C=O	1700–1750
C≡C	2100–2260
O-H (in carboxylic acids)	2500–3000
C-H	2850–3090
O-H (in alcohols)	3200–3600

Wavenumber is the reciprocal of wavelength.

Infrared spectrum of ethanol CH_3CH_2OH

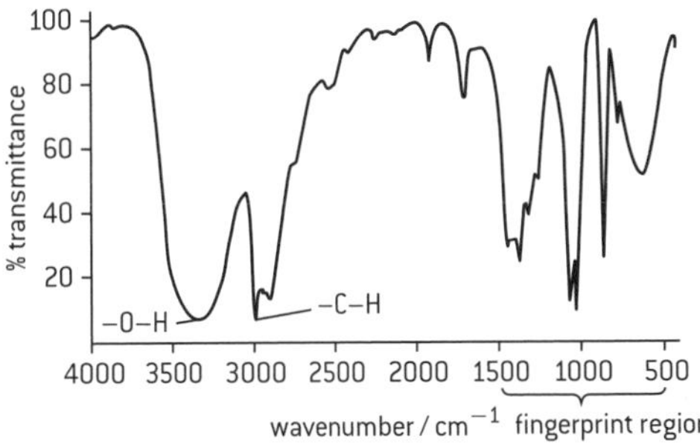

Infrared spectrum of ethyl ethanoate $CH_3-C(=O)-O-C_2H_5$

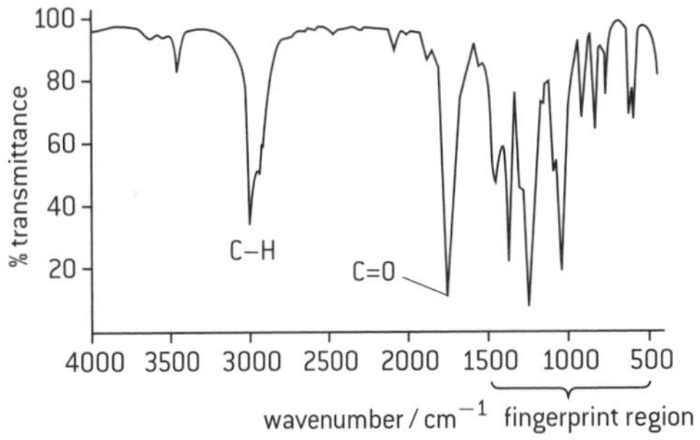

PROTON NUCLEAR MAGNETIC RESONANCE SPECTROSCOPY (^{1}H NMR)

Whereas infrared spectroscopy gives information about the types of bonds in a molecule, ^{1}H NMR spectroscopy provides information on the chemical environment of all the hydrogen atoms in the molecule. The nuclei of hydrogen atoms possess spin and can exist in two possible states of equal energy. If a strong magnetic field is applied the spin states may align themselves either with the magnetic field, or against it, and there is a small energy difference between them. The nuclei can absorb a photon of energy when transferring from the lower to the higher spin state. The photon's energy is very small and occurs in the radio region of the spectrum. The precise energy difference depends on the chemical environment of the hydrogen atoms.

The position in the ^{1}H NMR spectrum where the absorption occurs for each hydrogen atom in the molecule is known as the chemical shift, and is measured in parts per million (ppm). The area under each peak corresponds to the number of hydrogen atoms in that particular environment.

A simplified correlation chart

Type of proton	Chemical shift / ppm
$R-CH_3$	0.9–1.0
$-CH_2-R$	1.3–1.4
$RO-C(=O)-CH_2-$	2.0–2.5
$R-C(=O)-CH_2-$	2.2–2.7
$R-O-H$	1.0–6.0 (note the large range)
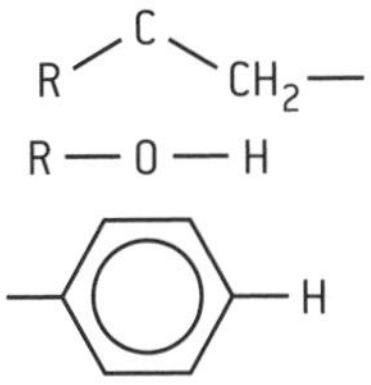	6.9–9.0
$R-C(=O)-H$	9.4–10.0

^{1}H NMR spectrum of ethyl ethanoate $CH_3-C(=O)-O-CH_2CH_3$

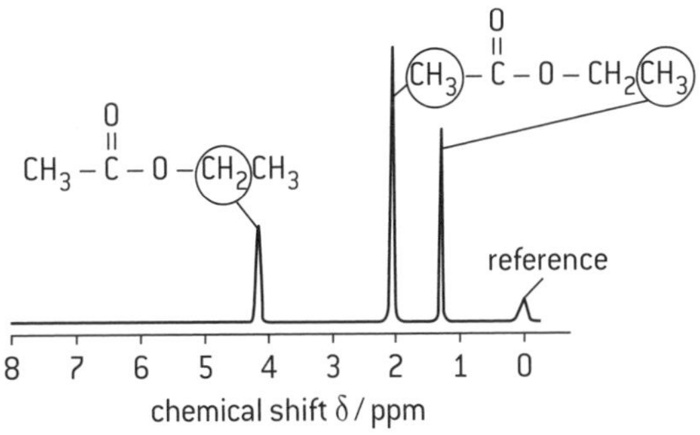

^{1}H NMR spectrum of ethanol CH_3CH_2OH

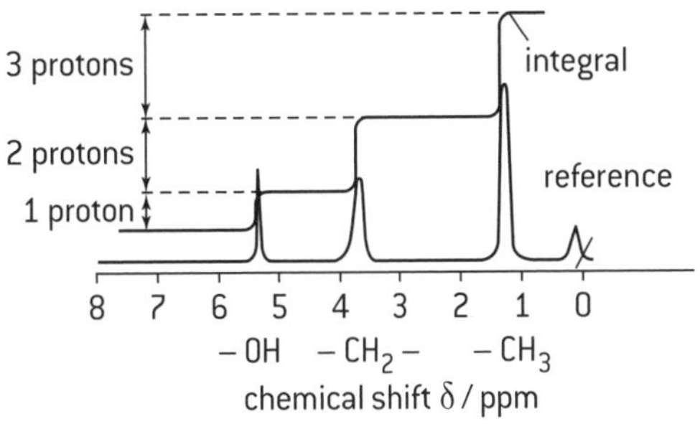

The additional trace integrates the area under each peak. The height of each section is proportional to the number of hydrogen atoms in each chemical environment.

Nuclear magnetic resonance (NMR) spectroscopy

THE MAIN FEATURES OF ^{1}H NMR SPECTRA

1. **The number of different absorptions (peaks)**
   Each proton in a particular chemical environment absorbs at a particular frequency. The number of peaks thus gives information as to the number of different chemical environments occupied by the protons.
2. **The area under each peak**
   The area under each absorption peak is proportional to the number of hydrogen atoms in that particular chemical environment. Normally each area is integrated and the heights of the integrated traces can be used to obtain the ratio of the number of hydrogen atoms in each environment.
3. **The chemical shift**
   Because spinning electrons create their own magnetic field the surrounding electrons of neighbouring atoms can exert a shielding effect. The greater the shielding the lower the frequency for the resonance to occur. The 'chemical shift' (δ) of each absorption is measured in parts per million (ppm) relative to a standard. The normal standard is tetramethylsilane (TMS) which is assigned a value of 0 ppm. A table of chemical shifts is to be found in the IB data booklet.
4. **Splitting pattern**
   In ^{1}H NMR spectroscopy the chemical shift of protons within a molecule is slightly altered by protons bonded to adjacent carbon atoms. This spin–spin coupling shows up in high resolution ^{1}H NMR as splitting patterns. If the number of adjacent equivalent protons is equal to n then the peak will be split into (n + 1) peaks.

TMS AS THE REFERENCE STANDARD

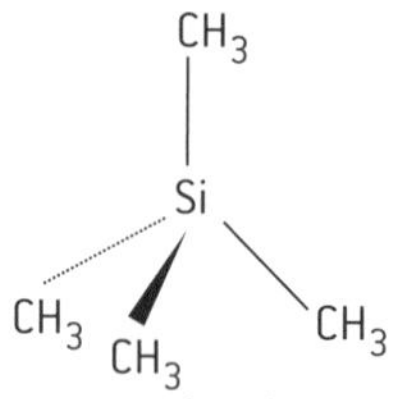

tetramethylsilane

The advantages of using tetramethylsilane $Si(CH_3)_4$ as the standard reference are:

- All the protons are in the same environment so it gives a strong single peak.
- It is not toxic and is very unreactive (so does not interfere with the sample).
- It absorbs upfield well away from most other protons.
- It is volatile (has a low boiling point) so can easily be removed from the sample.

SPIN–SPIN COUPLING

Splitting patterns are due to spin–spin coupling. For example, if there is one proton (n = 1) adjacent to a methyl group then it will either line up with the magnetic field or against it. The effect will be that the methyl protons will thus experience one slightly stronger and one slightly weaker external magnetic field resulting in an equal splitting of the peak. This is known as a doublet (n + 1 = 2).

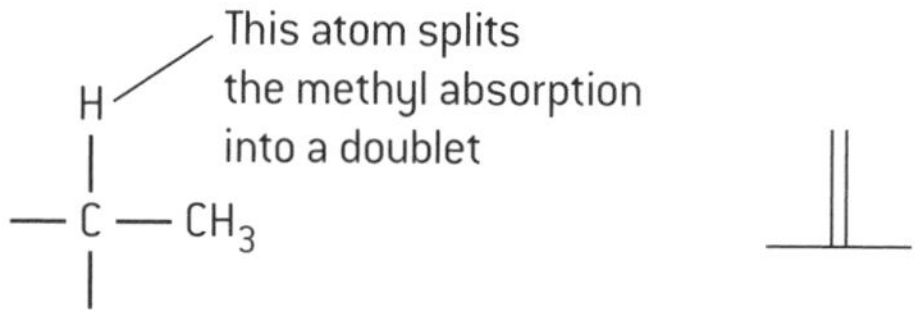

If there is a $-CH_2-$ group (n = 2) adjacent to a methyl group then there are three possible energy states available.

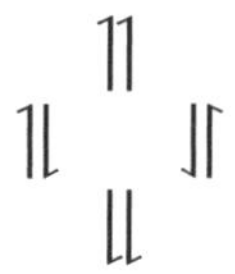

1. Both proton spins are aligned with the field
2. One is aligned with the field and one against it (2 possible combinations)
3. Both are aligned against the field.

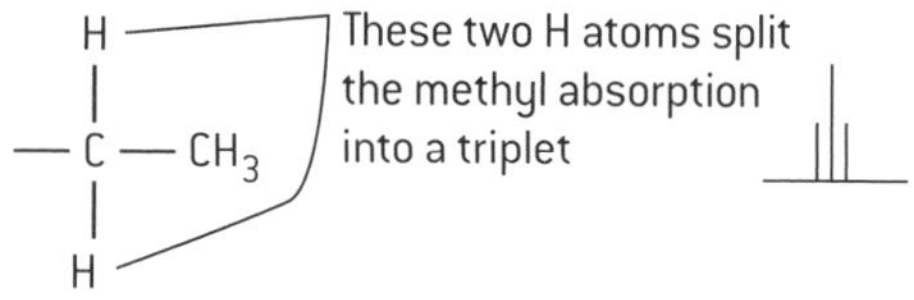

This results in a triplet with peaks in the ratio of 1:2:1.

The pattern of splitting can always be predicted using Pascal's triangle to cover all the possible combinations.

Number of adjacent protons (n)	Splitting pattern	Type of splitting
0	1	singlet
1	1 1	doublet
2	1 2 1	triplet
3	1 3 3 1	quartet
4	1 4 6 4 1	quintet

Thus a methyl group (n = 3) next to a proton will result in the absorption for that proton being split into a quartet with peaks in the ratio 1:3:3:1.

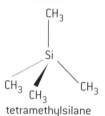

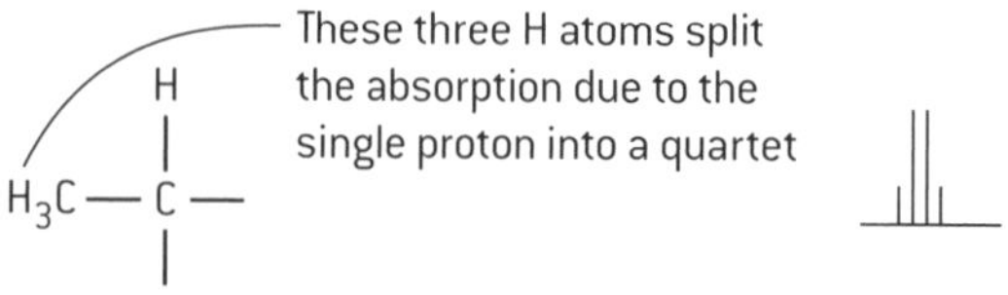

HL Applications of ^{1}H NMR spectroscopy

INTERPRETING A ^{1}H NMR SPECTRUM

1. The three different peaks show that the hydrogen atoms within the molecule are in three different chemical environments.
2. The integrated trace shows that the hydrogen atoms are in the ratio 2:3:3.
3. The chemical shifts of the three peaks identify them as:

   $R{-}CH_3$ 0.9 ppm, $CH_3{-}C(=O){-}$ 2.0 ppm and $R{-}CH_2{-}C(=O){-}$ 2.3 ppm
4. The $-CH_2-$ group has three adjacent protons so is split into a quartet, (n + 1 = 4).

   The $CH_3{-}C(=O){-}$ protons contain no adjacent protons so no splitting occurs.

   The CH_3- group next to the $-CH_2-$ group is split into a triplet (n + 1 = 3).

^{1}H NMR spectrum of butanone

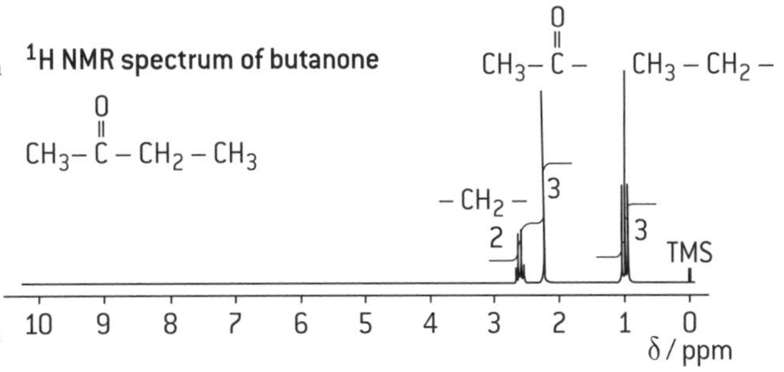

FURTHER EXAMPLES OF ^{1}H NMR SPECTRA INVOLVING SPLITTING PATTERNS

^{1}H NMR spectrum of phenylpropanone $C_6H_5{-}C(=O){-}CH_2{-}CH_3$

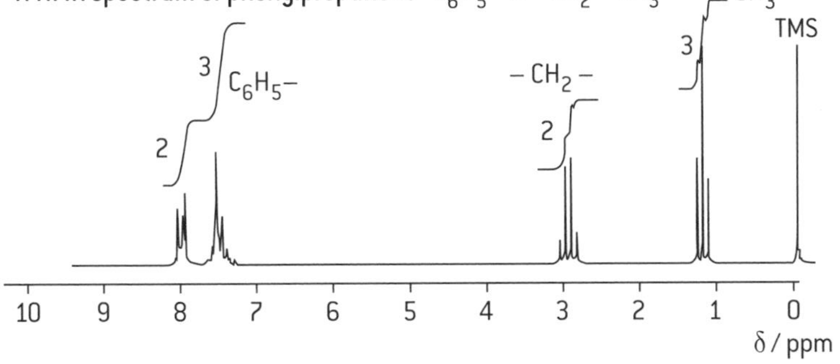

^{1}H NMR spectrum of 1,1-dichloroethane $CHCl_2CH_3$

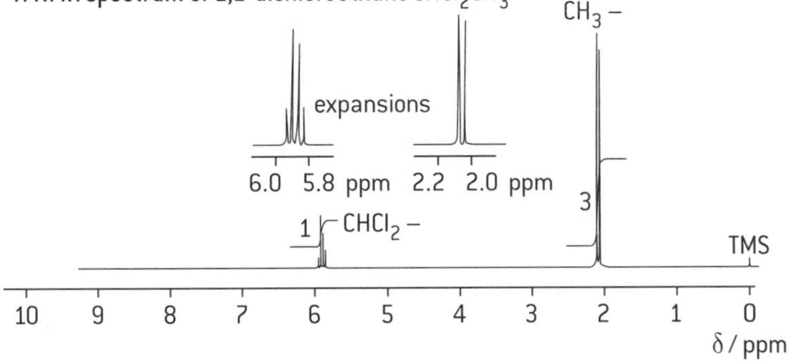

USES OF NMR SPECTROSCOPY

1. **Structural determination**
   ^{1}H NMR is a particularly powerful tool in structural determination as it enables information to be gained on the precise chemical environment of all the protons in the molecule. Similarly ^{13}C and other forms of NMR can also provide very detailed structural information including, for example, distinguishing between cis- and trans-isomers in organometallic compounds.
2. **Medicinal uses**
   NMR is particularly useful in medicine as the energy of the radio waves involved is completely harmless and there are no known side effects. ^{31}P is particularly useful in determining the extent of damage following a heart attack and in monitoring the control of diabetes. ^{1}H NMR is used in body scanning. The whole body of the patient can be placed inside the magnet of a large NMR machine. Protons in water, lipids and carbohydrates give different signals so that an image of the body can be obtained. This is known as MRI (magnetic resonance imaging). The image can be used to diagnose and monitor conditions, such as cancer, multiple sclerosis and hydrocephalus.

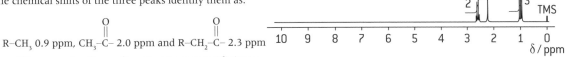

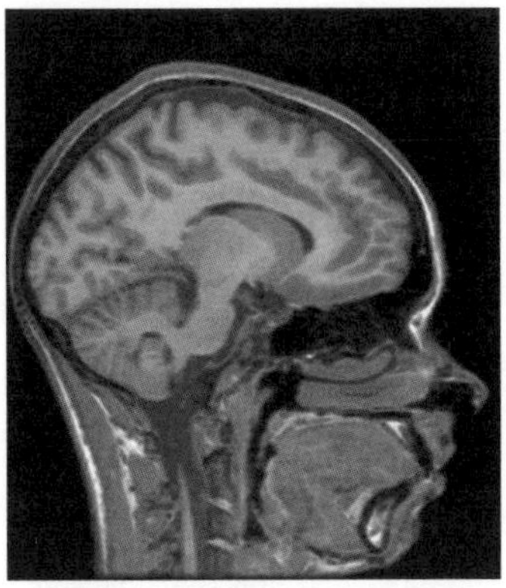

An MRI image of the human brain

Combination of different analytical techniques to determine structure

The determination of the organic structure of an unknown compound is usually achieved by combining the information from several different analytical techniques. This is illustrated by the following worked example for **Compound X**.

ELEMENTAL ANALYSIS

Compound X was found to contain 48.63% carbon, 8.18% hydrogen, and 43.19% oxygen by mass.

From this information the empirical formula of **Compound X** can be deduced as $C_3H_6O_2$.

Element	Amount/mol	Simplest ratio
C	48.63/12.01 = 4.05	3
H	8.18/1.01 = 8.10	6
O	43.19/16.00 = 2.70	2

INFRARED SPECTROSCOPY

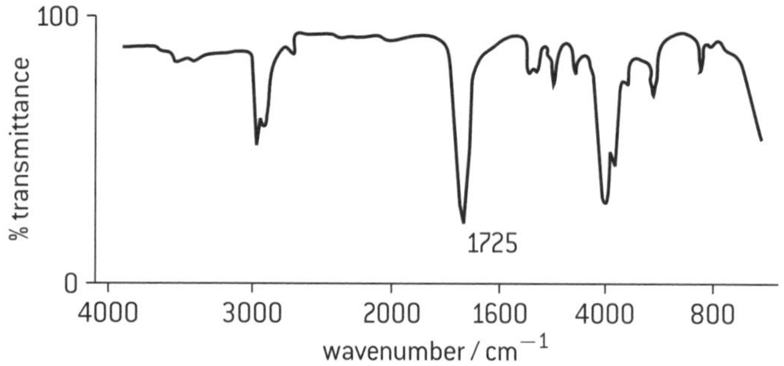

IR spectrum of compound X

Information available from the infrared spectrum:

- Absorption at 2980 cm^{-1} due to presence of C–H in **Compound X**.
- Absorption at 1725 cm^{-1} due to presence of C=O in **Compound X**.
- Absorption at 1200 cm^{-1} due to presence of C–O in **Compound X**.
- Absence of broad absorption at 3300 cm^{-1} indicates **Compound X** does not contain O–H.

MASS SPECTROMETRY

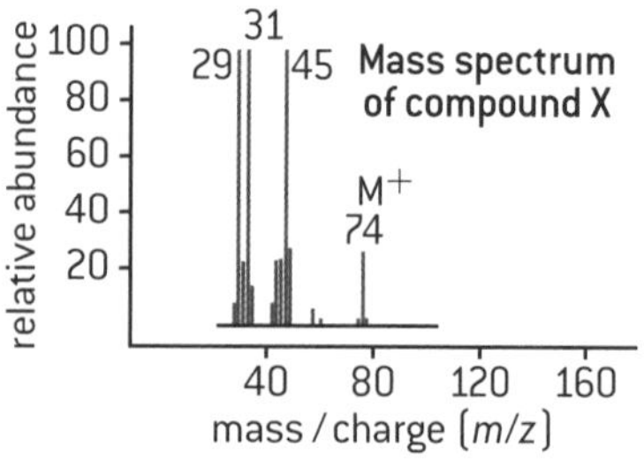

Information available from the mass spectrum:

- Since M^+ occurs at 74 the relative molecular mass of **Compound X** = 74.
- From this and the empirical formula it can be deduced that the molecular formula of **Compound X** is $C_3H_6O_2$.
- Fragment at 45 due to $(M - 29)^+$ so **Compound X** may contain C_2H_5– and/or CHO–.
- Fragment of 31 due to $(M - 43)^+$ so **Compound X** may contain C_2H_5O–
- Fragment at 29 due to $(M - 45)^+$ so **Compound X** may contain

  HOOC– or $H{-}\overset{\overset{O}{\|}}{C}{-}O{-}$.
- Peak at 75 due to the presence of ^{13}C.

^1H NMR SPECTROSCOPY

Information available from the ^1H NMR spectrum:

- Number of separate peaks is three so **Compound X** contains hydrogen atoms in three different chemical environments.
- From the integration trace the hydrogen atoms are in the ratio of 3:2:1 for the peaks at 1.3, 4.2 and 8.1 ppm respectively. Since there are six hydrogen atoms in the molecule this is the actual number of protons in each environment.

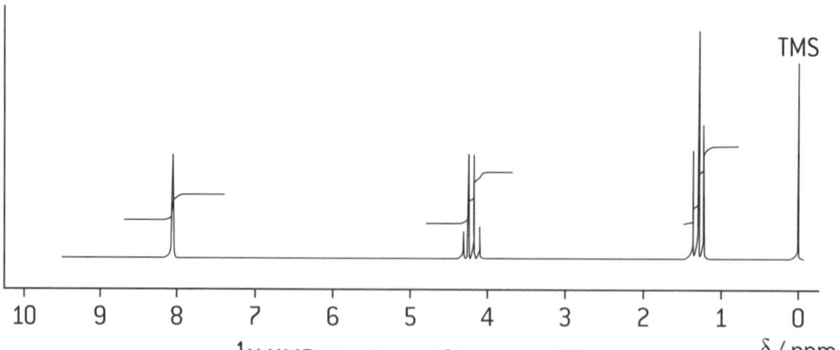

^1H NMR spectrum of compound X

- From the IB data booklet the chemical shift may be attributed to following types of proton.

  1.3 ppm $R{-}CH_3$ (cannot be $R{-}CH_2{-}R$ as it is for three protons)

  4.2 ppm $R{-}\overset{\overset{O}{\|}}{C}{-}O{-}CH_2{-}R$

  8.1 ppm not in IB data booklet but consistent with $H{-}\overset{\overset{O}{\|}}{C}{-}O{-}R$.
- From the splitting patterns the number of adjacent hydrogen atoms can be determined.

1.3 ppm	triplet	two adjacent hydrogen atoms
4.2 ppm	quartet	three adjacent hydrogen atoms
8.1 ppm	singlet	no adjacent hydrogen atoms

IDENTIFICATION

All the above information is consistent with only one definitive structure:

Compound X is: $H{-}\overset{\overset{O}{\|}}{C}{-}O{-}CH_2CH_3$
ethyl methanoate

HL X-ray crystallography

X-RAY CRYSTALLOGRAPHY

Atoms or ions can never be 'seen' in the conventional sense as atomic radii and bond lengths are smaller than the wavelength of visible light. However X-rays, which have much smaller wavelengths than visible light, can be used to determine the molecular and atomic structure of crystals where the particles are arranged in a regular array. If the distance between two layers in a crystal structure is d, then by relatively simple mathematics it can be seen that an incident ray hitting the surface of a plane at an angle θ has to travel a distance equal to 2d sinθ further if it is diffracted from the layer beneath. For the X-rays to remain in phase the extra distance must be equal to a whole number of wavelengths. This gives what is known as the Bragg equation:

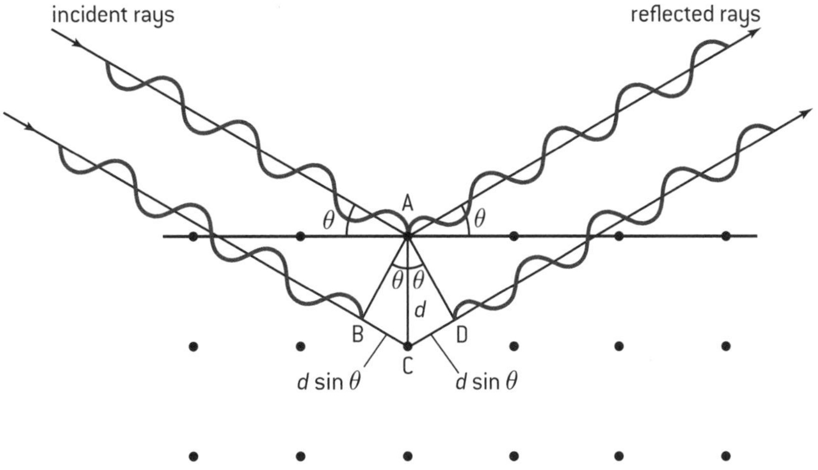

$n\lambda = 2d \sin\theta$ (where n is an integer)

By measuring the angles and intensities of these diffracted beams a three-dimensional picture of the position of all the atoms within the crystal can be obtained. This definitively gives the distances between all the atoms (bond lengths) and the bond angles and hence the unambiguous molecular or ionic structure. As well as for simple molecules such as ice, X-ray crystallography has been used to determine the structure of complex molecules such as DNA and proteins. Although the structure of compounds can be deduced from a combination of spectroscopic techniques, X-ray crystallography is still the main method used to confirm the exact structure of a new compound.

MULTIPLE CHOICE QUESTIONS – MEASUREMENT, DATA PROCESSING AND ANALYSIS

1. A 25.0 cm³ sample of a base solution of unknown concentration is to be titrated with a solution of acid of unknown concentration. Which of the following technique errors would give a value for the concentration of the base that is **too high**?

   I. The pipette that is used to deliver the base solution is rinsed only with distilled water before delivering the sample to be titrated.

   II. The burette that is used to measure the acid solution is rinsed with distilled water but not with the solution of the titration.

   A. I only C. Both I and II
   B. II only D. Neither I nor II

2. In a school laboratory, which of the items listed below has the greatest relative uncertainty in a measurement?

   A. A 50 cm³ burette when used to measure 25 cm³ of water
   B. A 25 cm³ pipette when used to measure 25 cm³ of water
   C. A 50 cm³ graduated cylinder when used to measure 25 cm³ of water
   D. An analytical balance when used to weigh 25 g of water

3. A piece of metallic indium with a mass of 15.456 g was placed in 49.7 cm³ of ethanol in a graduated cylinder. The ethanol level was observed to rise to 51.8 cm³. From these data, the best value one can report for the density of indium is

   A. 7.360 g cm^{-3} C. 1.359×10^{-1} g cm^{-3}
   B. 7.4 g cm^{-3} D. 32.4 g cm^{-3}

4. A mixture of sodium chloride and potassium chloride is prepared by mixing 7.35 g of sodium chloride with 6.75 g of potassium chloride. The total mass of the salt mixture should be reported to __________ significant figures; the mass ratio of sodium chloride to potassium chloride should be reported to __________ significant figures, and the difference in mass between sodium chloride and potassium chloride should be reported to __________ significant figures. The numbers required to fill the blanks above are, respectively,

   A. 4, 3, 2 C. 3, 3, 1
   B. 4, 2, 2 D. 4, 3, 1

5. Repeated measurements of a quantity can reduce the effects of

   A. both random and systematic errors
   B. neither random nor systematic errors
   C. only systematic errors
   D. only random errors

6. A 50.0 cm³ pipette with an uncertainty of 0.1 cm³ is used to measure 50.0 cm³ of 1.00 ± 0.01 mol dm^{-3} sodium hydroxide solution. The amount in moles of sodium hydroxide present in the measured volume is

   A. 0.0500 ± 0.0010
   B. 0.0500 ± 0.0001
   C. 0.0500 ± 0.0012
   D. 0.0500 ± 0.0006

7. Consider the following three sets each of five measurements of the same quantity which has an accurate value of 20.0.

I.	II.	III.
19.8	19.2	20.0
17.2	19.1	19.9
18.3	19.3	20.0
20.1	19.2	20.1
18.4	19.2	20.0

Which results can be described as precise?

A. I, II and III
B. II only
C. II and III
D. III only

8. A thermometer with an accuracy of ± 0.2 °C was used to record an initial temperature of 20.2 °C and a final temperature of 29.8 °C. The temperature rise was

A. 9.6 ± 0.4 °C
B. 9.6 ± 0.2 °C
C. 10 °C
D. 9.6 ± 0.1 °C

9. An experiment to determine the molar mass of solid hydrated copper(II) sulfate, $CuSO_4.5H_2O$ gave a result of 240 g. The experimental error was

A. 0.04%
B. 4%
C. 10%
D. 50%

10. Which sketch graph shows interpolation to find an unknown value.

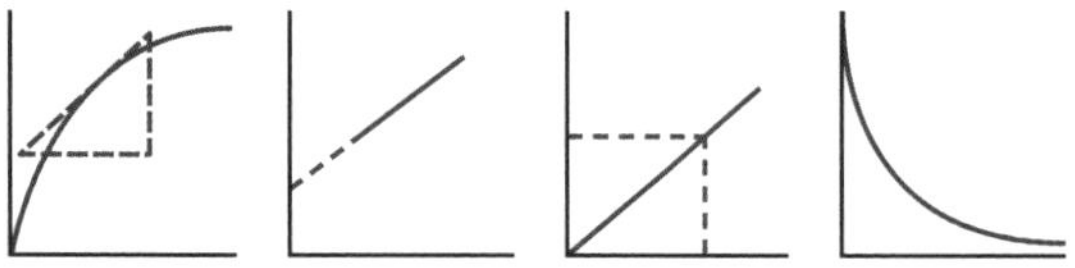

A. B. C. D.

11. What information can be obtained from the number of peaks in the ^1H NMR spectrum of a compound?

A. The number of different chemical environments occupied by the protons in one molecule of the compound.

B. The number of hydrogen atoms in one molecule of the compound.

C. The number of different functional groups in one molecule of the compound.

D. The number of carbon atoms in one molecule of the compound.

12. What is the index of hydrogen deficiency (IHD) for a compound with the molecular formula C_7H_8?

A. 3
B. 4
C. 6
D. 8

13. The mass spectrum of a compound shows peaks with *m*/*z* values of 88, 73, 59 and 43. Which compound could give this spectrum?

A. $HCOCH_2CH_2CH_3$
B. $CH_3CH_2COOCH_2CH_3$
C. $CH(CH_3)_2COOH$
D. $CH_3CH_2CH_2COOH$

14. Which is correct about the regions of the electromagnetic spectrum?

	Region of spectrum	Energy	Wavelength
A.	Ultraviolet	Highest	Longest
B.	Infrared	Highest	Shortest
C.	Radio	Lowest	Longest
D.	Visible	Lowest	Shortest

HL

15. The ^1H NMR spectrum of a particular compound shows three separate peaks. The relative height of each integration trace and the splitting patterns are

Peak number	Integration trace	Splitting pattern
1	3	triplet
2	2	quartet
3	1	singlet

Which compound could give this spectrum?

A. $HCOOCH_2CH_3$
B. $CH_3CH_2CH_2COOH$
C. $CH_3CH_2COOCH_3$
D. $CH_3CH(OH)CH_3$

16. Which are advantages for using tetramethylsilane as a standard reference in ^1H NMR?

I. All its protons are in the same chemical environment.

II. It is chemically unreactive.

III. It does not absorb energy in the same region as most other protons.

A. I and II only
B. I and III only
C. II and III only
D. I, II and III

17. Which compound will contain a peak with a triplet splitting pattern in its ^1H NMR spectrum?

A. CH_3OH
B. CH_3CH_2COOH
C. $CH_3CHClCOOH$
D. $(CH_3)_3COH$

18. Which can be determined by single crystal X-ray crystallography?

I. Bond lengths

II. Bond angles

III. Chemical structure

A. I and II only
B. I and III only
C. II and III only
D. I, II and III

SHORT ANSWER QUESTIONS – MEASUREMENT, DATA PROCESSING AND ANALYSIS

1. A simulated computer-based experiment was performed to determine the volume of nitrogen generated in an airbag from the decomposition of sodium azide, NaN_3.

   $2NaN_3(s) \rightarrow 2Na(s) + 3N_2(g)$

   The following data were entered into the computer.

Mass of NaN_3(s)/ kg($\pm$0.001 kg)	Temperature / °C ($\pm$0.50 °C)	Pressure / kPa $\pm$1 kPa
0.072	20.00	106

   a) State the number of significant figures for the mass, temperature and pressure data. [1]

   b) Calculate the amount (in mol) of sodium azide present. [2]

   c) Calculate the percentage uncertainty for each of the mass, temperature and pressure. [1]

   d) Determine the volume of nitrogen gas (in dm^3) produced under these conditions together with its uncertainty assuming complete decomposition of the sodium azide. [5]

2. a) Elemental analysis of Compound **X** shows that it contains 15.40% carbon, 3.24% hydrogen and 81.36% iodine by mass. Determine the empirical formula of Compound **X**. [1]

   b) The mass spectrum of Compound **X** is shown below.

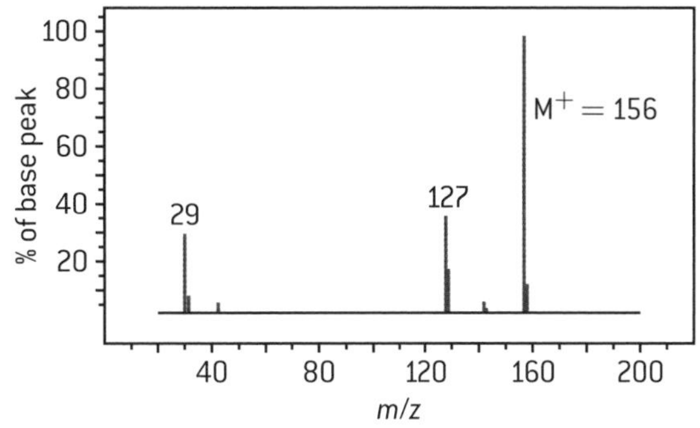

   (i) Deduce the molecular formula of Compound **X**. [1]

   (ii) Calculate the index of hydrogen deficiency (IHD) for Compound **X**. [1]

   (iii) Identify the fragments responsible for the peaks with *m/z* values of 127 and 29. [2]

   (iv) Deduce the identity of Compound **X**. [1]

   c) The ¹H NMR spectrum for Compound **X** shows two peaks with integration traces of two units and three units respectively. Explain how this information is consistent with your answer to (b) (iv). [2]

3. ¹H NMR spectroscopy can be used to distinguish between pentan-2-one and pentan-3-one. In each case state the number of peaks in their ¹H NMR spectra and state the ratios of the areas under each peak. [4]

HL

4. The ¹H NMR spectra of an unknown compound is given below:

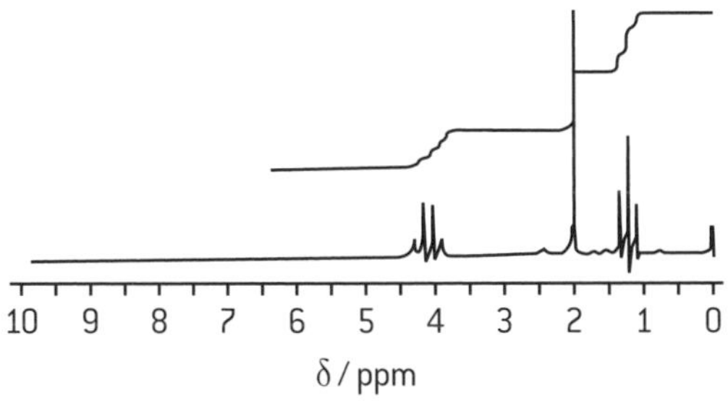

   a) Why is tetramethylsilane used in ¹H NMR spectroscopy? [1]

   b) What word is used to describe the multiplicity of the peaks centred at 4.1 ppm? [1]

   c) What is the ratio of the number of hydrogen atoms responsible for the chemical shifts centred at 1.2, 2.0 and 4.1 ppm respectively? [1]

   d) Two structures that have chemical shifts centred at 4.1 ppm are $RCOOCH_2R$ and $C_6H_5OCOCH_3$. From consideration of the rest of the spectra only one of these general structures is possible. Identify which one and explain your reasoning. [2]

   Below is the infrared spectrum of the same compound:

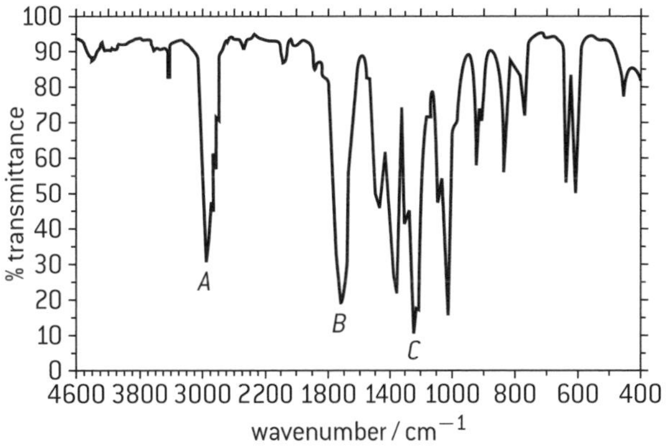

   e) Identify which particular vibrations are responsible for the peaks labelled *A*, *B* and *C*. [3]

   f) The mass specturm of the same compund shows a molecular ion peak at 88 *m/z*. The fragmentation pattern shows prominent peaks at 73 *m/z* and 59 *m/z* amongst others. Identify the ions responsible for these peaks. [3]

   g) Give the name and structural formula of the compound. [2]

Introduction to materials science

CLASSIFICATION OF MATERIALS AND TRIANGULAR BONDING DIAGRAM

Civilizations have been characterised by the materials they used, e.g. Stone Age, Iron Age, Bronze Age and now perhaps the Plastics Age. Materials can be classified in several different ways based upon their properties, their uses or on the type of bonding or structure they exhibit. Materials science investigates the relationship between the bonding and structures of materials on the atomic or molecular level and relates this to their macroscopic properties. The key to understanding the macroscopic properties is thus the identity of the type or types of bonding present in the material. Metallic substances tend to have very different properties to non-metals and the properties of ionic compounds differ greatly to covalent compounds. The degree of covalent, ionic or metallic character in a compound can be deduced from its position on a triangular bonding diagram – also known as a Van Arkel–Ketelaar diagram. Bond triangles show the three extremes of bonding in terms of electronegativity difference (ionicity) plotted against the average electronegativity (localization) for binary compounds with the three extreme cases of types of bonding – ionic, covalent and metallic – at the apices of an equilateral triangle.

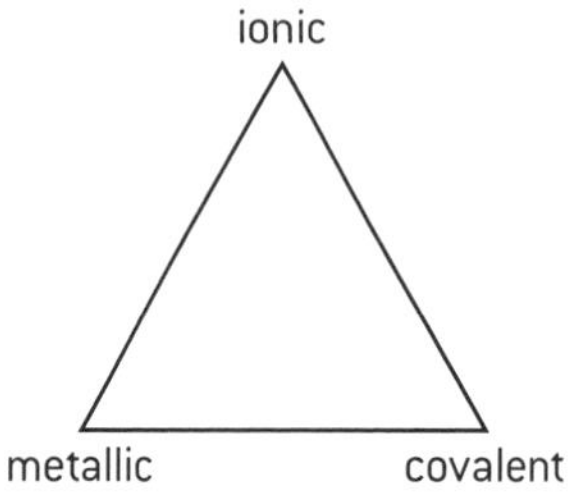

These three extreme cases are characterized by caesium (metallic), fluorine (covalent) and caesium fluoride (ionic). Other materials can be placed around the triangle based on their differences in electronegativities. The bottom side of the triangle, from metallic (low electronegativity) to covalent (high electronegativity) is for elements and compounds with varying degree of delocalization and directionality in the bond. Pure metals with only delocalized bonding occur on the extreme left (metallic corner) and at the other extreme are covalent compounds where all the electron orbitals overlap in a particular direction. The left side of the triangle (from ionic to metallic) is for delocalized bonds with varying electronegativity difference and the right side shows the change from covalent to ionic. The diagram below shows that compounds such as silicon tetrachloride and carbon dioxide are very much covalent whereas silicon dioxide is polar covalent but a compound such as aluminium chloride lies very close to the border between covalent and ionic. Similarly lithium oxide has a high degree of ionic character whereas lithium hydride lies close to the border of ionic and metallic.

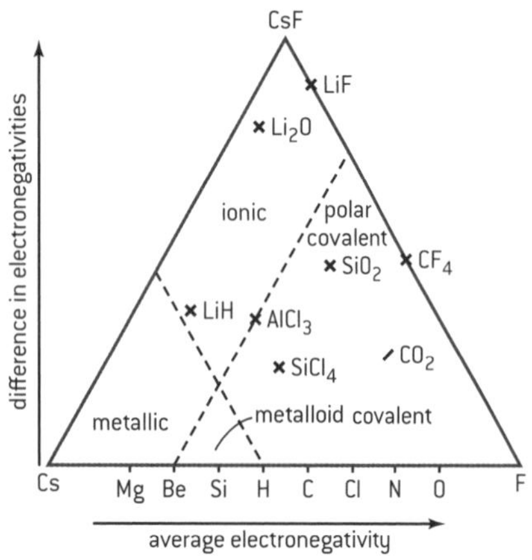

COMPOSITES

Composites are made from two or more materials with significantly different properties which when combined produce a material with different characteristics (e.g. lighter and stronger) to its constituent parts. The composite consists of two distinct phases, a reinforcing phase which is embedded in a matrix phase. Common composites include building materials such as concrete and cement and reinforced plastics and ceramics.

RELATIONSHIPS BETWEEN PHYSICAL CHARACTERISTICS AND BONDING AND STRUCTURE

Melting point. Melting points follow a similar trend to boiling points in that the weaker the attractive forces between the particles the lower the melting point. But they are also influenced by the type of packing between the particles in the solid state. Thus the melting point of metals can vary from quite low (caesium melts at 28.5 °C) to very high (copper melts at 1085 °C) even though they all contain delocalized electrons. Giant molecular covalent substances such as diamond and silicon dioxide have high melting points, whereas simple covalent molecules such as hydrogen gas and carbon dioxide have low melting points.

Conductivity. Materials containing delocalized electrons such as metals and graphite are good conductors of heat and electricity. Covalent compounds containing electrons localized in fixed orbitals are poor conductors. Ionic compounds can only conduct electricity when their ions are free to move in the molten state or in aqueous solution but are chemically decomposed in the process (see page 30).

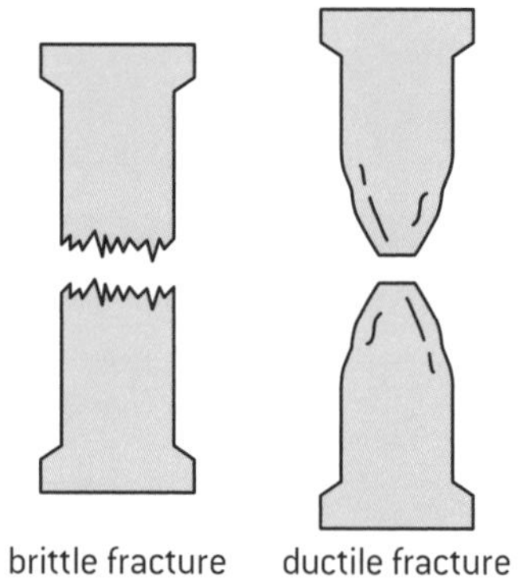

Brittleness. Many metals are malleable (the opposite of brittle) and ductile (can be drawn into a wire) as the close-packed layers of positive ions can slide over each other without breaking more bonds than are made. Impurities added to the metal disturb the lattice and so make the metal less malleable, which explains why most alloys are harder than the pure metal they are made from (see page 31). Many covalent substances such as sulfur are brittle, that is they fracture when subjected to stress rather than undergo deformation as the bonds cannot be reformed once broken. Because brittle substances do not undergo deformation when fractured the two parts can often be glued together to return to the original shape.

Permeability is defined as the ability of a porous material to allow fluids such as water to pass through it. This depends both upon the type of bonding in the material, which can attract or repel polar water molecules, and the spaces between the particles in the lattice.

Elasticity is the tendency of a solid material to return to its original shape after it has been deformed by the application of a force. In polymers such as rubber the polymer chain can be stretched without being broken when forces are applied. In metals the atoms can slide over each other so that the lattice changes size and shape and can return to the original shape when the force is removed.

Principles of extraction of metals from their ores

EXTRACTION OF METALS FROM THEIR ORES

The essential process involved in the extraction of metals form their ores is reduction. The ease with which this process occurs depends upon the position of the metal in the activity series. Metals very low in the activity series, such as gold, silver and copper may be found uncombined. Metals slightly higher in the series can be obtained either by simply heating the ore strongly or by using chemical reducing agents, such as more reactive metals, carbon or carbon monoxide. Generally metals that can be obtained by relatively simple chemical reduction have been known since ancient times. It is much harder and more expensive to reduce chemically the more reactive metals and it took the discovery of electricity in the 19th century before these could be produced commercially. It is still an expensive process as molten ores must be used. To reduce the energy required, and hence the cost of production, impurities are often added to the molten electrolyte to lower the melting point.

Methods of extracting some important metals

Activity series	Main ore	Method of extraction
Sodium	NaCl	Electrolysis of molten NaCl with $CaCl_2$ added to lower the melting point. $Na^+(l) + e^- \rightarrow Na(l)$
Aluminium	Al_2O_3	Electrolysis of Al_2O_3 in molten cryolite. $Al^{3+}(l) + 3e^- \rightarrow Al(l)$
Zinc	ZnS	Roast to form the oxide $2ZnS(s) + 3O_2(g) \rightarrow 2ZnO(s) + 2SO_2(g)$ then reduce chemically with carbon monoxide $ZnO(s) + CO(g) \rightarrow Zn(s) + CO_2(g)$ or electrolytically $Zn^{2+}(l) + 2e^- \rightarrow Zn(l)$
Iron	Fe_2O_3	Heat with carbon monoxide $Fe_2O_3(s) + 3CO(g) \rightarrow 2Fe(l) + 3CO_2$
Lead	PbS	Heat to form the oxide then reduce with carbon $PbO(s) + C(s) \rightarrow Pb(l) + CO(g)$
Copper	Cu or $CuFeS_2$	Heat in air to give copper and sulfur dioxide
Gold	Au	Metal found uncombined

PRODUCTION OF ALUMINIUM

Aluminium is primarily made by the electrolytic reduction of aluminium oxide.

The main ore of aluminium is bauxite. The aluminium is mainly in the form of the hydroxide $Al(OH)_3$ and the principal impurites are iron(III) oxide and titanium hydroxide. The impurities are removed by heating powdered bauxite with sodium hydroxide solution. The aluminium hydroxide dissolves because it is amphoteric.

$$Al(OH)_3(s) + NaOH(aq) \rightarrow NaAlO_2(aq) + 2H_2O(l)$$

The aluminate solution is filtered leaving the impurities behind. Seeding with aluminium hydroxide then reverses the reaction. The pure recrystallized aluminium hydroxide is then heated to produce aluminium oxide (alumina).

$$2Al(OH)_3(s) \rightarrow Al_2O_3(s) + 3H_2O(l)$$

In a separate process hydrogen fluoride is added to the aluminate solution followed by sodium carbonate to precipitate cryolite (sodium hexafluoroaluminate(III), Na_3AlF_6).

$$NaAlO_2(aq) + 6HF(g) + Na_2CO_3(aq) \rightarrow Na_3AlF_6(s) + 3H_2O(l) + CO_2(g)$$

The electrolysis of molten alumina takes places in an open-topped steel container lined with graphite. Alumina has a melting point of 2045 °C so it is mixed with cryolite. The resulting solution melts at about 950 °C so that much less energy is required. The aluminium is produced on the graphite lining which acts as the negative electrode (cathode). Molten aluminium is more dense than cryolite so it collects at the bottom of the cell where it can be syphoned off periodically.

$$Al^{3+}(l) + 3e^- \rightarrow Al(l)$$

The positive electrode is made of blocks of graphite. As the oxide ions are oxidized some of the oxygen formed reacts with the graphite blocks so that they have to be renewed regularly.

$$2O^{2-}(l) \rightarrow O_2(g) + 4e^-$$

$$C(s) + O_2(g) \rightarrow CO_2(g)$$

A modern cell can produce up to two tonnes of aluminium per day.

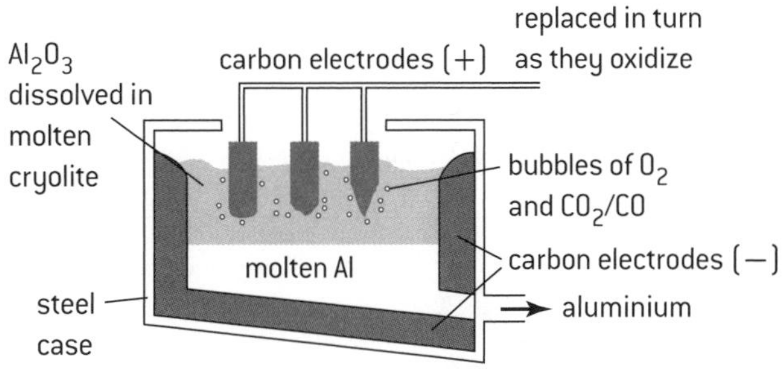

Production of aluminium by electrolysis

Faraday calculations and properties and analysis of alloys

DETERMINATION OF THE MASS OF METAL PRODUCED BY ELECTROLYSIS

The amount (in mol) of a metal produced by electrolysis depends upon the charge of electricity received by the molten electrolyte and the charge on the metal ion. Three times as much charge of electricity will be required to produce one mole of aluminium from Al^{3+} ions compared with one mole of sodium from Na^{+} ions. The charge passed through the electrolyte is equal to the size of the current flowing multiplied by the length of time it flows. A charge of one coulomb is produced by a current of one amp flowing for one second. The charge carried by one mole of electrons is equal to 96 500 coulombs and this value is known as one faraday (F).

Example 1: Calculate the mass of sodium produced at the negative electrode when a steady current of 15.2 A passes through a solution of molten sodium chloride for 3 hours 20 minutes.

$$\text{Charge passed} = 15.2 \times 200 \times 60 = 182\,400 \text{ C}$$

$$= \frac{182\,400}{96\,500} = 1.89 \text{ F}$$

$Na^{+}(l) + e^{-} \rightarrow Na\ (l)$ so one faraday required to produce one mole of sodium.

Amount of sodium produced = 1.89 mol

Mass of sodium produced = $1.89 \times 22.99 = 43.5$ g

Example 2: A typical aluminium cell produces 2000 kg of aluminium in 24 hours. Determine the average current that must be passed to achieve this.

$$2000 \text{ kg} = 2000 \times \frac{10^3}{26.98} = 7.413 \times 10^4 \text{ mol}$$

$Al^{3+}(l) + 3e^{-} \rightarrow Al(l)$

Charge required = $3 \times 7.413 \times 10^4 \times 96\,500 = 2.146 \times 10^{10}$ C

$$\text{Average current required for 24 hours} = \frac{2.146 \times 10^{10}}{(24 \times 60 \times 60)}$$

$$= 2.48 \times 10^5 \text{ A}$$

ALLOYS

Alloys are homogeneous mixtures of metals with other metals or non-metals. They have already been discussed together with some examples on page 31. Transition metals often form alloys with each other as their atoms have similar radii and the crystal structure is not seriously disrupted. The addition of other elements into the metallic structure alters the physical properties. Some properties such as density and electrical conductivity may not differ much from the constituent elements but others such as tensile strength and melting point may differ considerably. Steel is an alloy of iron, carbon and other metallic and non-metallic elements. It has a wide range of uses and by adjusting its composition it can be tailor-made with specific properties. For example, the addition of chromium increases the resistance of steel to corrosion. Stainless steel, used for kitchen knives and sinks contains about 18% chromium and 8% nickel. Toughened steel for use in drill bits, which need to retain a sharp cutting edge at high temperatures, contains up to 20% molybdenum.

DIAMAGNETISM AND PARAMAGNETISM

Metals and their compounds can exhibit different types of magnetism. Iron metal and some other metals (e.g. nickel and cobalt) show ferromagnetism. This is a permanent type of magnetism. In this type of magnetism unpaired electrons align parallel to each other in domains irrespective of whether an external magnetic or electric field is present. This property of iron has been utilized for centuries to make compasses which align with the Earth's magnetic field to point north.

Many other materials contain unpaired electrons. Unlike paired electrons, where the spins cancel each other out, the spinning unpaired electrons create a small magnetic field and will line up in an applied electric or magnetic field to make the material weakly magnetic when the field is applied, i.e. they reinforce the external magnetic field. This type of magnetism is known as **paramagnetism**. The more unpaired electrons there are in the material the more paramagnetic the material will be.

When all the electrons in a material are paired up the complex is said to be **diamagnetic**. Essentially the material is non-magnetic although the paired electrons do create a very small magnetic field in opposition to an externally applied field. Because the effect is so small, it is not observable in everyday life except for superconductors. Superconductivity occurs when certain materials are cooled below a characteristic critical temperature, which tends to be close to absolute zero.

DETECTION OF TRACE AMOUNTS OF METALS (ICP–OES AND ICP–MS)

Very small amounts of trace metals in alloys and other materials can be detected and determined quantitatively by Inductively Coupled Plasma (ICP) Spectroscopy together with either Optical Emission Spectroscopy (ICP–OES) or Mass Spectroscopy (ICP–MS) . Plasma is the fourth state of matter (the other three being solid, liquid and gas). A plasma contains charged particles and can be induced by a strong electromagnetic field. In ICP an intense electromagnetic field is created in a 'torch', which ionizes argon gas to form a high-temperature plasma. The sample of material to be analysed is introduced directly into the plasma flame where it is broken down into charged ions to release radiation of characteristic wavelengths, which can be detected in the optical spectrometer part of ICP–OES. By calibrating the instrument with specific amounts of known metals the values of the wavelengths emitted enable each metal to be identified. Since the concentration is proportional to the intensity of the peaks the concentration of each metal can also be determined by comparing the intensity of its peaks with those made by samples of known concentration. In ICP–MS the released ions are identified using a mass spectrometer rather than an optical spectrometer. As well as identifying the amounts of metals in alloys, these two techniques also have many uses in medicine and forensic science (e.g. heavy metal poisoning).

Catalysts

HOMOGENEOUS AND HETEROGENEOUS CATALYSIS

Catalysts function by providing an alternative reaction pathway with a lower activation energy so that more of the reactant particles will possess the necessary minimum energy to react when they collide. Catalysts may be in the same phase as the reactants (usually the same phase as the product(s) too) in which case they are known as **homogeneous catalysts**. A phase is similar to a state except that there is a physically distinct boundary between two phases. It is possible to have a single state but two phases. An example is two immiscible liquids such as oil and water. If the catalyst is in a different phase to the reactants then it is functioning as a **heterogeneous catalyst**.

CHOICE OF CATALYST

The advantage of a homogeneous catalyst compared with a heterogeneous catalyst is that all of the catalyst is exposed to the reactants whereas in heterogeneous catalysis the efficiency of the catalyst is dependent upon the surface area. The disadvantage of a homogenous catalyst is that it is usually harder to remove the catalyst after the reaction whereas a heterogeneous catalyst can be relatively easily removed by filtration. The most efficient catalysts of all are enzymes – biological catalysts – but these are very specific for a particular biological reaction due to the shape of the active site. When choosing a catalyst for a chemical reaction the following factors should be considered.

- Selectivity – Will the catalyst produce only the desired product?
- Efficiency – Will the catalyst cause a considerable increase in the rate? Will it continue to work well under severe conditions, such as those experienced by catalytic converters in cars, as well as mild conditions?
- Environmental impact – Will it be easy to dispose of the catalyst without causing harm to the environment? Many transition metals are classed as heavy metals and can cause problems if they enter the soil or ground water.
- Potential for poisoning – Catalysts rely on reactants occupying the active site reversibly. A poison will occupy the active site irreversibly so blocking access to reactants. Poisons include carbon monoxide, cyanide ions and sulfur.
- Cost – Industry is profit-based. Many transition metals such as rhodium, platinum and palladium are expensive and the cost to benefit ratio needs to be carefully calculated.

MODES OF ACTION OF CATALYSTS

Although homogenous catalysts can be recovered chemically unchanged at the end of the reaction they can form intermediate compounds during the reaction.

For example when nitrogen monoxide catalyses the oxidation of sulfur dioxide it is thought that nitrogen dioxide is formed as an intermediate compound.

$$2NO(g) + O_2(g) \rightarrow 2NO_2(g)$$

This is a redox reaction and the oxidation state of nitrogen has increased from +2 to +4. In the second step the nitrogen dioxide is reduced back to nitrogen monoxide by the sulfur dioxide.

$$2SO_2(g) + 2NO_2(g) \rightarrow 2SO_3(g) + 2NO(g)$$

So that nitrogen monoxide is unchanged at the end of the reaction and the overall equation is

$$2SO_2(g) + O_2(g) \rightarrow 2SO_3(g)$$

Because transition metal ions can exist in more than one oxidation state they tend to make good homogenous catalysts. In the reaction between peroxodisulfate(VI) ions and iodide ions it is thought that the $Fe^{2+}(aq)$ ions are oxidized to $Fe^{3+}(aq)$ ions and then back to $Fe^{2+}(aq)$ ions.

$$S_2O_8^{\ 2-}(aq) + 2Fe^{2+}(aq) \rightarrow 2SO_4^{\ 2-}(aq) + 2Fe^{3+}(aq)$$

$$2I^-(aq) + 2Fe^{3+}(aq) \rightarrow 2Fe^{2+}(aq) + I_2(aq)$$

Heterogeneous catalysts tend to function by adsorbing reactant molecules onto the surface of the catalyst (the active site) and bringing them into close contact with each other in the correct orientation. Many transition metals and their compounds have the ability to physically adsorb large amounts of gases on their surface which makes them particularly good heterogeneous catalysts. For example nickel or palladium can adsorb ethene and hydrogen so that they can react to form ethane.

The reaction of ethene with hydrogen to form ethane

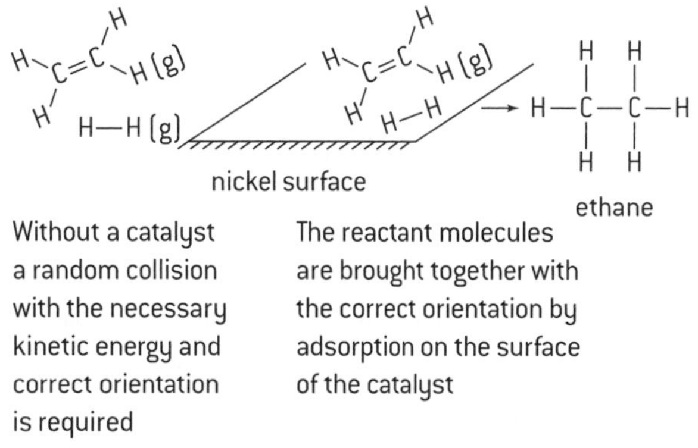
H—H (g)
nickel surface
ethane
Without a catalyst a random collision with the necessary kinetic energy and correct orientation is required
The reactant molecules are brought together with the correct orientation by adsorption on the surface of the catalyst

Two particular types of heterogeneous catalysts are zeolites and carbon nanotube support catalysts. Zeolites are microporous cage-like aluminosilicate compounds with complex formulas. They occur naturally and can also be made synthetically. They work by adsorbing reactant molecules and confining them in a small space so they can react together. They are particularly used in the petrochemical industry, for example as catalysts for cracking larger hydrocarbons into smaller more useful hydrocarbons and alkenes. Carbon nanocatalysts involve using carbon nanotubes (rather than silicon or alumina) as support for metals or metal oxide catalysts. Due to their small size the nanotubes provide a large surface area and other properties such as excellent electron conductivity and chemical inertness help to promote the action of the catalyst and make it much more efficient.

Liquid crystals (1)

LIQUID CRYSTALS

Liquid crystals are a phase or state of matter that lies between the solid and liquid state. In a liquid crystal the molecules tend to retain their orientation as in a solid but they can also move to different positions as in a liquid. The physical properties of liquid crystals (such as electrical conductivity, optical activity and elasticity) depend upon the orientation of the molecules relative to some fixed axis in the material. Examples of substances which can behave as liquid crystals under certain conditions include DNA, soap solution, graphite and cellulose together with some more specialized substances such as biphenyl nitriles.

An example of a biphenyl nitrile is 4-pentyl-4'-cyanobiphenyl (known as 5CB)

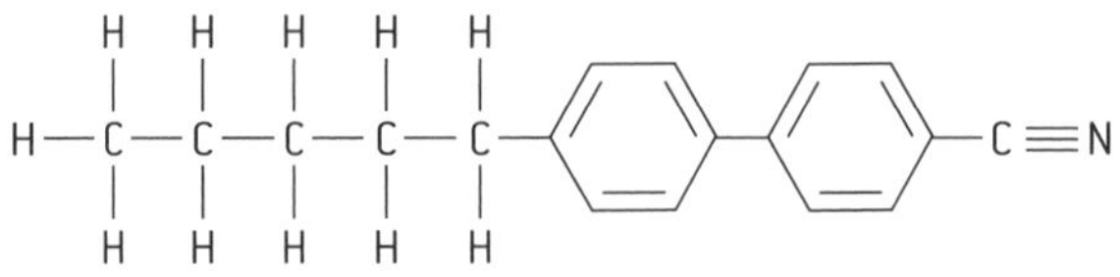

LYOTROPIC AND THERMOTROPIC LIQUID CRYSTALS

Liquid crystals only show liquid crystal properties under certain conditions. They are very sensitive to temperature and concentration. Essentially liquid crystals can be divided into two main types.

Lyotropic liquid crystals are solutions that show the liquid crystal phase at certain concentrations. An example of lyotropic liquid crystals is soap solution. At low dilution the polar soap molecules behave randomly but at higher concentrations they group together into larger units called micelles which in the liquid crystal phase are ordered in their orientation.

Thermotropic liquid crystals are pure substances that show liquid crystal behaviour over a range of temperature between the solid and liquid states. Examples of a thermotropic liquid crystal are biphenyl nitriles used in liquid crystal displays (LCDs).

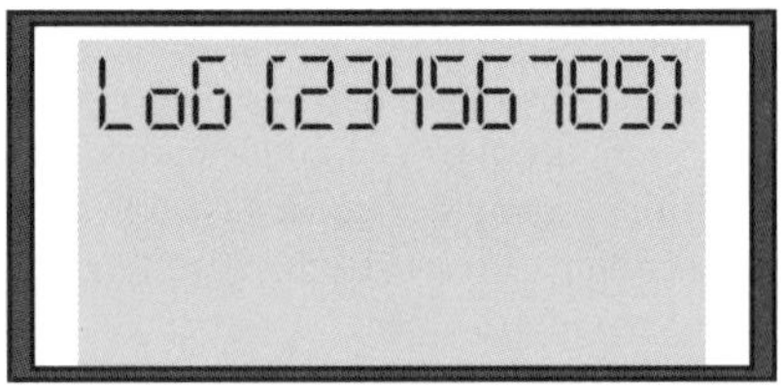

The use of thermotropic liquid crystals in a calculator screen

Within the thermotropic liquid crystal phase the rod-shaped molecules which are typically about 2.5×10^{-9} metres in length exist in groups or domains. The molecules can flow and are randomly distributed as in a liquid but within each domain they all point in the same direction. This is known as the **nematic phase**. As the temperature increases the orientation becomes increasingly more disrupted until eventually the directional order is lost and the normal liquid phase is formed.

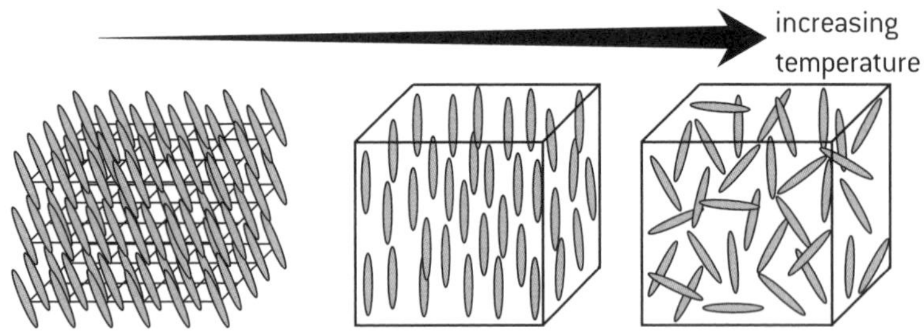

solid state molecules held in rigid 3-D structure

liquid crystal phase molecules within each domain oriented in the same direction

liquid crystal state molecules all randomly distributed

PRINCIPLES OF LCD DEVICES

Since liquid crystal molecules are polar their orientation can be controlled by an applied electric field. The orientation of the molecules affects the ability of the liquid crystal molecules to transmit light. In liquid crystal displays used in digital watches, calculators and laptops a small voltage is applied across a thin film of the material.

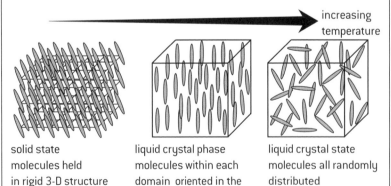

The flat screens used for computer monitors or televisions use liquid crystals

This controls the areas of the display that are light and dark and hence gives the characteristic readings of the pictures or letters. The great advantage of LCDs over other types of electronic display is that they use extremely small electric currents. The disadvantage is that they only work within a certain temperature range which explains why a digital watch or laptop screen may give a strange display in very hot or cold temperatures. Hence for use in an LCD a liquid crystal should:

- be a chemically stable compound
- contain polar molecules
- remain stable in the liquid crystal phase over a suitable range of temperature
- be able to orientate quickly (rapid switching speed).

Liquid crystals (2)

STRUCTURAL FEATURES OF BIPHENYL NITRILES

When the structure of 4-pentyl-4′-cyanobiphenyl is compared with other compounds that have good liquid crystal properties it can be seen that they have several features in common.

- The nitrile group in the biphenyl nitrile (and the –NN$^+$(O$^-$)– and –C=N– groups in the other two molecules) is polar. This ensures that the intermolecular forces are strong enough to align in a common direction.
- The two benzene rings in the molecules ensure that the molecules are rigid and therefore more rod-shaped.
- The long alkane chain group on the end of the molecule ensures that the molecules cannot pack so closely together and so helps to maintain the liquid crystal state.

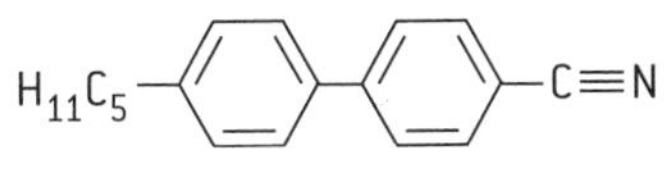

4-pentyl-4'-cyanobiphenyl 5CB

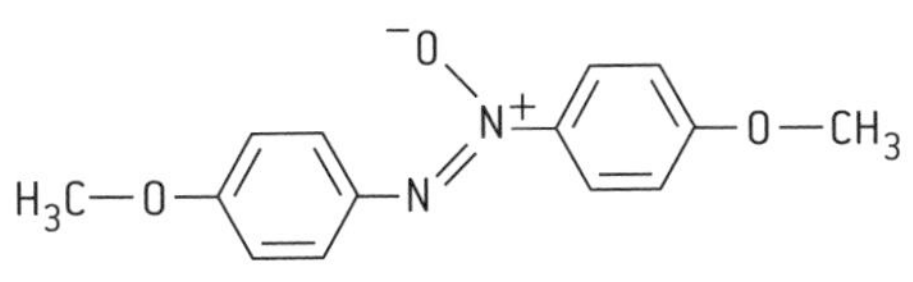

4-azoxyanisole PAA

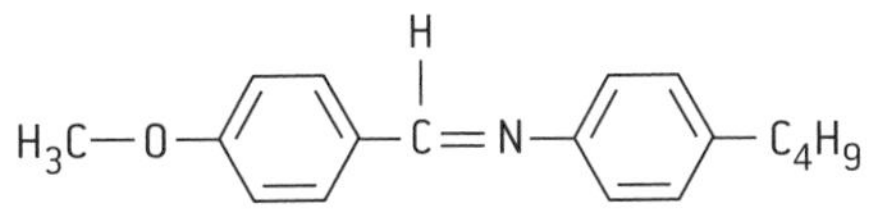

4′-methoxybenzylidene-4-butylaniline
MBBA

THE WORKINGS OF THE LCD DEVICE

In a liquid crystal display each pixel contains a liquid crystal film sandwiched between two glass plates. The plates have many very fine scratches at right angles to each other and have the property of polarizing light. The liquid crystal molecules in contact with the glass line up with the scratches and form a twisted arrangement between the plates due to intermolecular forces. This is known as **twisted nematic geometry**. The property utilized by the liquid crystals is their ability to interact with plane-polarized light (see page 96) which is rotated through 90° by the molecules as it passes through the film. When the two polarizers are aligned with the scratches, light will pass through the film and the pixel will appear bright. When a potential difference is applied across the film, the polar molecules will align with the film thus losing their twisted structure and ability to interact with the light. Plane-polarized light will now no longer be rotated so that the pixel appears dark.

the operation of the twisted nematic liquid crystal display

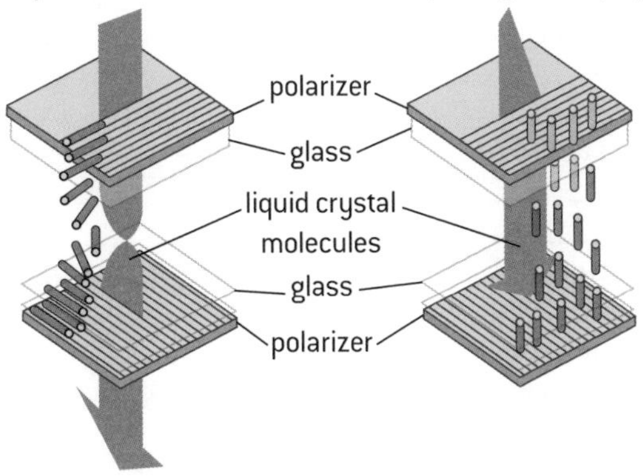

(a)off (b)on

Polymers

TYPES OF POLYMERS AND STRUCTURAL FEATURES

Polymers consist of many repeating units (monomers) joined together to form macromolecules with high molar masses – a theory first proposed by Herman Staudinger in 1920, which has since been substantiated by X-ray diffraction and scanning tunnelling electron microscopy. By understanding the molecular structure of polymers, chemists have been able to manipulate their properties and develop new polymers.

Polymers can be subdivided into thermoplastics and thermosets. Many alkenes polymerize to form thermoplastics that soften when heated and harden when cooled so that they can be remoulded each time they are heated. Thermosetting polymers, such as bakelite, polyurethanes and vulcanized rubber form prepolymers in a soft solid or viscous state that change irreversibly into hardened thermosets by curing, so that once shaped they cannot be remoulded. Some polymers, such as rubber, are flexible polymers known as elastomers. Elastomers can be deformed under force but will return to nearly their original shape once the stress is released. Generally the longer the chain length of a polymer the higher the strength and melting point, but cross-linking (or branching) and the orientation of the substituent groups can also affect particular properties.

Branching

Depending on the reaction conditions ethene can form high density or low density polythene. In high density poly(ethene), HDPE, there is little branching. This gives long chains that can fit together closely making the polymer stronger, denser, and more rigid than low density poly(ethene), LDPE. The presence of side chains in low density poly(ethene) results in a more resilient and flexible structure making it ideal for the production of film products, such as food wrappings.

Orientation of alkyl groups

In poly(propene) the methyl groups can all have the same orientation along the polymer chain – **isotactic**. Due to the regular structure isotactic polymers are more crystalline and tough. Isotactic poly(propene) is a thermoplastic and can be moulded into objects, such as car bumpers, and drawn into fibres for clothes and carpets. In **atactic** poly(propene) the chains are more loosely held so the polymer is soft and flexible, making it suitable for sealants and roofing materials.

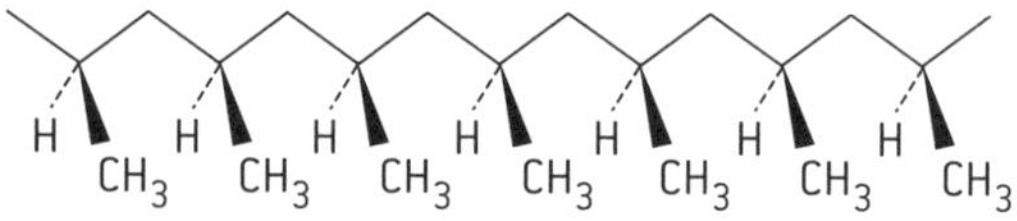

isotactic poly(propene) – all methyl groups orientated in same direction

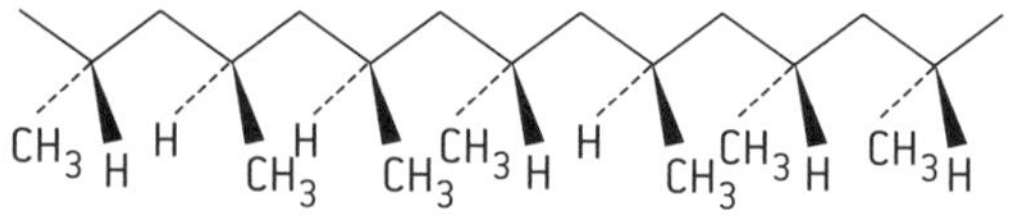

atactic poly(propene) – methyl groups arranged randomly

MODIFICATIONS TO POLYMERS

Plasticizers

Plasticizers are small molecules that can fit between the long polymer chains. They act as lubricants and weaken the attraction between the chains, making the plastic more flexible. By varying the amount of plasticizer added PVC can form a complete range of polymers from rigid to fully pliable.

Volatile hydrocarbons

If pentane is added during the formation of polystyrene and the product heated in steam the pentane vaporizes producing expanded polystyrene. This light material is a good thermal insulator and is also used as packaging as it has good shock-absorbing properties.

GREEN CHEMISTRY AND ATOM ECONOMY

Polymer development and use has grown rapidly in the past 60 years. The environmental impact of the industrial production of polymers is huge. Chemists try to find ways of reducing this by applying the principles of Green chemistry. One measure of the efficiency of the production is the atom economy which has been explained with an example on page 6. Generally addition polymers tend to have very high atom economies as all the monomers are converted into polymers. Condensation polymerization reactions, such as the formation of nylon, tend to have lower atom economies.

POLYMERIZATION OF 2-METHYLPROPENE

Addition polymers are formed by the addition reaction of alkenes to themselves. One such example is the polymerization of 2-methylpropene to form poly (2-methylpropene). The product, which is also known as butyl rubber or polyisobutylene, is a good elastomer and has many uses, such as car tyre inner tubes and cling film.

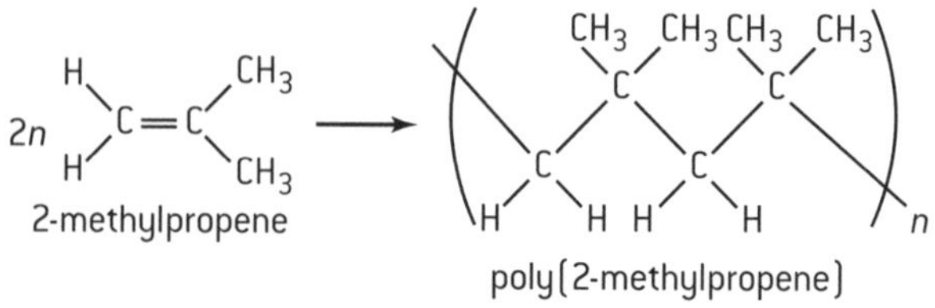

poly(2-methylpropene)

Nanotechnology

NANOTECHNOLOGY

Nanotechnology is defined as the research and technology of compounds within the range of one to one hundred nanometres (1.0×10^{-9} m to 1.0×10^{-7} m) in length, i.e. on the atomic scale. It creates and uses structures that have novel properties based on their small size. There are two main approaches. The 'bottom-up' approach involves building materials and devices from individual atoms, molecules or components. The 'top-down' approach involves constructing nano-objects from larger entities. Sometimes physical techniques are used that allow atoms to be manipulated and positioned to specific requirements. For example, a process known as dip-pen nanolithography can be used to place atoms in specific positions using an atomic force microscope. It is also possible to use chemical reactions such as in DNA nanotechnology where the specific base-pairing due to hydrogen bonding can be utilized to build desired molecules and structures.

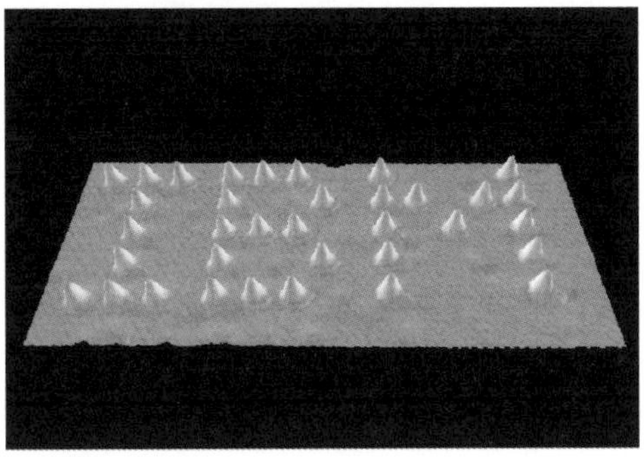

An image of IBM 'written' with xenon atoms on nickel using dip-pen nanolithography and 'seen' using scanning tunnelling microscopy.

IMPLICATIONS OF NANOTECHNOLOGY

Nanotechnology is concerned with the ability to control or manipulate on the atomic scale. It has the potential to solve many problems such as increase food production, prevent, monitor and cure diseases, and improve information and communication technology, although most of these benefits still probably lie somewhere far in the future. However, little is known about the potential risks associated with developing this technology. The hazards associated with small airborne particles are not properly known or covered by current toxicity regulations. The human immune system may be defenceless against new nano-scale products. There may also be social problems too as poorer societies may suffer as established technologies become redundant and demands for commodities change rapidly.

NANOTUBES

Nanotubes are tubes with a diameter in the region of just a few nanometres. They are made using only carbon atoms. The basic building block is a tube with the walls made from graphene and a hemisphere of buckminsterfullerene to close the ends. Both single and multi-walled graphene tubes, made from concentric nanotubes, have been formed. A wide variety of different materials, including elements, metal oxides and even small proteins have been inserted inside the tubes. Because they have a greatly increased ratio of surface area to volume they can act as extremely efficient and highly selective catalysts. They also have huge tensile strength and good thermal and electrical conductivity due to delocalized electrons.

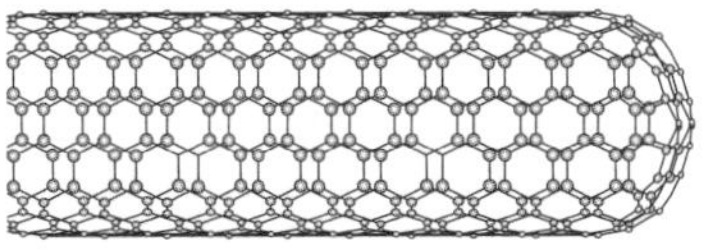

A diagram of part of a nanotube

Among the methods by which nanotubes can be prepared are arc discharge, chemical vapour deposition (CVD) and high pressure carbon monoxide (HIPCO). Arc discharge, which can give yields of up to 30%, involves either vaporizing the surface of one of the carbon electrodes, or discharging an arc through metal electrodes submersed in a hydrocarbon solvent, which forms a small rod-shaped deposit due to oxidation occurring on the anode. Chemical vapour deposition (CVD) is a commonly used method for the commercial production of nanotubes. In this process a carbon-containing gas, such as ethene in the plasma phase due to the influence of a strong electric field, is decomposed in the presence of an inert carrier gas, such as nitrogen, over a heated substrate formed from layers of metal particles. The carbon-containing gas is broken apart at the surface of the catalyst particle, and the resulting carbon is transported to the edges of the particles, where it forms the nanotubes. The HIPCO process creates single-walled carbon nanotubes from the reaction between high pressure carbon monoxide with iron pentacarbonyl, $Fe(CO)_5$ at temperatures between 1273 K and 1473 K. During this process the iron pentacarbonyl reacts to produce iron nanoparticles.

$$Fe(CO)_5(g) \rightarrow Fe(s) + 5CO(g)$$

The iron nanoparticles provide a nucleation surface for the transformation of carbon monoxide into carbon during the growth of the nanotubes.

$$xCO(g) \rightarrow CNT(s) + \tfrac{1}{2}xCO_2(g)$$

where x is typically 6000 giving a carbon nanotube (CNT) containing 3000 carbon atoms.

Environmental impact – plastics

ADVANTAGES AND DISADVANTAGES OF PLASTICS

Since the discovery of polymers such as nylon, polystyrene and polyurethanes in the 1930s, society has gained hugely from the use of plastics. Polymers can be tailor-made to perform a variety of functions based on properties such as strength, density, thermal and electrical insulation, flexibility, and lack of reactivity. There are, however, some disadvantages.

1. **Depletion of natural resources:** The majority of polymers are carbon-based. Currently oil is the major source of carbon although in the past it was coal. Both are fossil fuels and are in limited supply.
2. **Disposal:** Because of their lack of reactivity due to strong covalent bonds, plastics are not easily disposed of. Some, particularly PVC and poly(propene), can be recycled and others (e.g. nylon) are weakened and eventually decomposed by ultraviolet light. Plastics can be burned but if the temperature is not high enough poisonous dioxins can be produced along with toxic gases, such as hydrogen cyanide, hydrogen chloride and incomplete hydrocarbon combustion products.
3. **Biodegradability:** Most plastics do not occur naturally and are not degraded by micro-organisms. By incorporating natural polymers, such as starch, into plastics, they can be made more biodegradable. However, in the anaerobic conditions present in landfills biodegradation is very slow or will not occur at all.

DIOXINS AND POLYCHLORINATED BIPHENYLS

Dioxins can be formed when polymers are combusted unless the temperature is extremely high. They do not decompose in the environment and can be passed on in the food chain. Many dioxins, particularly chlorinated dioxins, are highly carcinogenic as they can disrupt the endocrine system (hormone action) and lead to cellular and genetic damage. Dioxins were used as defoliants present in Agent Orange during the Vietnam war. Dioxins contain unsaturated six-membered heterocyclic rings with two oxygen atoms, usually in positions 1 and 4. Examples of dioxins and dioxin-like substances include 1,4-dioxin, polychlorinated dibenzodioxins (PCDDs) and polychlorinated biphenyls (PCBs).

The general formulas of PCDDs and PCBs are given in Section 31 of the IB data booklet. Some specific examples are:

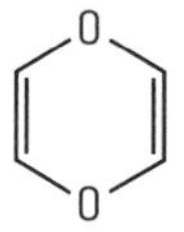

1, 4-dioxin

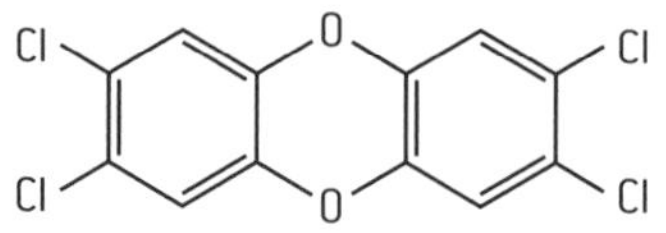

2, 3, 7, 8-tetrachlorodibenzodioxin (an example of a PCDD)

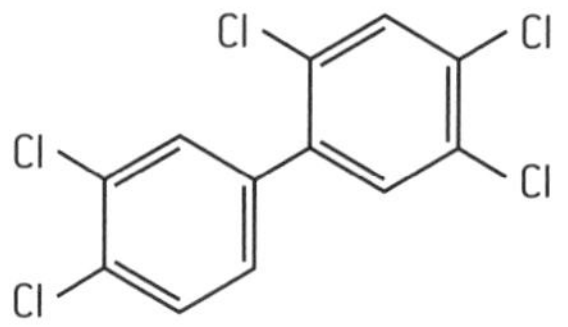

2, 3′, 4, 4′, 5-pentachlorobiphenyl (an example of a PCB)

PCBs contain from one to ten chlorine atoms attached to a biphenyl molecule. They are chemically stable and have high electrical resistance so were used in transformers and capacitors. Although not strictly dioxins (as they contain no oxygen atoms) they also persist in the environment and have carcinogenic properties.

RESIN IDENTIFICATION CODES

Different types of plastic need different types of treatment to enable them to be recycled or disposed of safely. The Resin Identification Code was developed to aid efficient separation by identifying each specific plastic with a numbered label in order to help process them before recycling. Which plastics can be recycled depends upon the recycling policy of the local community. Some are able to recycle all seven types of plastic whereas some may only be able to recycle PETE and HDPE types of plastic. Use of infrared spectroscopy can also be used to identify plastics that contain specific functional groups. For example, PETE can be recognized by the strong C=O absorption in the region of 1700–1750 cm^{-1} and PVC shows an absorption due to the C–Cl bond between 600 and 800 cm^{-1}.

Resin Identification Code (RIC)	Plastic types	Resin Identification Code (RIC)	Plastic types
1 PETE	polyethylene terephthalate	5 PP	polypropylene
2 HDPE	high-density polyethylene	6 PS	polystyrene
3 PVC	polyvinyl chloride	7 OTHER	other
4 LDPE	low-density polyethylene		

Table of Resin Identification Codes

For plastics that cannot be recycled, special incinerators need to be built to ensure the high temperatures needed to avoid dioxins and other toxic pollutants being formed when they are combusted. These temperatures are not reached in house fires where there are real dangers from the toxic gases emitted. To help reduce the risk, low smoke zero halogen cabling is often used in electrical wiring.

The use of toxic volatile phthalate esters as plasticizers is being phased out in many countries as they have been associated with birth defects. Because there are no covalent bonds between the phthalate esters and the plastics they are slowly released into the environment where they enter the atmosphere and the food chain. This process accelerates as the plastics deteriorate with age.

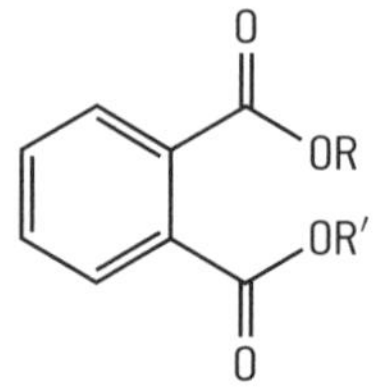

general formula of phthalate esters

Superconducting metals and X-ray crystallography

SUPERCONDUCTORS

Superconductors are materials (elements or compounds) that will conduct electric currents with no resistance below a critical temperature. As the material cools below the critical temperature to reach the superconducting state the material creates a mirror image magnetic field which cancels out an externally applied magnetic field. This is known as the Meissner effect. The mirror image magnetic field is formed by the generation of electric currents near the surface of the material, which cancel the applied magnetic field within the body of the superconductor. Above the critical temperature, electrical resistance is caused by collisions between electrons and positive ions in the lattice. Below the critical temperature the Bardeen–Cooper–Schrieffer (BCS) theory explains that electrons in superconductors form Cooper pairs which are able to move freely through the superconductor. Normally electrons behave as free particles and are repelled by other electrons. At low temperatures the positive ions in the lattice are distorted slightly by a passing electron. A second electron is attracted to this slight positive deformation and a coupling of these two electrons occurs to form a Cooper pair that acts as a single entity and causes no resistance.

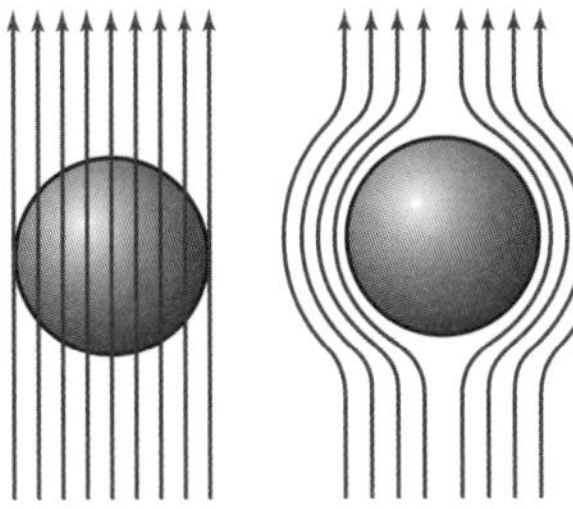

The Meissner effect. Magnetic fields are excluded from a superconductor below its critical temperature, T_c

TYPE 1 AND TYPE 2 SUPERCONDUCTORS

Superconductors can be divided into two types. Type 1 superconductors were discovered first and require the coldest temperatures to become superconductive. They exhibit a very sharp transition to a superconducting state and show complete diamagnetism so can repel a magnetic field completely.

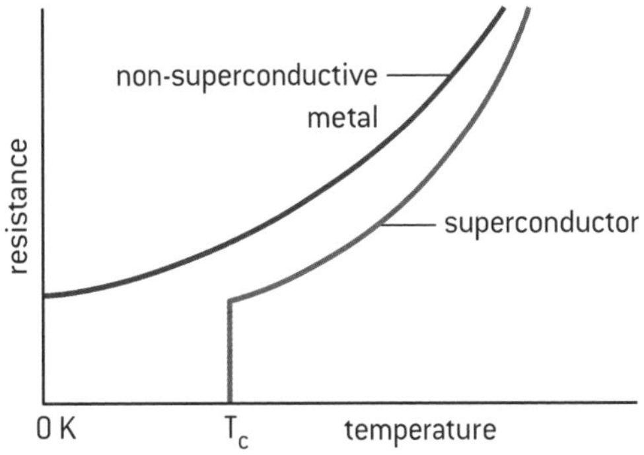

Type 1 superconductors show a sharp transition to a superconducting state.

Type 1 superconductors tend to be metals and metalloids. Type 2 super conductors show a gradual transition to the superconducting state and tend to be metallic compounds or alloys. They tend to have higher critical temperatures than Type 1 superconductors. Some recently-discovered Type 2 superconducting 'perovskites' (metal-oxide ceramics that normally have a ratio of 2 metal atoms to every 3 oxygen atoms) have critical temperatures above 0 °C.

STRUCTURES OF METALS AND IONIC COMPOUNDS

X-ray diffraction can be used to analyse the structures of metals and ionic compounds including the more complex perovskites. By using the Bragg equation, $n\lambda = 2d \sin\theta$ (see page 107) the precise location of all the metal atoms or ions in the crystal lattice can be determined. Crystal lattices contain simple repeating unit cells and the number of nearest neighbours of an atom or ion is known as its coordination number. Many metals form cubic structures. Polonium forms a simple cubic cell with a coordination number of six.

Group 1 metals form a body-centred cubic structure with the atoms filling 68% of the available space. Each metal atom is surrounded by eight other atoms as its nearest neighbours so the coordination number is 8.

Many other metals pack more efficiently using 74% of the available space with a coordination number of 12. In face-centred cubic close packing the unit cell has one atom in the centre of each face rather than one in the centre.

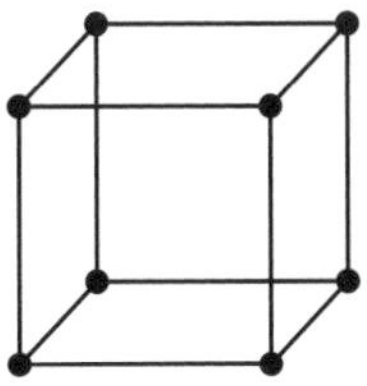

Simple cubic unit cell. Each atom contributes $\frac{1}{8}$ to the cell so the cell contains $8 \times \frac{1}{8} = 1$ atom.

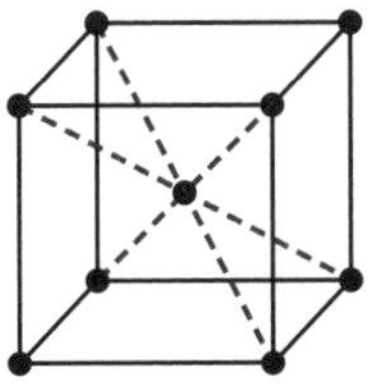

Body-centred cubic unit cell. Each corner atom contributes $\frac{1}{8}$ to the cell so the cell contains $1 + \left(8 \times \frac{1}{8}\right) = 2$ atoms.

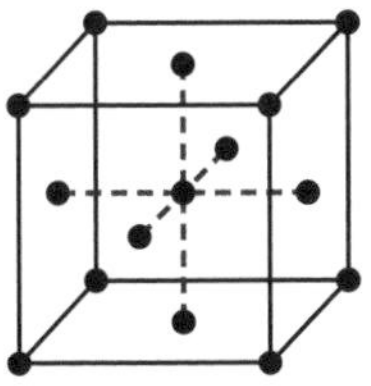

Face-centred cubic unit cell. Each corner atom contributes $\frac{1}{8}$ to the cell, each side atom contributes $\frac{1}{2}$ to the cell so the cell contains $\left(8 \times \frac{1}{8}\right) + \left(6 \times \frac{1}{2}\right) = 4$ atoms.

Worked example

Potassium has a metallic radius of 2.31×10^{-10} m and forms a body-centred cubic unit cell. Determine the density of solid potassium.

Step 1. Find the volume of one potassium atom

$V = \frac{4}{3}\pi r^3 = \frac{4}{3} \times 3.14 \times (2.31 \times 10^{-10})^3$
$= 5.16 \times 10^{-29}\ m^3$

Step 2. Find the volume of one mole of potassium atoms

$V = 5.16 \times 10^{-29} \times 6.02 \times 10^{23}$
$= 3.11 \times 10^{-5}\ m^3\ mol^{-1}$

Step 3. Find the volume of a lattice containing one mole of potassium atoms

Since it is body-centred cubic the atoms occupy 68% of the available space so the volume of the lattice $= 3.11 \times 10^{-5} \times \frac{100}{68} = 4.57 \times 10^{-5}\ m^3\ mol^{-1}$

Step 4. Determine the density knowing that one mole of potassium has a mass of 39.10 g

Density $= 39.10/(4.57 \times 10^{-5}) = 8.56 \times 10^5\ g\ m^{-3} = 0.856\ g\ cm^{-3}$

Condensation polymers

FORMATION OF CONDENSATION POLYMERS

Addition polymers are usually made from alkenes whereas condensation polymers are made from monomers that contain at least two reactive functional groups (or the same functional group at least twice). When the monomers condense to form the polymer, small molecules, such as water or hydrogen chloride, are also produced. For example, polyamides, such as nylon, can be made by condensing diamines with dicarboxylic acids to form a carboxamide (amide) link (also known as a peptide bond).

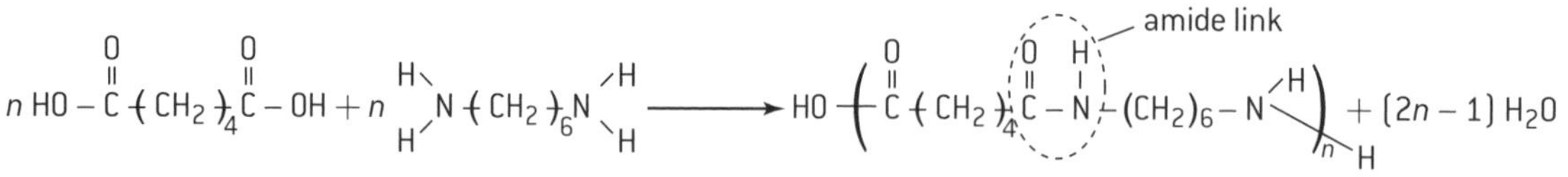

hexane-1, 6-dioic acid 1,6-diaminohexane repeating unit nylon 6,6

(6, 6 because each monomer contains 6 carbon atoms)

An example of a polyester is polyethene terephthalate (known as Terylene in the UK and as Dacron in the USA), which is made from benzene-1,4-dicarboxylic acid and ethane-1,2-diol. It is used for textiles.

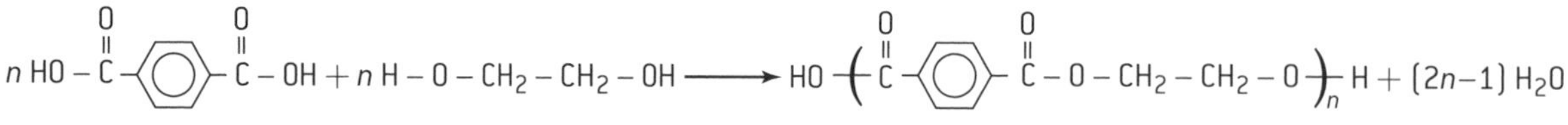

benzene-1, 4-dicarboxylic acid ethane-1, 2-diol repeating unit 'Terylene' or 'Dacron'

KEVLAR

The properties of condensation polymers depend upon their structural features. Cross-linking generally adds considerable strength to the polymer. A good example of this is Kevlar, the material from which lightweight bullet-proof vests, composites for motorcycle helmets and armour are made. Kevlar is a polyamide made by condensing 1,4,-diaminobenzene (*para*-phenylenediamine) with benzene-1,4-dicarbonyl chloride (terephthaloyl dichloride).

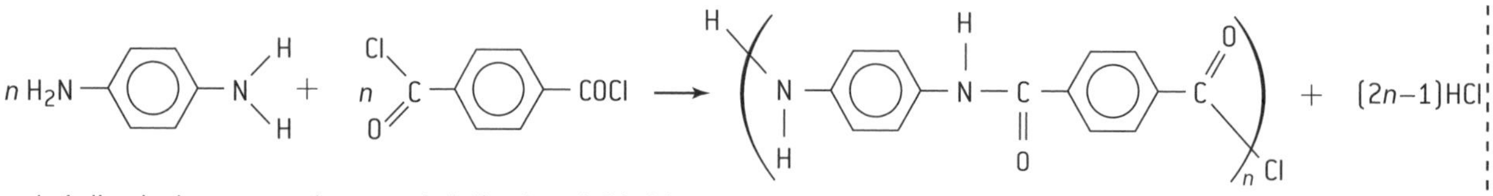

1, 4-diaminobenzene benzene-1, 4-dicarbonyl chloride 'Kevlar'

Kevlar consists of rigid, rod-shaped polar molecules with cross-linking formed by strong intramolecular hydrogen bonding between the chains. This gives a very ordered and strong three-dimensional structure, hence its use as a protective material.

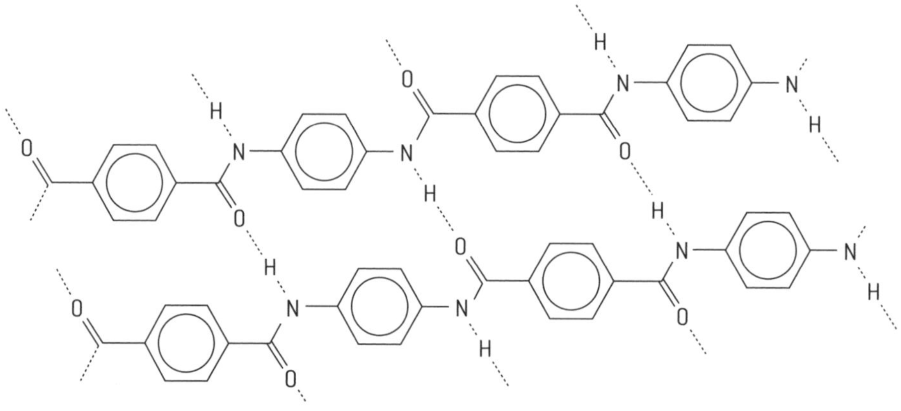

Three-dimensional structure of Kevlar

In acidic solution, Kevlar can act as a lyotropic liquid crystal. It is lyotropic because the alignment of its molecules depends upon the concentration of the solution. In strong acid solution the oxygen and nitrogen atoms in the amide linkage become protonated and this breaks the hydrogen bonding between the chains. In very strong acidic solution, such as concentrated sulfuric acid, the structure is broken down completely (as the amide links can also decompose) and Kevlar loses its liquid crystal properties and also its protective properties. Thus, equipment made of Kevlar should be stored well away from acids.

MODIFICATION OF POLYMERS

The addition of plasticizers to PVC and volatile hydrocarbons to addition polymers has already been covered. Expanded polystyrene is formed by adding pentane during the formation of styrene and the product heated in steam. This light material is a good thermal insulator and is also used in packaging as it has good shock-absorbing properties.

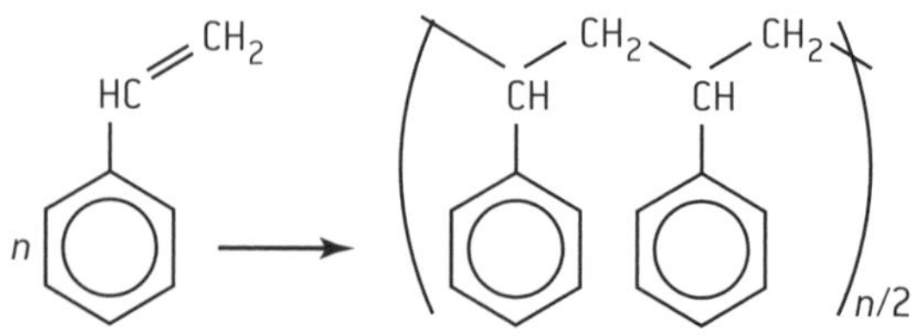

styrene polystyrene

Air can be blown into polyurethanes to make polyurethane foams for use as cushions and thermal insulation. The fibres of polyesters can be blended with other manufactured or natural fibres for making clothes which are dye-fast and more comfortable than pure polyesters.

Like addition polymers, the disadvantages of condensation polymers include their disposal and the effect on the environment. Condensation polymers are not generally biodegradable and when polyurethanes are burned they can release hydrogen cyanide gas.

Environmental impact – heavy metals (1)

HEAVY METALS

Heavy metals that exist as ions in polluted water include cadmium, mercury, lead, chromium, nickel, copper and zinc. Their sources are varied. Cadmium is found in the effluent near zinc mining and in batteries and paints. Mercury is used a fungicide in seed dressings and used to be used as an electrode in the electrolysis of sodium chloride. Lead used to be used in paints and leaded gasoline and is still used in car batteries and roofing material. Other heavy metals originate from specific industrial processes (e.g. chromium used to be used for tanning leather), in batteries (e.g. NiCd cells) and biocides (e.g. copper in woodworm treatment). Some methods of removing heavy metals from polluted water are precipitation, adsorption and chelation.

PRECIPITATION OF HEAVY METAL IONS FROM WATER

Even 'insoluble' salts are still very slightly soluble in water. For a salt formed from a metal M with a non-metal X:

$$MX(s) \rightleftharpoons M^+(aq) + X^-(aq)$$

The equilibrium expression for this heterogeneous process will be $K_{sp} = [M^+(aq)] \times [X^-(aq)]$ where the equilibrium constant, K_{sp}, is known as the solubility product. (Note that MX(s) does not appear in the expression as a solid has no concentration.)

Many metal sulfides have very low solubility products so one effective way of precipitating heavy metal ions is by bubbling hydrogen sulfide through polluted water. The solubility product can be used to calculate the amount of a metal ion that will remain in solution after it has been precipitated. For example, the solubility product of lead sulfide, PbS is 1.30×10^{-28}.

$$Pb(s) \rightleftharpoons Pb^{2+}(aq) + S^{2-}(aq)$$

$$K_{sp} = [Pb^{2+}(aq)] \times [S^{2-}(aq)] \quad \text{but } [Pb^{2+}(aq)] = [S^{2-}(aq)]$$

Therefore $K_{sp} = [Pb^{2+}(aq)]^2 = 1.30 \times 10^{-28}$ at 298 K

and $[Pb^{2+}(aq)] = (1.30 \times 10^{-28})^{½} = 1.14 \times 10^{-14}$ mol dm^{-3}

Hence the concentration of lead ions in the aqueous solution is 1.14×10^{-14} mol dm^{-3}. The relative atomic mass of lead is 207.19, the mass of lead ions that dissolves in one litre of water at 298 K is therefore $207.19 \times 1.14 \times 10^{-14}$ g which is equal to only 2.36×10^{-12} g. Hence the precipitation as the sulfide is very efficient. Even though this is very small it can be reduced even further by adding more sulfide ions to the solution. Now the concentration of the lead ions is not the same as the concentration of the sulfide ions. This is known as the **common ion effect**. If the concentration of the sulfide ions is made to be 1.00 mol dm^{-3} then since

$$K_{sp} = [Pb^{2+}(aq)] \times [S^{2-}(aq)] \text{ and } [S^{2-}(aq)] = 1.00 \text{ mol dm}^{-3}$$

$$K_{sp} = [Pb^{2+}(aq)] = 1.30 \times 10^{-28} \text{ at 298 K.}$$

The concentration of lead ions remaining in the solution at 298 K is now only 1.30×10^{-28} mol dm^{-3} and the mass of lead ions remaining in one litre is just 2.69×10^{-26} g.

For many heavy metal ions found in waste water it is slightly more complicated as the salts formed are not binary but the principle is the same. For example, nickel ions in waste water can be precipitated by adding hydroxide ions.

$$Ni^{2+}(aq) + 2OH^-(aq) \rightleftharpoons Ni(OH_2)(s)$$

The solubility product for nickel(II) hydroxide at 298 K is 6.50×10^{-18}.

Hence $\quad K_{sp} = [Ni^{2+}(aq)] \times [OH^-(aq)]^2$

Assuming that all the hydroxide ions in the solution come just from the dissolved nickel(II) hydroxide then

$$[OH^-(aq)] = 2[Ni^{2+}(aq)].$$

Therefore $K_{sp} = 6.50 \times 10^{-18} = [Ni^{2+}(aq)] \times (2[Ni^{2+}(aq)])^2 = 4[Ni^{2+}(aq)]^3$

Hence $[Ni^{2+}(aq)] = \left(\frac{6.50}{4} \times 10^{-18}\right)^{\frac{1}{3}} = (1.63 \times 10^{-18})^{\frac{1}{3}} = 1.18 \times 10^{-6}$ mol dm^{-3}.

Since the relative atomic mass of nickel is 58.69, the mass of nickel ions remaining in one litre of water at 298 K is 6.93×10^{-5} g. This small amount can be reduced even more by using the common ion effect, i.e. by adding more hydroxide ions. However care must be taken to ensure that soluble complex ions are not formed. For example, zinc hydroxide redissolves in excess hydroxide ions to form $[Zn(OH)_4]^{2-}(aq)$.

Environmental impact – heavy metals (2)

CHELATION

Chelation occurs when a molecule or ion containing non-bonding pairs of electrons bonds to a metal atom or ion forming two or more coordinate bonds. In other words chelating agents are polydentate ligands. Usually chelating agents are organic compounds and are sometimes known as sequestering agents. Polydentate ligands form more stable complexes than similar monodentate ligands. One explanation for this is that the equilibrium constants for chelation reactions are high. From the expression $\Delta G = -RT\ln K$ it can be seen that a high value for the equilibrium constant will give a large negative value for the free entropy change, i.e. the reaction will be highly spontaneous. Since ΔG is also equal to $\Delta H - T\Delta S$ the large negative value for ΔG can also be explained by the large positive entropy change, ΔS, as the complex contains just one ligand the disorder of the system increases, as six monodentate ligands are released into the system.

Two important chelating agents are ethane-1,2-diamine and the ethylenediaminetetraacetate ion, $EDTA^{4-}$. Ethane-1,2-diamine contains two non-bonding pairs of electrons, one on each of the nitrogen atoms so acts as a bidentate ligand. $EDTA^{4-}$ contains six non-bonding pairs of electrons and can use all six of these to form a hexadentate ligand.

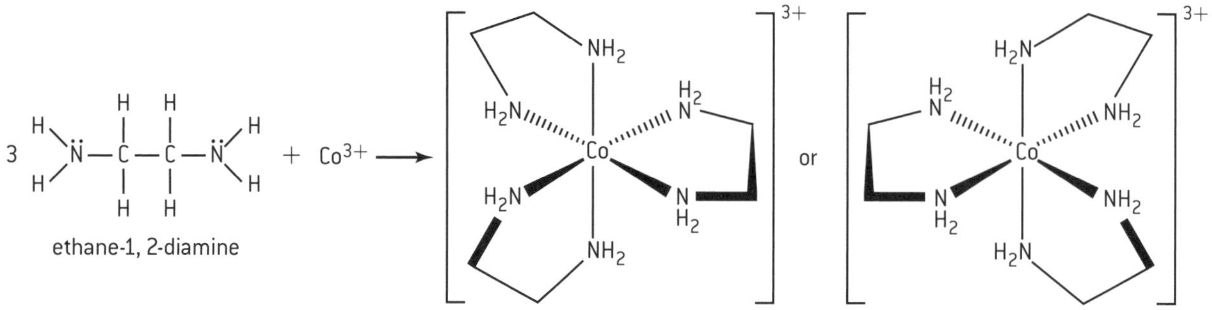

The three bidentate ligands can form two enantiomers with the Co^{3+} ion

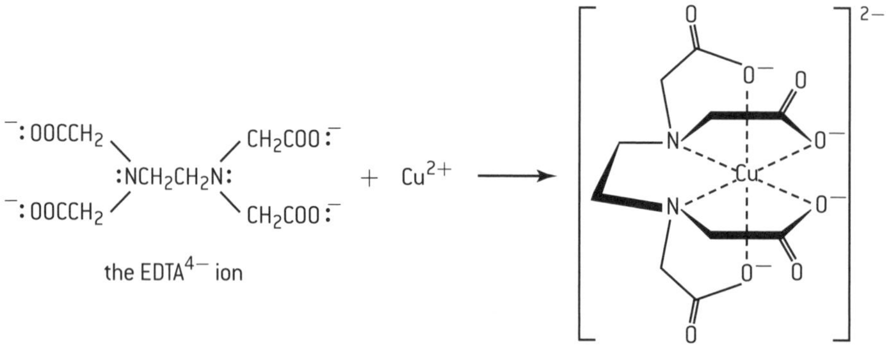

the hexadentate $[Cu(EDTA)]^{2-}$ complex

THE HABER–WEISS AND FENTON REACTIONS

Toxic doses of transition metals can disturb the normal oxidation–reduction balance in cells through various mechanisms. They can promote the formation of free radicals, such as the superoxide anion, O_2^- and the hydroxyl radical, HO· which can damage macromolecules, including DNA, proteins and lipids. Hydroxyl radicals can be generated in cells by the reaction of hydrogen peroxide with the superoxide ion. This is known as the Haber–Weiss reaction. The reaction is slow but can be catalysed by iron. The overall process is

$$O_2^-(aq) + H_2O_2(aq) \rightarrow HO\cdot(aq) + OH^-(aq) + O_2(g)$$

The first step involves the catalytic reduction of iron(III) to iron(II)

$$Fe^{3+}(aq) + O_2^-(aq) \rightarrow Fe^{2+}(aq) + O_2(aq)$$

The iron(II) is then oxidized back to iron(III) by hydrogen peroxide to form a hydroxide ion and a hydroxyl radical.

$$Fe^{2+}(aq) + H_2O_2(aq) \rightarrow Fe^{3+}(aq) + OH^-(aq) + HO\cdot(aq)$$

This second step is known as the Fenton reaction. Fenton's reagent is a solution of hydrogen peroxide and iron(II) sulfate used to purify contaminated water. The iron(II) ions act as a catalyst. In this reaction the iron(II) is first oxidized to iron(III) by acidified hydrogen peroxide to form a hydroxyl radical and then reduced back to iron(II) by more hydrogen peroxide. The radicals generated purify the water by destroying living organisms. Oxidation of organic compounds by Fenton's reagent is both rapid and exothermic and results in the aerobic oxidation of the carbon and hydrogen content of contaminants to form ultimately carbon dioxide and water.

SHORT ANSWER QUESTIONS – OPTION A – MATERIALS

1. a) Salt (sodium chloride) and haematite, iron(III) oxide, are two minerals that have been known to humans for millennia. Explain why elemental iron has been known since the Iron Age and yet elemental sodium was only discovered some 200 years ago. [2]

   b) Some compounds of iron are paramagnetic whereas other compounds of iron are diamagnetic. Explain the difference between *paramagnetic* and *diamagnetic* on a molecular level. [2]

   c) Stainless steel contains iron alloyed with carbon, chromium and some other metals. In addition to making the steel resistant to rusting, state two other changes that occur when iron is alloyed with carbon and other metals. [2]

2. a) Determine the time taken to produce 1000 kg of aluminium when a steady current of 2.00×10^5 A is passed through a solution of alumina in molten cryolite, Na_3AlF_6. [3]

   b) Discuss why the production of aluminium makes a significant contribution to the greenhouse effect. [3]

3. Liquid crystals are sometimes used in the construction of 'smart windows'. Smart windows are milky white as their randomly arranged liquid crystals scatter light. When a voltage is applied, the liquid crystals align in the same direction. The light then passes through them without scattering, making the windows transparent.

   a) State the property of the liquid crystal molecules that allows them to align when a voltage is applied. [1]

   b) List two substances that can behave as liquid crystals. [1]

   c) Distinguish between *thermotropic* and *lyotropic* liquid crystals. [2]

4. a) Distinguish between *homogeneous* and *heterogeneous* catalysts and explain their different mode of action. [4]

   b) List three factors that should be considered when choosing a catalyst for a particular process. [3]

   c) Iron is a good catalyst for the Haber Process. Suggest why it is used in a finely divided (powdered) form. [1]

5. a) Propene can polymerize to form either isotactic or atactic poly(propene). Draw a section of isotactic poly(propene) containing four methyl groups. [2]

   b) Describe how the properties of isotactic poly(propene) differ from those of atactic poly(propene). [2]

   c) Explain why low density poly(ethene), LDPE, is more flexible than high density poly(ethene), HDPE. [2]

   d) Deduce the atom economy for the conversion of ethene to poly(ethene). [1]

6. a) One of the methods for making nanotubes is chemical vapour deposition, CVD. Explain why an inert gas such as nitrogen and not oxygen must be used in the process. [2]

   b) Explain why nanotubes are good conductors of electricity. [1]

HL

7. Kevlar can be made by condensing 1,4,-diaminobenzene with benzene-1,4-dicarbonyl chloride.

   a) Draw the repeating unit of Kevlar. [2]

   b) Explain why Kevlar is such a strong polymer. [2]

   c) Explain why a bullet proof vest made of Kevlar should not be allowed to come into contact with sulfuric acid. [2]

8. Cadmium ions, $Cd^{2+}(aq)$ are extremely poisonous. They can be removed from solution by adding hydroxide ions to precipitate cadmium hydroxide, $Cd(OH)_2$. The solubility product of cadmium hydroxide is 7.2×10^{-15} at 298 K.

   a) Calculate the solubility of cadmium hydroxide in mol dm^{-3} at 298 K. [3]

   b) Explain why the solubility of cadmium ions in solution can be reduced even further by adding excess hydroxide ions. [2]

   c) Heavy metals can also be removed from water by chelation. Explain why ethane-1,2-diamine is a good chelating agent. [2]

9. A simple cubic structure contains atoms placed at the corners of a cube.

   a) Determine the number of atoms in one unit cell. [1]

   b) Determine the coordination number of each atom in a simple cubic structure. [1]

   c) Explain how the distance between the atoms in a simple cubic structure can be determined. [2]

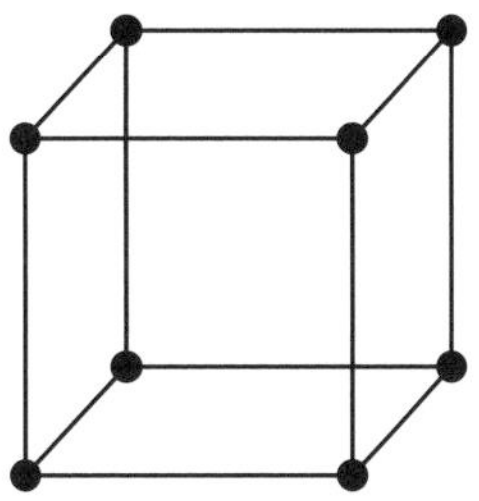

Introduction to biochemistry

METABOLISM

The chemical reactions that occur in living organisms are collectively known as metabolism. Essentially these enzyme-catalysed reactions allow organisms to grow and reproduce, maintain their structures and respond to their environment. Reactions in which organic matter is broken down to produce energy involve the process of *catabolism* whereas reactions which use energy to synthesize larger molecules such as proteins and nucleic acids are known as *anabolism*.

The chemical reactions involved in metabolism are organized into metabolic pathways. Each step is catalysed by an enzyme and normally occurs in a controlled aqueous environment. Biological molecules are diverse in nature and their functions depend upon their precise structure and shape.

Typical examples of biological molecules include:

- Proteins – made from linking amino acids by peptide bonds.
- Lipids – these include fats and steroids.
- Carbohydrates – they have the general formula $C_x(H_2O)_y$ and include monosaccharides (e.g. glucose and fructose) and polysaccharides (e.g. starch and cellulose).
- Nucleotides – consist of a phosphate group, a ribose sugar group and a nitrogenous base. Polymers of nucleotides include DNA and RNA.
- Co-enzymes – compounds such as adenosine triphosphate (ATP) that are metabolic intermediates whose function is to carry chemical groups (or energy) between different reactions.

PHOTOSYNTHESIS AND RESPIRATION

Photosynthesis is the process used by plants and other organisms to synthesize energy-rich molecules such as carbohydrates from carbon dioxide and water using light energy. During photosynthesis oxygen is released. This is the origin of the oxygen in the atmosphere and photosynthesis is responsible for maintaining atmospheric oxygen levels. Essentially photosynthesis is the process of converting carbon into biomass. The process is complicated and needs the presence of chlorophyll but can be simplified by the overall equation:

$$6CO_2(g) + 6H_2O(l) \xrightarrow{\text{light}} C_6H_{12}O_6(aq) + 6O_2(g) \quad \Delta H \text{ positive}$$

Respiration is also a complex set of reactions in which energy-rich molecules such as carbohydrates are broken down in the presence of oxygen to provide energy for cells. Ultimately the products of respiration are carbon dioxide and water, which are released into the atmosphere. The overall equation is the reverse of photosynthesis.

$$C_6H_{12}O_6(aq) + 6O_2(g) \rightarrow 6CO_2(g) + 6H_2O(l) \quad \Delta H \text{ negative}$$

CONDENSATION AND HYDROLYSIS REACTIONS

Biological polymers (biopolymers) are formed by condensation reactions. These involve the reaction between two smaller molecules to form one larger molecule with the evolution of a small molecule such as water. For condensation polymerization to occur each reacting molecule must possess at least two reactive functional groups. Classic examples include the condensation of amino acids to form proteins and the condensation of sugars to form starch. For example, the two amino acids alanine, $H_2N–CHCH_3–COOH$ and cysteine, $H_2N–CH(CH_2SH)–COOH$ can condense together to form two different possible dipeptides if each dipeptide contains one of each of the two acid residues.

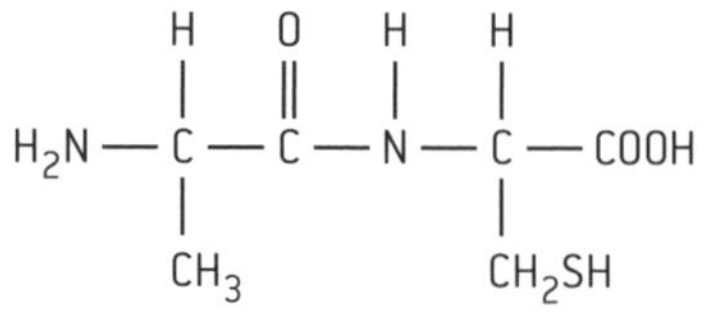

H_2N-Ala-Cys-COOH

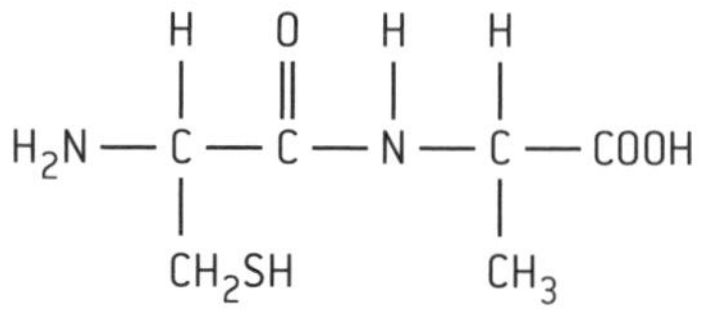

H_2N-Cys-Ala-COOH

Each end of the dipeptides contains a reactive group so can undergo further condensation reactions with more amino acids to produce polypeptides.

Hydrolysis is the reverse of condensation. A molecule is hydrolysed when a water molecule (often in the presence of acid or alkali) reacts with a larger molecule to break a bond and form two smaller molecules. The hydrolysis of proteins produces amino acids and the hydrolysis of starch (polysaccharide) produces sugars (monosaccharides). For example, sucrose, a disaccharide, can be hydrolysed to form glucose and fructose.

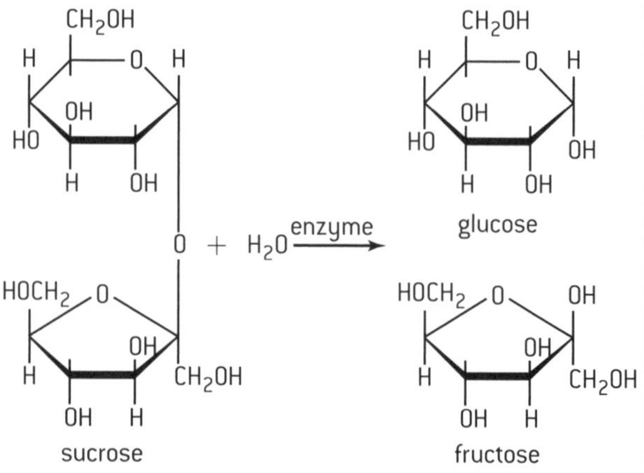

Structure of proteins

AMINO ACIDS AND THE STRUCTURE OF PROTEINS

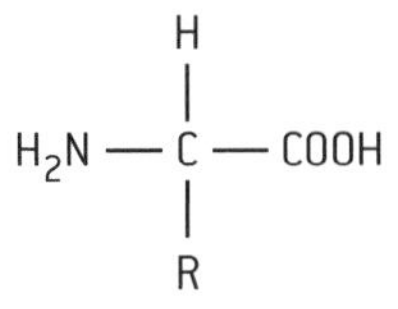

2-amino acid

Amino acids contain both an amine functional group and a carboxylic acid functional group. When both are attached to the same carbon atom they are known as 2-amino acids (or α-amino acids). They are solids at room temperature and have quite high melting points. This is because they can exist as zwitterions in which the hydrogen atom from the carboxylic acid group protonates the amine group to form a carboxylate anion and a substituted ammonium cation within the same compound (see page 126).

Proteins are large macromolecules made up of chains of 2-amino acids. There are about twenty 2-amino acids that occur naturally. A full list of the 2-amino acids together with their common names, abbreviations and structural formulas is given in Section 33 of the IB data booklet. The amino acids bond to each other through condensation reactions resulting in the formation of a polypeptide, in which the amino acid residues are bonded to each other by a carboxamide link (peptide bond).

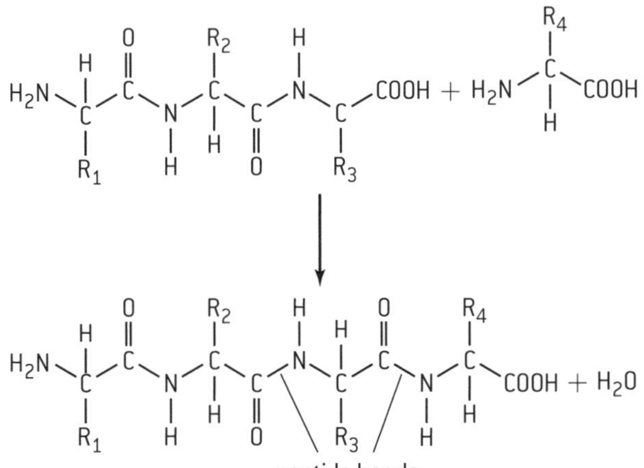

Each protein contains a fixed number of amino acid residues connected to each other in strict sequence. This sequence, e.g. gly-his-ala-ala-leu- ... is known as the primary structure of proteins. The secondary structure describes the way in which the chain of amino acids folds itself due to intramolecular hydrogen bonding. The folding can either be α-helix in which the protein twists in a spiralling manner rather like a coiled spring, or β-pleated to give a sheet-like structure.

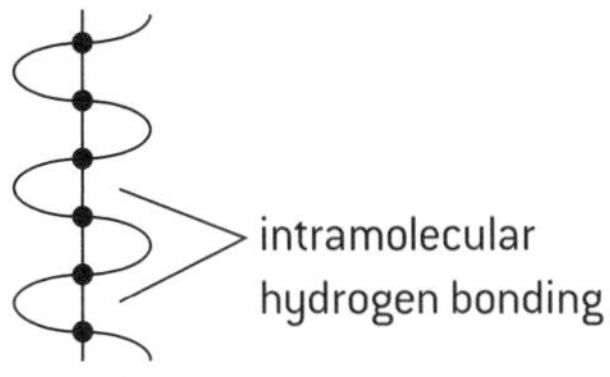

α-helix

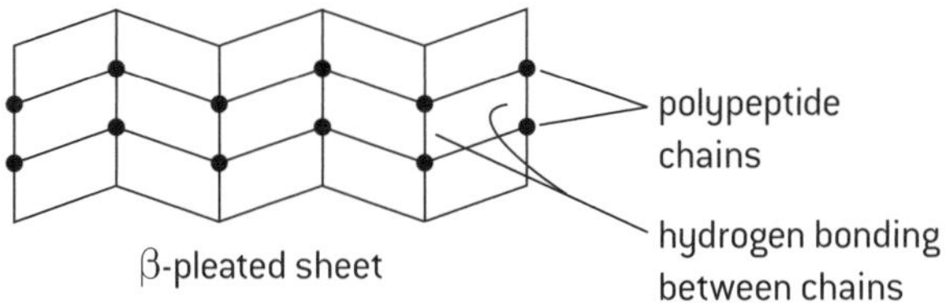

β-pleated sheet

The tertiary structure describes the overall folding of the chains by interactions between distant amino acids to give the protein its three-dimensional shape. These interactions may be due to hydrogen bonds, London dispersion forces between non-polar side groups, and ionic attractions between polar groups. In addition two cysteine residues can form **disulfide bridges** when their sulfur atoms undergo oxidation.

Examples of interactions between side groups on polypeptide chains:

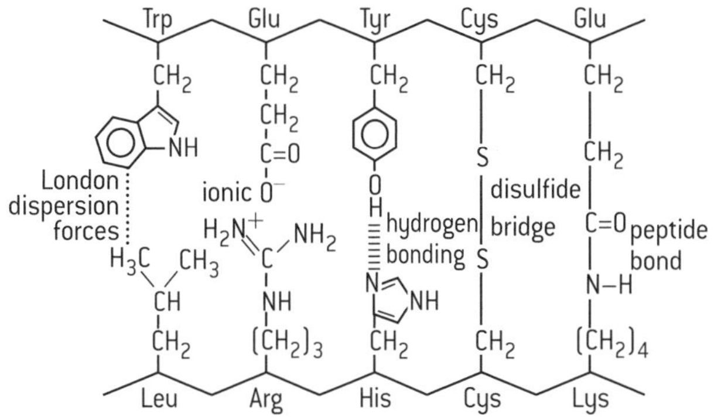

Separate polypeptide chains can interact together to give a more complex structure – this is known as the quaternary structure. Haemoglobin has a quaternary structure that includes four protein chains (two α-chains and two β-chains) grouped together around four haem groups.

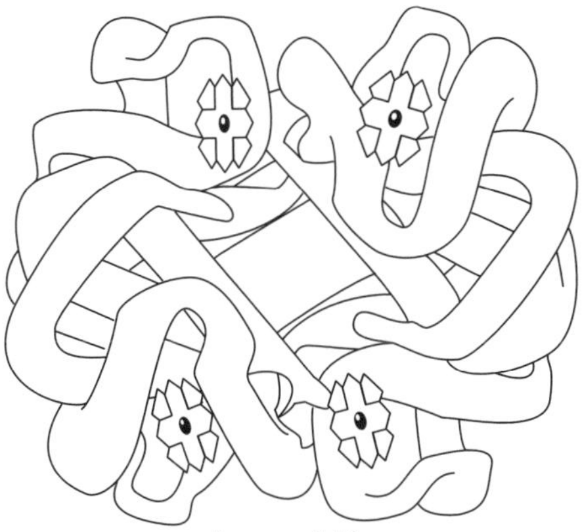

haemoglobin

Fibrous and globular proteins

Haemoglobin is an example of a globular protein. Globular proteins have complex tertiary and sometimes quaternary structures (e.g. haemoglobin) folded into spherical (globular) shapes. They are usually soluble to some extent in water as the hydrophobic side chains tend to be in the centre of the structure. Fibrous proteins, such as collagen, have little or no tertiary structure and form long parallel polypeptide chains. Fibrous proteins have cross-linking at intervals to form long fibres or sheets and have mainly structural roles such as keratin in hair and collagen, which is found in skin and the walls of blood vessels and acts as connective tissue.

Analysis of proteins

The primary structure of proteins can be determined either by paper chromatography or by electrophoresis. In both cases the protein must first be hydrolysed by hydrochloric acid and heat to successively release the amino acids. The three-dimensional structure of the complete protein can be confirmed by X-ray crystallography.

PAPER CHROMATOGRAPHY

A small spot of the unknown amino acid sample is placed near the bottom of a piece of chromatographic paper. Separate spots of known amino acids can be placed alongside. The paper is placed in a solvent (eluent), which then rises up the paper due to capillary action. As it meets the sample spots the different amino acids partition themselves between the eluent and the paper to different extents, and so move up the paper at different rates. When the eluent has nearly reached the top, the paper is removed from the tank, dried, and then sprayed with an organic dye (ninhydrin) to develop the chromatogram by colouring the acids. The positions of all the spots can then be compared.

If samples of known amino acids are not available the R_f value (retention factor) can be measured and compared with known values as each amino acid has a different R_f value. It is possible that two acids will have the same R_f value using the same solvent, but different values using a different solvent. If this is the case the chromatogram can be turned through 90° and run again using a second solvent.

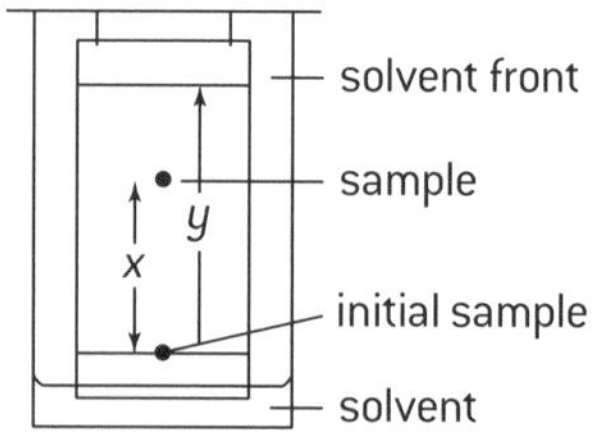

$$R_f \text{ value} = \frac{\text{distance travelled by sample}}{\text{distance travelled by solvent}} = \frac{x}{y}$$

Rotate chromatogram and use a second solvent with a different polarity.

first solvent direction

second solvent direction

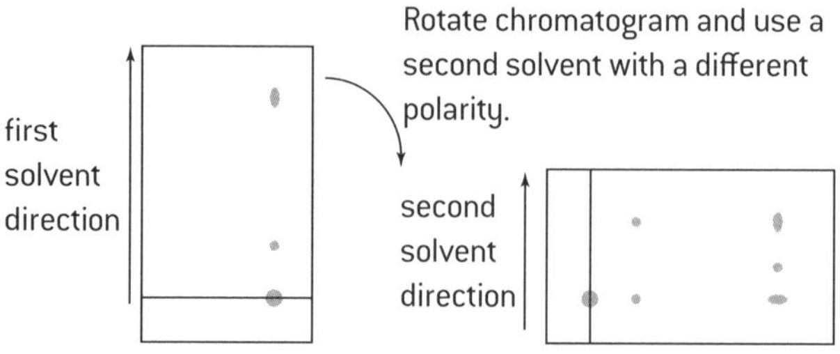

Three-component mixture split into two by first solvent.

Separation of the three components is achieved.

ELECTROPHORESIS

The structure of amino acids alters at different pH values. At low pH (acid medium) the amine group will be protonated. At high pH (alkaline medium) the carboxylic acid group will lose a proton. This explains why amino acids can function as buffers. If H^+ ions are added they are removed as $-NH_4^+$ and if OH^- ions are added the $-COOH$ loses a proton to remove the OH^- ions as water. For each amino acid there is a unique pH value (known as the **isoelectric point**) where the acid will exist as the zwitterion.

The medium on which electrophoresis is carried out is usually a polyacrylamide gel. So the process is known as PAGE (polyacrylamide gel electrophoresis). The sample is placed in the centre of the gel and a potential difference applied across it. Depending on the pH of the buffer the different amino acids will move at different rates towards the positive and negative electrodes. At its isoelectric point a particular amino acid will not move as its charges are balanced. When separation is complete the acids can be sprayed with ninhydrin and identified by comparing the distance they have travelled with standard samples, or from a comparison of their isoelectric points.

increasing pH →

$$H_3N^+ - \overset{\overset{\displaystyle H}{|}}{\underset{\underset{\displaystyle R}{|}}{C}} - COOH$$

pH < isoelectric point

$$H_3N^+ - \overset{\overset{\displaystyle H}{|}}{\underset{\underset{\displaystyle R}{|}}{C}} - COO^-$$

isoelectric point zwitterion

$$H_2N - \overset{\overset{\displaystyle H}{|}}{\underset{\underset{\displaystyle R}{|}}{C}} - COO^-$$

pH > isoelectric point

(−) initial mixture of five amino acids (+)

Arg His Ser Phe Glu

buffer solution of pH = 5.7

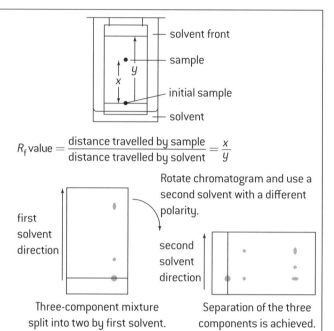

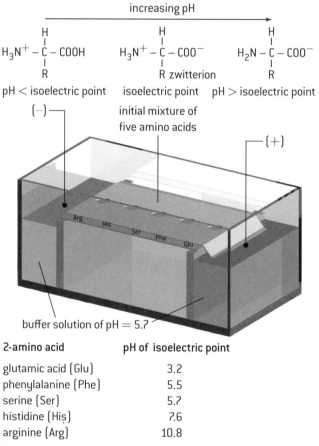

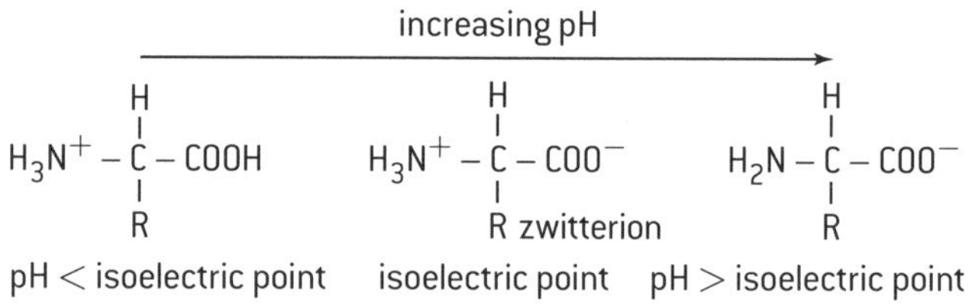

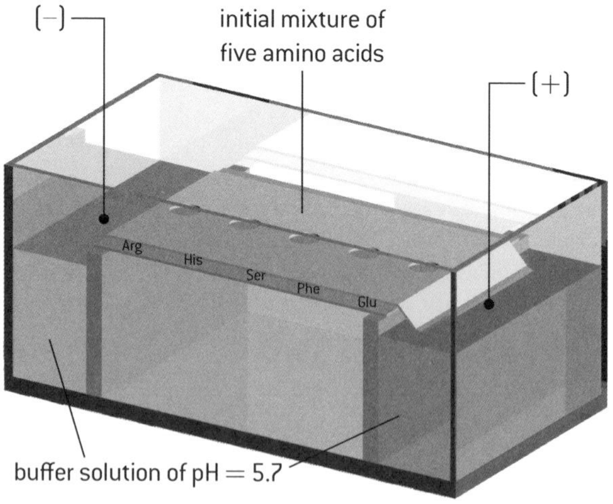

2-amino acid	pH of isoelectric point
glutamic acid (Glu)	3.2
phenylalanine (Phe)	5.5
serine (Ser)	5.7
histidine (His)	7.6
arginine (Arg)	10.8

Separation of a mixture of five amino acids by electrophoresis
Serine does not move as its isoelectric point is the same pH as the buffer. Histidine and arginine contain $-NH_3^+$ at pH 5.7, so move towards the negative electrode. Glutamic acid and phenylalanine contain $-COO^-$ at pH 5.7, so move towards the positive electrode.

Enzymes

USES OF PROTEINS

Proteins have many different functions in the body. They can act as biological catalysts for specific reactions (enzymes). They can give structure (e.g hair and nails consist almost entirely of polypeptides coiled into α-helices), and provide a source of energy. Some hormones are proteins, e.g. FSH (follicle stimulating hormone), responsible for triggering the monthly cycle in females.

CATALYTIC ACTIVITY OF ENZYMES AND ACTIVE SITE

Enzymes are protein molecules that catalyse biological reactions. Each enzyme is highly specific for a particular reaction, and extremely efficient, often being able to increase the rate of reaction by a factor greater than 10^8. Enzymes work by providing an alternative pathway for the reaction with a lower activation energy, so that more of the reactant particles (substrate) will possess the necessary minimum activation energy.

Effect of adding an enzyme on activation energy

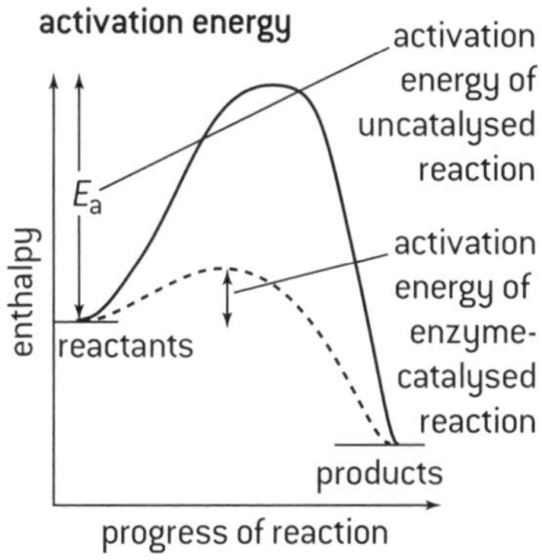

The specificity of enzymes depends on their tertiary and quaternary structure. The part of an enzyme that reacts with the substrate is known as the active site. This is a groove or pocket in the enzyme where the substrate will bind. The site is not necessarily rigid but can alter its shape to allow for a better fit – known as the induced fit theory.

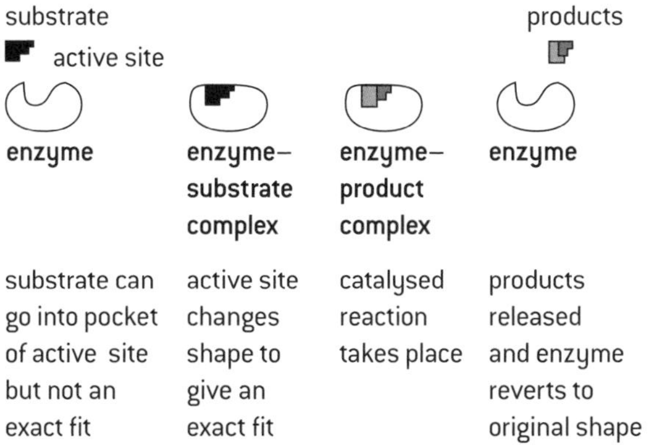

Induced fit theory of enzyme catalysis

The induced fit theory replaces the old 'lock and key' theory which assumes that enzymes have a fixed shape into which the substrate fits.

ENZYME KINETICS

At low substrate concentrations the rate of the enzyme-catalysed reaction is proportional to the concentration of the substrate. However at higher concentrations the rate reaches a maximum. This can be explained in terms of enzyme saturation. At low substrate concentrations there are enough active sites present for the substrate to bind and react. Once all the sites are used up the enzyme can no longer work any faster.

Effect of concentration of substrate on rate of enzyme-catalysed reaction

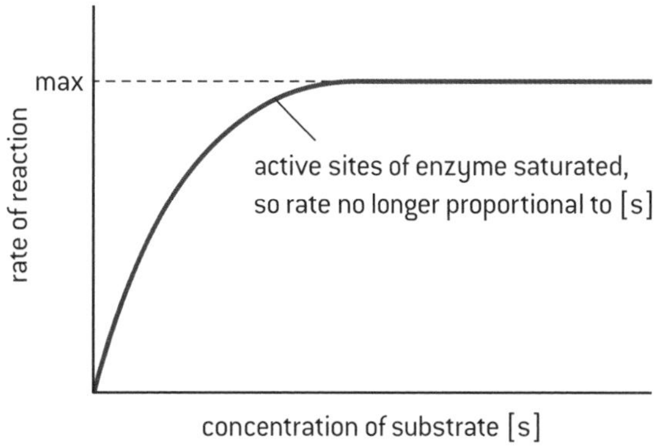

EFFECT OF TEMPERATURE, PH AND HEAVY METAL IONS ON ENZYME ACTIVITY

The action of an enzyme depends on its specific shape. Increasing the temperature will initially increase the rate of enzyme-catalysed reactions, as more of the reactants will possess the minimum activation energy. The optimum temperature for most enzymes is about 40°C. Above this temperature enzymes rapidly become denatured as the weak bonds holding the tertiary structure together break.

At different pH values the charges on the amino acid residues change affecting the bonds between them, and so altering the tertiary structure and making the enzyme ineffective. Heavy metals can poison enzymes by reacting with –SH groups replacing the hydrogen atom with a heavy metal atom or ion so that the tertiary structure is altered.

Effect of temperature on enzyme activity

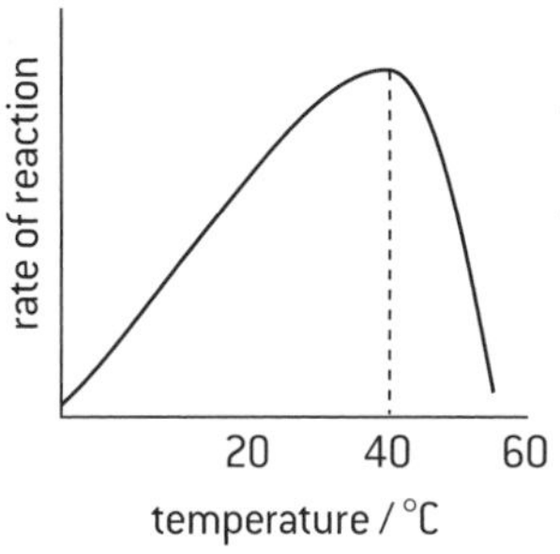

Effect of pH on enzyme activity

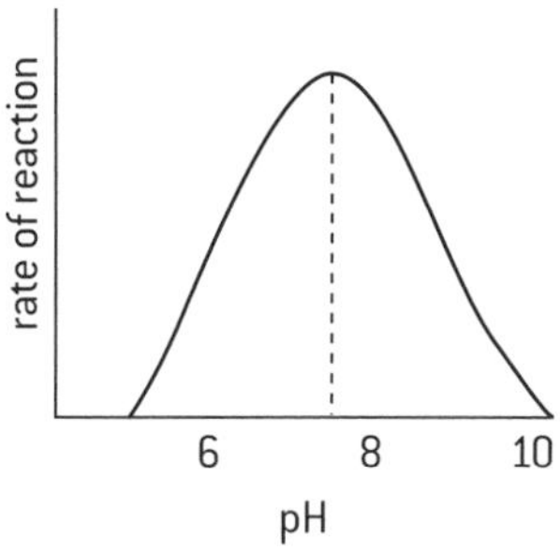

Lipids (1)

Lipids are organic molecules with long hydrocarbon chains that are soluble in non-polar solvents. They are mainly used for energy storage, insulating and protecting vital organs, forming cell membranes and, in some cases, acting as hormones. Three important types of lipids are triglycerides (fats and oils), phospholipids (lecithin) and steroids (cholesterol).

FATS AND OILS

Fats and oils are triesters (triglycerides) formed from the condensation reaction of propane-1,2,3-triol (glycerol) with long chain carboxylic acids (fatty acids).

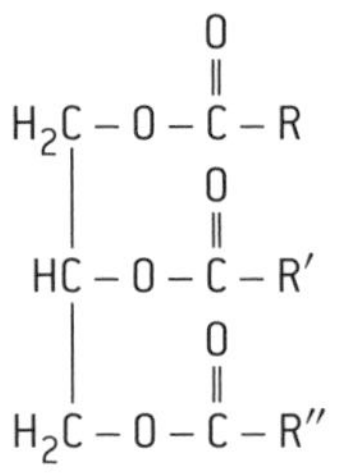

R, R′, and R″ are long chain hydrocarbons formed from carboxylic acids which may be the same or different.

General formula of a fat or oil.

Fats are solid triglycerides; examples include butter, lard and tallow. Oils are liquid at room temperature and include castor oil, olive oil and linseed oil. The essential chemical difference between them is that fats contain saturated carboxylic acid groups (i.e. they do not contain C=C double bonds). Oils contain at least one C=C double bond and are said to be unsaturated. Most oils contain several C=C double bonds and are known as polyunsaturated.

PHOSPHOLIPIDS

Phospholipids form an integral part of all cell membranes. They are essentially made of four components. A backbone such as propane-1,2,3-triol (glycerol), linked by esterification to two fatty acids and a phosphate group which is itself condensed to a nitrogen-containing alcohol. There are many different phospholipids. They can be exemplified by phosphatidyl choline – the major component of lecithin, present in egg yolk.

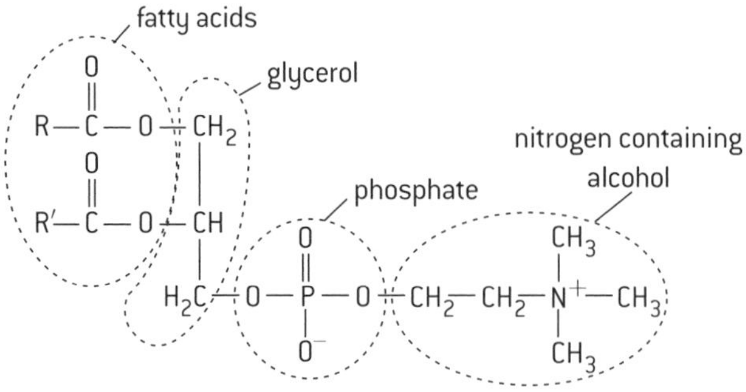

The structure of phosphatidyl choline showing the origins of the four distinct components.

CHOLESTEROL

Cholesterol has the characteristic four-ring structure possessed by all steroids.

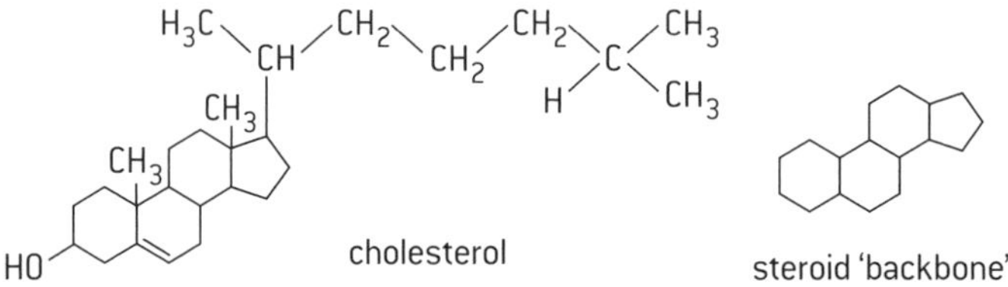

It is transported around the body by lipoproteins. **Low density lipoproteins** (LDL) are in the order of 18–25 nm and transport cholesterol to the arteries where it can line the walls of the arteries leading to cardiovascular diseases. The major source of these low density lipoproteins are saturated fats, in particular those derived from lauric (C_{12}), myristic (C_{14}) and palmitic (C_{16}) acids. Smaller lipoproteins, in the order of 8–11 nm, known as **high density lipoproteins** (HDL) can remove the cholesterol from the arteries and transport it back to the liver.

FATTY ACIDS

Stearic acid (m.pt 69.6 °C) and linoleic acid (m.pt −5.0 °C) both contain the same number of carbon atoms and have similar molar masses. However, linoleic acid contains two double bonds. Generally the more unsaturated the fatty acid the lower its melting point.

The regular tetrahedral arrangement of saturated fatty acids means that they can pack together closely, so the London dispersion forces holding molecules together are stronger as the surface area between them is greater. As the bond angle at the C=C double bonds changes from 109.5° to 120° in unsaturated acids it produces a 'kink' in the chain. They are unable to pack so closely and the London dispersion forces between the molecules become weaker, which results in lower melting points. This packing arrangement is similar in fats and explains why unsaturated fats (oils) have lower melting points.

$CH_3(CH_2)_{16}COOH$

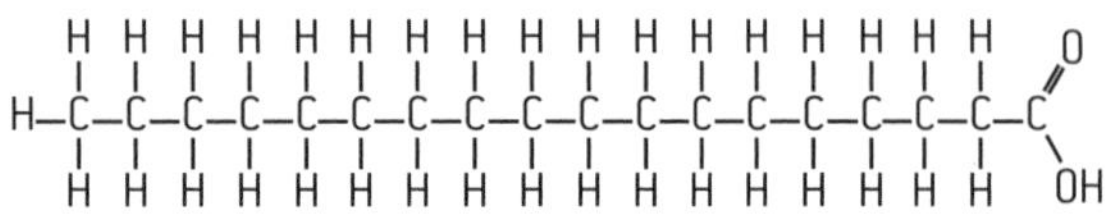

stearic acid – a saturated fatty acid

Name	Number of C atoms per molecule	Number of C=C bonds	Melting point/°C
saturated fatty acids			
lauric acid $CH_3(CH_2)_{10}COOH$	12	0	44.2
myristic acid $CH_3(CH_2)_{12}COOH$	14	0	54.1
palmitic acid $CH_3(CH_2)_{14}COOH$	16	0	62.7
stearic acid $CH_3(CH_2)_{16}COOH$	18	0	69.6
unsaturated fatty acids			
oleic acid $CH_3(CH_2)_7CH=CH(CH_2)_7COOH$	18	1	10.5
linoleic acid $CH_3(CH_2)_4CH=CHCH_2CH=CH(CH_2)_7COOH$	18	2	−5.0

Lipids (2)

HYDROGENATION

Oils naturally contain only cis-unsaturated fatty acids. These are generally healthier than saturated fats as they increase HDL cholesterol. Unsaturated fats can be hydrogenated to saturated fats with a higher melting point by adding hydrogen under pressure in the presence of a heated nickel catalyst. However during the hydrogenation process, partial hydrogenation can occur and the trans-isomers may be formed. Trans-unsaturated fatty acids are present in fried foods such as French fries and some margarines. Unlike natural mono- and poly-unsaturated oils, trans-unsaturated fats increase the formation of LDL cholesterol ('bad' cholesterol) and thus increase the risk of heart disease.

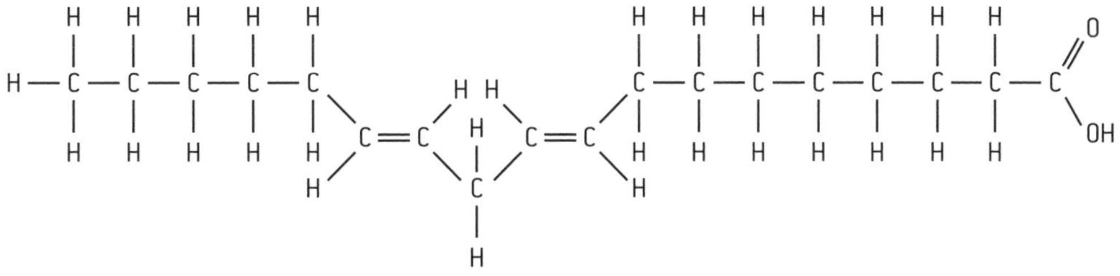

the structure of the trans, trans- form of linoleic acid

HYDROLYSIS AND RANCIDITY OF FATS

Fats and oils are hydrolysed in the body by enzymes, known as lipases, to glycerol and fatty acids. These in turn are broken down by a series of redox reactions to produce ultimately carbon dioxide, water and energy. Because they are essentially long-chain hydrocarbons with only two oxygen atoms each on the three carboxyl atoms fats are in a less oxidized form than carbohydrates so weight for weight produce more energy.

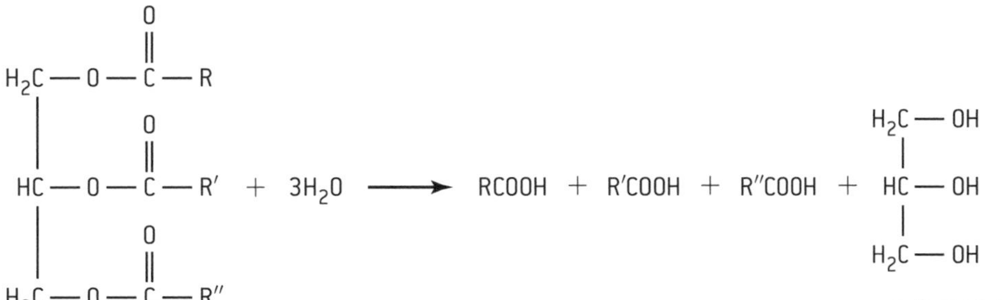

hydrolysis of a fat

propane-1, 2, 3-triol (glycerol)

Lipids (fats and oils) in food become rancid when our senses perceive them to have 'gone off' due to a disagreeable smell, texture or appearance. This may be caused either by hydrolysis of the triesters (hydrolytic rancidity) as shown above to produce disagreeable smelling fatty acids or by oxidation of the fatty acid chains.

Oxidative rancidity is typically due to the addition of oxygen across the C=C double bonds in unsaturated fatty acids. Oily fishes, such as mackerel, contain a high proportion of unsaturated fatty acids and are prone to oxidative rancidity. The process proceeds by a free radical mechanism catalysed by light in the presence of enzymes.

IODINE NUMBER

Unsaturated fats can undergo addition reactions. The addition of iodine to unsaturated fats can be used to determine the number of C=C double bonds, since one mole of iodine will react quantitatively with one mole of C=C double bonds. Iodine is coloured. As the iodine is added to the unsaturated fat the purple colour of the iodine will disappear as the addition reaction takes place. Once the colour remains the amount of iodine needed to react with all the C=C double bonds can be determined. Often fats are described by their iodine number, which is the number of grams of iodine that add to 100 g of the fat.

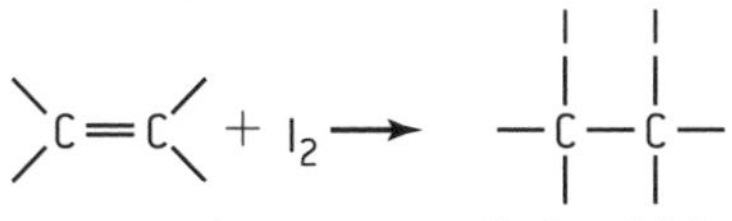

unsaturated fat

diiodo-addition product

THE ROLES OF LIPIDS IN THE BODY

- Energy storage. Because they contain proportionally less oxygen than carbohydrates they release more energy when oxidized.
- Insulation and protection of organs. Fats are stored in adipose tissue, which provides both insulation and protection to parts of the body.
- Steroid hormones. Examples include female and male sex hormones such as progesterone and testosterone and the contraceptive pill. Sometimes steroids are abused. Anabolic steroids have similar structures to testosterone and are taken to build up muscle.
- Cell membranes. Lipids provide the structural component of cell membranes.

More controversially, lipids are thought to affect health, particularly heart disease. Although the evidence is disputed by many, some think that saturated fatty acids, particularly lauric (C_{12}), myristic (C_{14}) and palmitic (C_{16}) acids increase LDL, as do trans-unsaturated fats, causing heart problems. Conversely omega-3-polyunsaturated fatty acids such as natural unsaturated fats (e.g. olive oil) are thought to lower the level of LDL and consequently are thought to be good for you.

Carbohydrates

MONOSACCHARIDES

All monosaccharides have the empirical formula CH_2O. In addition they contain a carbonyl group (C=O) and at least two -OH groups. If the carbonyl group is an aldehyde (RCHO) they are known as an aldose, if the carbonyl group is a ketone (RCOR') they are known as a ketose. Monosaccharides have between three and six carbon atoms.

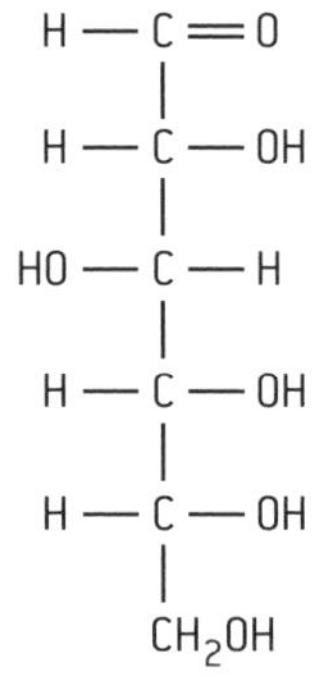

Straight chain glucose (an aldose)

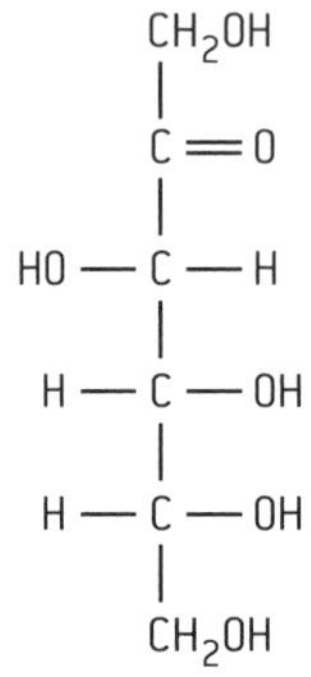

Straight chain fructose (a ketose)

Monosaccharides with the general formula $C_5H_{10}O_5$ are known as pentoses (e.g. ribose) and monosaccharides with the general formula $C_6H_{12}O_6$ are known as hexoses (e.g. glucose).

Many structural isomers of monosaccharides are possible. In addition several carbon atoms are chiral (asymmetric) and give rise to optical isomerism. As well as this, open chain structures and ring structures are possible. The form of glucose that is found in nature is known as D-glucose. Note that the ring structures are cyclic ethers as they contain an oxygen atom bonded on either side by a carbon atom within the ring.

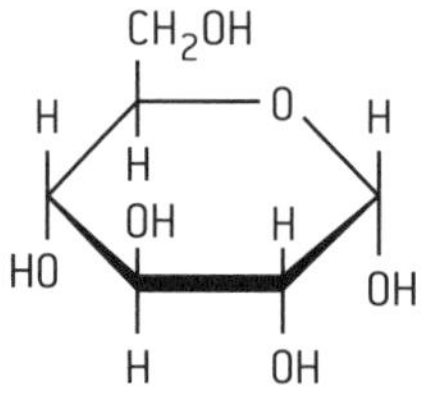

D-glucose

Six-membered ring monosaccharides are known as pyranoses. Hexoses can also have a furanose structure where they have a five-membered ring containing an oxygen atom.

MAJOR FUNCTIONS OF POLYSACCHARIDES IN THE BODY

Carbohydrates are used by humans:

- **to provide energy:** foods such as bread, biscuits, cakes, potatoes and cereals are all high in carbohydrates.
- **to store energy:** starch is stored in the livers of animals in the form of glycogen – also known as animal starch. Glycogen has almost the same chemical structure as amylopectin.
- **as precursors** for other important biological molecules, e.g. they are components of nucleic acids and thus play an important role in the biosynthesis of proteins.
- **as dietary fibre:** dietary fibre is mainly plant material that is not hydrolysed by enzymes secreted by the human digestive tract but may be digested by microflora in the gut. Examples include cellulose, hemicellulose, lignin and pectin. It may be helpful in preventing conditions such as diverticulosis, irritable bowel syndrome, obesity, Crohn's disease, haemorrhoids and diabetes mellitus.

POLYSACCHARIDES

Monosaccharides can undergo condensation reactions to form disaccharides and eventually polysaccharides. For example, sucrose, a disaccharide formed from the condensation of D-glucose in the pyranose form and D-fructose in the furanose form.

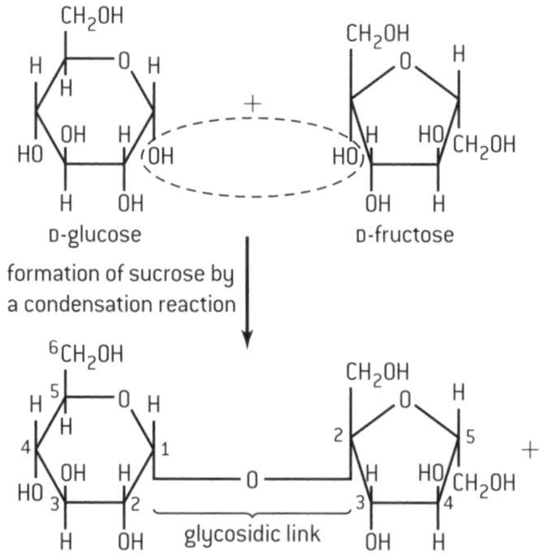

sucrose

The link between the two sugars is known as a glycosidic link. In the case of sucrose the link is between the C-1 atom of glucose and the C-2 atom of fructose. The link is known as a 1,2 glycosidic bond. Maltose, another disaccharide is formed from two glucose molecules condensing to form an 1,4 glycosidic bond. Lactose is a disaccharide in which the D-galactose is linked at the C-1 atom to the C-4 atom of D-glucose to form a 1,4 glycosidic bond.

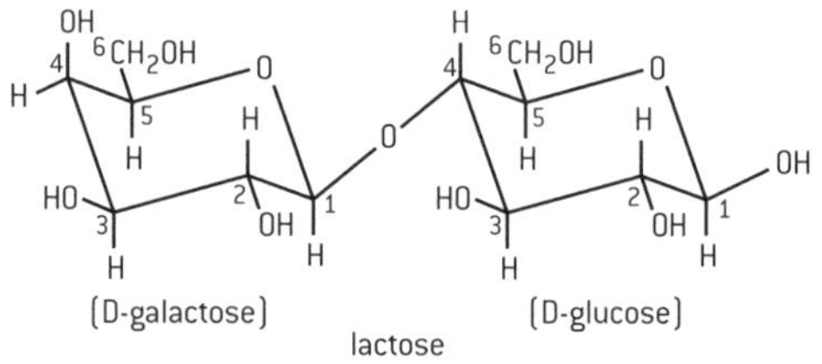

lactose

One of the most important polysaccharides is starch. Starch exists in two forms: amylose, which is soluble in hot water, and amylopectin, which is more soluble in water. Amylose is a straight chain polymer of D-glucose units with 1,4 glycosidic bonds:

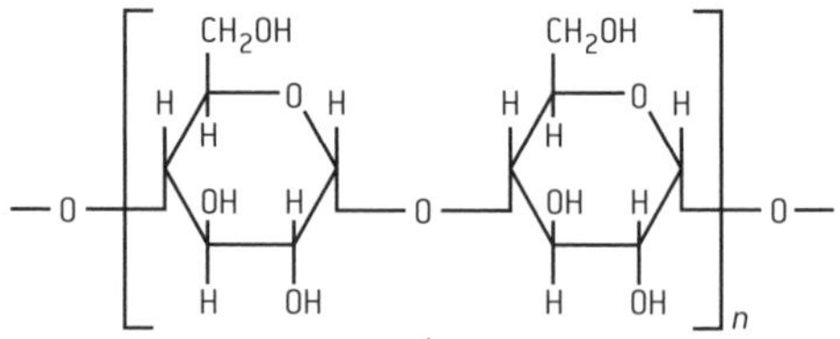

amylose

Amylopectin also consists of D-glucose units but it has a branched structure with both 1,4 and 1,6 glycosidic bonds:

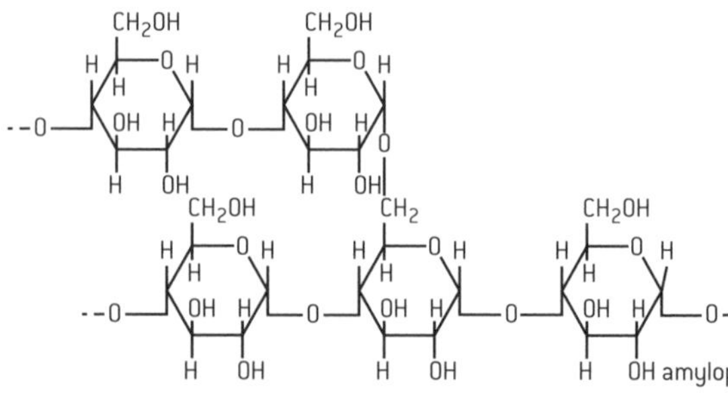

amylopectin

Most plants use starch as a store of carbohydrates and thus energy. Cellulose, a polymer of D-glucose contains 1,4 linkages. Cellulose, together with lignin, provides the structure to the cell walls of green plants. Most animals, including all mammals, do not have the enzyme cellulase so are unable to digest cellulose or other dietary fibre polysaccharides.

Vitamins

VITAMINS

Vitamins are micro-nutrients. Micro-nutrients are substances required in very small amounts (mg or µg). They mainly function as a co-factor of enzymes and include not only vitamins but also trace minerals such as Fe, Cu, F, Zn, I, Se, Mn, Mo, Cr, Co and B.

Vitamins can be classified as fat soluble or water soluble. The structure of fat soluble vitamins is characterized by long, non-polar hydrocarbon chains or rings. These include vitamins A, D, E, F and K. They can accumulate in the fatty tissues of the body. In some cases an excess of fat soluble vitamins can be as serious as a deficiency. The molecules of water soluble vitamins, such as vitamin C and the eight B-group vitamins, contain hydrogen attached directly to electronegative oxygen or nitrogen atoms that can hydrogen bond with water molecules. They do not accumulate in the body so a regular intake is required. Vitamins containing C=C double bonds and -OH groups are readily oxidized and keeping food refrigerated slows down this process.

VITAMIN A (RETINOL)

Although it does contain one -OH group, vitamin A is fat soluble due to the long non-polar hydrocarbon chain. Unlike most other vitamins it is not broken down readily by cooking. Vitamin A is an aid to night vision.

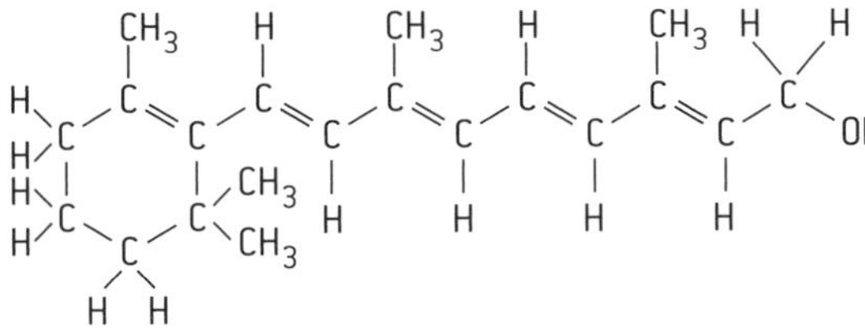

vitamin A (retinol)

VITAMIN C (ASCORBIC ACID)

Due to the large number of polar -OH groups vitamin C is soluble in water so is not retained for long by the body. The most famous disease associated with a lack of vitamin C is scorbutus ('scurvy'). The symptoms are swollen legs, rotten gums and bloody lesions. It was a common disease in sailors, who spent long periods without fresh food, until the cause was recognized.

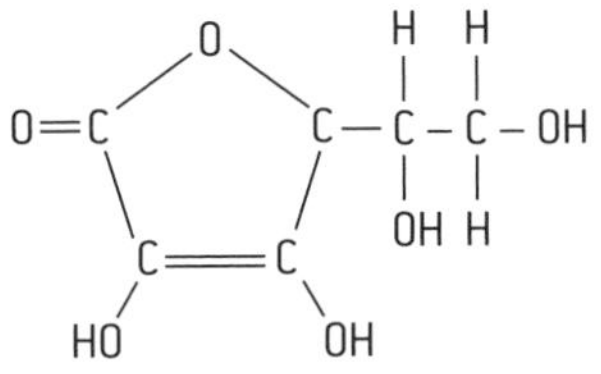

vitamin C (ascorbic acid)

VITAMIN D (CALCIFEROL)

Vitamin D is essentially a large hydrocarbon with one –OH group and is fat soluble.

A deficiency of vitamin D leads to bone softening and malformation – a condition known as rickets.

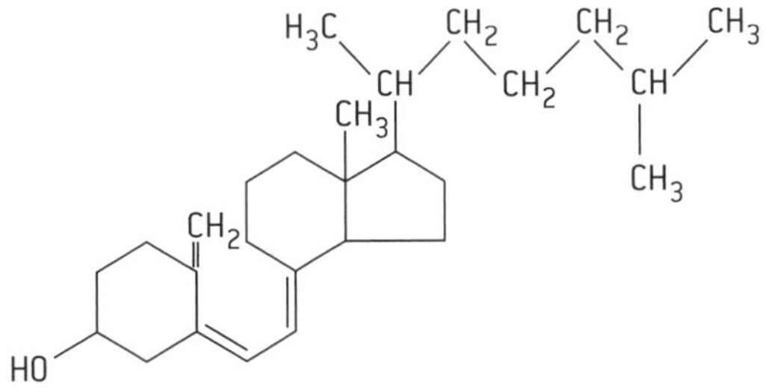

vitamin D

MALNUTRITION

Malnutrition occurs when either too much food is consumed, which leads to obesity, or the diet is lacking in one or more essential nutrients. Specific micro-nutrient deficiencies include:

- Fe – anaemia
- I – goitre
- vitamin A (retinol) – xerophthalmia, night blindness
- vitamin B_3 (niacin) – pellagra
- vitamin B_1 (thiamin) – beriberi
- vitamin C (ascorbic acid) – scurvy
- vitamin D (calciferol) – rickets.

Solutions to combat malnutrition include:

- eating fresh food rich in vitamins and minerals
- adding nutrients that are missing in commonly consumed foods
- genetic modification of food
- providing nutritional supplements.

Biochemistry and the environment

GREEN CHEMISTRY

Our increasing knowledge and use of biochemistry has led to solutions to some issues but has also caused environmental problems in other areas. Scientists have a responsibility to be aware of the impact of their research on the environment and should actively find ways to counter any negative impact their work may have on the environment. Examples of negative impact include the use of enzymes in biological detergents and the overuse of antibiotics in animal feed. Green chemistry, which is sometimes also known as sustainable chemistry, encourages the reduction and prevention of pollution at source. It does this by trying to minimize the use and formation of substances harmful to the environment. One way in which this can be achieved is to make use of atom economy (see page 6).

BIODEGRADABILITY

Although most plastics are organic in origin they are petroleum-based so cannot easily be broken down by natural organisms and cause big pollution problems. Biodegradable plastics are plastics capable of being broken down by bacteria or other organisms, ultimately to carbon dioxide and water. They are based on natural renewable polymers containing ester or glycosidic links, such as starch, that can be hydrolysed. In theory, starch-based bioplastics produced as biomass could be almost carbon neutral but there are problems such as using land that could otherwise be used for growing food and the release of the greenhouse gas methane if the plastics are decomposed anaerobically in landfill sites.

Enzymes can also be used to biodegrade pollutants. Enzymes are used to aid the breakdown and dispersal of oil spills. This reduces the effect of dispersal agents, such as 2-butoxyethanol, but does not replace them completely. The oil still needs to be broken into smaller droplets before the microbes containing the enzymes can be effective. The use of enzymes in biological detergents is also well known. This has the advantage to the environment of lowering the temperature at which clothes need to be washed so making the process more efficient, i.e. saving on energy and fossil fuels.

HOST–GUEST CHEMISTRY

Host–guest complexes are made of two or more molecules or ions bonded together through non-covalent bonding, which is critical in maintaining the 3-D structure of the molecule. Non-covalent interactions include hydrogen bonds, ionic bonds and van der Waals' forces. These forces, which are weaker than covalent bonding, allow large molecules to bind specifically but transiently to one another to form supramolecules. They work by mimicking some of the actions performed by enzymes by selectively binding to 'guest' species. For example, they have been used to deliver drugs more effectively in humans by increasing the solubility and availability of the drug and reducing drug resistance. They are also used to remove toxic materials (xenobiotics) from the environment. For example, radioactive 137-caesium from nuclear waste and carcinogenic amines from polluted water.

XENOBIOTICS

Chemicals found in organisms, which are not normally present or produced by the organism, or are present in organisms in abnormally high amounts, are known as xenobiotics. Examples of xenobiotics include drugs in animals. Antibiotics, for example, are not produced by animals nor are they part of a normal diet. The use of antibiotics in animal feed and in sewage plants has meant that they pass through into the human food chain and increase resistant strains of bacteria. Some xenobiotics may be natural compounds but most are pollutants. Two classic xenobiotics are dioxins and polychlorinated biphenyls (PCBs).

Dioxins can be formed when polymers are combusted, unless the temperature is extremely high. They do not decompose in the environment and can be passed on in the food chain. Many dioxins, particularly chlorinated dioxins, are highly carcinogenic as they can disrupt the endocrine system (hormone action) and lead to cellular and genetic damage. Examples of dioxins and dioxin-like substances include 1,4-dioxin, polychlorinated dibenzodioxins (PCDDs) and polychlorinated biphenyls (PCBs). The general formulas of PCDDs and PCBs are given in Section 31 of the IB data booklet. Some specific examples are:

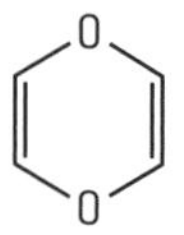

1, 4-dioxin

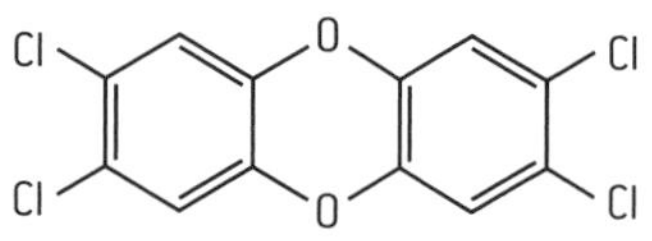

2, 3, 7, 8-tetrachlorodibenzodioxin
(an example of a PCDD)

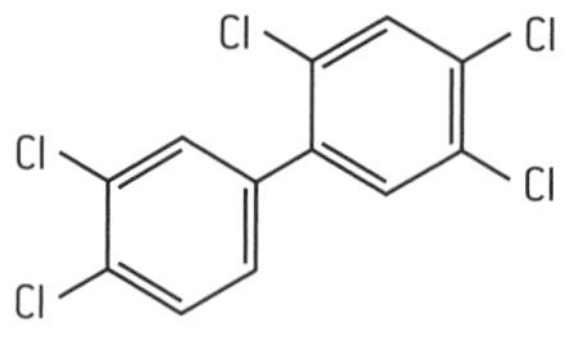

2, 3′, 4, 4′, 5-pentachlorobiphenyl
(an example of a PCB)

PCBs contain from one to ten chlorine atoms attached to a biphenyl molecule. They are chemically stable and have high electrical resistance so were used in transformers and capacitors. Although not strictly dioxins (as they contain no oxygen atoms) they also persist in the environment and have carcinogenic properties.

One of the problems associated with xenobiotics is biomagnification. As the xenobiotic passes through the food chain its concentration increases in higher species. DDT (dichlorodiphenyltrichloroethane) is an effective insecticide particularly against the malaria mosquito but its use is now banned as it accumulates at high levels in birds of prey, which threatens their survival.

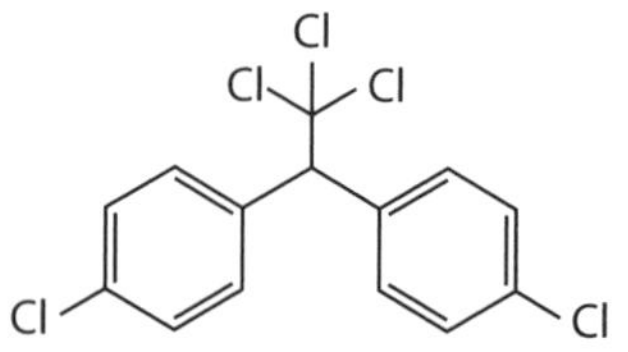

DDT dichlorodiphenyltrichlorethane

Proteins and enzymes (1)

V_{MAX} AND THE MICHAELIS-MENTEN CONSTANT, K_M

We have seen already that at low substrate concentrations the rate of an enzyme-catalysed reaction is proportional to the concentration of the substrate but it reaches a maximum at higher substrate concentrations. This maximum is known as V_{max}.

Effect of concentration of substrate on rate of enzyme-catalysed reaction

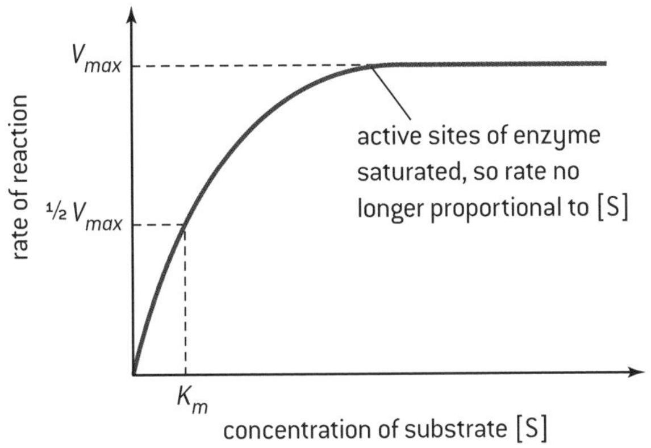

The Michaelis-Menten constant, K_{m} is the substrate concentration when the rate of the reaction is ½ V_{max}. A particular enzyme with the same substrate will always have the same value for K_{m}. It indicates whether the enzyme functions efficiently at low substrate concentrations, or whether high substrate concentrations are necessary for efficient catalysis.

COMPETITIVE AND NON-COMPETITIVE INHIBITION

Inhibitors are substances that slow down the rate of enzyme-catalysed reactions. Competitive inhibitors resemble the substrate in shape, but cannot react. They slow down the reaction because they can occupy the active site on the enzyme thus making it less accessible to the substrate. Non-competitive inhibitors also bind to the enzyme, but not on the active site. They bind at the allosteric site, which causes the enzyme to change its shape so that the substrate cannot bind. As the substrate concentration is increased the effect of competitive inhibitors lessens, as there is increased competition for the active sites by the substrate. With non-competitive inhibitors increasing the substrate concentration has no effect, as the enzyme's shape still remains altered.

Effect of substrate concentration on inhibitors

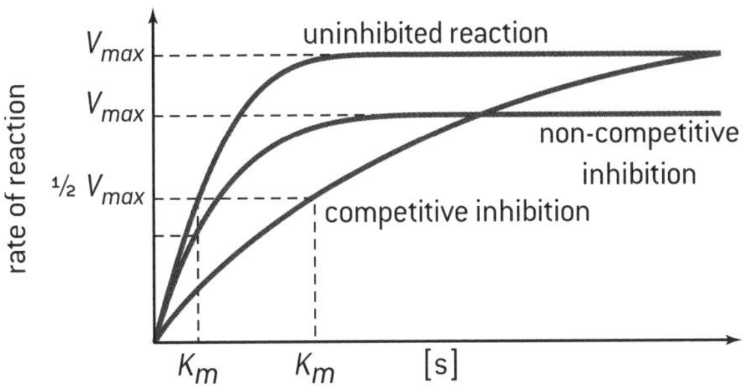

For non-competitive inhibitors, V_{max} is lower but K_m is the same. For competitive inhibitors, V_{max} is the same but K_m is increased.

BUFFER SOLUTIONS

Enzymes only function efficiently within a narrow pH region. Outside of this region the structure is altered and the enzyme becomes denatured, hence the need for buffering. A buffer solution resists changes in pH when small amounts of acid or alkali are added to it.

An acidic buffer solution can be made by mixing a weak acid together with the salt of that acid and a strong base. An example is a solution of ethanoic acid and sodium ethanoate. The weak acid is only slightly dissociated in solution, but the salt is fully dissociated into its ions, so the concentration of ethanoate ions is high.

$$NaCH_3COO(aq) \rightarrow Na^+(aq) + CH_3COO^-(aq)$$

$$CH_3COOH(aq) \rightleftharpoons CH_3COO^-(aq) + H^+(aq)$$

If an acid is added the extra H^+ ions coming from the acid are removed as they combine with ethanoate ions to form undissociated ethanoic acid, so the concentration of H^+ ions remains unaltered.

$$CH_3COO^-(aq) + H^+(aq) \rightleftharpoons CH_3COOH(aq)$$

If an alkali is added the hydroxide ions from the alkali are removed by their reaction with the undissociated acid to form water, so again the H^+ ion concentration stays constant.

$$CH_3COOH(aq) + OH^-(aq) \rightarrow CH_3COO^-(aq) + H_2O(l)$$

In practice acidic buffers are often made by taking a solution of a strong base and adding excess weak acid to it, so that the solution contains the salt and the unreacted weak acid.

$$\underset{\text{limiting reagent}}{NaOH(aq)} + CH_3COOH(aq) \rightarrow \underbrace{\underset{\text{salt}}{NaCH_3COO(aq)} + H_2O(l) + \underset{\text{excess weak acid}}{CH_3COOH(aq)}}_{\text{Buffer solution}}$$

An alkali buffer with a fixed pH greater than 7 can be made from a weak base together with the salt of that base with a strong acid. An example is ammonia with ammonium chloride.

$$NH_4Cl(aq) \rightarrow NH_4^+(aq) + Cl^-(aq)$$

$$NH_3(aq) + H_2O(l) \rightleftharpoons NH_4^+(aq) + OH^-(aq)$$

If H^+ ions are added they will combine with OH^- ions to form water and more of the ammonia will dissociate to replace them. If more OH^- ions are added they will combine with ammonium ions to form undissociated ammonia. In both cases the hydroxide ion concentration and the hydrogen ion concentration remain constant.

Proteins and enzymes (2)

BUFFER CALCULATIONS

The equilibrium expression for weak acids also applies to acidic buffer solutions,

e.g. ethanoic acid/sodium ethanoate solution.

$$K_a = \frac{[H^+] \times [CH_3COO^-]}{[CH_3COOH]}$$

The essential difference is that now the concentrations of the two ions from the acid will not be equal.

Since the sodium ethanoate is completely dissociated the concentration of the ethanoate ions in solution will be almost the same as the concentration of the sodium ethanoate, as very little will come from the acid.

If logarithms are taken and the equation is rearranged then:

$$pH = pK_a + \log_{10}\frac{[CH_3COO^-]}{[CH_3COOH]}$$

This is known as the Henderson–Hasselbalch equation (the general formula can be found in Section 1 of the IB data booklet).

Two facts can be deduced from this expression. Firstly the pH of the buffer does not change on dilution, as the concentration of the ethanoate ions and the acid will be affected equally.

Secondly the buffer will be most efficient when $[CH_3COO^-] = [CH_3COOH]$. At this point, which equates to the half equivalence point when ethanoic acid is titrated with sodium hydroxide, the pH of the solution will equal the pK_a value of the acid.

Calculate the pH of a buffer containing 0.200 mol of sodium ethanoate in 500 cm^3 of 0.100 mol dm^{-3} ethanoic acid (given that K_a for ethanoic acid = 1.8×10^{-5} mol dm^{-3}).

$[CH_3COO^-] = 0.400$ mol dm^{-3}; $[CH_3COOH] = 0.100$ mol dm^{-3}

$$K_a \approx \frac{[H^+] \times 0.400}{0.100} = 1.8 \times 10^{-5} \text{ mol dm}^{-3}$$

$[H^+] = 4.5 \times 10^{-6}$ mol dm^{-3}

pH = 5.35

Calculate what mass of sodium propanoate must be dissolved in 1.00 dm^3 of 1.00 mol dm^{-3} propanoic acid (pK_a = 4.87) to give a buffer solution with a pH of 4.5.

$$[C_2H_5COO^-] = \frac{K_a \times [C_2H_5COOH]}{[H^+]} = \frac{10^{-4.87} \times 1.00}{10^{-4.5}}$$

$= 0.427$ mol dm^{-3}

Mass of NaC_2H_5COO required $= 0.427 \times 96.07 = 41.0$ g

BLOOD

An important buffer is blood, which only functions correctly within a very narrow pH range. Blood is a complex buffering system, which is responsible for carrying oxygen around the body. One of the components of the system is that the oxygen adds on reversibly to the haemoglobin in the blood.

$$HHb + O_2 \rightleftharpoons H^+ + HbO_2^-$$

If the pH increases ($[H^+]$ falls) the equilibrium will move to the right and the oxygen will tend to be bound to the haemoglobin more tightly. If the pH decreases ($[H^+]$ increases) the oxygen will tend to be displaced from the haemoglobin. Both of these processes are potentially life threatening.

PROTEIN ASSAY BY UV-VIS SPECTROSCOPY

Determining the concentration of a protein in solution by UV-VIS spectroscopy depends essentially on two relationships. The first is that the protein needs to be made into a coloured compound such that the intensity of the colour depends upon the concentration of the protein in the solution. One way in which this can be done is to add a dye called Coomassie Brilliant Blue. The coloured complex with the dye absorbs light at a particular wavelength. In the case of Coomassie Brilliant Blue this wavelength is 595 nm. This can be shown by running a spectrum of the solution containing the complex and seeing that the maximum absorption (known as λ_{max}) occurs at 595 nm. The second relationship required involves the Beer–Lambert Law. This states that for dilute solutions at a fixed wavelength

$$\log_{10}\frac{I_o}{I} = \varepsilon lc$$

where: I_o is the intensity of the incident radiation and I is the intensity of the transmitted radiation.

ε is the molar absorption coefficient (a constant for each absorbing substance).

l is the path length of the absorbing solution (usually 1.0 cm) and c is the concentration.

Most spectrometers measure $\log_{10} I_o/I$ directly as absorbance. If the path length is kept the same by using the same cuvette (sample tube) and all the readings are taken at λ_{max} then it is easy to see that the measured absorbance is directly proportional to the concentration. Using Coomassie Brilliant Blue the Beer–Lambert Law holds true for solutions of protein covering the range from 0 to approximately 1500 μg cm^{-3}.

To find the concentration of the solution of the protein with unknown concentration, it is therefore necessary to first obtain a calibration curve by using a range of known concentrations of protein and measuring the associated absorbance. A line of best fit is obtained and once the absorbance of the unknown sample has been measured its concentration can be determined by interpolation of the graph.

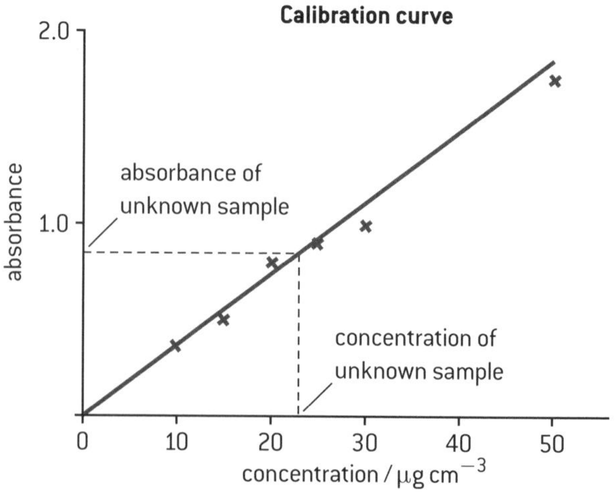

Nucleic acids

STRUCTURE OF NUCLEOTIDES AND NUCLEIC ACIDS

Almost all cells in the human body contain DNA (deoxyribonucleic acid). DNA and a related material RNA (ribonucleic acid) are macromolecules with relative molar masses of up to several million. Both nucleic acids are made up of repeating base-sugar-phosphate units called nucleotides. A nucleotide of DNA contains the condensation products of deoxyribose (a pentose sugar), phosphoric acid, and one of four nitrogen-containing bases, adenine (A), guanine (G), cytosine (C) or thymine (T). RNA contains a different sugar, ribose, but also contains a phosphate group and four nitrogen-containing bases. Three of the bases are the same as those in DNA but the fourth, uracil (U), replaces thymine.

In DNA, the polynucleotide units are wound into a helical shape with about 10 nucleotide units per complete turn. Two helices are then held together by hydrogen bonds between the bases to give the characteristic double helix structure. The stability of this double helix structure is due to the base-stacking interactions between the hydrophilic and hydrophobic components as well as hydrogen bonding between the nucleotides. The hydrogen bonds are very specific. Cytosine can only hydrogen bond with guanine and adenine can only hydrogen bond with thymine (uracil in RNA). Unlike DNA, RNA normally exists as single polynucleotide chains.

sugars

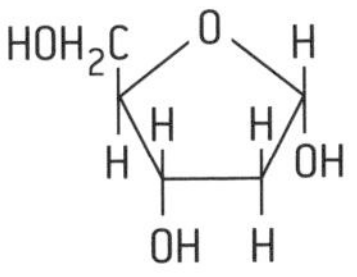

deoxyribose (used in DNA)

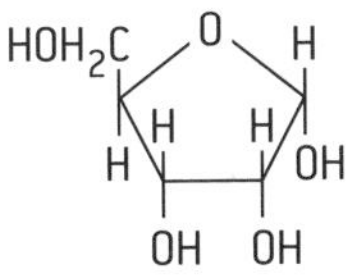

ribose (used in RNA)

bases (showing complementary hydrogen bonding)

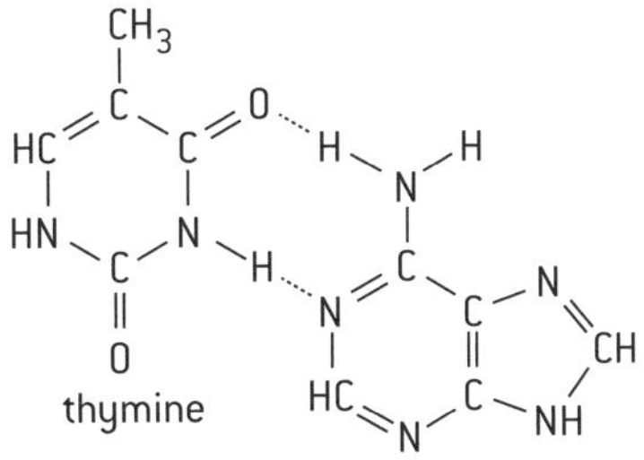

thymine adenine

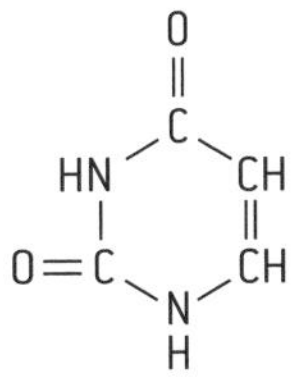

uracil (replaces thymine in RNA)

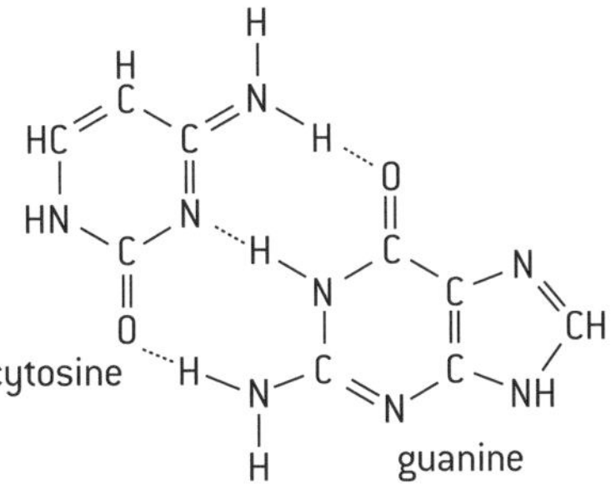

cytosine guanine

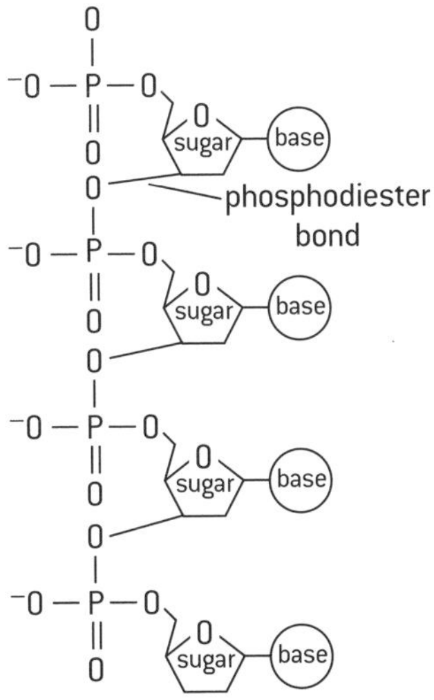

Nucleotides condense to form a polynucleotide. Each nucleotide is joined by a phosphodiester bond between C_3 of the sugar and the neighbouring phosphate group.

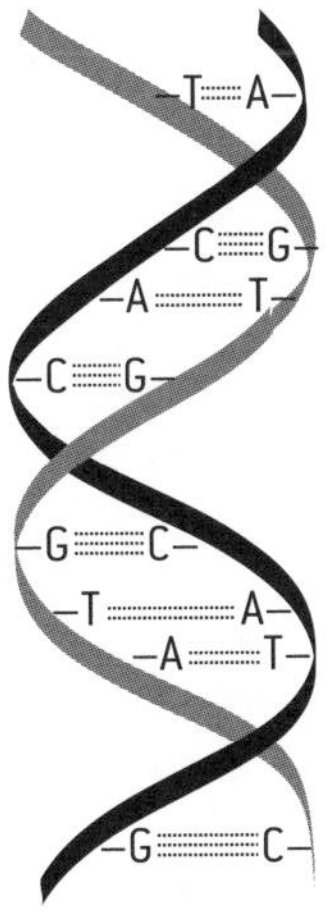

The double helix structure of DNA is shown here. Note the hydrogen bonds between the two different strands of polynucleotides.

The genetic code

THE GENETIC CODE

When cells divide, the genetic information has to be replicated intact. The genetic information is stored in chromosomes found inside the nucleus. In humans there are 23 pairs of chromosomes. Chromosomes are effectively a very long DNA sequence. The DNA is compacted efficiently in the eukaryotic nucleus by forming DNA-protein complexes with histones. Histones are positively charged proteins that bond tightly to the negatively charged phosphate groups in the DNA's phosphate-sugar backbone. The DNA in the cell starts to partly unzip as hydrogen bonds between the bases break. Sugar-base units will be picked up from the aqueous solution to form a complementary new strand. Because adenosine can only hydrogen bond with thymine (A–T) and cytosine can only hydrogen bond with guanine (C–G) the new strand formed will be identical to the original. Thus if the sequence of bases in one strand is -C-G-A-T-T-A- the complementary strand will have the sequence -G-C-T-A-A-T-.

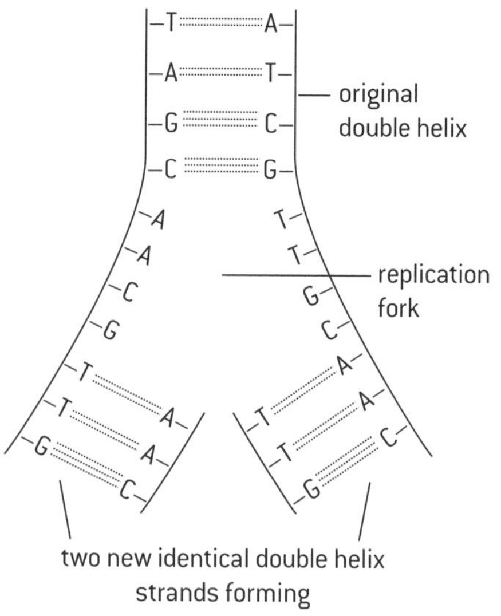

Replication of DNA

The information required to make complex proteins is passed from the DNA to messenger RNA by a similar unzipping process, known as transcription, except that the new strand of mRNA contains a different sugar and uracil in place of thymine.

The coded information held in the mRNA is then used to direct protein synthesis using a triplet code by a process known as translation. Each sequence of three bases represents one amino acid and is known as the triplet code. The triplet code allows for up to 64 permutations known as codons. This is more than sufficient to represent the 20 amino acids and several different codons may represent the same amino acid. Consecutive DNA codons of AAA, TAA, AGA, GTG, and CTT will transcribe to RNA codons of UUU, AUU, UCU, CAC, and GAA which will cause part of a strand of a protein to be formed that contains the amino acid residues –Phe-Ile-Ser-His-Glu-.

UUU	Phe	UCU	Ser	UAU	Tyr	UGU	Cys
UUC	Phe	UCC	Ser	UAC	Tyr	UGC	Cys
UUA	Leu	UCA	Ser	UAA	Terminator	UGA	Terminator
UUG	Leu	UCG	Ser	UAG	Terminator	UGG	Trp
CUU	Leu	CCU	Pro	CAU	His	CGU	Arg
CUC	Leu	CCC	Pro	CAC	His	CGC	Arg
CUA	Leu	CCA	Pro	CAA	Gln	CGA	Arg
CUG	Leu	CCG	Pro	CAG	Gln	CGG	Arg
AUU	Ile	ACU	Thr	AAU	Asn	AGU	Ser
AUC	Ile	ACC	Thr	AAC	Asn	AGC	Ser
AUA	Ile	ACA	Thr	AAA	Lys	AGA	Arg
AUG	Met	ACG	Thr	AAG	Lys	AGG	Arg
GUU	Val	GCU	Ala	GAU	Asp	GGU	Gly
GUC	Val	GCC	Ala	GAC	Asp	GGC	Gly
GUA	Val	GCA	Ala	GAA	Glu	GGA	Gly
GUG	Val	GCG	Ala	GAG	Glu	GGG	Gly

The genetic code carried by RNA

GENETICALLY MODIFIED FOODS

Genetic engineering involves the process of selecting a single gene for a single characteristic and transferring that sequence of DNA from one organism to another. Thus a genetically modified (GM) food can be defined as one derived or produced from a genetically modified organism. The GM food can be substantially different or essentially the same in composition, nutrition, taste, smell, texture and functional characteristics to the conventional food. An example of genetically modified food is the FlavrSavr tomato. In normal tomatoes, a gene is triggered when they ripen to produce a substance that makes the fruit go soft and eventually rot. In the FlavrSavr tomato the gene has been inhibited to produce a tomato with a fuller taste and a longer shelf life.

Benefits of GM foods

- With crops, it can enhance the taste, flavour, texture and nutritional value and also increase the maturation time.
- Plants can be made more resistant to disease, herbicides and insect attack.
- With animals, GM foods can increase resistance to disease, increase productivity and feed efficiency to give higher yields of milk and eggs.
- Anti-cancer substances and increased amounts of vitamins (such as vitamin A in rice) could be incorporated and exposure to less healthy fats reduced.
- Environmentally 'friendly' bio-herbicides and bio-insecticides can be formed. GM foods can lead to soil, water and energy conservation and improve natural waste management.

Potential concerns of GM foods

- The outcome of alterations is uncertain as not enough is known about how genes operate.
- They may cause disease as antibiotic-resistant genes could be passed to harmful microorganisms.
- Genetically engineered genes may escape to contaminate normal crops with unknown effects.
- They may alter the balance of delicate ecosystems as food chains become damaged.
- There are possible links to an increase in allergic reactions (particularly with those involved in food processing).

Biological pigments (1)

ULTRAVIOLET AND VISIBLE ABSORPTION IN ORGANIC MOLECULES

Organic compounds containing unsaturated groups such as C=C, C=O, –N=N–, $-NO_2$ and the benzene ring can absorb in the ultraviolet or visible part of the spectrum. Such groups are known as chromophores and the precise energy of absorption is affected by the other groups attached to the chromophore. The absorption is due to electrons in the bond being excited to an empty orbital of higher energy, usually an anti-bonding orbital. The energy involved in this process is relatively high and most organic compounds absorb in the ultraviolet region and thus appear colourless. For example ethene absorbs at 185 nm. However, if there is extensive conjugation of double bonds (i.e. many alternate C–C single bonds and C=C double bonds) in the molecule involving the delocalization of pi electrons then less energy is required to excite the electrons and the absorption occurs in the visible region. Biological pigments are coloured compounds produced by metabolism. Good examples include anthocyanins, carotenoids, chlorophyll and haem. One other obvious example is the pigment melanin, which is responsible for different tones of skin, eye and hair colour.

ANTHOCYANINS

Anthocyanins are aromatic, water-soluble pigments widely distributed in plants. They contain the flavonoid $C_6C_3C_6$ skeleton.

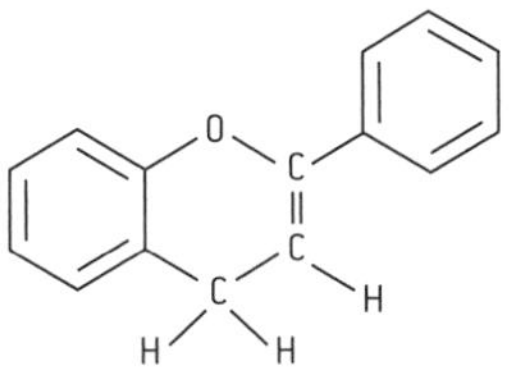

The flavonoid $C_6C_3C_6$ backbone

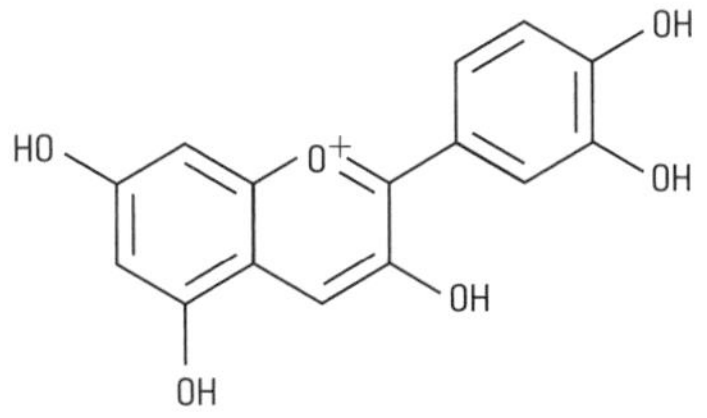

Structure of cyanidin in acidic solution. Less conjugation so absorbs in blue-green region and transmits red light.

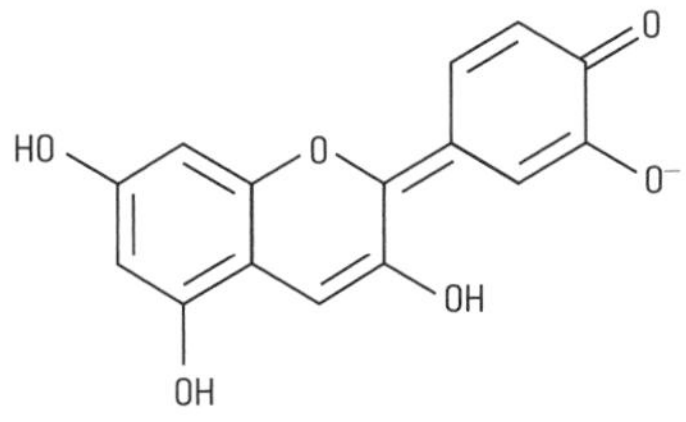

Structure of cyanidin in alkaline solution. More conjugation so absorbs in the orange region of the spectrum and transmits blue light.

It is the conjugation of the pi electrons contained in this structure that accounts for the colour of anthocyanins. The more extensive the conjugation, the lower the energy (longer the wavelength) of the light absorbed. This can be exemplified using cyanidin. In acidic solution it forms a positive ion and there is less conjugation than in alkaline solution where the pi electrons in the extra double bond between the carbon and oxygen atom are also delocalized.

This difference in colour depending on pH explains why poppies that have acidic sap are red whereas cornflowers, which also contain cyanidin but have alkaline sap, are blue. Other anthocyanins differ in the number and types of other groups such as hydroxyl or methoxy groups, which affect the precise wavelength of the light absorbed and hence the colour transmitted. Because their precise colour is so sensitive to pH changes anthocyanins can be used as indicators in acid–base titrations. The addition of other groups also affects other properties of anthocyanins. The basic flavonoid $C_6C_3C_6$ backbone is essentially non-polar. As more polar hydroxyl groups are added the potential for them to form hydrogen bonds with water molecules increases and many anthocyanins, such as cyanidin with several –OH groups, are appreciably soluble in water for this reason.

CAROTENOIDS

Carotenoids are lipid-soluble pigments, and are involved in harvesting light in photosynthesis. The conjugation in carotenoids is mainly due to a long hydrocarbon chain (as opposed to the ring system in anthocyanins) consisting of alternate single and double carbon to carbon bonds.

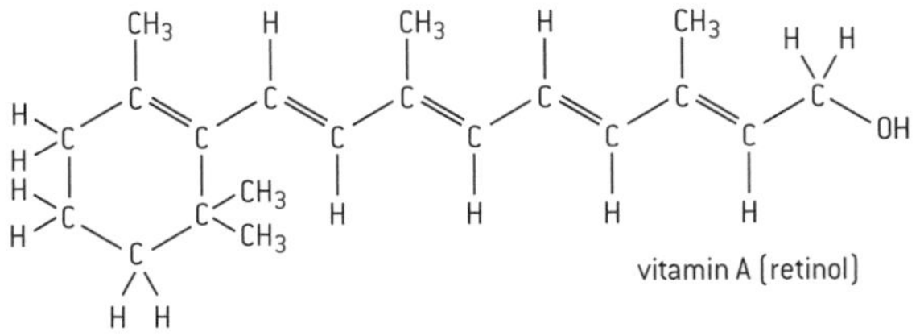

vitamin A (retinol)

The majority of carotenoids are derived from a (poly)ene chain containing forty carbon atoms, which may be terminated by cyclic end groups and may also be complemented with oxygen-containing functional groups. The hydrocarbon carotenoids are known as xanthophylls. Examples include α-carotene, β-carotene and vitamin A. α- and β-carotene and vitamin A are all lipid-soluble and not water-soluble. Although vitamin A does contain one polar hydroxyl group the rest of the molecule is a large non-polar hydrocarbon. Because of the unsaturation in the double bond carotenoids are susceptible to oxidation. This oxidation process can be catalysed by light, metals and hydroperoxides. It results in a change of colour, loss of activity in vitamin A and is the cause of bad smells.

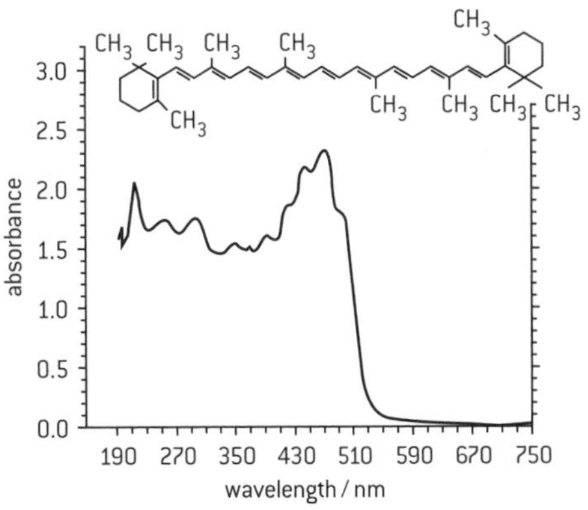

β-carotene is found in carrots and has a characteristic orange colour. It contains eleven conjugated double bonds and absorbs strongly in the violet-blue (400–510 nm) region.

Biological pigments (2)

CHLOROPHYLL AND HAEM

Porphyrin compounds, such as haemoglobin, myoglobin, chlorophyll and many cytochromes are chelates of metals with large nitrogen-containing macrocyclic ligands. Porphyrins contain a cyclic system in which all the carbon atoms are sp^2 hybridized. This results in a planar structure with extensive pi conjugation. The non-bonding pairs of electrons on the four nitrogen atoms enable the porphyrin to form coordinate bonds with metal ions. Chlorophyll contains a magnesium ion and its structure is given in Section 35 of the IB data booklet. It is found in two closely related forms. In chlorophyll a, the -R group is a methyl group, $-CH_3$, and in chlorophyll b the -R group is an aldehyde group, $-CHO$. Chlorophyll is essential for photosynthesis. Its function is to absorb light energy and undergo a redox reaction to donate an electron through a series of intermediates in an electron transport chain. Cytochromes in the electron transport chain contain haem groups in which the iron ion interconverts between iron(II) and iron(III) during redox reactions.

Haemoglobin, which is found in the blood, carries oxygen from the respiratory organs to the rest of the body. Like myoglobin (which is found in muscles) and some of the cytochromes, it contains haem (also spelt heme) groups with the porphyrin group bound to an iron(II) ion. When oxygen binds to one of the iron atoms in the complex to form HbO_2 it causes the iron atom to move towards the centre of the porphyrin ring and at the same time the imidazole side-chain of a histidine residue is pulled towards the porphyrin ring. This produces a strain that is transmitted to the remaining three monomers in the quaternary haemoglobin. This brings about a similar conformational change in the other haem sites so that it is easier for a second oxygen molecule to bind to a second iron atom. Each time the haem group's affinity to attract oxygen increases as the remaining sites become filled. Haem becomes fully saturated with oxygen when all four iron atoms have been utilized forming HbO_8. The binding of oxygen is, thus, a cooperative process. Various factors affect the amount of oxygen that binds. Low pH and a relatively high pressure of carbon dioxide (i.e. during exhalation) cause oxygen to be released from the haemoglobin into the tissues. Conversely at lower carbon dioxide pressure (which causes the pH to rise) more oxygen is taken up by the haemoglobin. Temperature also has an effect. At higher temperatures more oxygen is released from haemoglobin. When muscles are metabolically active they emit energy (heat) and the haemoglobin provides them with the increased oxygen required.

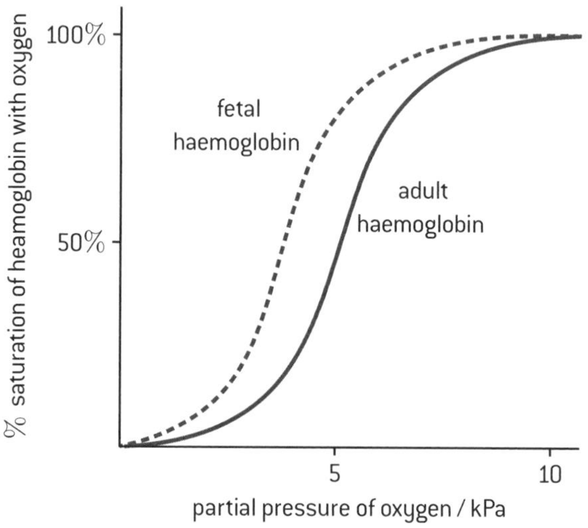

The partial pressure also affects the oxygen uptake. Because of the cooperative process achieved through the induced changes of the haemoglobin protein complex the oxygen binding curve of saturation of haemoglobin with oxygen against partial pressure is sigmoidal in shape compared with the normal hyperbolic curve expected if no cooperative binding takes place.

Haemoglobin exists in a slightly different form in fetal blood. It has a greater affinity for oxygen than normal haemoglobin so more oxygen is bound to the haemoglobin at lower partial pressures. This enables the fetal blood in the placenta to take up oxygen from the mother's blood.

Carbon monoxide is a dangerous poison as carbon monoxide is a stronger ligand than oxygen and forms an irreversible complex with the iron in haemoglobin. It thus acts as a competitive inhibitor and prevents the haemoglobin from binding with oxygen.

PAPER AND THIN LAYER CHROMATOGRAPHY

The technique of paper chromatography used to identify amino acids has already been discussed on page 126. It is ideal to use to separate and identify biological pigments by measuring R_f values as no dye or stain is needed to see the spots. In paper chromatography the stationary phase is the water contained in the cellulose fibres in the paper. Thin layer chromatography can also be used. This is similar to paper chromatography but uses a thin layer of a solid, such as alumina, Al_2O_3, or silica, SiO_2, on an inert support such as glass. When absolutely dry it works by adsorption but, like paper, silica and alumina have a high affinity for water, therefore the separation occurs more by partition with water as the stationary phase. The choice of a suitable solvent depends on the polarity or otherwise of the particular pigments. One real advantage of thin layer chromatography over paper chromatography is that each of the separated components can be recovered pure. The section containing the component is scraped off the glass and then dissolved in a suitable solvent. The solution is then filtered to remove the solid support and the solvent can then be evaporated to leave just the pure component.

Stereochemistry in biomolecules

CHIRALITY

A chiral carbon atom is asymmetric, i.e. contains four different atoms or groups attached to it. Many important biological molecules are chiral and only one of the particular enantiomers is normally active in nature, although sometimes a different enantiomer may have a detrimental effect. This was the case with the drug thalidomide, which was prescribed in the 1950s and 1960s. One enantiomer alleviated the effects of morning sickness in pregnant women, the other enantiomer caused severe defects in the fetus.

All amino acids apart from glycine, $H_2N–CH_2–COOH$, exist as enantiomers but only the L configuration is found in proteins.

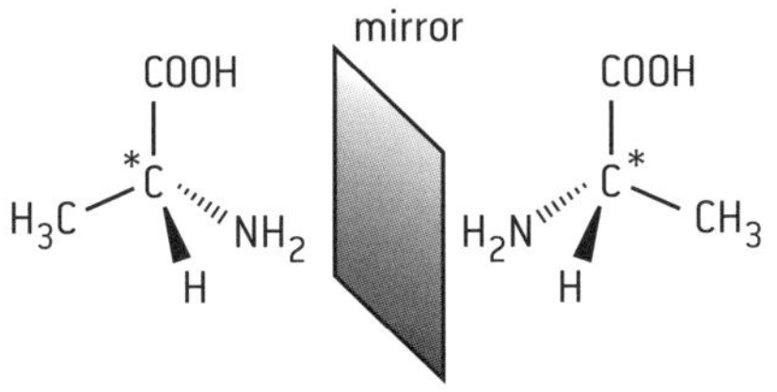

enantiomers of alanine
(* asymmetric carbon/chiral carbon)

Sugars contain several chiral carbon atoms. The D and L stereoisomers of sugars refer to the configuration of the chiral carbon atom furthest from the aldehyde or ketone group. The D forms occur most frequently in nature. The stereochemistry of sugars is further complicated by the position of the hydroxyl groups. The ring forms of sugars have isomers, known as α and β, depending on whether the position of the hydroxyl group at carbon 1 (glucose) or carbon 2 (fructose) lies below the plane of the ring (α) or above the plane of the ring (β).

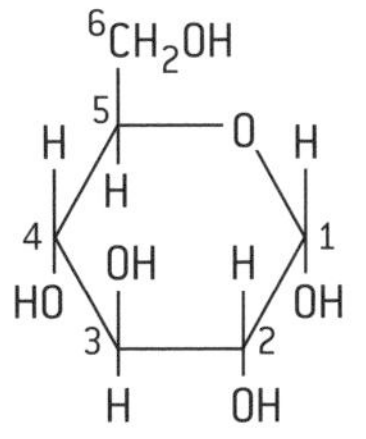

α-D-glucose

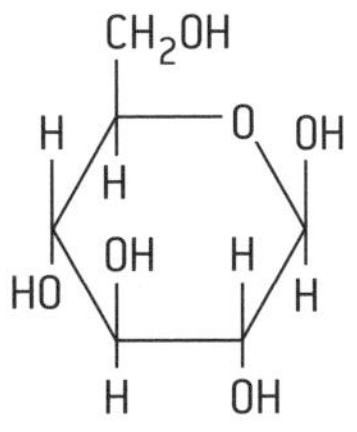

β-D-glucose

Polysaccharides formed from glucose can be very different depending upon whether the α- or β- form is involved. Starch, which can be digested by humans, is formed from polymerizing α-D-glucose. Amylose is a straight-chain polymer of α-D-glucose with α-1,4 glycosidic bonds and amylopectin is also derived from α-D-glucose with both α-1,4 and α-1,6 glycosidic bonds. Cellulose is also a polymer of glucose but it is formed from β-D-glucose with β-1,4 linkages.

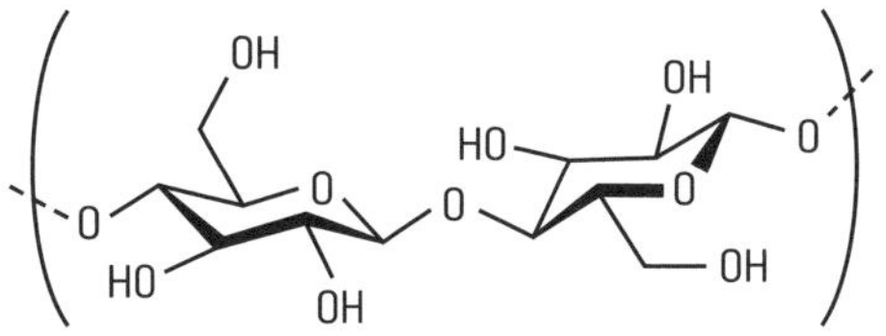

A repeating unit of cellulose showing the β-1,4-linkage

Cellulose, together with lignin, provides the structure to the cell walls of green plants. Most animals, including all mammals, do not have the enzyme cellulose so are unable to digest cellulose. Known as 'roughage', dietary fibre does play an important role in the diet as it aids digestion and makes defecation easier.

CIS- AND TRANS- ISOMERISM

Fatty acids occur naturally as the cis-isomers but as described on page 129 trans-isomers, which can increase the risk of heart disease, can be formed during the partial hydrogenation of unsaturated fats. Like saturated fatty acids, the trans- acids are straighter than their bent cis- isomers. This means that they can pack together more easily and so have higher melting points. For example elaidic acid (trans-9-octadecenoic acid) melts at 45 °C whereas oleic acid (cis-9-octadecenoic acid) melts at 13 °C.

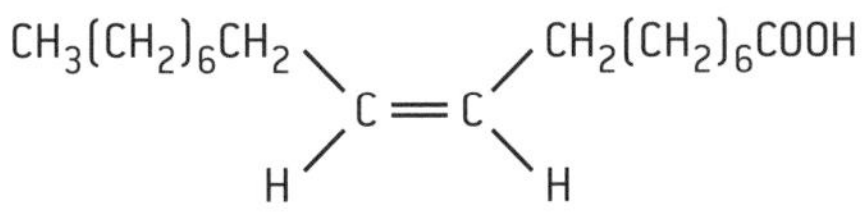

oleic acid (melting point 13°C)

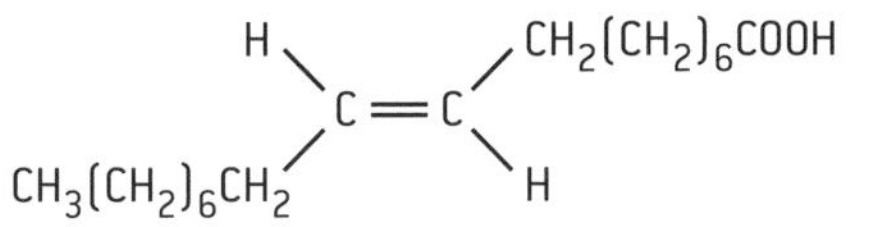

elaidic acid (melting point 45°C)

The process whereby photons of light are converted into electrical signals in the retina at the back of the eye is called the visual cycle. Rhodopsin in the retina consists of a protein, opsin and a covalently bonded co-factor retinal, which is produced in the retina from vitamin A. When light falls on the retina it converts the carotenoid retinal from the cis- to the trans- form and as this happens a nerve impulse is transmitted via an interaction, which causes a conformational change in the structure of opsin to send a signal along the optic nerve to the brain.

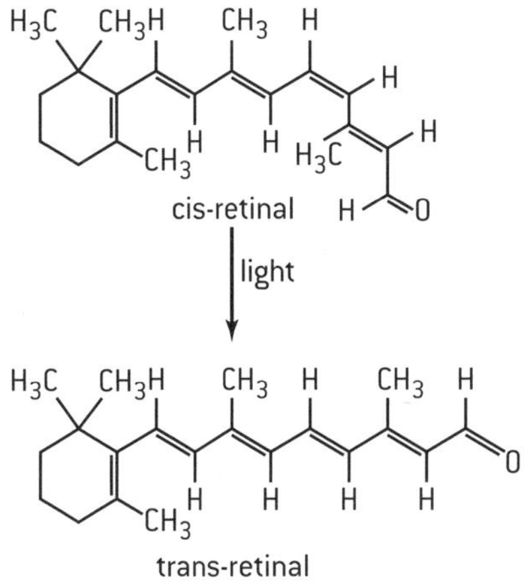

The effect of light on cis-retinal

SHORT ANSWER QUESTIONS – OPTION B – BIOCHEMISTRY

1. a) Deduce the structure of methionine in
   (i) acid solution ($pH < 4$)
   (ii) at the isoelectric point ($pH = 5.7$)
   (iii) alkaline solution ($pH > 9$) [3]

   b) Draw the two dipeptides that can be formed when one molecule of methionine condenses with one molecule of alanine. [2]

   c) Determine the number of different tripeptides that could be formed from methionine, alanine and cysteine if each tripeptide contains one residue from each of the three amino acids. [1]

   d) Design an experiment you could use in a school laboratory to determine whether a given protein contains a methionine residue. [4]

2. The diagram below shows a triacyclglycerol (triglyceride).

$$\begin{array}{r} CH_3(CH_2)_4CH{=}CHCH_2CH{=}CH(CH_2)_7CO{-}O{-}CH_2 \\ | \\ CH_3(CH_2)_4CH{=}CHCH_2CH{=}CH(CH_2)_7CO{-}O{-}CH \\ | \\ CH_3(CH_2)_7CH{=}CH(CH_2)_7CO{-}O{-}CH_2 \end{array}$$

   a) State the main dietary group this compound belongs to. [1]

   b) Deduce whether this compound is likely to have come from an animal or vegetable source. [2]

   c) Deduce whether this compound is likely to be a solid or a liquid at room temperature. [2]

   d) List two major functions of this class of compounds in the body. [2]

3. The structures of vitamin A (retinol) and vitamin C (ascorbic acid) are given in Section 35 of the IB data booklet.

   a) (i) Identify two functional groups present in retinol. [2]
   (ii) Classify vitamin A and vitamin C as water or fat soluble and justify the difference on the molecular level. [3]
   (iii) State one physical symptom for each of vitamin A and vitamin C deficiency and state the common name given to vitamin C deficiency. [3]

   b) Interpret the information that 0.014 moles of a particular oil was found to react exactly with 14.2 g of iodine. [3]

4. a) A mixture of the amino acids serine (Ser), glutamic acid (Glu) and lysine (Lys) was separated using electrophoresis and a buffer of pH 5.7. A drop containing the mixture was placed in the centre of the paper and a potential difference was applied. The amino acids were developed and the following results were obtained.

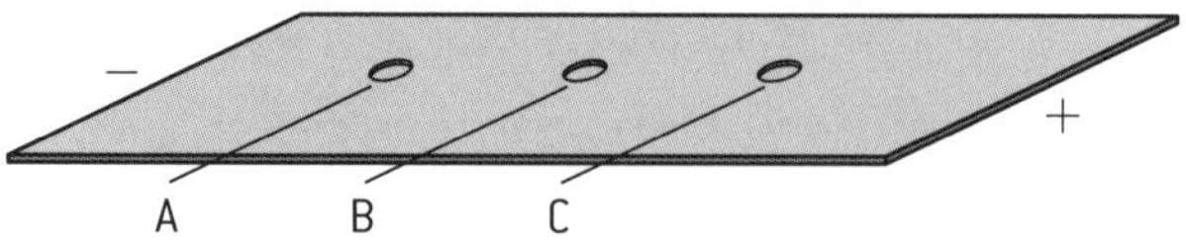

   (i) Describe how the amino acid spots may have been developed. [1]
   (ii) Predict which amino acid is present at spot C. Explain your answer. [3]
   (iii) The amino acid at spot B is at its isoelectric point. Describe one characteristic of an amino acid at its isoelectric point. [1]

   b) Explain, using equations, how the amino acid glycine (Gly) can act as a buffer. [2]

HL

5. a) Iron combines reversibly with oxygen in haemoglobin. Other than variable oxidation states, state two typical characteristics of transition metals that are shown by iron in haemoglobin. [2]

   b) The ability of haemoglobin to carry oxygen at body temperature depends on the concentration of oxygen, the concentration of carbon dioxide and on the pH. The graph shows how the percentage saturation of haemoglobin with oxygen changes with pH at different partial pressures of oxygen.

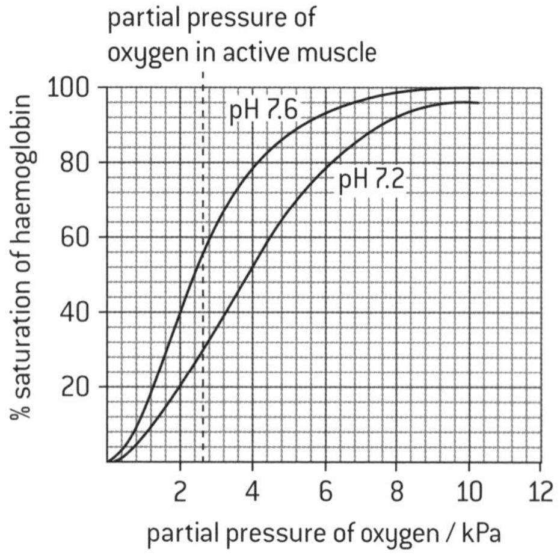

   (i) The partial pressure of oxygen in active muscle is shown by the dotted line at 2.8 kPa. Calculate the difference in the percentage saturation of haemoglobin with oxygen in active muscle when the pH changes from 7.6 to 7.2. [1]
   (ii) When the cells in muscles respire they excrete carbon dioxide and sometimes lactic acid as waste products. Explain how this affects the ability of haemoglobin to carry oxygen. [2]

6. The acid dissociation constant, K_a, for lactic acid, $CH(CH_3)(OH)COOH$, is 1.38×10^{-4} at 298 K.

   a) Explain why lactic acid can exist in two different enantiomeric forms. [1]

   b) Calculate the pH of 0.100 mol dm^{-3} lactic acid solution at 298 K. [3]

   c) Determine the mass of lactic acid that must be added together with 2.00 g of sodium hydroxide to make 500 cm^3 of a buffer solution at 298 K with a pH of 4.00. [4]

Energy sources

INTRODUCTION

Energy is the ability to do work, and is measured in joules (J). The first Law of Thermodynamics states that energy within a closed system can neither be created nor destroyed. Living organisms require energy to remain alive. This energy is obtained from the metabolism of food. Societies also need energy to function. Useful sources of energy must release energy at reasonable rates and produce minimal pollution. Energy sources are either renewable, that is naturally replenished, or non-renewable, that is finite. Apart from tidal energy and nuclear energy, the Sun is the ultimate source of energy for the planet. The Sun is also responsible for the Earth's climate and ecosystems. Humans' use of energy, particularly when derived from fossil fuels, interferes with both the climate and ecosystems. Types of energy include kinetic and potential but 'energy sources' refers specifically to ways of obtaining energy to produce electrical power.

ENERGY SOURCES

At the moment, oil is the top energy source used in the world. The order for the main energy sources currently used in the world in terms of producing energy is: crude oil > coal > natural gas > hydroelectric > nuclear fission > wind > biofuels > solar > geothermal (source: oilprice.com). Crude oil, coal and natural gas are non-renewable fossil fuels and contain stored energy from photosynthesis trapped millions of years ago. Nuclear fission is non-renewable as it depends upon a supply of fissionable material. Hydroelectric, wind, biofuels, solar and geothermal are all renewable energy sources. Other forms of renewable energy that may be used more in the future are tidal and wave energy. When the energy from any of these sources is converted into electrical energy, some of the energy is lost as heat to the surroundings so none of the processes is 100% efficient. Heat is also lost as the generated electricity is transported by power lines.

The choice of fuel depends on many factors including availability, ease of extraction, storage and the environmental and social effects of extracting and using the energy as well as physical factors. These physical factors include:

$$\text{The energy density} = \frac{\text{energy released from the fuel}}{\text{volume of fuel consumed}}$$

$$\text{The specific energy} = \frac{\text{energy released from the fuel}}{\text{mass of fuel consumed}}$$

$$\text{The efficiency of an energy transfer} = \frac{\text{useful output energy}}{\text{total input energy}} \times 100$$

(expressed as a percentage)

For example, the specific energy and energy density of ethanol (biofuel) and gasoline (fossil fuel) can be compared using their enthalpies of combustion, molar masses and densities.

	Ethanol C_2H_5OH	**Gasoline (octane) C_8H_{18}**
M_r	**46.08**	**114.26**
$\Delta H_c^{\ominus}$ / kJ mol^{-1}	**−1367**	**−5470**
Density / g cm^{-3}	**0.789**	**0.703**
Specific energy / kJ kg^{-1}	**29670**	**47890**
Energy density / kJ dm^{-3}	**23410**	**33650**

Clearly on all physical counts gasoline (octane) is the better fuel. It produces 1.6 times as much energy as ethanol by mass and 1.4 times as much energy by volume. Both produce carbon dioxide (a greenhouse gas) when burned. Gasoline adds to the total amount of carbon dioxide in the atmosphere as it is a fossil fuel whereas ethanol is formed from crops grown now so does not significantly alter the amount of carbon dioxide in the atmosphere and so is more environmentally friendly. Weighed against this is the fact that the large scale use of ethanol uses up land to grow crops that could otherwise be used to grow food.

Nuclear fuel is a very efficient fuel in terms of specific energy density but has many safety and pollution issues associated with it. Hydrogen also has a high specific energy (143 000 kJ kg^{-1}) but there are problems with the transport and storage of hydrogen as it cannot be liquefied by pressure alone and the methods of manufacturing hydrogen are also energy intensive. Wind and solar energy depend upon weather conditions and so the ways of storing the energy they produce are also important.

The generation of electric power involves using mechanical energy to drive electrical generators. For energy sources such as coal, oil, natural gas and biofuels, the mechanical energy is obtained by using the heat produced by combustion. Nuclear fuels use the heat generated by the nuclear reactions. Wind and hydroelectric power can bypass this step as they produce mechanical energy by driving a turbine directly. Coal and oil-fired power stations typically can only convert about 33% of their energy output into electricity, i.e. their efficiency of energy transfer is 33%. By using a combined cycle involving synthesis gas this can be increased to about 60%.

Fossil fuels (1)

FORMATION OF FOSSIL FUELS

Coal is fossilized plant material containing mainly carbon together with hydrogen, nitrogen and sulfur. Most coal was formed during the Carboniferous period (286–360 million years ago). The action of pressure and heat through geological forces converted the plant material in stages from peat to lignite to bituminous soft coal to hard coal (anthracite). At each stage the percentage of carbon increases. Coal contains between 80 and 90% carbon by mass.

Crude oil was formed from the remains of marine organisms mainly during the Paleozoic era, up to 600 million years ago. Thick sediments built up on top of the organic layers and under the action of high pressures and biochemical activity crude oil was formed. The oil migrated through rocks due to earth movements and collected in traps. Crude oil is a complex mixture of straight-chain, branched, cyclic and aromatic hydrocarbons, although it consists mainly of alkanes.

Natural gas was formed at the same time as crude oil and the two are often found together, although it may occur on its own or with coal. It consists mainly of methane (85–95%) with varying amounts of ethane, propane, butane and other gases such as hydrogen sulfide.

COMPOSITION AND CHARACTERISTICS OF CRUDE OIL FRACTIONS

Sulfur must first be removed from crude oil before it is refined as it can poison catalysts by blocking their active sites. The oil then undergoes fractional distillation to separate it into different boiling fractions. The number of carbon atoms, boiling ranges and the uses of the different fractions are summarized in the table:

Fraction	Carbon chain length	Boiling range / °C	Main uses
Refinery gas	1–4	<30	Used as fuel on site, gaseous cooking fuel and as feedstock for chemicals e.g. methane used to provide hydrogen gas for the Haber process.
Gasoline and naphtha	5–10	40–180	Gasoline (petrol) for cars. Feedstock for organic chemicals (by steam cracking).
Kerosene	11–12	160–250	Fuel for jet engines; domestic heating; cracked to provide extra gasoline (petrol).
Gas oil (diesel oil)	13–25	220–350	Diesel engines and industrial heating; cracked to produce extra gasoline (petrol).
Residue	>20	>350	Fuel for large furnaces; vacuum distilled to make lubricating oils and waxes. Residue of bitumen and asphalt used to surface roads and waterproof roofs.

CRACKING AND REFORMING

The performance of hydrocarbons as fuels is improved by cracking and catalytic reforming reactions.

Cracking is the process conducted at high temperatures whereby large hydrocarbons are broken down into smaller more useful molecules. The products are usually alkanes and alkenes. For example decane can be broken down to form octane and ethene.

$$C_{10}H_{22}(g) \rightarrow C_8H_{18}(g) + C_2H_4(g)$$

The alkanes are usually branched isomers (e.g. 2,2,4-trimethylpentane) and are added to octane to improve the octane rating. The alkenes are used to make other chemicals, particularly addition polymers.

More useful branched alkanes can be obtained from straight-chain hydrocarbons by mixing them with hydrogen and heating them at 770 K over a platinum/alumina catalyst at high pressure. This process, known as isomerization, is a particular form of reforming. Other reforming processes in which straight-chain alkanes are reformed into molecules with the same number of carbon atoms include cyclization to make ring molecules and aromatization to make benzene.

Examples of reforming

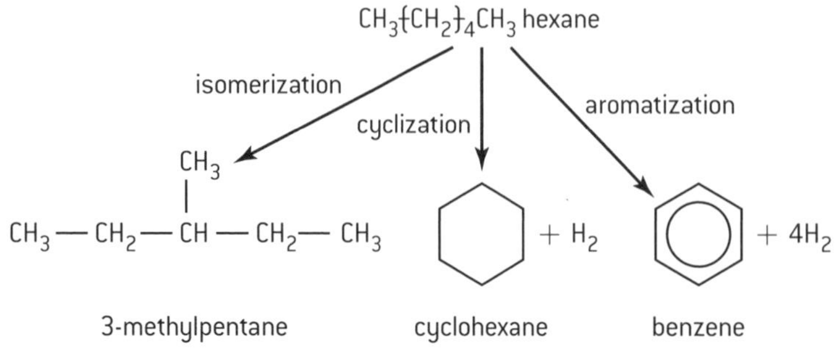

CARBON FOOTPRINT

A carbon footprint is the total amount of greenhouse gases (CO_2 and CH_4) produced during human activities. Carbon footprints can be estimated using online calculators and are generally expressed in equivalent tons of carbon dioxide. For most people the majority of their carbon footprint comes indirectly from the fuel used to produce and transport goods compared with the emissions which come directly from using gasoline in their own car or taking a flight. Fossil fuels have by far the highest carbon footprint whereas nuclear power and renewable sources of energy such as wind and hydroelectric power all contribute little towards the carbon footprint. Consumers can opt to offset their carbon footprint by, for example, planting trees. The best ways to lower carbon footprints are to reduce the use of fossil fuels and materials in general, reuse materials and recycle.

Fossil fuels (2)

OCTANE RATING

In an efficient internal combustion engine a spark ignites the fuel-air mixture just before the piston reaches 'top dead centre' so that the full force of the explosion pushes the piston down just as it reaches the top of the cylinder. Under the conditions of high temperature and pressure the reaction may start before the spark and the engine will be less efficient. This is known as pre-ignition or **knocking**. The more straight-chain the alkane the higher the tendency for knocking.

Fuels are classified according to their **octane number**. Generally the more branched the alkane the higher the octane number. Pure heptane is assigned an octane number of zero and an isomer of octane, 2,2,4-trimethylpentane, has an octane rating of 100. Thus gasoline (petrol) with an octane rating of 95 will burn as efficiently as a mixture of 95% 2,2,4-trimethylpentane and 5% heptane. In the past tetraethyllead $Pb(C_2H_5)_4$ was added to petrol to raise the octane rating. Lead-free gasoline (petrol) contains added aromatic hydrocarbons such as benzene and more branched hydrocarbons obtained through cracking.

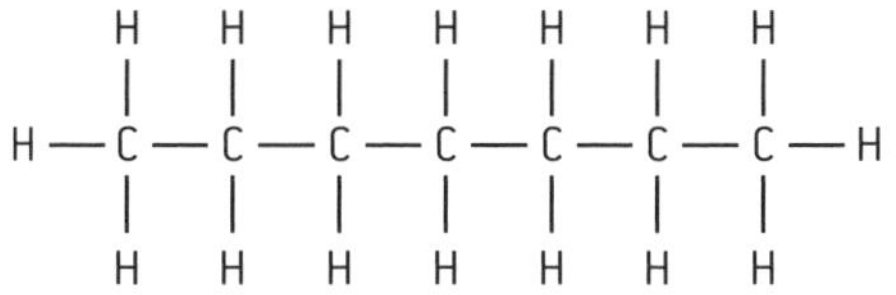

heptane: octane rating 0

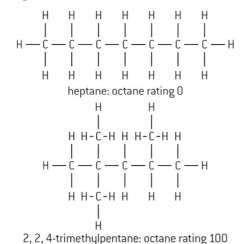

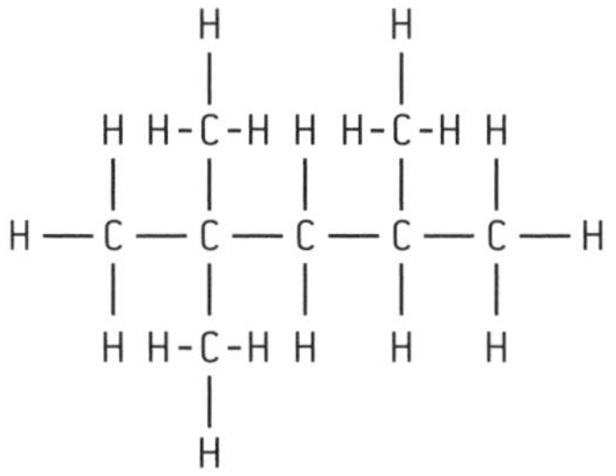

2, 2, 4-trimethylpentane: octane rating 100

COAL GASIFICATION AND LIQUEFACTION

Before the advent of plentiful supplies of natural gas, coal was routinely turned into 'coal gas'. As supplies of natural gas diminish, interest in coal gasification may revive. Coal gas (also known as synthesis gas, water gas or town gas) contains a mixture of hydrogen and carbon monoxide and is made by heating coal in the presence of steam.

$$C(s) + H_2O(g) \rightarrow CO(g) + H_2(g)$$

Reacting coal gas with more hydrogen in the presence of a heated catalyst converts it into SNG (substitute or synthetic natural gas).

$$CO(g) + 3H_2(g) \rightarrow CH_4(g) + H_2O(g)$$

SNG can also be made by heating crushed coal in steam at 700 °C using potassium hydroxide as a catalyst.

$$2C(s) + 2H_2O(g) \rightarrow CH_4(g) + CO_2(g)$$

SNG is a cleaner gas (as it removes pollution due to sulfur dioxide), which is easier to transport but the process is less efficient as it uses up some 30% of the available energy during the conversion process.

In Germany in the 1930s and in South Africa, where coal is abundant, coal has been converted into a liquid fuel. As the price of oil increases this process may become more important economically. The method is known as the Fischer–Tropsch process. Synthesis gas is reacted with more steam to increase the proportion of hydrogen in the mixture.

$$CO(g) + H_2O(g) \rightarrow CO_2(g) + H_2(g)$$

The hydrogen and carbon monoxide are then passed into a fluidized bed reactor containing iron or cobalt catalysts to produce a mixture of hydrocarbons that can be separated by fractional distillation.

RELATIVE ADVANTAGES AND DISADVANTAGES OF FOSSIL FUELS

Fossil fuel	Advantages	Disadvantages
Coal	1. Present in large quantities and distributed throughout the world. 2. Can be converted into synthetic liquid fuels and gases. 3. Feedstock for organic chemicals. 4. Has the potential to yield vast quantities of energy compared with renewable sources and safer than nuclear power. 5. Longer lifespan (350 years?) compared with oil or gas	1. Contributes to acid rain and global warming. 2. Not so readily transported (no pipelines). 3. Coal waste (slag heaps) lead to ground acidity and visual and chemical pollution. 4. Mining is dangerous – cave-ins, explosions and long term effect of coal dust on miners. 5. Dirty (produces dust, smoke and particulates).
Oil	1. Easily transported in pipelines or by tankers. 2. Convenient fuel for use in cars, lorries, etc. 3. Feedstock for organic chemicals.	1. Contributes to acid rain and global warming. 2. Limited lifespan (30–50 years?) and uneven distribution worldwide. 3. Risk of pollution associated with transportation by tankers.
Natural gas	1. Clean fuel 2. Easily transported in pipelines and pressurized containers. 3. Does not contribute to acid rain. 4. Releases a higher quantity of energy per kg than coal or oil.	1. Contributes to global warming. 2. Limited lifespan (30 years?) and uneven distribution worldwide. 3. Greater risk of explosions due to leaks.

Nuclear fusion and nuclear fission (1)

NUCLEAR REACTIONS

In a normal chemical reaction, valence shell electrons are rearranged as bonds are broken and new bonds formed. There is no change in the nucleus and no new elements are formed. In nuclear reactions the nucleus itself rearranges. The protons and neutrons in the nucleus are held together by strong forces known as the binding energy.

A graph of binding energy per nucleon against mass number shows that atomic nuclei with a mass number of approximately 56 (i.e. the nucleus of iron) have the maximum binding energy and are thus the most stable. Nuclei to the left or right of this maximum will undergo nuclear change in such a way that, as they approach the maximum, energy will be released.

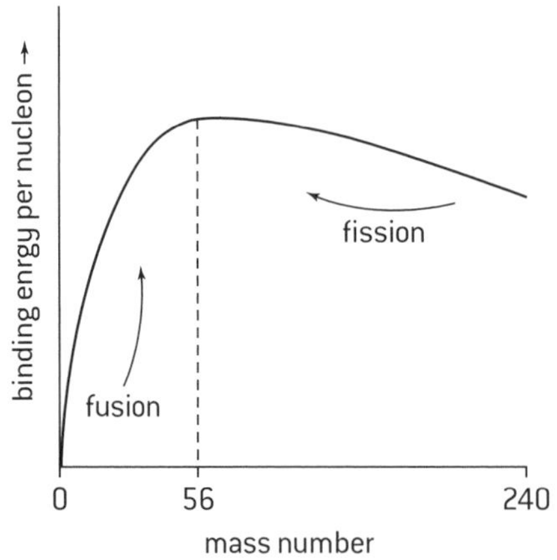

How binding energy varies with mass number

This explains nuclear fusion whereby small atoms combine to form heavier nuclei and nuclear fission – the splitting of heavy nuclei to form two or more lighter nuclei. In both cases the total mass of the products is less than the mass of the initial nucleus or nuclei. During nuclear reactions, mass is converted into energy according to Einstein's equation $E = mc^2$ (where c is the velocity of light). Thus in a nuclear reaction new elements are formed and the energy change is potentially much greater than in a chemical reaction. When an unstable radioactive isotope decays it can emit three different types of radiation. Alpha particles which are helium nuclei, ${}^{4}_{2}He^{2+}$, beta particles which are electrons, ${}^{0}_{-1}e^{-}$, and gamma radiation, which is high energy electromagnetic radiation.

NUCLEAR EQUATIONS

Nuclear equations must balance. The total mass numbers and the nuclear charge numbers (atomic numbers) must be equal on both sides of the equation. During α particle emission the new element will have a mass of four less than the original element and an atomic number that is two less.

e.g. ${}^{238}_{92}U \rightarrow {}^{234}_{90}Th + {}^{4}_{2}He$

During β emission when an electron is ejected from the nucleus at a velocity approaching the speed of light, the new element will have the same mass number but the atomic number will have increased by one.

e.g. ${}^{14}_{6}C \rightarrow {}^{14}_{7}N + {}^{0}_{-1}e$

Nuclear reactions also occur artificially when nuclei are bombarded with other small particles such as α particles or neutrons. In each case the total mass numbers and nuclear charges on both sides of the equation must still balance.

e.g. ${}^{235}_{92}U + {}^{1}_{0}n \rightarrow {}^{144}_{56}Ba + {}^{90}_{36}Kr + 2{}^{1}_{0}n$

Other small particles that may be involved in nuclear reactions include protons ${}^{1}_{1}p$ and positrons ${}^{0}_{+1}e^{+}$. Positrons are positive electrons, sometimes called β^{+} particles to distinguish them from electrons (β^{-} particles).

HALF-LIFE $t_{1/2}$

It is impossible to state when an individual unstable isotope will decay as it occurs spontaneously. However when a large number of atoms are together in a sample of the isotope the rate of decay depends on the amount of atoms present. The time taken for any specified amount to decrease by exactly one half remains constant and is independent of pressure and temperature. The half-life for a particular isotope is defined as the time taken for a given sample to decay to one half of the **mass** of the **original isotope**. This is different to a chemical reaction where half-life is defined as the time taken for the **concentration** of a reactant to decrease to one half of its initial value. The half-life of ${}^{32}_{15}P$ is 14.3 days. After 14.3 days a 1.0 g sample of ${}^{32}_{15}P$ will have decayed to 0.50 g and after a further 14.3 days only 0.25 g of ${}^{32}_{15}P$ will be remaining. Questions are often set the other way round. For example, calculate the half-life of ${}^{131}I$ if a sample of ${}^{131}I$ is found to contain $\frac{1}{32}$ of the original amount after 40.30 days.

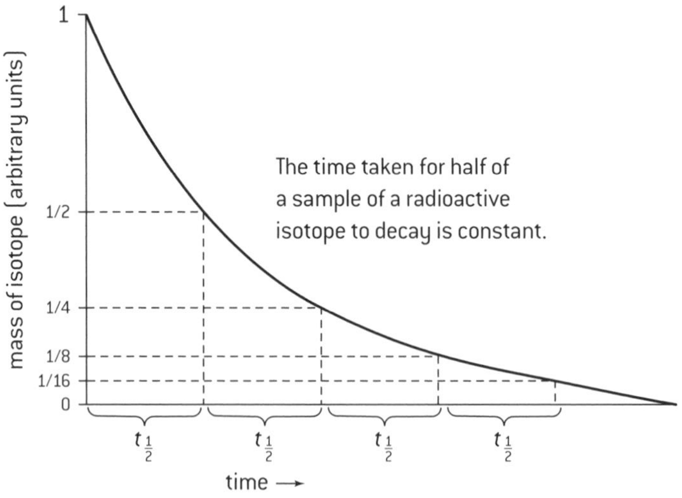

$$1 \xrightarrow{t_{1/2}} \frac{1}{2} \xrightarrow{t_{1/2}} \frac{1}{4} \xrightarrow{t_{1/2}} \frac{1}{8} \xrightarrow{t_{1/2}} \frac{1}{16} \xrightarrow{t_{1/2}} \frac{1}{32}$$

The total number of half-lives is five so the half-life $= \frac{40.30}{5} = 8.06$ days. This was solved by inspection. It can also be solved mathematically. The number of atoms remaining N is related to the original number of atoms N_0 by the equation $N = N_0 e^{-\lambda t}$ where λ is known as the decay constant.

$$\text{At } t_{1/2},\ N = \frac{N_0}{2} \quad \text{so} \quad t_{1/2} = \frac{\ln 2}{\lambda}$$

In the example above N is equal to $\frac{N_0}{32}$ so $\frac{1}{32} = e^{-\lambda t}$ and $\ln\left(\frac{1}{32}\right) = -40.03 \times \lambda$ which makes λ equal to 0.0860 days^{-1}. Substituting this value into the second equation gives $t_{1/2} = \ln \frac{2}{0.0860} = 8.06$ days.

Radioactive isotope	Half-life
${}^{212}_{84}Po$	3×10^{-7} seconds
${}^{221}_{87}Fr$	4.8 minutes
${}^{222}_{86}Rn$	3.8 days
${}^{14}_{6}C$	5730 years
${}^{238}_{92}U$	4.5×10^{9} years

Nuclear fusion and nuclear fission (2)

POWER FROM NUCLEAR FISSION

A nuclear power station essentially contains two main components. The reactor to produce the heat from a nuclear reaction and a turbine to drive a generator to produce electricity. The nuclear reactor uses a fuel of uranium or plutonium. The uranium used is ^{235}U. This reacts with neutrons to form smaller nuclei and more neutrons. A typical reaction is:

$$^{1}_{0}n + ^{235}_{92}U \rightarrow ^{141}_{56}Ba + ^{92}_{36}Kr + 3^{1}_{0}n + \text{energy}$$

Because more neutrons are produced than are used a chain reaction is possible. The critical mass is when this becomes self-sustaining and the neutrons need to be controlled. The mass of the products is less than the reactants and the mass defect is converted into energy. Natural uranium only contains a small percentage of ^{235}U, most of it is ^{238}U. In a breeder reactor neutrons react with ^{238}U to form 239-plutonium, which is fissionable.

$$^{238}_{92}U + ^{1}_{0}n \rightarrow ^{239}_{92}U \rightarrow ^{0}_{-1}e + ^{239}_{93}Np \rightarrow ^{0}_{-1}e + ^{239}_{94}Pu$$

$$^{1}_{0}n + ^{239}_{94}Pu \rightarrow ^{90}_{38}Sr + ^{147}_{56}Ba + 3^{1}_{0}n + \text{energy}$$

NUCLEAR SAFETY

Arguments against the use of nuclear energy include the risk of terrorist attack, an accident, the disposal of waste and the promotion of less polluting forms of alternative energy. Accidents at nuclear power plants in Chernobyl in Ukraine in 1986 and in Fukushima in Japan in 2011 have heightened the concerns about the safety of nuclear power stations.

1. The possibility of a meltdown. A meltdown occurs when a nuclear reactor becomes out of control and essentially becomes a nuclear bomb. A nuclear power station requires the slow release of energy so the fuel contains much less of the fissionable isotope ($^{235}_{92}U$) than a nuclear bomb. The neutrons emitted by the fissionable material in a power station are absorbed by the non-fissionable isotope ($^{238}_{92}U$) so in theory cannot build up enough momentum to establish a spontaneously explosive chain reaction.
2. Escape of radioactive material. This can occur as the fuel is being transported or while it is being used. In the Chernobyl disaster a fire ignited the graphite moderator and a cloud of radioactive gas spread across much of Europe. The accident in Japan was caused by an earthquake and tsunami. Ordinary materials such the surrounding air and clothes worn by workers also have the potential to transfer low-level waste outside the plant. Concern has also been expressed about the risk of plutonium or other radioactive material falling into the hands of terrorists.
3. High-level nuclear waste. There are considerable problems associated with the disposal of high-level nuclear waste from spent fuel rods. These must be stored for very long periods (hundreds if not thousands of years) before they become relatively harmless. Current methods include vitrifying the waste in glass and burying it deep underground or in ocean trenches. Humankind has no experience of storing such materials safely for this length of time. The possibilities of disruption by earthquakes or the slow seepage of the waste into the water table make this one of the strongest arguments against using nuclear energy.

4. Low-level nuclear waste. This not only comes from power plants but also from research laboratories or hospitals.

   It includes items such as rubber gloves, paper towels and protective clothing that have been used in areas where radioactive materials are handled. The level of activity is low and the half-lives of the radioactive isotopes are generally short. It can be stored in vast tanks of cooled water called 'ponds' where it can lose much of its activity. Before it is then discharged into the sea it is filtered through an ion-exchange resin which removes strontium and caesium, the two elements responsible for much of the radioactivity. Other methods of disposal include keeping the waste in steel containers inside concrete-lined vaults.

POWER FROM NUCLEAR FUSION

Nuclear fusion offers the possibility of an almost unlimited source of energy, which would almost be pollution free since the main fuel, deuterium, is inexpensive, abundant in sea water and no radioactive waste would remain. The essential reaction is:

$$^{2}_{1}H + ^{2}_{1}H \rightarrow ^{3}_{2}He + ^{1}_{0}n + \text{energy}$$

Other reactions involving tritium, $^{3}_{1}H$, another isotope of hydrogen could also be used but tritium is much less readily available and would need to be bred.

$$^{2}_{1}H + ^{3}_{1}H \rightarrow ^{4}_{2}He + ^{1}_{0}n + \text{energy}$$

However the problems of controlling nuclear fusion reactions have yet to be overcome. Essentially, the intensely hot reaction mixture known as plasma has to be contained and maintained for long enough to fuse the nuclei together. Since this involves temperatures approaching forty million degrees (4×10^7 °C) the problems are considerable.

ABSORPTION SPECTRA OF STARS

The energy from the Sun and other stars comes from uncontrolled nuclear fusion reactions. Helium was in fact first discovered by looking at the absorption spectrum of the Sun. The electrons are excited and absorb energy as they are promoted from lower to higher levels. Each electronic transition shows up as a black line on what is otherwise a continuous spectrum. From the positions of the lines, the elemental composition of stars can be ascertained as each element has its own unique spectrum. An absorption spectrum is the opposite of an emission spectrum where energy is emitted as excited electrons drop down to lower energy levels but the lines have the same wavelengths and energy associated with them. The emission spectrum of hydrogen is discussed on page 11.

Solar energy

PHOTOSYNTHESIS

Approximately 5.6×10^{21} kJ of energy reaches the surface of the planet from the Sun each year. About 0.06% of this is used to store energy in plants. This is achieved through photosynthesis – a complex process summarized by the reaction of carbon dioxide and water to form carbohydrates in the presence of chlorophyll.

$$6CO_2(g) + 6H_2O(l) \rightarrow C_6H_{12}O_6(s) + 6O_2(g)$$

The process is endothermic requiring 2816 kJ of energy per mole of glucose. The green pigment chlorophyll contains an extended system of alternating single and multiple bonds known as conjugation. Substances such as chlorophyll and α-carotene, which contain extended conjugation, are able to absorb visible light by exciting an electron within this conjugated system. The energy which is absorbed by chlorophyll when it interacts with light is able to drive photosynthesis. The products of photosynthesis are used as food to provide energy for animals through the reverse process – respiration. Wood is mainly cellulose – a polymer made up of repeating glucose units.

CONVERSION OF BIOMASS INTO ENERGY

The energy stored in biomass can be released in a variety of ways. For example:

1. Direct combustion of plant material or combustion of waste material derived from plants such as animal dung.

   e.g. $C_6H_{12}O_6(s) + 6O_2(g) \rightarrow 6CO_2(g) + 6H_2O(l)$ $\Delta H = -2816$ kJ mol^{-1}

2. Biogas. The anaerobic decay of organic matter by bacteria produces a mixture of mainly methane and carbon dioxide known as biogas. The manure from farm animals can generate enough methane to provide for the heating, cooking and refrigeration needs of rural communities.

3. Fermentation to produce ethanol. Carbohydrates can be fermented by enzymes in yeast.

   $$C_6H_{12}O_6(s) \rightarrow 2C_2H_5OH(l) + 2CO_2(g)$$

   The ethanol can then be burned to produce energy.

   $$C_2H_5OH(l) + 3O_2(g) \rightarrow 2CO_2(g) + 3H_2O(l) \quad \Delta H = -1371 \text{ kJ mol}^{-1}$$

   By combining ethanol with gasoline a fuel called gasohol can be produced, which can be used by unmodified cars making them less reliant on the supply and cost of pure gasoline (petrol).

   Biofuels are renewable, readily available and relatively non-polluting. In many hot countries animal dung is dried and used as a fuel for heating and cooking. Garbage consisting of animal and vegetable waste is burned in incinerators in several cities to provide heat and electricity. This has the added advantage of reducing the amount of waste that has to be dumped in landfill sites. However disadvantages of biofuels include the fact that they are widely dispersed, they take up land where food crops can be grown and they remove nutrients from the soil.

TRANSESTERIFICATION

Vegetable oils are derived from glycerol (an alcohol containing three hydroxyl groups) and unsaturated fatty acids. They are triglycerides and contain three ester groups.

Vegetable oils contain a considerable amount of energy but they cannot be used in internal combustion engines as they are too viscous. This is because they have a high molar mass. However they can be converted into a usable fuel by a process known as transesterification. In this process an ester can react with an alcohol to be converted into a different ester in the presence of a strong acid or strong base.

RCOOR + R'OH → RCOOR' + ROH (where R' is different to R)

Biodiesel for use in diesel engines can be made from vegetable oils and waste cooking oils by heating them with an excess of an alcohol such as methanol or ethanol in the presence of sodium hydroxide. The alcohol is in excess in order to drive the position of equilibrium towards the transester side although some fatty acids may also be present in the products.

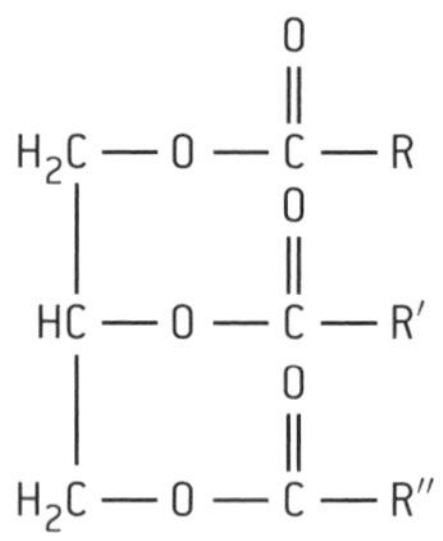

General formula of a vegetable oil where R, R' and R" are long chain hydrocarbons derived from fatty acids which may be the same or different.

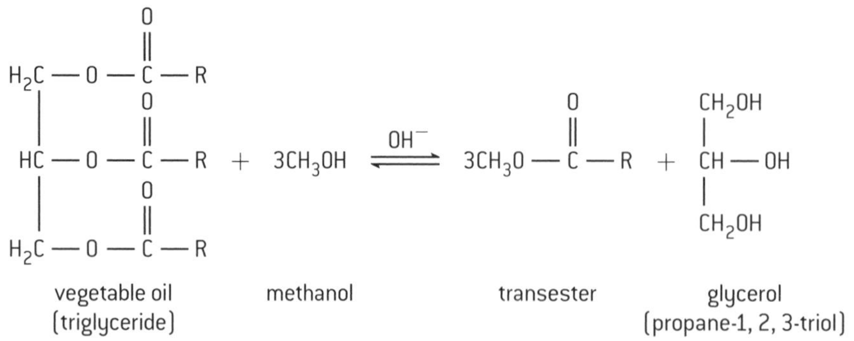

The three alkyl ester molecules formed have a similar energy content in total to the triglyceride from which they were derived but are much more volatile as they only contain one ester group each and have a much lower molar mass.

Environmental impact — global warming

THE GREENHOUSE EFFECT

A steady state equilibrium exists between the energy reaching the Earth from the Sun and the energy reflected by the Earth back into space. This regulates the mean average temperature of the Earth's surface. The incoming radiation is shortwave ultraviolet and visible radiation. Some is reflected back into space and some is absorbed by the atmosphere before it reaches the surface. The energy reflected back from the Earth's surface is longer wavelength infrared radiation. Not all of the radiation escapes. Greenhouse gases in the atmosphere allow the passage of the incoming shortwave radiation but absorb some of the reflected infrared radiation and re-radiate it back to the Earth's surface. They can absorb infrared radiation provided there is a change in dipole moment as the bonds in the molecule stretch and bend.

Vibrations of H_2O and CO_2

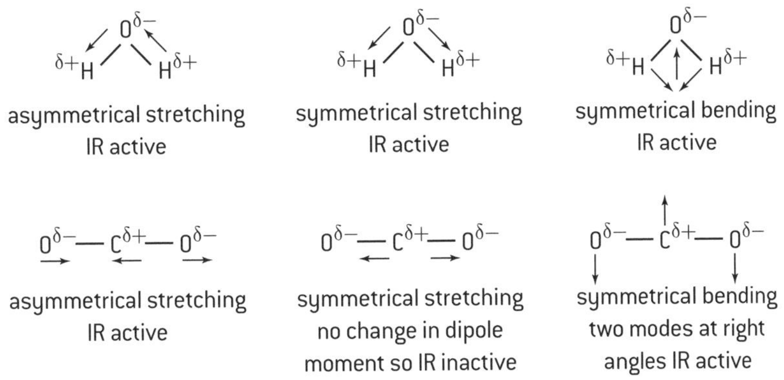

The contribution to global warming made by different greenhouse gases will depend both on their concentration in the atmosphere (abundance) and on their ability to absorb infrared radiation. Apart from water, carbon dioxide contributes about 50% to global warming and methane 18%. Other greenhouse gases such as CFCs, N_2O and O_3 are many times better at absorbing heat than carbon dioxide but because their concentration is so low their effect is limited.

Gas	Main source	Heat trapping effectiveness compared to CO_2	Overall contribution to increased global warming
H_2O	Evaporation of oceans	0.1	–
CO_2	Combustion of fossil fuels and biomass	1	50%
CH_4	Anaerobic decay of organic matter caused by intensive farming	30	18%

INFLUENCE OF GREENHOUSE GAS EMISSIONS ON GLOBAL WARMING

Most of the greenhouse gases have natural as well as human-made sources. As humans have burned more fossil fuels, the concentration of carbon dioxide in the air has risen steadily. Readings taken from Mauna Loa in Hawaii show a steady increase in the concentration of carbon dioxide by about 1 ppm (0.0001%) each year for the past 50 years. During the same period measurements of the Earth's mean temperature also show a general increase. During the past 100 years the mean temperature of the Earth has increased by about 1 °C although there were some years where the temperature fell rather than rose. Evidence from ice core samples from Greenland shows that there have also been large fluctuations in global temperature in the past; however most scientists now accept that the current global warming is a direct consequence of the increased emission of greenhouse gases. The predicted consequences of global warming are complex and there is not always agreement. The two most likely effects are:

1. Changes in agriculture and biodistribution as the climate changes.
2. Rising sea-levels due to thermal expansion and melting polar ice caps and glaciers.

CARBON DIOXIDE

The increase in carbon dioxide in the atmosphere in the past 150 years is mainly due to the increased use of fossil fuels for transport, heating and industrial processes. The Intergovernmental Panel on Climate Change (IPCC) and the Kyoto Protocol (later extended in Qatar) set targets to mitigate against climate change. Ways to limit human-made emissions of carbon dioxide include reducing dependency on fossil fuels by moving to renewable energy sources as well as carbon capture and storage. Carbon dioxide can be captured chemically using sodium hydroxide or calcium oxide and stored in certain types of silicon-based rocks. There is a natural heterogeneous equilibrium between the carbon dioxide gas present in the atmosphere and the aqueous carbon dioxide dissolved in the oceans. The increase in carbon dioxide emissions has also caused the concentration of dissolved carbon dioxide in the oceans to increase. Because it forms carbonic acid, H_2CO_3, in sea-water this has resulted in a drop in pH, i.e. an increase in the acidity of the oceans. One of the effects of this is to destroy coral reefs, which are essentially made of limestone.

$$CaCO_3(s) + 2H^+(aq) \rightarrow Ca^{2+}(aq) + CO_2(g) + H_2O(l)$$

PARTICULATES

Particulates, such as soot and volcanic dust, can have the opposite effect to greenhouse gases. They cool the Earth by scattering the shortwave radiation from the Sun and reflecting it back into space. Clouds do this as well. The lowering of mean global temperatures during the 1940s and 1960s has been attributed to the increased volcanic activity during these periods.

Electrochemistry, rechargeable batteries and fuel cells (1)

BATTERIES

A battery is a general term for an electrochemical cell in which chemical energy is converted into electrical energy. The electrons transferred in the spontaneous redox reaction taking place in the voltaic cell produce the electricity. Batteries are a useful way to store and transport relatively small amounts of energy. Some batteries (primary cells), such as the carbon–zinc dry cell, can only be used once as the electrochemical reaction taking place is irreversible. Secondary cells can be recharged as the redox reactions involved can be reversed using electricity. Examples of rechargeable batteries include the lead–acid battery and nickel–cadmium and lithium-ion batteries.

The voltage of a battery is essentially the difference in electrode potential between the two half-cells. It primarily depends only on the nature of the chemical components of the two half-cells. By combining two cells in series the voltage can be doubled. Three cells in series will triple the voltage of a cell. The power of a battery (i.e. the total work that can be obtained from the cell) is the rate at which it can deliver energy and is measured in joules per second. It is affected both by the size of the cell and the physical quantities of the materials present. An electrochemical cell has internal resistance due to the finite time it takes for ions to diffuse. The maximum current of a cell is limited by its internal resistance.

CARBON–ZINC DRY CELL

The common dry cell contains a paste of ammonium chloride and manganese(IV) oxide as the electrolyte. The positive electrode (cathode) is made of graphite (often with a brass cap) and the zinc casing is the negative electrode (anode). A porous separator acts as a salt bridge. When the cell is working electrons flow from the zinc to the graphite.

Oxidation at the anode (–) $Zn \rightarrow Zn^{2+} + 2e^-$

Reduction at the cathode (+) $MnO_2 + NH_4^+ + e^- \rightarrow MnO(OH) + NH_3$

The ammonia then reacts with the zinc ions to form $[Zn(NH_3)_4]^{2+}$. This prevents a build up of ammonia gas.

A carbon–zinc dry cell produces 1.5 volts when new. In addition to having to replace the battery when it wears out it has a poor shelf-life and the acidic ammonium chloride can corrode the zinc casing.

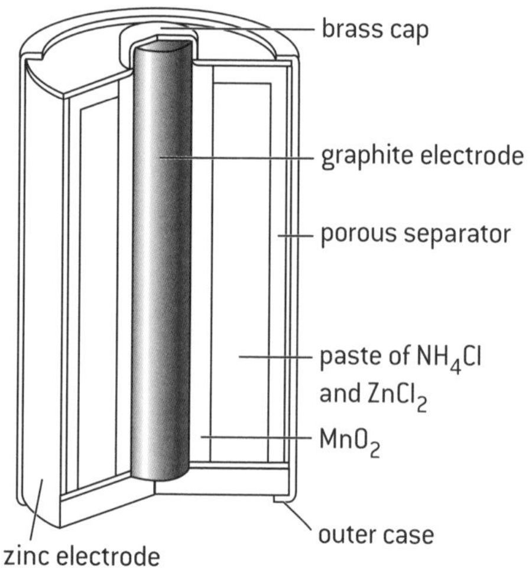

carbon–zinc dry cell

LEAD–ACID BATTERY

The lead–acid battery is used in automobiles and is an example of a secondary cell. Usually it consists of six cells in series producing a total voltage of 12 V. The electrolyte is an aqueous solution of sulfuric acid. The negative electrodes are made of lead and the positive electrodes are made of lead(IV) oxide.

Oxidation at anode (–) $Pb + SO_4^{2-} \rightarrow PbSO_4 + 2e^-$

Reduction at cathode (+) $PbO_2 + 4H^+ + SO_4^{2-} + 2e^- \rightarrow PbSO_4 + 2H_2O$

The overall reaction taking place is thus:

$$Pb + PbO_2 + 4H^+ + 2SO_4^{2-} \rightarrow 2PbSO_4 + 2H_2O$$

The reverse reaction takes place during charging. This can be done using a battery charger or through the alternator as the automobile is being driven. As sulfuric acid is used up during discharging, the density of the electrolyte can be measured using a hydrometer to give an indication of the state of the battery. The disadvantages of lead–acid batteries are that they are heavy and both lead and sulfuric acid are potentially polluting.

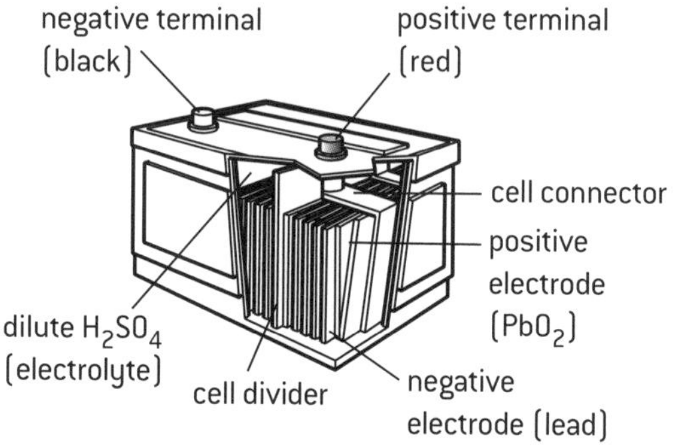

Lead–acid battery – during recharging hydrogen and oxygen are evolved from the electrolysis of dilute H_2SO_4 so it needs topping up occasionally with distilled water.

NICKEL–CADMIUM AND LITHIUM-ION BATTERIES

Rechargeable nickel–cadmium (NiCd) batteries are used in electronics and toys. They have a cell potential of 1.2 V. The positive electrode is made of nickel hydroxide which is separated from the negative electrode made of cadmium hydroxide. The electrolyte is potassium hydroxide. During discharge the following reaction occurs:

$$2NiO(OH) + Cd + 2H_2O \rightarrow 2Ni(OH)_2 + Cd(OH)_2$$

This process is reversed during charging. One of the disadvantages of NiCd batteries is that cadmium is an extremely toxic heavy metal so the batteries need to be disposed of responsibly. Laptops, tablets, smart phones and other handheld devices often use lithium-ion batteries. These contain lithium atoms complexed to other ions, e.g. Li_xCoO_2, in the positive electrode and it is these ions rather than lithium itself that undergo the redox reactions. The negative electrode is made of graphite. Lithium-ion batteries are much lighter than NiCd batteries and produce a higher voltage, 3.6–4.2 V, but they do not have such a long lifespan.

Electrochemistry, rechargeable batteries and fuel cells (2)

FUEL CELLS

A hydrogen fuel cell utilizes the reaction between oxygen and hydrogen to produce water. Unlike combustion, the energy is given out not as heat but as electricity. As reactants are used up, more are added so a fuel cell can give a continuous supply of electricity. They are used in spacecraft as, unlike rechargeable batteries, they do not need an external source of electricity for recharging. The electrolyte is aqueous sodium hydroxide. It is contained within the cell using porous electrodes that allow the passage of water, hydrogen and oxygen.

Oxidation at anode (−) $H_2 + 2OH^- \rightarrow 2H_2O + 2e^-$

Reduction at cathode (+) $O_2 + 2H_2O + 4e^- \rightarrow 4OH^-$

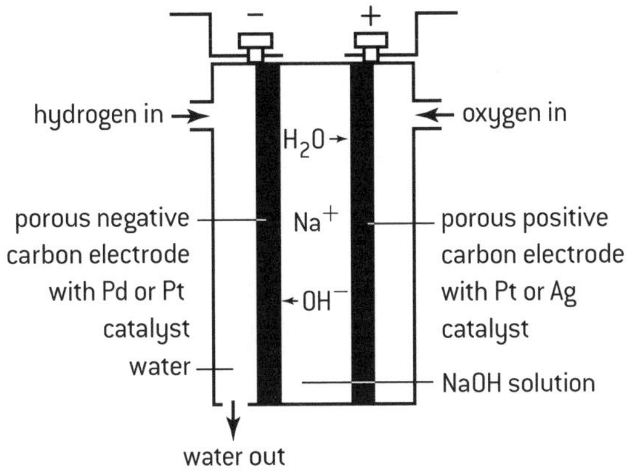

hydrogen–oxygen fuel cell with an electrolyte of NaOH

Hydrogen fuel cells can also be made using phosphoric acid as the electrolyte in which case the equations are:

Oxidation at anode (−) $H_2 \rightarrow 2H^+ + 2e^-$

Reduction at cathode (+) $O_2 + 4H^+ + 4e^- \rightarrow 2H_2O$

In both cases the overall equation is $2H_2(g) + O_2(g) \rightarrow 2H_2O(l)$ and the entropy change is negative.

Other fuels apart from hydrogen can be used. For example, the methanol fuel cell produces carbon dioxide and water as the products.

Oxidation at anode (−) $CH_3OH + H_2O \rightarrow CO_2 + 6H^+ + 6e^-$

Reduction at cathode (+) $\frac{3}{2}O_2 + 6H^+ + 6e^- \rightarrow 3H_2O$

In both hydrogen and methanol fuels cells the electrolyte can be supported on a solid polymer membrane. These are known as proton-exchange membrane (PEM) cells. They are used to produce electricity at relatively low temperatures, and can vary output quickly to meet shifts in power demand. The advantage of a hydrogen fuel cell is that it does not pollute, as water is the only product. Currently fuel cells are very expensive to produce.

The thermodynamic efficiency of a fuel cell is given by $\Delta G^\ominus/\Delta H^\ominus$. For the hydrogen fuel cell $\Delta H^\ominus$ has the value −286 kJ mol^{-1} and $\Delta G^\ominus$ has the value of −237 kJ mol^{-1} ($\Delta G^\ominus$, the energy available for work, is less than $\Delta H^\ominus$ as some energy has been used to overcome the negative entropy change as the system becomes more ordered). This gives a thermodynamic efficiency of 83%. Although real fuel cells do not approach this ideal efficiency it is still much higher than the 30–60% efficiency obtained by combusting fuels in power stations.

MICROBIAL FUEL CELLS

Microbial fuel cells (MFCs) use bacteria to generate an electric current. They are able to tap into the electron transport chain of cells and liberate electrons that normally would be taken up by oxygen or other intermediates. The fuel is oxidized at the anode by microorganisms, generating carbon dioxide, electrons and protons. Electrons are transferred to the cathode through an external electric circuit, while protons are transferred to the cathode through a membrane. Electrons and protons react with oxygen at the cathode to form water. MFCs have the potential to be a sustainable energy source using different carbohydrates or substrates present in waste waters as the fuel. The *Geobacter* species of bacteria, for example, can be used in some cells to oxidize ethanoate ions under anaerobic conditions.

$CH_3COO^- + 2H_2O \rightarrow 2CO_2 + 7H^+ + 8e^-$

NERNST EQUATION AND CONCENTRATION CELLS

When a copper half-cell is connected to a zinc half-cell the total emf produced is 1.10 V under standard conditions (see page 76).

The reaction taking place is $Zn(s) + Cu^{2+}(aq) \rightleftharpoons Zn^{2+}(aq) + Cu(s)$.

It is an equilibrium reaction although the position of equilibrium lies far on the product side. From Le Chatelier's Principle it can be seen that either increasing the concentration of the copper ions or decreasing the concentration of the zinc ions will move the equilibrium further to the right and increase the voltage. The dependence of voltage on concentration is given by the Nernst equation.

$$E = E^\ominus - \left(\frac{RT}{nF}\right)\ln Q$$

Where R is the gas constant, F is Faraday's constant, T is the temperature in Kelvin and n is the number of electrons transferred. Q is the reaction quotient, i.e. the equilibrium expression, in this case the concentration of the aqueous zinc ions divided by the concentration of the aqueous copper ions. If both concentrations are 1.00 mol dm^{-3} as they are under standard conditions then $E^\ominus = E$ as $\ln 1 = 0$. However if the concentration of the copper ions is doubled and the concentration of the zinc ions is halved then Q will equal $\frac{1}{4}$.

Then E will be equal to $E^\ominus - \frac{RT}{nF}\ln 0.25 = 1.10 - \frac{(8.31 \times 298)}{(2 \times 96500)}\ln 0.25 = 1.12$ V

One application of the Nernst equation is that a cell can be made by combining two half-cells with the same element but with solutions of different concentrations. This is known as a concentration cell. For example a Cu(s) / Cu^{2+}(1.00 mol dm^{-3}) half-cell connected to a Cu(s) / Cu^{2+}(1.00×10^{-2} mol dm^{-3}) half-cell will produce an emf of $-\frac{RT}{2F}\ln 0.01 = 0.059$ V.

Although concentration cells produce a very small voltage they are important in biological systems. Nerve impulses are produced because nerve cells contain different concentrations of potassium ions on the inside and outside of the cell.

Nuclear fusion and nuclear fission (3)

MASS DEFECT AND NUCLEAR BINDING ENERGY

Protons and neutrons in the nucleus of an atom are held together by very strong forces. A measure of these forces is known as the **binding energy**. This can be defined as the energy that must be supplied to one mole of the atoms to break down the nuclei into separate neutrons and protons *or* the energy released when separate neutrons and protons combine to form one mole of the atomic nuclei. It can be calculated from the **mass defect**.

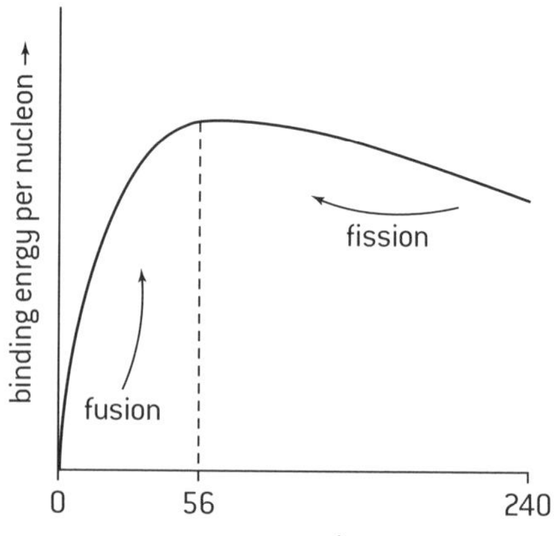

How binding energy varies with mass number

The mass defect is the difference in the combined mass of all the separate protons and neutrons compared with the actual mass of the nucleus. Strictly speaking, it also includes the mass of the electrons too but these are so small they can be ignored. Consider a helium nucleus. It contains two neutrons (relative mass = 2 × 1.0087) and two protons (relative mass = 2 × 1.0078) giving a total relative mass of 4.0330. However the actual relative atomic mass of helium is 4.0039. The relative mass defect is thus 0.0291. By using Einstein's equation $E = mc^2$ this converts to a binding energy of approximately 2.7×10^9 kJ mol^{-1}. A graph of binding energy per nucleon against mass number shows that atomic nuclei with a mass number of approximately 56 (i.e. the nuclei of iron) have the maximum binding energy and are thus the most stable. Nuclei to the left or right of this maximum will undergo nuclear change in such a way that, as they approach the maximum, energy will be released. This explains nuclear fusion, whereby small atoms combine to form heavier nuclei, and nuclear fission – the splitting of heavy nuclei to form two or more lighter nuclei.

CALCULATING THE ENERGY RELEASED IN A NUCLEAR REACTION

The mass numbers and atomic numbers balance in a nuclear equation. However if accurate values are used for the masses it can be seen that a small decrease in mass takes place during the reaction. The amount of energy that is formed by this decrease in mass can be calculated using Einstein's equation $E = mc^2$.

e.g. relative mass

$$^{235}_{92}\text{U} + {}^{1}_{0}\text{n} \rightarrow {}^{90}_{36}\text{Kr} + {}^{144}_{56}\text{Ba} + 2{}^{1}_{0}\text{n}$$

$$\underbrace{235.0439 + 1.0087}_{236.0526} \qquad \underbrace{89.9470 + 143.8810 + 2.0174}_{235.8454}$$

The relative mass loss = 236.0526 – 235.8454 = 0.2072 which is approximately 0.1% of the initial uranium. If the masses are measured in grams then a loss of 0.2072 g (2.072×10^{-4} kg) is equivalent to $2.072 \times 10^{-4} \times (2.998 \times 10^8)^2 = 1.862 \times 10^{13}$ J (or 1.862×10^{10} kJ). This amount of energy is the theoretical amount that one mole of uranium-235 (235 g) can produce if it all reacted according to the above equation.

Similar calculations can be performed for the mass loss during nuclear fusion.

e.g. relative mass

$$^{2}_{1}\text{H} + {}^{3}_{1}\text{H} \rightarrow {}^{4}_{2}\text{He} + {}^{1}_{0}\text{n}$$

$$\underbrace{2.014102 + 3.016049}_{5.030151} \qquad \underbrace{4.002602 + 1.008665}_{5.011267}$$

The relative mass loss = 5.030151 – 5.011267 = 0.018884. This equates to $1.8884 \times 10^{-5} \times (2.998 \times 10^8)^2 = 1.697 \times 10^{12}$ J (or 1.697×10^9 kJ). This is the theoretical energy released by 5 g of hydrogen isotopes. If this is scaled up to 235 g to compare it with the fission energy released from the same mass of uranium-235 it becomes 7.977×10^{10} kJ which shows the greater potential that nuclear fusion has as an energy source.

ENRICHED URANIUM

Naturally occurring uranium contains 99.3% uranium-238 and only 0.7% by mass of uranium-235. To obtain enriched uranium the isotopes have to be separated so that the percentage of uranium-235 is increased. This is because uranium-235 is the only naturally occurring isotope present in appreciable amounts that is fissile with neutrons.

Uranium occurs naturally in several different ores. The ore uraninite contains uranium(IV) oxide, UO_2. This is a crystalline ionic solid with the same structure as calcium fluorite, CaF_2. It has a high melting point (2865 °C) which makes it difficult to separate the different isotopes. However it can be converted into uranium hexafluoride. Uranium hexafluoride is a gas at ambient temperatures (it sublimes at 65.5 °C) even though it has a very high relative molar mass. This is because the six fluorine atoms surround the central uranium atom to form a non-polar molecule with an octahedral shape which can only form weak intermolecular forces with other uranium hexafluoride molecules. Because fluorine only exists as a single isotope the slight difference in mass between $^{238}UF_6$ and $^{235}UF_6$ is due only to the different isotopes of uranium. This means that $^{238}UF_6$ and $^{235}UF_6$ can be separated by either centrifugation or by gaseous diffusion. The more modern method of gas centrifugation uses a large number of rapidly rotating cylinders to create a strong centrifugal force so the heavier $^{238}UF_6$ gas molecules move towards the outside of the cylinder and the lighter gas molecules of $^{235}UF_6$ concentrate closer to the centre as enriched uranium. Gas diffusion is based on Graham's Law which is explained on the following page.

HL Nuclear fusion and nuclear fission (4)

RATE OF RADIOACTIVE DECAY

The decay of radioactive isotopes is a first-order reaction i.e.

rate = $\lambda[N]$ where λ is the rate constant (also known as the decay constant) and N is the concentration at time t.

If the rate is expressed as $-\frac{dN}{dt}$

then $\frac{dN}{dt} = \lambda[N]$

Integration of this expression gives $\ln\frac{[N_0]}{[N]} = \lambda t$ where $[N_0]$ is the initial concentration when $t = 0$ and ln is the natural logarithm.

This can also be expressed in the form $N = N_0e^{-\lambda t}$, which is how it is given in the IB data booklet.

Since radioactive decay depends on the nucleus it does not matter how the isotope is chemically bound or whether it is present as the free element. The rate of decay will be the same and it is more usual to refer to $[N]$ as the amount of isotope present rather than concentration. This can be expressed in moles or as mass.

At $t_{1/2}$ the amount of N will be half the initial amount. Thus $\frac{[N_0]}{[N]}$ will equal 2. The integrated expression then becomes $\ln 2 = \lambda t_{1/2}$ or $t_{1/2} = \frac{0.693}{\lambda}$.

From this expression it can be seen that the half-life of radioactive decay is independent of the amount of isotope present as stated earlier in this option.

Worked examples

1. The half-life of radium-226 is 1622 years. Calculate how long it will take for a sample of radium-226 to decay to 10% of its original radioactivity.

   Step 1. Use $t_{1/2}$ to find the rate constant from the equation $\lambda t_{1/2} = 0.693$

   $$\lambda = \frac{0.693}{1622} = 4.27 \times 10^{-4}\ \text{year}^{-1}$$

   Step 2. Insert the value for λ into the integrated form of the rate equation $\lambda t = \ln\left(\frac{[N_0]}{[N]}\right)$.

   $$4.27 \times 10^{-4} \times t = \ln\left(\frac{100\%}{10\%}\right)$$

   $$=> t = 5392\ \text{years}$$

   It will take 5392 years to decay to 10% of its original activity.

2. A piece of old wood was found to give 10 counts per minute per gram of carbon when subjected to ^{14}C analysis. New wood has a count of 15 cpm g^{-1}. The half-life of ^{14}C is 5570 years. Calculate the age of the old wood.

   Step 1. Use $t_{1/2}$ to find the rate constant.

   $$\lambda = \frac{0.693}{5570} = 1.24 \times 10^{-4}\ \text{years}^{-1}.$$

   Step 2. Insert the value of λ into the integrated form of the rate equation.

   $$1.24 \times 10^{-4} \times t = \ln\frac{[N_0]}{[N]} = \ln\frac{^{14}\text{C content in new wood}}{^{14}\text{C content in old wood}}$$

   $$= \ln\frac{15}{10} = \ln 1.5$$

   $$=> t = 3270\ \text{years}$$

   The wood is 3270 years old.

GRAHAM'S LAW OF EFFUSION

Graham's Law is based on the fact that two different gases at the same temperature have the same kinetic energy. Since kinetic energy can be expressed as ½ mv^2 where v is the velocity of the gaseous molecules it follows that $m_1v_1^2 = m_2v_2^2$ where m_1 and m_2 are the molar masses of the two different gases travelling at velocities of v_1 and v_2 respectively. The lighter the gas the faster it is travelling compared with heavier gases at the same temperature.

A gas **diffuses** when it evenly fills its container and this occurs because, according to the kinetic theory of gases, gaseous molecules move with rapid random motion. **Effusion** occurs when gas molecules escape from a small hole in the container. Because lighter gases have a greater velocity they effuse more quickly than heavier gases. Graham's law of effusion states that the rate of effusion of a gas is inversely proportional to the square root of the molar mass and is usually expressed as $\frac{\text{Rate}_1}{\text{Rate}_2} = \sqrt{\frac{M_2}{M_1}}$.

Thus if a mixture of sulfur dioxide ($M_r = 64$) and methane ($M_r = 16$) molecules is allowed to effuse the methane molecules will effuse twice as fast. This process of effusion can be used to enrich uranium as $^{235}UF_6$ will effuse faster than $^{238}UF_6$.

NUCLEAR SAFETY

The problems of dealing with low-level and high-level nuclear waste and of ensuring nuclear safety have already been highlighted on page 145. The main health hazard caused by radioactive isotopes in nuclear material is due to the type of radiation they emit. This radiation is ionizing in nature and can produce free radicals such as the superoxide radical, O_2^- and the hydroxyl radical, OH. These radicals can initiate free-radical chain reactions that can damage the DNA and the enzymes in living cells and thus lead to mutations and cancer.

Photovoltaic cells and dye-sensitized solar cells (DSSC)

ELECTRICAL CONDUCTIVITY OF SILICON

Metals conduct electricity because they have relatively low ionization energies and contain delocalized electrons. Non-metals (apart from graphite) are poor conductors because the electrons are normally held in fixed positions. Silicon is a semiconductor. Unlike metals, the conductivity of semiconductors increases with increased temperature.

A crystal of silicon contains a lattice of silicon atoms bonded to each other by shared pairs of electrons. These electrons are in fixed positions so silicon is a poor conductor under normal conditions. However the energy required to excite an electron and free it from its bonding position is equivalent to the energy of light with a wavelength of 1.1×10^{-6} m. Visible light has a shorter wavelength in the range of $4–7 \times 10^{-7}$ m. This is higher in energy and so photons of sunlight are able to excite a valence electron in silicon. The electron is then free to move through the crystal lattice in the conduction band making it an electrical conductor. This is the basis of the photoelectric effect and is the theory behind solar powered batteries. In practice, the process is not very efficient and the cost of purifying the silicon is high. However solar cells are not polluting and do not use up valuable fossil fuel reserves.

One method of improving the efficiency of the photoelectric effect is by doping. This process involves adding very small amounts of atoms of other elements usually from group 13 (Al, Ga or In) or from group 15 (P or As). When a group 15 element is added the extra electron can move easily throughout the crystal lattice making it a better conductor compared with pure silicon. Such doping produces an n-type semiconductor because the conductivity is due to negative electrons. When a group 13 element is added the element now has one less electron than silicon. This produces a 'hole' in the lattice. When a free electron moves into this hole it produces a new hole where the electron was formerly located. The hole can be regarded as a positive carrier so the semiconductor is known as a p-type.

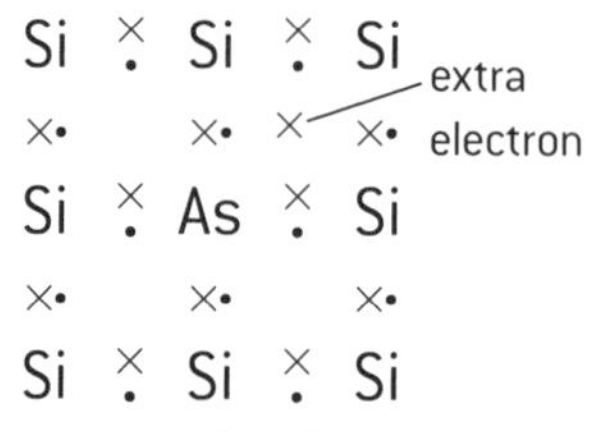

n-type semiconductor

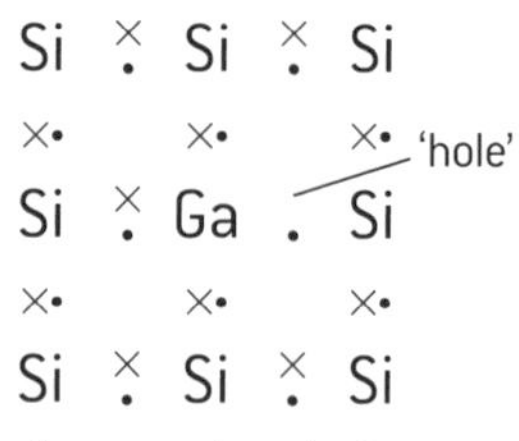

p-type semiconductor

DYE-SENSITIZED SOLAR CELLS (DSSCs)

Dye-sensitized solar cells work in a different way to silicon-based photoelectric cells. In the silicon-based cell the silicon provides both the source and the means of conductivity of the electrons. In a DSSC the photoelectrons originate from the dye when it absorbs light and are then transported through a semiconductor.

A normal C=C double bond in an organic compound absorbs high energy radiation in the ultraviolet region and in the process an electron is excited to a higher energy level. If the compound contains many alternate double and single carbon to carbon bonds then it is said to be conjugated and the energy required to excite an electron is lower. The more conjugated the molecule is, the lower the energy of light required. Molecules such as chlorophyll and α-carotene, which are highly conjugated, absorb in the visible region. As their name suggests, dyes also absorb in the visible region and sunlight interacts with dyes such as ruthenium–polypyridine complexes, which are used in Grätzel DSSCs to produce the photoelectrons.

The Grätzel DSSC essentially consists of three parts. The anode is made of tin(IV) oxide, SnO_2, doped with fluoride ions on a transparent plate. The back of this plate is covered in a thin layer of nanoparticles of titanium(IV) oxide, TiO_2 which provides a large surface area upon which the photosensitive dye is covalently bonded. This plate is sealed together with a separate plate consisting of an iodide electrolyte on a thin sheet of a conducting metal such as platinum which acts as the cathode.

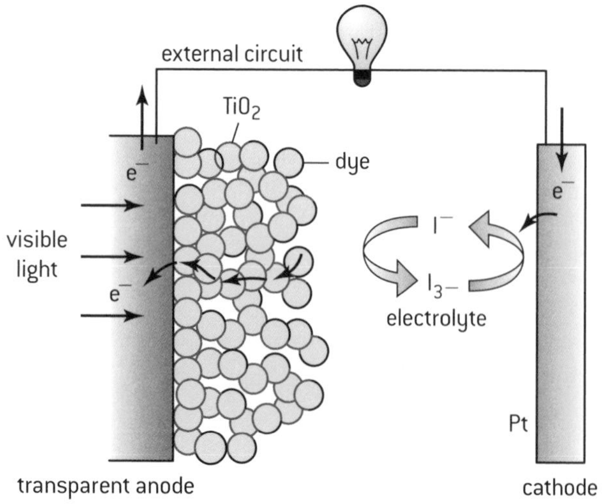

Representation of a Grätzel DSSC

When light passes through the transparent plate it is absorbed by the dye, which releases electrons into the titanium(IV) oxide layer, which acts as the semiconductor. During this process the dye becomes oxidized. The electrons flow toward the transparent tin-based anode and through the external circuit. They return to the platinum cathode and flow into the iodide electrolyte. The electrolyte then transports the electrons back to the dye molecules.

The action of the iodide electrolyte is to reduce the dye back to its normal state during which iodide ions are oxidized to triiodide ions, I_3^-.

$$3I^- \rightarrow I_3^- + 2e^-$$

Note that the triiodide ion can be thought of as a complex of iodine and iodide ions so effectively iodide ions are being reduced to iodine $2I^- \rightarrow I_2 + 2e^-$.

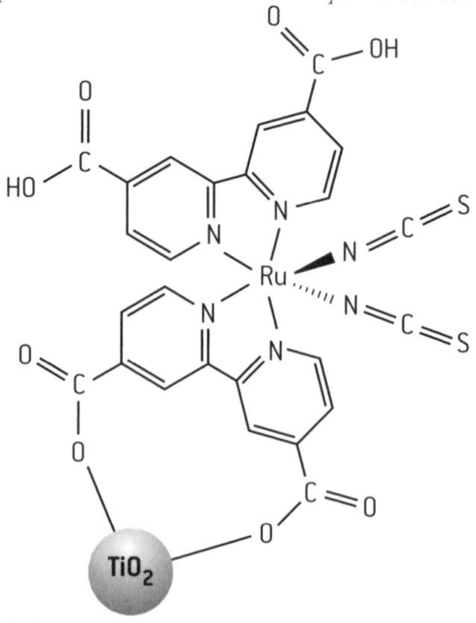

A highly conjugated ruthenium–polypyridine dye showing how it bonds to TiO_2 in Grätzel DSSCs

The returning electrons from the external circuit reduce the I_3^- ions back to iodide ions on the platinum cathode.

$$I_3^- + 2e^- \rightarrow 3I^-$$

DSSCs have several advantages over silicon-based photoelectric cells. They are simple to manufacture, semi-flexible and semi-transparent and relatively cheap. They also absorb visible light over a larger range of wavelengths.

SHORT ANSWER QUESTIONS – OPTION C – ENERGY

1. a) In the context of nuclear reactions explain the meaning of *fusion*. [1]

   b) Explain why a fusion reaction results in the release of a large quantity of energy. [2]

   c) Explain why the energy released during a fusion reaction is not currently an energy source used by society to generate electricity. [2]

   d) The main reaction in a uranium reactor is fission of ^{235}U. A side reaction is caused when neutrons react with ^{238}U.

$$^{238}_{a}U + ^{b}_{c}X \rightarrow ^{239}_{92}U \rightarrow ^{239}_{94}Y + 2^{d}_{e}Z$$

   Deduce the identities of X,Y,Z and a,b,c,d and e. [4]

X		b	
Y		c	
Z		d	
a		e	

2. Plutonium-238 has a half-life of 88 years and emits alpha particles.

   a) Deduce the equation for the natural radioactive decay of ^{238}Pu when it emits one alpha particle. [2]

   b) Explain why it is impossible to state how long an individual plutonium-238 atom will take to decay. [2]

   c) Determine how long will it take for 2.00 g of plutonium-238 to reduce to 0.3125 g of plutonium-238? [2]

3. a) Distinguish between *cracking* and *reforming* reactions used in crude oil refining. [2]

   b) Two products obtained from crude oil are ethene, C_2H_4 and octane, C_8H_{18}. Compare the specific energies (in kJ kg^{-1}) of these two substances. [2]

   c) Gasoline (petrol) for use in cars is given an octane rating. Explain what is meant by an octane number of 95. [2]

   d) Suggest two reasons why ethene is not used as a fuel in cars. [2]

4. a) The structure of chlorophyll is given in section 35 of the IB data booklet. State the essential feature of this molecule that enables it to absorb light in the visible region of the spectrum. [1]

   b) Wood is a biofuel. The cellulose contained within wood is made up of repeating units of glucose, $C_6H_{12}O_6$. State the equation for the formation of glucose in plants by photosynthesis. [2]

   c) Vegetable oils can be converted into diesel fuel by a process known as transesterification.

      (i) Explain why vegetable oils themselves are not used as diesel fuel. [2]

      (ii) State the equation for the reaction of ethanol with a vegetable oil represented by the structure below to form biodiesel. [2]

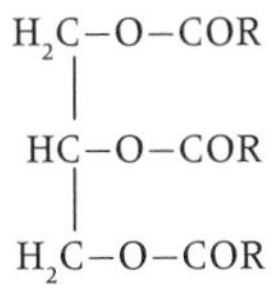

      (iii) Suggest one reason why excess ethanol is used in this reaction. [1]

5. a) Nitrogen, oxygen, carbon dioxide and water vapour are all naturally present in the atmosphere. Describe the greenhouse effect and explain at the molecular level why only some of these gases are responsible for it. [4]

   b) Explain why the concentration of carbon dioxide has increased during the last 100 years and why this has increased the acidity of the oceans. [3]

   c) List two ways in which governments can meet their pledges to reduce carbon dioxide emissions. [2]

HL

6. a) The standard electrode potential of the $Cu(s)/Cu^{2+}(aq)$ half-cell is given in Section 24 of the IB data booklet. Calculate the value of the electrode potential if the concentration of the aqueous copper(II) is reduced to 1.00×10^{-2} mol dm^{-3} and all the other variables remain constant. [2]

   b) Calculate the cell emf when a $Cu(s)/Cu^{2+}(aq)$ half-cell with a copper(II) concentration of 1.00×10^{-2} mol dm^{-3} is connected via a salt bridge to a standard $Zn(s)/Zn^{2+}(aq)$ half-cell. [2]

   c) Compare and contrast the voltage and power that can be obtained from the cell in b) and a copper half-cell connected to a zinc half-cell where both half-cells are operating under standard conditions. [3]

7. a) The mass of a proton and the mass of a neutron can be found in section 4 of the IB data booklet. The accurate relative atomic mass of ^{238}U is 238.050789. Calculate the binding energy in the nucleus of one atom of ^{238}U (ignore the contribution made by the electrons to the relative atomic mass). [2]

   b) Naturally occurring uranium is largely made up of ^{238}U with a small amount of ^{235}U. Describe how enriched uranium can be obtained. [3]

   c) An isotope in some radioactive nuclear waste has a half-life of 69 years. Calculate how long it would take for the activity of this particular isotope in the waste to decrease by 99%. [2]

8. a) List three advantages that dye-sensitized solar cells have compared with silicon-based photovoltaic cells. [3]

   b) An electrolyte that is used in dye-sensitized solar cells is a solution of iodine in iodide ions. Explain with equations how the electrolyte functions when the cells are working. [4]

   c) Explain the function of titanium(IV) oxide in a DSSC and why it is present in nanosized particles. [3]

Pharmaceutical products and drug action

THE EFFECTS OF DRUGS AND MEDICINES

For centuries people have used natural materials to provide relief from pain, heal injuries and cure disease. Many of these folk remedies have been shown to be very effective and the active ingredients isolated and identified. Morphine was extracted from the poppy *Papaver somniferum* early in the 19th century and later salicylic acid, the precursor of aspirin, was isolated from willow bark.

The words 'drug' or 'medicine' have different connotations in different countries and are difficult to define precisely. Generally a drug or medicine is any chemical (natural or human-made) that does one or more of the following:

- alters incoming sensory sensations
- alters mood or emotions
- alters the physiological state (including consciousness, activity level or coordination).

Drugs and medicines are commonly (but not always) taken to improve health. They accomplish this by assisting the body in its natural healing process. The different mechanisms of drug action are still not fully understood and there is evidence that the body can be 'fooled' into healing itself through the 'placebo' effect.

METHODS OF ADMINISTERING DRUGS

In order to reach the site where their effects are needed, the majority of drugs must be absorbed into the bloodstream. The method of administering the drug determines the route taken and the speed with which it is absorbed into the blood. The four main methods are: by mouth (oral); inhalation; through the anus (rectal) and by injection (parenteral).

Drugs may also be applied topically so that the effect is limited mainly to the site of the disorder such as the surface of the skin. Such drugs may come in the form of creams, ointments, sprays and drops.

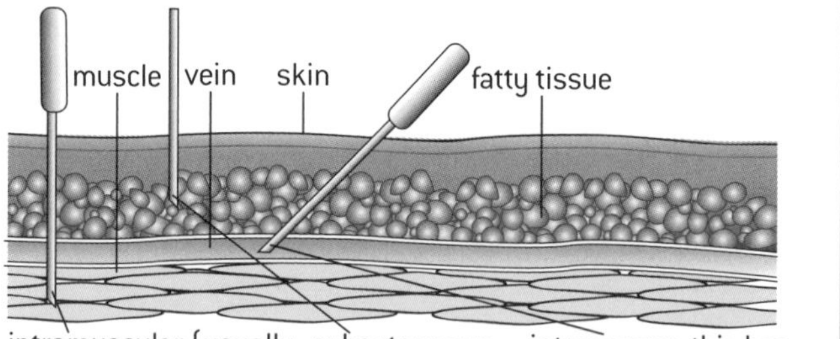
muscle
vein
skin
fatty tissue
intramuscular (usually injected into arm, leg or buttock muscles)
subcutaneous: injected directly under the skin
intravenous: this has the most rapid effect as the drug enters the bloodstream directly

RESEARCH, DEVELOPMENT AND TESTING OF NEW PRODUCTS

The research and development of new drugs is a long and expensive process. Traditionally, a new product is isolated from an existing species or synthesized chemically. The process needs to identify the need, the structure of the drug, possible methods of synthesis, extraction of the product and the yield. It is then subjected to thorough laboratory and clinical pharmacological studies to demonstrate its effectiveness. Before studies are allowed on humans it must be tested on animals to determine the **lethal dose** required to kill fifty percent of the animal population, known as the LD_{50}. The **effective dose** required to bring about a noticeable effect in 50% of the animal population, ED_{50}, is also obtained.

A factor known as the **therapeutic index**, can then be calculated.

$$\text{Therapeutic index for } animals = \frac{LD_{50}}{ED_{50}}$$

There are ethical issues about using animals for drug research and the lethal dose cannot be determined experimentally on humans.

$$\text{Therapeutic index for } humans = \frac{TD_{50}}{ED_{50}}$$

where TD_{50} is the dose that causes a toxic effect in 50% of the human population.

Drugs with a low therapeutic index need to be controlled very carefully as exceeding an effective dose could produce serious toxicity or even death. Ideally the minimum amount of the drug required to achieve the desired effect should be administered. The **therapeutic window** is the range of dosages between the minimum amounts of the drug that produce the desired effect and a medically unacceptable adverse effect.

BIOAVAILABILITY OF A DRUG

Drug-receptor interactions depend upon the structure of the drug and the site of activity. Bioavailability is the fraction of the administered dosage that reaches the target part of the human body. The polarity of the drug, the types of functional groups present in the drug and the method of administration can all affect bioavailability. For example, the early penicillins could not be taken orally as they were destroyed by acids in the stomach before they could be effective. Many drugs, such as morphine and soluble aspirin are administered in an ionic form to make them more soluble. They then revert to the undissociated form in the body. Some modern anti-cancer drugs have extremely high bioavailability. They contain radioactive nanoparticles embedded in the drug molecule. When the drug binds to the cancerous cells they effectively deliver the radioactivity directly to the target and destroy the cancerous cells without causing damage elsewhere in the body.

Once the therapeutic index is determined the drug can then be used in an initial clinical trial on humans. This is usually on volunteers as well as on patients, half of whom are given the drug and half of whom are given a placebo. This initial trial is closely monitored to establish the drug's safety and possible side effects. Drugs usually have unwanted **side effects**, for example aspirin can cause bleeding of the stomach and morphine, which is normally used for pain relief, can cause constipation. Side effects can be relative depending on why the drug is taken. People with diarrhoea are sometimes given a kaolin and morphine mixture and people who have suffered from a heart attack are advised to take aspirin as it is effective as an anti-clotting agent. The severity of the complaint will determine an acceptable **risk to benefit ratio**. If an effective treatment is found for a life-threatening disease then a high risk from side effects will be more acceptable. The **tolerance** of the drug is also determined. Drug tolerance occurs as the body adapts to the action of the drug. A person taking the drug needs larger and larger doses to achieve the original effect. The danger with tolerance is that as the dose increases so do the risks of dependence and the possibility of reaching the lethal dose. Some drugs can cause **addiction,** also known as substance dependence, where reducing or stopping the drug causes withdrawal symptoms. If the drug passes the initial clinical trial it will then go through a rigorous series of further phases where its use is gradually widened in a variety of clinical situations. If it passes all these trials it will eventually be approved by the Drug Administration of a particular country for use either as an OTC (over the counter) drug or for use only through prescription by a doctor.

Aspirin

ASPIRIN AND OTHER MILD ANALGESICS

For a long time the bark of the willow tree (*Salix alba*) was used as a traditional medicine to relieve the fever symptoms of malaria. In the 1860s, chemists showed that the active ingredient in willow bark is salicylic acid (2-hydroxybenzoic acid) and by 1870 salicylic acid was in wide use as a pain killer (analgesic) and fever depressant (antipyretic). However salicylic acid has the undesirable side effect of irritating and damaging the mouth, oesophagus and stomach membranes. In 1899 the Bayer Company of Germany introduced the ethanoate ester of salicylic acid, naming it 'Aspirin'.

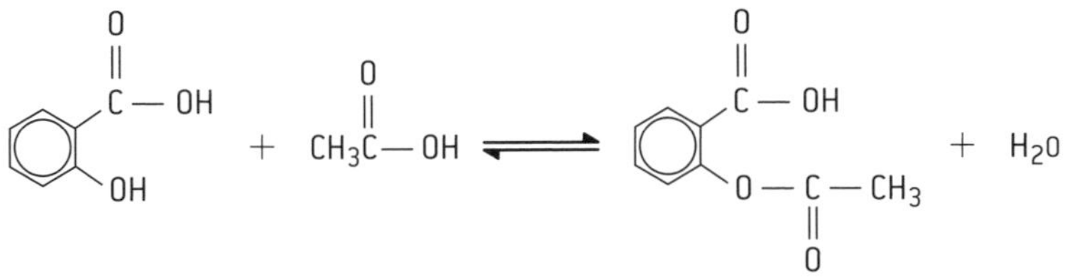

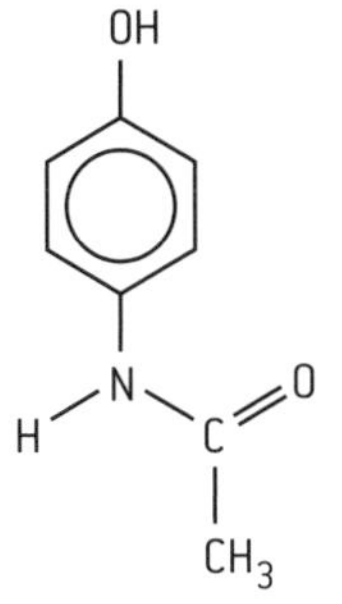

paracetamol

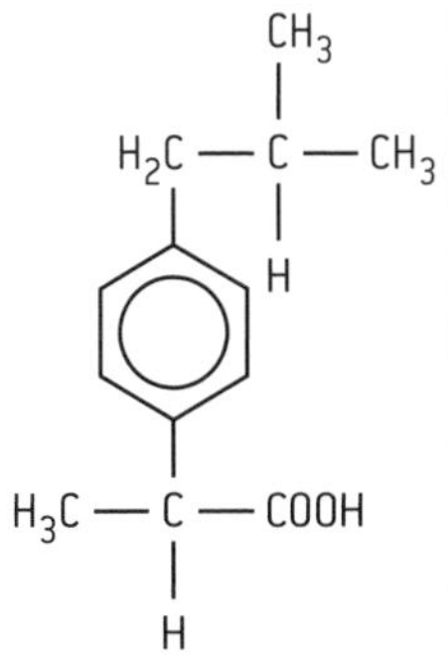

ibuprofen

Aspirin is less irritating to the stomach membranes and the above reaction is only reversed in the body to produce salicylic acid when the aspirin reaches the more alkaline conditions of the small intestine.

Aspirin is thought to work by preventing a particular enzyme, prostaglandin synthase, being formed at the site of the injury or pain. This enzyme is involved in the synthesis of prostaglandins which produce fever and swelling, and the transmission of pain from the site of an injury to the brain. Because of its anti-inflammatory properties, aspirin can also be taken for arthritis and rheumatism. Aspirin also has an ability to prevent blood clotting and is sometimes taken to prevent strokes or the recurrence of heart attacks.

The most common side effect of aspirin is that it causes bleeding in the lining of the stomach. This effect is increased by taking aspirin with alcohol as the alcohol has a **synergistic effect**. A few people are allergic to aspirin with just one or two tablets leading to bronchial asthma. The taking of aspirin by children under twelve has been linked to Reye's syndrome – a potentially fatal liver and brain disorder. Exceeding the safe dose of aspirin can be fatal as the salicylic acid leads to acidosis due to a lowering of the pH of the blood.

Other mild analgesics have structures that show some similarity with aspirin (all are built on a benzene ring with two substituents).

Paracetamol (known as acetaminophen in the USA) contains a hydroxyl group and an amide group. It is often preferred to aspirin as a mild pain reliever, particularly for young children, as its side-effects are less problematic, although in rare cases it can cause kidney damage and blood disorders. Serious problems can arise however if an overdose is taken. Even if the overdose does not result in death, it can cause brain damage and permanent damage to the liver and kidneys.

SYNTHESIS OF ASPIRIN

To obtain a good yield of aspirin, salicylic acid is reacted with ethanoic anhydride (rather than ethanoic acid) in the presence of an acid catalyst. The product can be recrystallized from hot water to obtain pure aspirin.

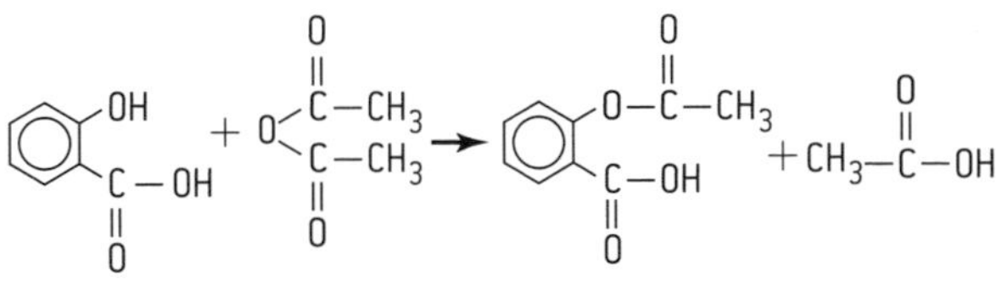

salicylic acid | ethanoic anhydride | aspirin | ethanoic acid

SOLUBLE ASPIRIN

Although aspirin is slightly polar due to the carboxylic acid group it is not very soluble in water. The solubility can be increased dramatically by reacting it with sodium hydroxide to turn it into an ionic salt.

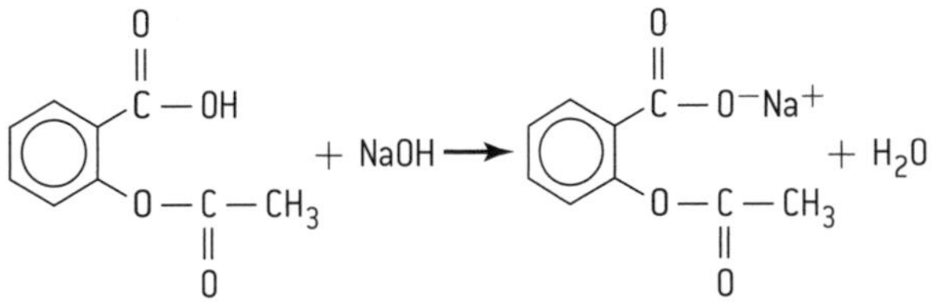

aspirin | soluble aspirin

Once the aspirin anion reaches the strongly acidic environment of the stomach it reverts back to the un-ionized form.

The purity of the product can be determined from its melting point as impurities lower the melting point and cause it to melt over a wider range. The melting point of pure aspirin is 138–140 °C.

The purity can also be tested by looking at its infrared spectrum. Aspirin shows two peaks at 1750 and 1680 cm^{-1} due to the two C=O bonds, and a very broad absorption between 2500 and 3500 cm^{-1} due to the carboxylic acid group.

Infrared spectrum of aspirin

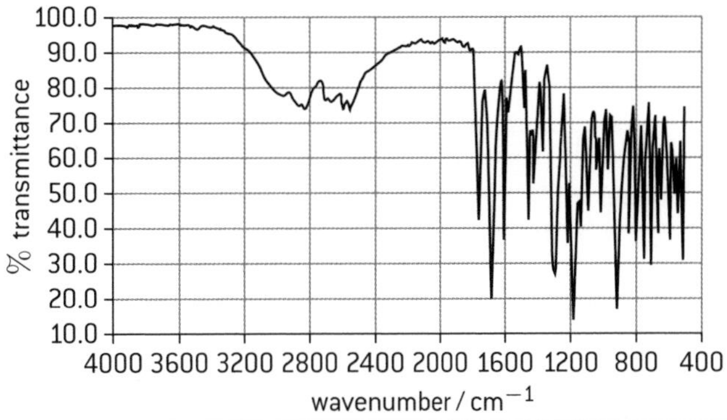

Penicillin

DISCOVERY AND GENERAL STRUCTURE

One of the main reasons for low life expectancy in the early part of the 20th century was death due to bacterial infections such as septicaemia. Penicillin was first discovered in the 1920s but was not brought into production until the Second World War. The original penicillin occurs naturally in a mould (*Penicillium notatum*) and had to be laboriously grown in large tanks containing corn-steep liquor. In the 1950s the structure of penicillin was determined and this enabled chemists to synthesize different types of penicillin and other antibiotics (antibacterials originating from moulds) in the laboratory without recourse to moulds.

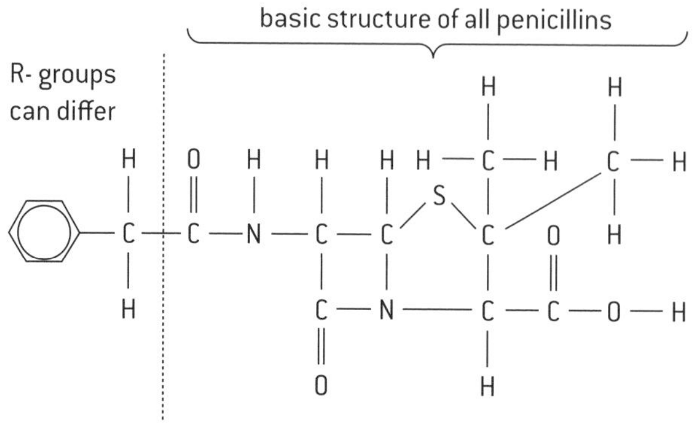

benzylpenicillin (penicillin G)
The original penicillin was broken down by stomach acid so needed to be injected.

MODE OF ACTION

A typical bacterium consists of a single cell with a protective wall made up of a complex mixture of proteins, sugars and lipids. Inside the cell wall is the cytoplasm, which may contain granules of glycogen, lipids and other food reserves. Each bacterial cell contains a single chromosome consisting of a strand of deoxyribonucleic acid (DNA). Not all bacteria cause disease and some are beneficial.

Penicillins work by preventing bacteria from making normal cell walls.

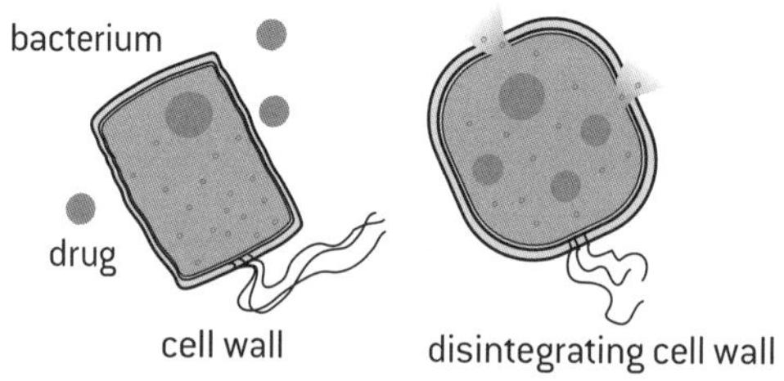

Penicillin interferes with cell-wall formation. As the cell swells, the osmotic pressure causes the wall to disintegrate and the bacterium dies.

All penicillins contain a four-membered beta-lactam ring. Because of the restrictions of the ring, the normal bond angles of 109.5° and 120° are not able to be obtained. This makes the carboxamide group in the ring highly reactive as the ring can readily break due to the strained angles. The group containing the beta-lactam ring is very similar to a combination of the two amino acids cysteine and valine. When the ring opens these parts of the penicillin become covalently bonded to the enzyme that synthesizes the cell walls of the bacterium, thus blocking its action.

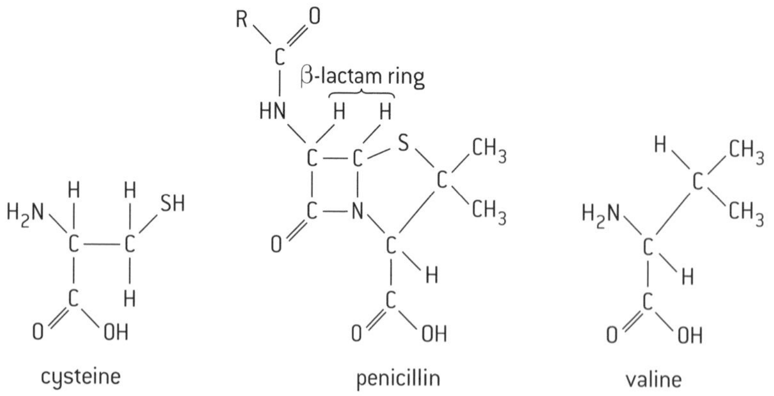

RESISTANCE TO PENICILLINS

When penicillin became readily available to doctors it was often overprescribed to cure minor illnesses such as a sore throat. Certain bacteria were resistant to penicillin and were able to multiply. Their resistance was due to the presence of an enzyme called penicillinase, which could deactivate the original penicillin, penicillin G. To combat this, chemists developed other penicillins whereby the active part of the molecule is retained but the side chain is modified. However, as bacteria multiply and mutate so fast, it is a continual battle to find new antibiotics that are effective against an ever more resistant breed of 'super bugs'. These include the methicillin-resistant *Staphylococcus aureus* (MRSA) and strains of *Mycobacterium tuberculosis* which cause tuberculosis, TB. To treat these infections a strict adherence to a treatment regime often involving a 'cocktail' of different antibiotics is required to prevent the risk of further resistance developing.

Patients not completing a prescribed course of antibiotics and the use of antibiotics in animal feedstocks has also contributed to this problem. Healthy animals are given antibiotics to prevent risk of disease but the antibiotics are passed on through the meat and dairy products to humans, increasing the development of resistant bacteria.

Opiates

STRONG ANALGESICS

Strong analgesics are only available on prescription and are given to relieve the severe pain caused by injury, surgery, heart attack, or chronic diseases, such as cancer. They work by interacting temporarily with receptor sites in the brain without depressing the central nervous system, with the result that pain signals within the brain and spinal cord are blocked. The most important naturally occurring strong analgesics are morphine and codeine found in the opium poppy. These are known as opiates or narcotics (although originally narcotic referred to any sleep-inducing drug).The active part of the morphine molecule has been identified. Codeine and semi-synthetic (obtained by simple structural modifications to morphine) opiates, such as heroin, and totally synthetic compounds (e.g. 'demerol') all possess this basic structure and function as strong analgesics.

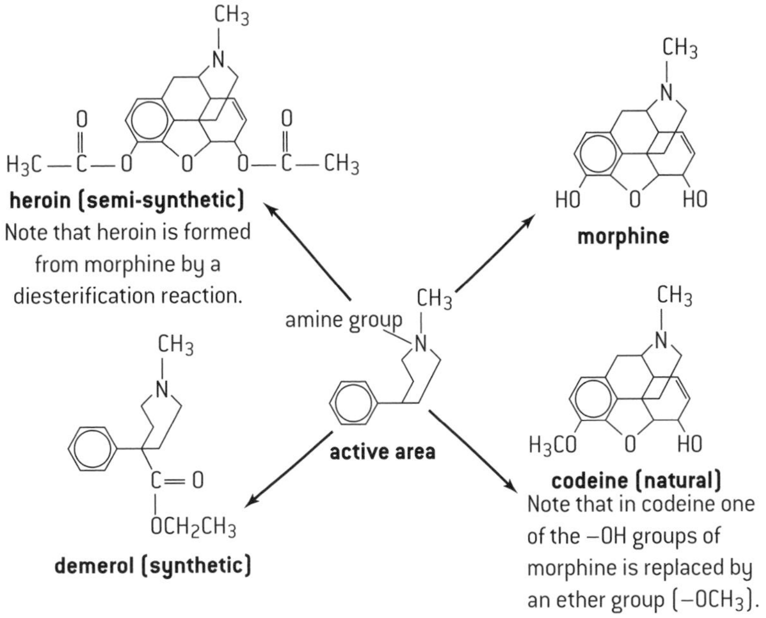

EFFECTS OF OPIATES

Diamorphine (heroin) is a more powerful painkiller than morphine but also more addictive. All opiates cause addiction and also lead to tolerance where more of the drug is required to give the same effect and there is a danger of reaching the lethal limit.

Short-term effects

- Induce a feeling of euphoria (sense of well-being)
- Dulling of pain
- Depress nervous system
- Slow breathing and heart rate
- Cough reflex inhibited
- Nausea and vomiting (first time users)
- High doses lead to coma and/ or death

Long-term effects

- Constipation
- Loss of sex drive
- Disrupts menstrual cycle
- Poor eating habits
- Risk of AIDS, hepatitis, etc. through shared needles
- Social problems e.g. theft, prostitution

Withdrawal symptoms occur within 6 to 24 hours for addicts if the supply of the drug is stopped. These include hot and cold sweats, diarrhoea, anxiety and cramps. One treatment to wean addicts off their addiction is to use methadone as a replacement for heroin. Methadone is also an amine and functions as an analgesic but does not produce the euphoria craved by addicts.

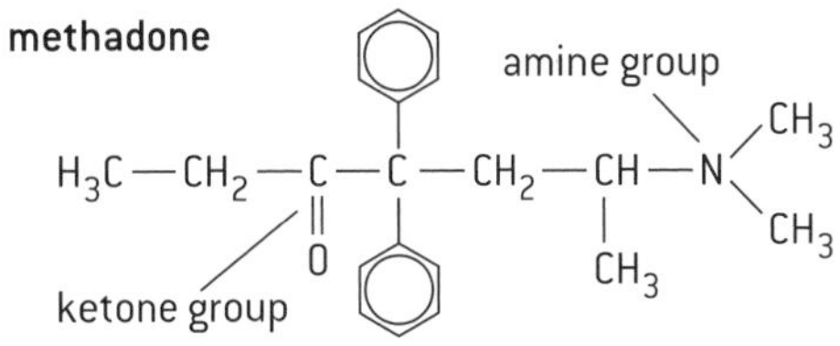

POTENCY OF HEROIN COMPARED WITH MORPHINE

Heroin (diamorphine) is much more potent and produces a much greater feeling of euphoria than morphine. This can be explained by the difference in the polarity of the two substances. Morphine molecules contain two polar hydroxyl (–OH) groups. When morphine is converted into heroin to form the diester these are replaced by much less polar ethanoate groups. This makes heroin much more soluble in lipids that are non-polar. Heroin is thus able to rapidly penetrate the lipid-based blood–brain barrier and reach the brain in higher concentrations than morphine.

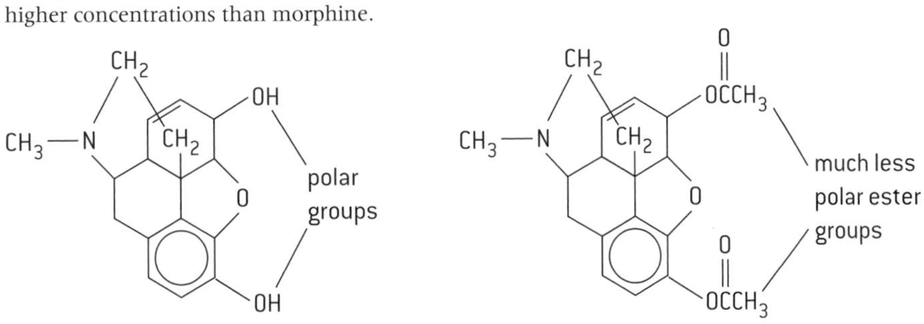

morphine heroin

Both morphine and heroin contain a tertiary amine group. They can be converted into ionic salts by reacting with hydrochloric acid (this is similar to the reaction of ammonia with hydrochloric acid to form ammonium chloride). Thus for severe pain following surgery or to alleviate the pain due to terminal cancer, diamorphine is often injected in the form of diamorphine hydrochloride to increases its bioavailability. Once it is transported in the blood to the brain it reverts back to the undissociated form to cross the lipid-based blood–brain barrier.

pH regulation of the stomach

ANTACIDS

Antacids are used to reduce excess stomach acid. They are most effective if taken between one and three hours after eating, as food typically remains in the stomach for up to four hours after a meal. Antacids are essentially simple bases such as metal oxides, hydroxides, carbonates or hydrogen-carbonates. They work in a non-specific way by neutralizing the acid, preventing inflammation, relieving pain and discomfort and allow the mucus layer and stomach lining to mend. When used in the treatment of ulcers they prevent acid from attacking the damaged stomach lining and so allow the ulcer to heal. Common examples include $Al(OH)_3$, $NaHCO_3$, Na_2CO_3, $Ca(OH)_2$ and 'milk of magnesia' which is a mixture of MgO and $Mg(OH)_2$. Typical neutralization reactions are:

$$NaHCO_3(s) + HCl(aq) \rightarrow NaCl(aq) + CO_2(g) + H_2O(l)$$

$$Na_2CO_3(s) + 2HCl(aq) \rightarrow 2NaCl(aq) + CO_2(g) + H_2O(l)$$

$$Mg(OH)_2(s) + 2HCl(aq) \rightarrow MgCl_2(aq) + 2H_2O(l)$$

$$Al(OH)_3(s) + 3HCl(aq) \rightarrow AlCl_3(aq) + 3H_2O(l)$$

Worked example

Which would be the most effective in combating indigestion – a spoonful of liquid containing 1.00 g of magnesium hydroxide or a spoonful of liquid containing 1.00 g of aluminium hydroxide?

M_r for $Mg(OH)_2 = 24.31 + (2 \times 17.01) = 58.33$

M_r for $Al(OH)_3 = 26.98 + (3 \times 17.01) = 78.01$

Amount of $Mg(OH)_2$ in $1.00\,g = \frac{1.00}{58.33} = 0.0171\,mol$

Amount of $Al(OH)_3$ in $1.00\,g = \frac{1.00}{78.01} = 0.0128\,mol$

$Mg(OH)_2(s) + 2HCl(aq) \rightarrow MgCl_2(aq) + 2H_2O(l)$

Amount of HCl neutralized by 1.00 g of $Mg(OH)_2 = 2 \times 0.0171 = 0.0342\,mol$

$Al(OH)_3(s) + 3HCl(aq) \rightarrow AlCl_3(aq) + 3H_2O(l)$

Amount of HCl neutralized by 1.00 g of $Al(OH)_3 = 3 \times 0.0128 = 0.0384\,mol$

Therefore the aluminium hydroxide would be slightly more effective.

SIDE EFFECTS

Like all drugs, antacids and drugs that inhibit or suppress acid production in the stomach may have side effects. These may be caused by the drugs themselves or by **active metabolites**, the active form of the drug after it has been processed by the body. Usually the effect of active metabolites is weaker than those of the parent drug although in some cases it may actually be stronger and be responsible for the main therapeutic effect of the parent drug.

The most common side effects of antacids are belching to release gas and diarrhoea from magnesium-containing antacids or constipation from aluminium-containing antacids. Side effects of ranitidine and the PPIs include diarrhoea, headache and dizziness. Long term use of PPIs may lead to a higher risk of osteoporosis and increase the risk of developing food and drug allergies.

INHIBITION OF STOMACH ACID PRODUCTION

Much of the digestive process takes place in the stomach, a collapsible muscular bag that can hold between 2 and 4 litres of food. The walls of the stomach are lined with a layer of cells that secrete mucus, pepsinogen (a precursor for the enzyme pepsin that breaks down proteins into peptides) and hydrochloric acid, collectively known as gastric juices. The hydrogen ion concentration of the hydrochloric acid normally lies between $3 \times 10^{-2}\,mol\,dm^{-3}$ and $3 \times 10^{-3}\,mol\,dm^{-3}$ giving a pH value between 1.5 and 2.5. The wall of the stomach is protected from the action of the acid by a lining of mucus. Problems can arise if the stomach lining is damaged or when too much acid is produced, which can eat away at the mucus lining.

Certain drugs can inhibit the production of stomach acid by targeting specific processes. Histamine is produced by certain cells in the lining of the stomach, called the enterochromaffin-like cells (ECL cells). The histamine released from ECL cells then stimulates the acid-making cells (parietal cells) in the lining of the stomach to release acid. H2-receptor antagonists (H2RA), such as ranitidine (Zantec), are a class of drugs used to block the action of histamine on the histamine H2-receptors of parietal cells and thus inhibit acid production.

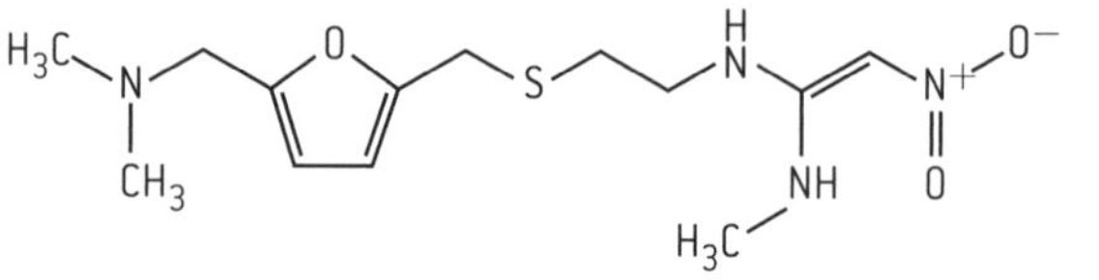

ranitidine (trade name Zantec®)
an H2 receptor antagonist or 'H2 blocker'

Other drugs used to suppress acid secretion in the stomach include proton pump inhibitors such as omeprazole (Prilosec) and esomeprazole (Nexium). PPIs cause a long-lasting reduction of gastric acid production and have largely superseded H2-receptor antagonists. They work by interacting with a cysteine residue in a chemical system called the hydrogen-potassium adenosine triphosphatase enzyme system, H^+/K^+ ATPase (known as the 'proton pump') found in the parietal cells in the stomach lining that make stomach acid. This inhibits the ability of these cells to secrete acid and helps to prevent ulcers from forming or assist the healing process. They can also help to reduce acid reflux-related symptoms such as heartburn.

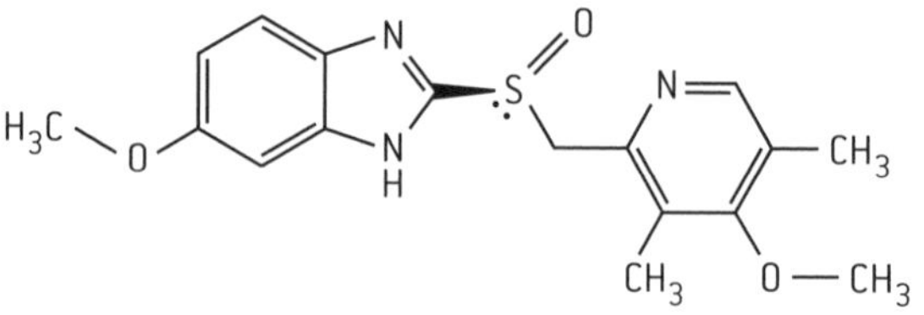

esomeprazole (Nexium®)

The formula of omeprazole is given in section 37 of the IB data booklet. Esomeprazole is an enantiomer of omeprazole. Omeprazole actually consists of a racemic mixture of the two isomers and undergoes a chiral shift (on the S atom) in the body that converts the inactive enantiomer into the active enantiomer (esomeprazole).

Buffer solutions

BUFFER SOLUTIONS

It is important to maintain the pH in the body within certain strict limits to ensure enzymes function correctly. This is done by buffer solutions, which resist changes in pH when small amounts of acid or alkali are added to them.

An acidic buffer solution can be made by mixing a weak acid together with the salt of that acid and a strong base. An example is a solution of ethanoic acid and sodium ethanoate. The weak acid is only slightly dissociated in solution, but the salt is fully dissociated into its ions, so the concentration of ethanoate ions is high.

$$NaCH_3COO(aq) \rightarrow Na^+(aq) + CH_3COO^-(aq)$$

$$CH_3COOH(aq) \rightleftharpoons CH_3COO^-(aq) + H^+(aq)$$

If an acid is added the extra H^+ ions coming from the acid are removed as they combine with ethanoate ions to form undissociated ethanoic acid, so the concentration of H^+ ions remains unaltered.

$$CH_3COO^-(aq) + H^+(aq) \rightleftharpoons CH_3COOH(aq)$$

If an alkali is added the hydroxide ions from the alkali are removed by their reaction with the undissociated acid to form water, so again the H^+ ion concentration stays constant.

$$CH_3COOH(aq) + OH^-(aq) \rightarrow CH_3COO^-(aq) + H_2O(i)$$

In practice acidic buffers are often made by taking a solution of a strong base and adding excess weak acid to it, so that the soluton contains the salt and the unreacted weak acid.

$$NaOH(aq) + CH_3COOH(aq) \rightarrow NaCH_3COO(aq) + H_2O(l) + CH_3COOH(aq)$$

limiting reagent — salt — excess weak acid

(salt + excess weak acid) = buffer solution

An alkali buffer with a fixed pH greater than 7 can be made from a weak base together with the salt of that base with a strong acid. An example is ammonia with ammonium chloride.

$$NH_4Cl(aq) \rightarrow NH_4^+(aq) + Cl^-(aq)$$

$$NH_3(aq) + H_2O(l) \rightleftharpoons NH_4^+(aq) + OH^-(aq)$$

If H^+ ions are added they will combine with OH^- ions to form water and more of the ammonia will dissociate to replace them. If more OH^- ions are added they will combine with ammonium ions to form undissociated ammonia. In both cases the hydroxide ion concentration and the hydrogen ion concentration remain constant.

BUFFER CALCULATIONS

The equilibrium expression for weak acids also applies to acidic buffer solutions,

e.g. ethanoic acid/sodium ethanoate solution.

$$K_a = \frac{[H^+] \times [CH_3COO^-]}{[CH_3COOH]}$$

The essential difference is that now the concentrations of two ions from the acid will not be equal.

Since the sodium ethanoate is completely dissociated the concentration of the ethanoate ions in solution will be all the same as the concentration of the sodium ethanoate; only a very little will come from the acid.

If logarithms are taken and the equation is rearranged the

$$pH = pK_a + \log_{10}\frac{[CH_3COO^-]}{[CH_3COOH]}$$

This is known as the Henderson–Hasselbalch equation (the general formula can be found in section 1 of the IB data booklet).

Two facts can be deduced from this expression. Firstly the pH of the buffer does not change on dilution, as the concentration of the ethanoate ions and the acid will be affected equally. Secondly the buffer will be most efficient when $[CH_3COO^-] = [CH_3COOH]$. At this point, which equates to the half-equivalence point when ethanoic acid is titrated with sodium hydroxide, the pH of the solution will equal the pK_a value of the acid.

Calculate the pH of a buffer containing 0.200 mol of sodium ethanoate in 500 cm^3 of 0.100 mol dm^{-3} ethanoic acid (given that K_a for ethanoic acid $= 1.8 \times 10^{-5}$ mol dm^{-3}).

$[CH_3COO^-] = 0.400$ mol dm^{-3}; $[CH_3COOH] = 0.100$ mol dm^{-3}

$$K_a \approx \frac{[H^+] \times 0.400}{0.100} = 1.8 \times 10^{-5}\ \text{mol dm}^{-3}$$

$[H^+] = 4.5 \times 10^{-6}$ mol dm^{-3}

pH = 5.35

Calculate what mass of sodium propanoate must be dissolved in 1.00 dm^3 of 1.00 mol dm^{-3} propanoic acid ($pK_a = 4.87$) to give a buffer solution with a pH of 4.5.

$$[C_2H_5COO^-] = \frac{K_a \times [C_2H_5COOH]}{[H^+]} = \frac{10^{-4.87} \times 1.00}{10^{-4.5}}$$

$$= 0.427\ \text{mol dm}^{-3}$$

Mass of NaC_2H_5COO required $= 0.427 \times 96.07 = 41.0$ g

Antiviral medications

BACTERIA AND VIRUSES

Bacteria are single cell organisms. Inside the cell wall is the cytoplasm, which may contain granules of glycogen, lipids and other food reserves. Each bacterial cell contains a single chromosome consisting of a strand of deoxyribonucleic acid, DNA.

There are many different types of virus and they vary in their shape and structure. All viruses, however, have a central core of DNA or RNA surrounded by a coat (capsid) of regularly packed protein units called capsomeres each containing many protein molecules. They have no nucleus or cytoplasm and therefore, unlike bacteria, they are not cells. They are much smaller than bacteria and they do not feed, excrete or grow and they can only reproduce inside the cells of living organisms using the materials provided by the host cell.

Athough viruses can survive outside the host they can only replicate by penetrating the living host cell and injecting their DNA or RNA into the cell's cytoplasm. The virus then 'takes over' the biochemical machinery inside the cell. This causes the cell to die or become seriously altered and causes the symptoms of the viral infection. The cell is made to produce new DNA or RNA and forms large numbers of new viruses. These are then released and move on to infect other healthy cells. Because they lack a cell structure and because they multiply so quickly they are difficult to target with drugs. The two main ways in which antiviral drugs may work is by altering the cell's genetic material so that the virus cannot use it to multiply, or by blocking enzyme activity within the host cell so that the virus cannot multiply or is prevented from leaving the cell. One of the problems with developing antiviral drugs is that the viruses themselves are regularly mutating – this is particularly true with the Human Immunodeficiency Virus (HIV).

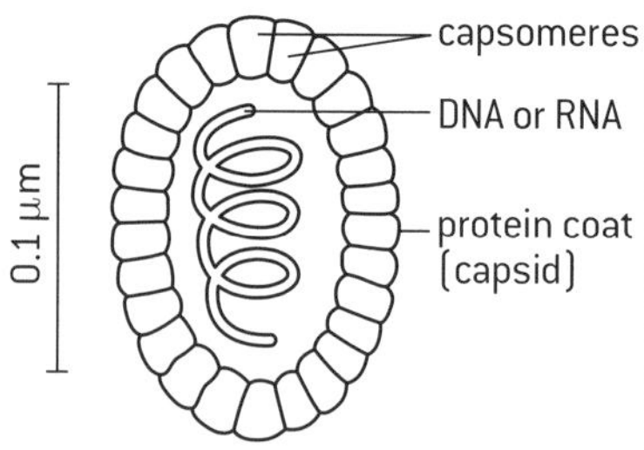

generalized diagram of a virus

TWO SPECIFIC ANTIVIRAL DRUGS

Common viral infections include the common cold, influenza and childhood diseases such as mumps and chicken pox. One drug that is effective against both the influenza A and B viruses is oseltamivir which has the trade name Tamiflu. One of the enzymes used by all influenza viruses to stick to the host cell wall as it leaves is called neuraminidase and the drug works by inhibiting the active site on this enzyme. In the liver, oseltamivir, which is an ester, is hydrolysed to its carboxylate anion which inhibits the viral neuraminidase and prevents it from acting on sialic acid, an acid found on the proteins on the surface of the host cells. This blocking action prevents the new viral particles from being released and so the virus cannot infect other cells and multiply.

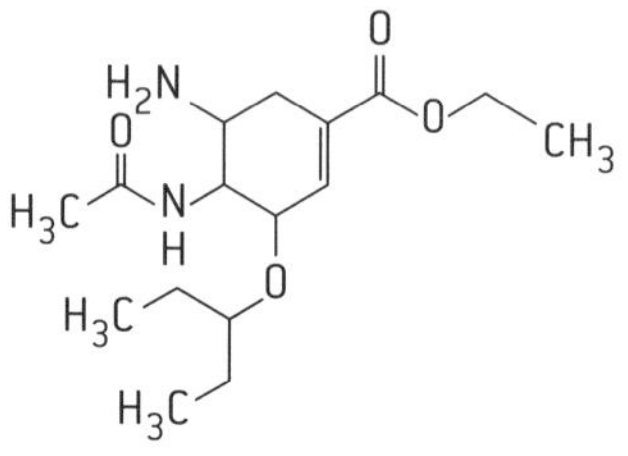
oseltamivir

Another influenza drug, zanamivir (trade name Relenza) also functions as a neuraminidase inhibitor. Both oseltamivir and zanamivir have some structural similarities to sialic acid (a derivative of neuraminic acid, which is a monosaccharide with a nine carbon backbone) and this may explain how they are able to inhibit the viral neuraminidase enzyme. Zanamivir contains many polar hydroxyl and amine groups together with a carboxylic acid group, which makes it much more soluble in polar solvents than oseltamivir.

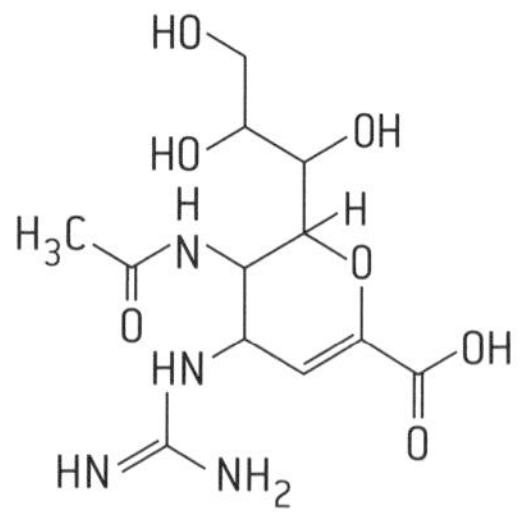
zanamivir

AIDS

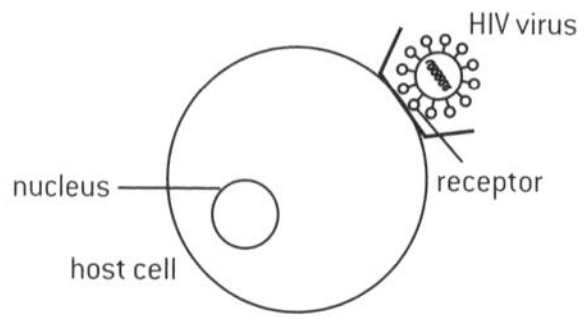

1. Virus binds to a receptor on the host cell.
Possible action of antiviral
Binding site could be altered to prevent virus attaching to host cell.

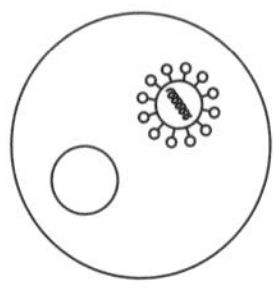

2. Virus enters host cell.
Possible action of antiviral
Cell wall could be altered to prevent virus entering cell.

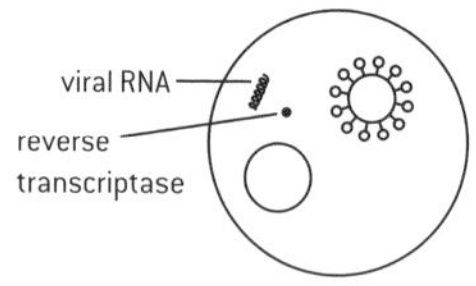

3. Virus loses its protective coat and releases RNA and reverse transcriptase.
Possible action of antiviral
Drugs might be developed which would prevent the virus from losing its protective coat.

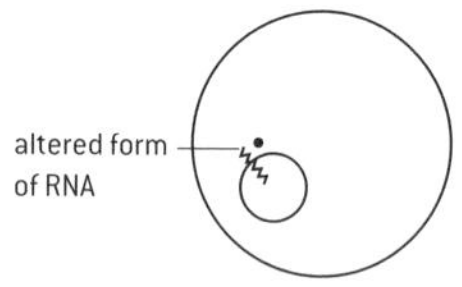

4. The reverse transcriptase converts the viral RNA into a form which can enter the nucleus of the host cell so it can integrate with the cell's DNA.
Possible action of antiviral
AZT works by blocking the action of reverse transcriptase.

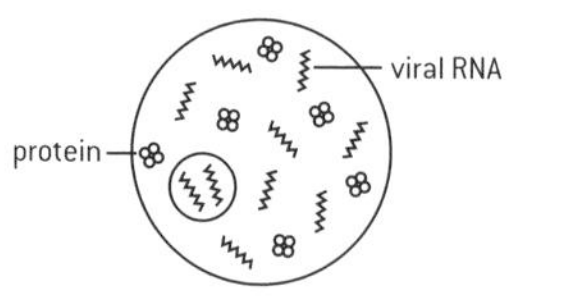

5. The host cell produces new viral RNA and protein.
Possible action of antiviral
May be able to inhibit the production of new viral RNA and proteins by altering the genetic material of the virus.

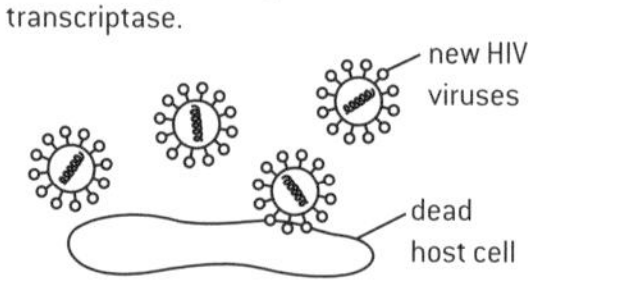

6. The new RNA and proteins form new viruses which then leave the host cell. The host cell dies.
Possible action of antiviral
Develop a drug which prevents the new viruses from leaving the cell. This is how amantadine works against the influenza virus.

AIDS (Acquired Immune Deficiency Syndrome) is caused by a retrovirus – that is it contains RNA rather than DNA. The virus invades certain types of cells, particularly the white blood cells that normally activate other cells in the immune system with the result that the body is unable to fight infection. Once it invades a host cell it makes viral-DNA from the RNA template using an enzyme called reverse transcriptase. This is the opposite process that takes place in normal cells in which RNA is made from a DNA template using transcriptase as the enzyme. Various drugs have met some success in treating the HIV virus but there is still no cure. One, AZT (zidovudine), is a reverse-transcriptase inhibitor and combines with the enzyme that the HIV virus uses to build DNA from RNA and inhibits its active site. Since it is only retroviruses that use this enzyme AZT does not affect normal cells. Other possible ways in which potential antiviral drugs might work are shown above.

Environmental impact of some medications

RADIOACTIVE WASTE

The treatment of cancer often involves radiation therapy. This may be from an external radioactive source or using radioisotope therapy where a radioisotope is attached to another molecule or antibody, which then guides it to the target tissue after being injected or taken orally. Radioactive waste can be divided into high-level waste (HLW) and low-level waste (LLW). Low-level waste includes items such as rubber gloves, paper towels and protective clothing that have been used in areas where radioactive materials are handled. The level of activity is low and the half-lives of the radioactive isotopes are generally short. High-level waste has high activity and generally the isotopes have long half-lives so the waste will remain active for a long period. Isotopes used in radioisotope therapy tend to be (relatively) low-level waste (typical isotopes used are ^{131}I, ^{89}Sr and ^{153}Sm). Any type of radioactive waste needs to be kept separate from other types of waste. LLW is usually disposed of in landfill or in the sea whereas HLW is vitrified (turned into a glass-like material) and stored underground in concrete bunkers.

ANTIBIOTIC AND SOLVENT WASTE

Microorganisms in water or the soil can take up waste antibiotics. Bacteria in the organisms can become resistant to antibiotics by developing enzymes that can break down the antibiotics. If the bacteria find their way into drinking water or food they then make antibiotics much less effective when they are required to fight disease. Antibiotic waste comes from the disposal of unused antibiotics, through the urine of people who are on antibiotics and from animals where the feedstock contains antibiotics, which often gets discharged into rivers.

Many different solvents may be used in the production of drugs. Most countries have strict guidelines about the safe disposal of solvents. They should be separated into chlorine containing and non-chlorine containing solvents. Chlorinated solvents must be incinerated at very high temperatures to prevent the formation of carcinogenic dioxins. Non-chlorinated solvents may be recycled (or their solutes such as heavy metals extracted and recycled), burned to provide energy or, if they are innocuous such as sodium chloride solution, disposed of in rivers or the sea.

GREEN CHEMISTRY

While the safe disposal of waste with minimal damage to the environment is important, green chemistry (also known as sustainable chemistry) aims to reduce harm to the environment by minimizing the use and generation of hazardous substances in the first place – in other words to reduce the pollution at its source and to conserve natural resources including energy. This is particularly important in the pharmaceutical industry where often the research and development of a new drug involves many steps, each involving many potentially polluting substances.

Important factors when designing and producing new drugs include:

1. Aiming for a high atom economy and a low environmental factor. The atom economy is the ratio of the total mass of the desired product(s) to the total mass of all the products. Essentially this is a measure of how much of the reactants remain in the final product (see page 6).

   The environmental-factor (E-factor) is defined as the mass of the total waste products divided by the mass of the desired product, which highlights the need to avoid producing waste products.

   $$E = \frac{\text{total waste (kg)}}{\text{product (kg)}}$$

   Pharmaceutical chemistry, where almost all the products contain carbon, also uses carbon efficiency.

   $$\text{Carbon efficiency} = \frac{\text{amount of carbon in product}}{\text{total amount of carbon present in reactants}}$$

2. The number of steps in a synthesis should be kept to a minimum. Generally the more separate steps required to reach the desired product the lower the percentage yield and the higher the amount of waste reactants and products and the more energy used. In the past, chemists looked for intellectually elegant syntheses. A modern elegant synthesis is one that uses less raw material and energy, and produces less waste.

3. Use greener and safer solvents and reactants. Solvents play an important role in many of the separate steps in a synthesis. The energy and materials needed to manufacture the solvent as well as the problems caused by the disposal of the solvents (if they cannot be recycled) all need to be taken into account. The safety of the workforce should also be considered so chlorinated solvents and benzene, which are potentially carcinogenic, should be avoided.

   In addition, green synthetic chemists will also consider using renewable feedstocks, using suitable catalysts to reduce energy demands by lowering operating temperatures and to consider the fate of the potential drug regarding its breakdown products and disposal after use.

   One example of the use of green chemistry in practice is the development of the influenza drug Tamiflu (oseltamivir). Tamiflu was first synthesized by chemists working at Monash University in Australia. Their technique only produced very small amounts of product and used unsafe lithium nitride, LiN_3. The first commercial scale production was done by Glaxo. They avoided the use of lithium nitrides but still used azides and chromatography as a purification technique rather than the greener recrystallization. Since then more and more ways of synthesizing Tamiflu have been developed, each with a more sustainable route. One of the problems is that one of the starting materials is a compound called shikimic acid, which is currently uneconomical to synthesize and is isolated from the Chinese star anise plant (*Illicium anisatum*). One green hope is that it can be obtained by bioengineering using fermenting bacteria.

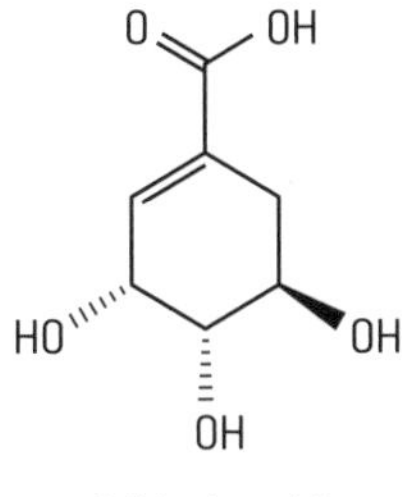

shikimic acid

Taxol—a chiral auxiliary case study

TAXOL

Taxol (also known as paclitaxel) is used to treat ovarian, breast and lung cancer. It is injected as a colourless fluid as part of chemotherapy treatment. It works by preventing cancer cells from dividing and replicating. It occurs naturally in the bark of the Pacific yew tree (*Taxus brevifolia*). Once it became established as an effective anti-cancer drug, concern was expressed about the damage being done to the yew tree population as the amount of bark required to make even small amounts of the drug is considerable and stripping the bark kills the trees. The challenge for chemists was to synthesize the drug from crude-oil based starting materials. Initially a semi-synthetic route was found starting from a compound found in the needles of a related yew tree (*Taxus baccata*). More recently, the total synthesis from crude-oil based reagents has been achieved but the process is not very green. Most of the commercially available taxol is still obtained either by a semi-synthetic route or by plant cell fermentation (PCF) technology, which uses much less energy and less hazardous chemicals.

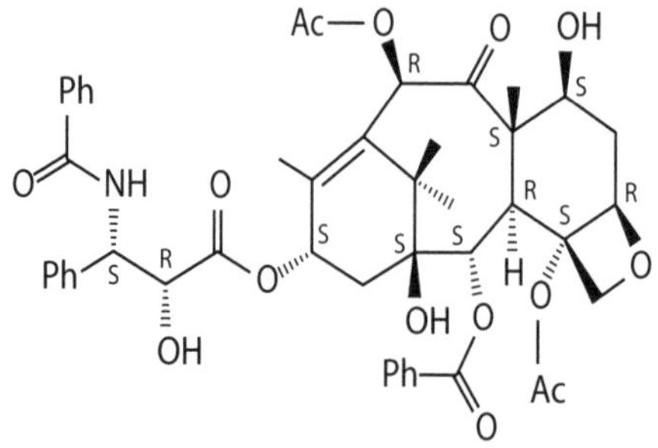

The structure of taxol showing the exact stereochemistry at each of the eleven chiral carbon atoms (Ph = phenyl; Ac = acetyl)

The structure of taxol is given in section 37 of the IB data booklet but this does not make clear the exact stereochemistry. The difficulty with the synthesis of taxol is that it contains no fewer than 11 separate chiral carbon atoms. This means that there is a very large number of possible enantioners and diastereoisomers of which only one corresponds to taxol itself with the correct R or S configuration on each of the 11 chiral carbon atoms.

USE OF CHIRAL AUXILIARIES

The traditional synthesis of an optically active compound normally produces a racemic mixture (50:50 mixture) of the two enantiomers, which then has to be separated into the two isomers by using chromatography together with another optically active compound. If this technique is used to make taxol then the yield of the correct enantiomer would be very small. A more recent technique using chiral auxiliaries now makes it possible to synthesize just the desired isomer.

Attaching an auxiliary, which is itself optically active, to the starting material creates the stereochemical conditions necessary for the reaction to form only one enantiomer, i.e. asymmetrically to form just the required isomer. After the desired product has been formed the auxiliary is removed and recycled. Taxol is too complicated to give as an example here but the reaction of propanoic acid to form just one of the two possible enantiomers of 2-hydroxypropanoic acid (lactic acid) illustrates the technique.

Use of a chiral auxilliary

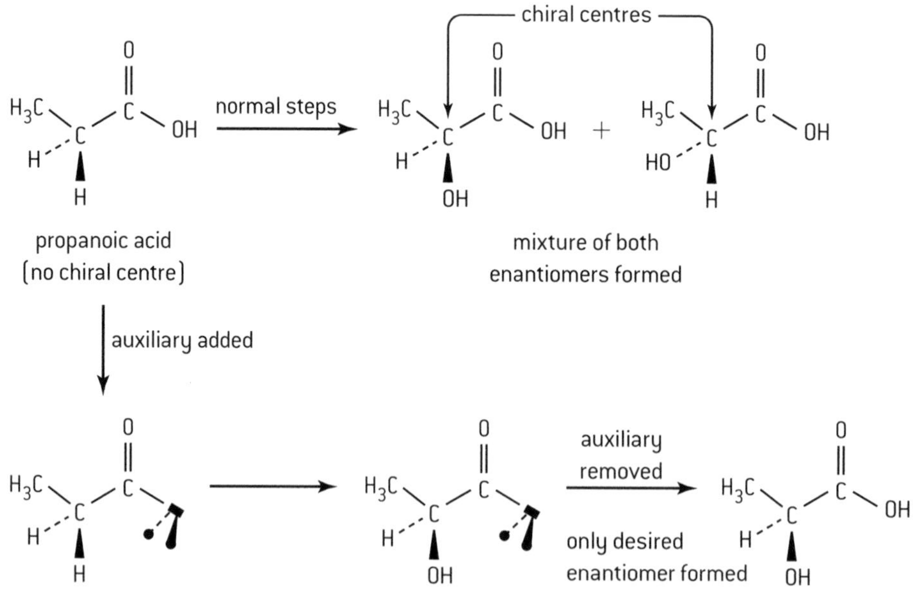

The particular enantiomer required can be distinguished from other enantiomers by using a polarimeter as it will rotate the plane of plane-polarized light by a fixed amount in either a clockwise or anti-clockwise direction. This technique is described on page 96.

Nuclear medicine

RADIOTHERAPY AND SIDE EFFECTS

Radiotherapy can involve many different types of emissions depending on the type of radiotherapy and the reason it is being used. These include alpha particles, ${}^{4}_{2}He$, beta particles, ${}^{0}_{-1}e^{-}$, protons, ${}^{1}_{1}H$, neutrons, n, positrons, ${}^{0}_{+1}e^{+}$ and gamma radiation, γ. The ionizing radiation may be used diagnostically or used to destroy cancer cells. Much of it is high-energy radiation and has the potential to cause harm as well as do good and the dosage and its use need to be carefully controlled. One notable exception to this is magnetic resonance imaging. The patient is placed inside a large magnetic field and very low energy radio waves are used to determine the environment of particular nuclei, e.g. ${}^{1}H$ or ${}^{31}P$, in the body to give a detailed image. This is a practical application of NMR spectroscopy (see page 105) and is non-invasive.

Side effects from radiotherapy can vary from person to person and depend upon the type of radiotherapy received. External radiotherapy tends to cause more side effects than the more targeted internal therapy. Hair loss may occur although, unlike chemotherapy, this is usually limited to the treatment area where the radiation enters and leaves the body. Other short-term effects may include nausea, fatigue, loss of appetite and skin disorders. Radiation may cause damage to the DNA in surrounding healthy cells and regenerating tissue and possibly cause some sterility.

DIAGNOSTIC RADIOGRAPHY

Tc-99m is a metastable radioactive isotope of technetium with a radioactive half-life of 6 h and a biological half-life of 1 day (i.e. half of it will be excreted from the body in 24 h) so it is ideal as a tracer. Because of its short half-life, it is generated in hospitals by the beta decay of molybdenum-99. It can be targeted at particular parts of the body and can be used diagnostically, for example to determine how well both kidneys are functioning. Its progress through the body can be followed by the gamma radiation it emits as the energy produced is roughly the same as that detected by conventional X-ray equipment. Because it is a gamma emitter there is a very small risk of tumours developing following exposure but the potential benefits make this a risk worth taking. The decay product, ground state Tc-99 is a beta emitter but the energy released is low as it has a long half-life and the amount in the body rapidly decreases over time.

$$ {}^{99}_{43}Tc \xrightarrow{t_{1/2} = \text{ 6 hours}} {}^{99}_{43}Tc + \gamma \xrightarrow{t_{1/2} = \text{ 211,000 years}} {}^{99}_{44}Ru + {}^{0}_{-1}e^{-} $$

Another radioisotope that can be used as a tracer in diagnostic medicine is iodine-131 which is a beta and gamma emitter with a half-life of eight days. However it remains in the body for some time and its use can lead to thyroid cancer developing so it tends to be used mainly in high doses to kill thyroid tumours (where it becomes incorporated in thyroxin) rather than diagnostically.

TARGETED THERAPY

Targeted alpha therapy (TAT)

TAT is used to direct radiation directly at cancerous cells and in particular to prevent micrometastatic cancerous cells (secondary tumour formation) in other parts of the body. The range of α particles in tissues is typically only 50–100 μm so when targeted at a tumour they have the ability to release a large amount of energy directly at the specific tumour and do much less damage to the surrounding area than beta emitters. A typical isotope used for TAT is lead-212.

Pb-212 has a half-life of 10.6 hours. It initially decomposes by beta emissions to give bismuth-212 which then rapidly decomposes by alpha-emission to give Tl-208.

$$ {}^{212}_{82}Pb \rightarrow {}^{212}_{83}Bi \rightarrow {}^{208}_{81}Tl $$

Boron neutron capture therapy (BNCT)

BNCT is used to treat brain tumours and recurrent head and neck cancers. Initially the patient is injected with a non-radioactive isotope of boron, ${}^{10}_{5}B$, which is selectively absorbed by the cancer cells. The patient is then irradiated with neutrons, which are captured by the boron-10. This causes a nuclear reaction and the boron is converted into an isotope of lithium, ${}^{7}_{3}Li$ with the emission of a high energy alpha particle that acts directly on the tumour while doing relatively little damage to the surrounding area.

$$ {}^{10}_{5}B + {}^{1}_{0}n \rightarrow {}^{7}_{3}Li + {}^{4}_{2}He $$

Other targeted therapy

Other examples of radioisotopes used for targeting radiation to specific areas are lutetium-177 and yttrium-90. Yttrium-90 is a beta emitter whereas lutetium-177 is both a gamma and beta emitter. Lutetium-177 has a shorter penetration range than yttrium-90 making it ideal for smaller tumours. It is added to a carrier called DOTA-TATE, which attaches to specific tumours such as neuroendocrine tumours and certain types of thyroid tumours.

HALF-LIFE CALCULATIONS

Radioactive nuclei have constant half-lives so the equation $N = N_o e^{-\lambda t}$ can be used where N is the mass or amount remaining after time t, N_o is the initial mass or amount and λ is the decay constant. Information on half-life can be found on page 144.

Worked example

Technetium-99m has a half-life of 6.01 hours. A fresh sample containing 5.00 g of technicium-99m is prepared from molybdenum-99 in a hospital at 0900 h. Calculate the mass of technetium-99m remaining in the 5.00 g sample at 1700 h later that day.

Step 1. Use $t_{1/2}$ to find the rate constant from the equation

$$ \lambda t_{1/2} = 0.693 $$

$$ \lambda = \frac{0.693}{6.01} = 0.115\text{ hour}^{-1} $$

Step 2. Insert the value for λ into the integrated form of the rate equation $\lambda t = \ln \frac{[N_o]}{[N]}$.

$0.115 \times 8.00 = \ln\left(\frac{5.00}{N}\right)$ where N = mass remaining after 8.00 hours

$\ln\left(\frac{5.00}{N}\right) = 0.92$ so $N = \frac{5.00}{2.51} = 1.99$ g

1.99 g of technetium-99m will be remaining.

Drug detection and analysis (1)

IDENTIFICATION OF DRUGS USING SPECTROSCOPY

This topic builds on what has already been covered in pages 102–106. Remember that mass spectroscopy can be used to identify not only the molar mass but also individual fragments with particular masses. Infrared spectroscopy can identify specific functional groups as these absorb in different regions of the infrared spectrum with the precise absorption depending upon neighbouring atoms. Proton NMR spectroscopy gives different peaks for hydrogen atoms that are in different chemical environments. The area under each peak is related to the number of hydrogen atoms in that particular environment and the splitting pattern is determined by the protons bonded directly to the adjacent carbon atoms. These techniques are now applied to three particular drugs, the mild analgesics aspirin, paracetamol and ibuprofen.

Mass spectroscopy

All three compounds have different molar masses so we would expect to see molecular ion peaks, M^+, with m/z values of 180 (aspirin), 151(paracetamol) and 206 (ibuprofen).

All three compounds contain methyl groups so we would expect peaks corresponding to $(M - 15)^+$ for all three compounds.

Only aspirin and ibuprofen contain a carboxylic acid group so they will give peaks due to loss of –COOH at $(M - 45)^+$ Aspirin contains an ethanoate ester group so we would expect a peak at (M – 59), i.e. with an m/z value of 121, due to loss of $-OCOCH_3$. Ibuprofen contains an isobutyl group, $-CH_2CH(CH_3)_2$, so we would expect a peak at $(M - 57)^+$, i.e. with an m/z value of 149. This logic can be extended to accommodate other fragments for example paracetamol would be expected to show a peak with an m/z value of 93 $(M - 58)^+$ due to loss of the $-NHCOCH_3$ group.

Infrared spectroscopy

All three compounds will show a characteristic spectrum in the 'fingerprint region' between 1400–400 cm^{-1}. Aspirin and paracetamol will give a strong very broad absorption between 2500–3300 cm^{-1} due to the COOH functional group whereas paracetamol will show a slightly less broad peak due to the OH group at 3230–3500 cm^{-1}. This broad peak may mask the peak due to the N–H absorption as it occurs in a similar region. Perhaps the easiest way to distinguish them is from the position of their carbonyl peaks. C=O for aldehydes, ketones, carboxylic acids and esters normally appears in the region of 1700–1750 cm^{-1}. Aspirin is the only one to show two carbonyl peaks as it contains two C=O groups. The ibuprofen carboxylic acid C=O absorption occurs at 1721 cm^{-1} and the C=O in the amide functional group in paracetamol absorbs at around 1650 cm^{-1}.

^1H NMR spectroscopy

This provides a very quick and easy way to distinguish between the three mild analgesics.

All will show a complex pattern with an area corresponding to four protons at about 7 ppm due to the hydrogen atoms attached to the phenyl group. Aspirin will just show two other peaks with areas corresponding to one and three hydrogen atoms respectively. The methyl group will appear as a singlet as there are no hydrogen atoms on adjacent carbon atoms and the carboxylic acid proton will have a large shift and also be a singlet.

Other than the peak due to the four phenyl protons, paracetamol will show three peaks, all singlets, in the ratio of 1:1:3 due to the phenol proton, the proton attached directly to the nitrogen atom and the methyl group respectively.

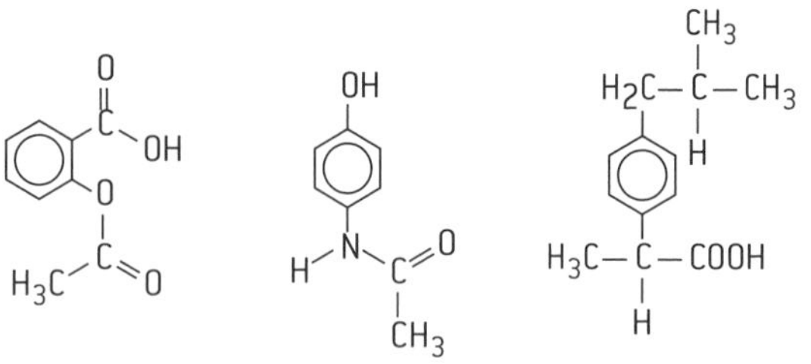

aspirin paracetamol ibuprofen

Ibuprofen is the most interesting. The spectrum gives the following information.

^1H NMR spectrum of ibuprofen

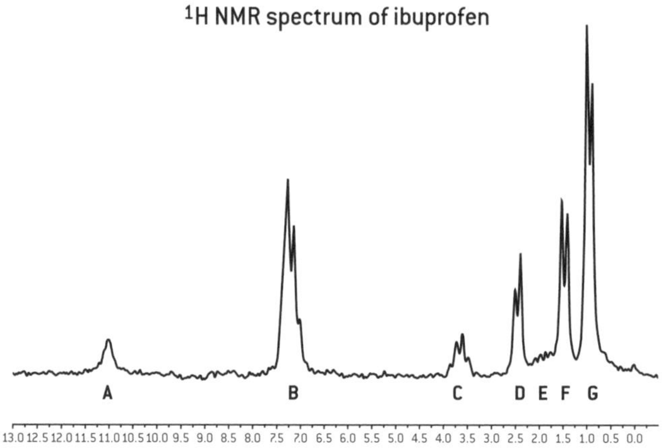

Peak	Chemical shift / ppm	Integration trace	Splitting pattern
A	11.0	1	singlet
B	7.1	4	complex
C	3.7	1	quartet
D	2.5	2	doublet
E	1.8	1	complex
F	1.5	3	doublet
G	0.8	6	doublet

The singlet with the largest shift at 11.0 ppm is due to the carboxylic acid proton. The four protons centred at 7.1 ppm are the ones attached to the phenyl ring. The quartet at 3.7 ppm must be due to the proton next to the –COOH group as it is split by the adjacent methyl protons. The doublet at 2.5 ppm must be due to the $-CH_2$ group attached to the benzene ring which is split by the one hydrogen atom on an adjacent carbon atom. The complex singlet at 1.8 ppm is due to the C–H hydrogen atom surrounded by two adjacent methyl groups and a $-CH_2$ group, which cause considerable splitting. The doublet at 1.5 ppm with an integration trace of three must be due to the three protons in the $-CH_3$ group on the carbon atom attached to the carboxylic acid group as it is split by the single hydrogen atom on the adjacent carbon atom. The doublet at 0.8 ppm with an integration trace of six must be due to the six hydrogen atoms in the two methyl groups attached to the same carbon atom.

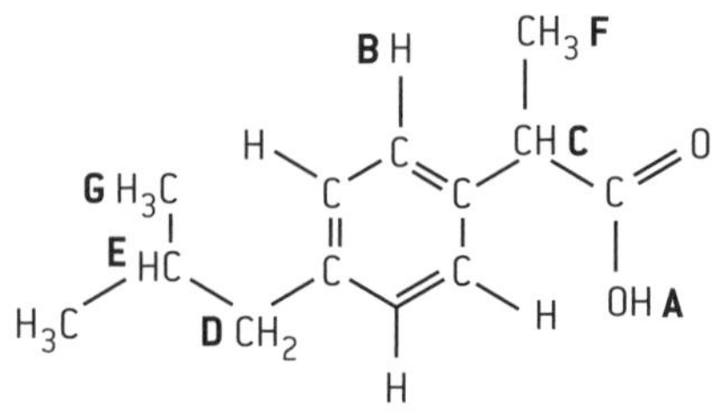

Drug detection and analysis (2)

PURIFICATION OF ORGANIC PRODUCTS

Spectroscopic identification can only be easily carried out on pure samples. There are several methods that can be used to obtain pure components from a mixture. For small amounts, chromatography can be used to separate and identify complex mixtures both quantitatively and qualitatively. It can also be used to determine how pure a substance is. There are several different types of chromatography. They include paper, thin layer (TLC), column (LC), gas–liquid (GLC) and high performance liquid chromatography (HPLC). In each case there are two phases: a **stationary** phase that stays fixed and a **mobile** phase that moves. Chromatography relies upon the fact that in a mixture, the components have different tendencies to adsorb onto a surface or dissolve in a solvent. This provides the means for separating them.

Techniques to extract larger amounts of a component of a mixture depend on the type of mixture. Solutes can be separated from solutions by recrystallization and solvents can be separated from solutions by distillation. For a mixture of liquids fractional distillation can be used. This is a practical application of Raoult's law. Raoult's law states that for an ideal solution the vapour pressure of each component equals the vapour pressure of the pure component multiplied by its mole fraction, where the mole fraction is the number of moles of the component divided by the total number of moles present. That is, for a solution containing two components A and B

$$P = P_A X_A + P_B X_B$$

where X_A and X_B are the mole fractions of A and B respectively and P is the vapour pressure above the solution.

So that when a solution boils, the vapour will be richer than the solution in the more volatile component. If component A is more volatile and this vapour is condensed and then the liquid formed boiled and condensed a large number of times. Then each time the vapour will become richer in A. This is what happens on the fractionating column to produce eventually A, the more volatile component, in a pure form.

Other methods of distillation include steam distillation – a technique useful for separating higher boiling point oils from plants by distilling them below their normal boiling point – and vacuum distillation, used when organic compounds decompose below their normal boiling point. Solvent extraction can be used for compounds that distribute between two solvents. This depends upon their relative solubilities between two immiscible liquids such as water and an organic solvent. The substance is extracted into the layer in which it is most soluble and the layers separated using a separating funnel.

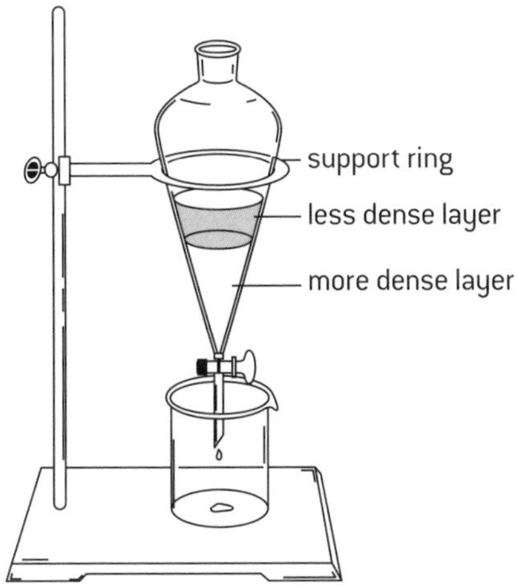

TESTING FOR DRUGS IN ATHLETES

The taking of anabolic steroids can increase muscle and enhance athletic performance as they have a similar effect to testosterone. As spectroscopic techniques have become more refined, even traces of illegal substances remaining in the body a long time after they were taken can be detected and the testing of athletes for the presence of anabolic steroids is routinely carried out. Analysis of urine samples from athletes by GC-MS detects the presence of steroids up to five days after their use but more modern techniques can also identify the long term metabolites (breakdown products) of steroids present in the urine in very low concentration, which persist for much longer than the steroids themselves.

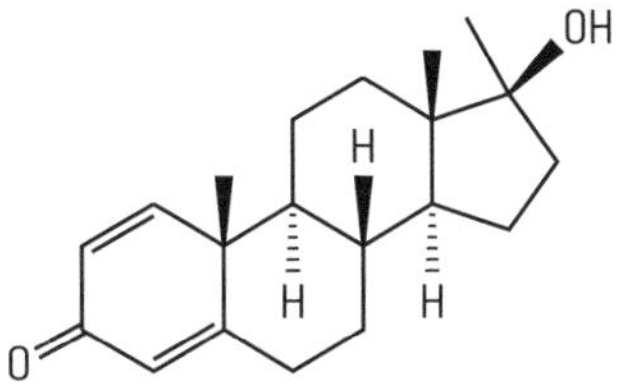

Methandrostenolone – a synthetic anabolic steroid taken by body builders

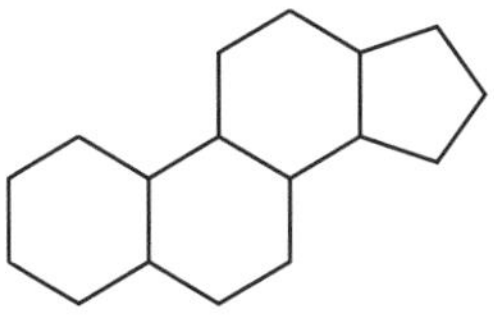

The four-ring structure common to all steroids

GC-MS

This method, which combines gas-liquid chromatography with mass spectrotrometry, is used to separate and identify mixtures of volatile liquids that do not decompose at temperatures at or near their boiling points. The stationary phase consists of a liquid (e.g. a long-chain alkane) coated onto a solid support in a long, thin capillary tube. The mobile phase is an inert gas such as nitrogen or helium. The sample is injected through a self-sealing cap into an oven for vaporization. The sample is then carried by the inert gas into the column, which is coiled and fitted into an oven. At the end of the column the separated components exit into the mass spectrometer.

The mass spectrometer is connected to a computer, which contains a library of the spectra of all known compounds. The computer matches the spectra and gives a print out of all the separate components and their concentrations. A similar technique (LC-MS) combines high performance liquid chromatography with mass spectrometry.

ALCOHOL BREATHALYSER

At the roadside, a motorist may be asked to blow into a breathalyser. This may involve acidified potassium (or sodium) dichromate(VI) crystals turning green as they are reduced by the alcohol to Cr^{3+} as the ethanol is oxidized. Modern hand-held intoximeters also involve a redox reaction using fuel cell technology. The fuel cell is a porous disk with platinum on both sides saturated with an electrolyte. When the breath sample passes across the cell the alcohol is converted into ethanoic acid releasing electrons.

$$C_2H_5OH + H_2O \rightarrow CH_3COOH + 4H^+ + 4e^-$$

The breath alcohol concentration is determined from the amount of the current produced.

SHORT ANSWER QUESTIONS – OPTION D – MEDICINAL CHEMISTRY

1. a) Outline why there is a different way of calculating the therapeutic index for humans than for animals. [1]

   b) Distinguish between *therapeutic window* and *therapeutic index* for humans. [2]

   c) Examine the concept of *side effects* using morphine as an example. Your answer should include two reasons why morphine is administered and two different side effects. [3]

   d) Explain the difference between tolerance and addiction. [2]

2. The general structure of penicillin is given below.

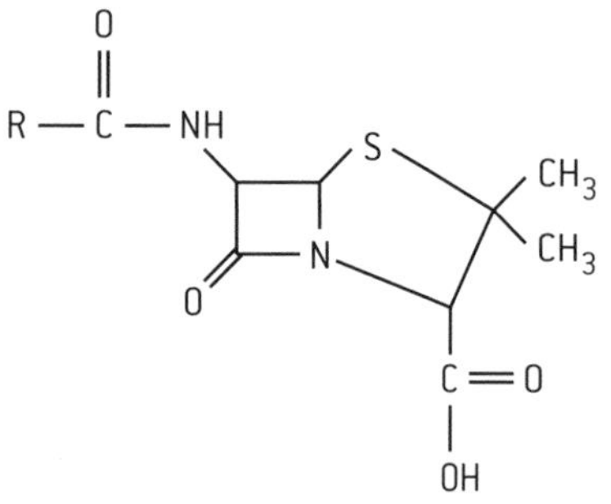

   a) The letter R in the above structure represents a side chain. State one reason why there are a number of different modifications of this side chain. [1]

   b) State why a prescribed course of penicillin should be completed. [1]

   c) Apart from the cost, what is an effect of the over-prescription of penicillins? [1]

   d) Explain the importance of the beta-lactam group in penicillin's ability to destroy bacteria. [3]

3. The mild analgesic aspirin can be prepared from reacting 2-hydroxybenzoic acid with either ethanoic anhydride or with ethanoyl chloride. The balanced equation for the reaction with ethanoyl chloride is:

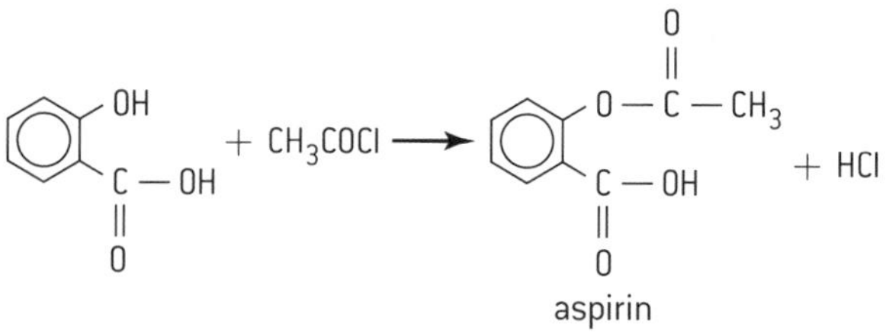

   a) Identify the type of reaction that is taking place. [1]

   b) Predict how the infrared spectrum of aspirin will differ to the infrared spectrum of 2-hydroxybenzoic acid in the region of 1700–1750 cm^{-1}. [2]

   c) Paracetamol can also be prepared by reacting a simple compound with ethanoyl chloride. Draw the structure of this compound. (The structure of paracetamol is shown in the IB data booklet.) [2]

   d) Both aspirin and paracetamol are important mild analgesics, but like all drugs they have some disadvantages. State **one** disadvantage in the use of each drug. [2]

4. a) The structures of oseltamivir and zanamivir are given in section 37 of the IB data booklet.

   (i) State the class of drugs to which they belong and state one illness they are used to treat. [2]

   (ii) Identify which one will be more soluble in water and explain why. [2]

   b) The structure of esomeprazole is also given in the IB data booklet.

   (i) State the class of drugs to which it belongs and state one condition it is used to treat. [2]

   (ii) Distinguish between the structures of esomeprazole and omeprazole. [1]

HL

5. a) Explain why it is important to carry out clinical trials on all the different enantiomers of a new drug. [2]

   b) Most reactions to form chiral compounds give a racemic mixture which then has to be separated into the two different enantiomers. Describe how a chiral auxiliary can be used to isolate the desired enantiomer of a particular drug. [3]

   c) The anti-cancer drug taxol can be synthesized using chiral auxiliaries. Part of its structure is shown below. Identify with an asterisk * two chiral centres.

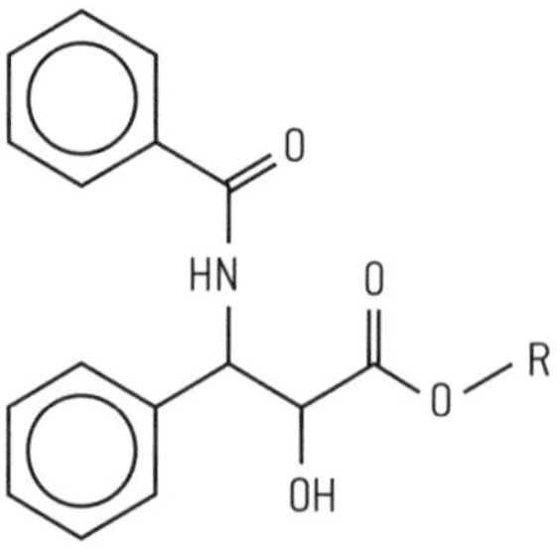

6. a) The radioactive isotope lead-212 is used in targeted alpha therapy (TAT).

   (i) Deduce the nuclear equation for the breakdown of lead-212 to form alpha particles. [2]

   (ii) Distinguish between targeted alpha therapy and boron neutron capture therapy (BNCT). [3]

   b) Iodine-131 is used both diagnostically and to treat tumours. It has a half-life of 8.02 days.

   (i) Deduce the nuclear equation for the disintegration of iodine-131 to form beta radiation. [2]

   (ii) Calculate the percentage of iodine-131 remaining in a sample of pure iodine-131 after 31.0 days has elapsed. [2]

   (iii) State **two** reasons why technetium-99m is often preferred to iodine-131 as a diagnostic isotope. [2]

Introduction

RELATIONSHIP TO THE SYLLABUS

The full IB Chemistry Guide is published by the International Baccalaureate and you should ensure that your teacher provides you with a copy of the parts that contain the detailed syllabus content relating to your specific course. For Standard level you need the Core and the Core topics of the one Option that you are studying. For Higher Level you require the Core, Additional Higher Level material (AHL) and the Core and AHL topics for the one Option you are studying. Chapters 1 to 11 of this book cover all you need for the chemistry content of the Core and AHL material and Chapters 12–15 cover all you need for the chemistry content of the four options.

If you look at each sub-topic then you will see that the chemistry content is listed under several different headings. This is because good teaching of IB Chemistry should be done holistically. It relates the chemical theory to wider issues including the underlying philosophy as to how the knowledge is obtained and how it relates to society. The different headings are:

Essential idea. This is listed at the beginning of each sub-topic and aims to give the overall interpretation of what the sub-topic is about in terms of the public understanding of science.

Nature of Science. This gives some specific examples of how the topic illustrates some of the specific aspects of the Nature of Science.

Understandings. This covers the main ideas that the teacher needs to cover.

Applications and skills. This outlines the specific skills and applications that need to be developed from the understandings.

Guidance. This is written basically for teachers and examiners but is also useful to students as it gives helpful information on the depth of treatment required, what examples to use and, in some cases, details of what does not need to be covered.

International-mindedness. This looks at how the topic can relate to the international nature of chemistry with some specific examples, although international-mindedness is not listed for all sub-topics.

Theory of Knowledge. This provides examples that may help you when it comes to writing your TOK essays or giving your TOK presentation.

Utilization. This shows how the topic may be linked to other parts of the syllabus or to other Diploma subjects or to real-life applications.

Aims. There are ten listed aims for all the science subjects and this refers to how one or more of the ten aims can be addressed through studying this sub-topic.

The IB summarizes this in the following table:

Essential idea: This lists the Essential idea for each sub-topic.

1.1 Sub-topic	
Nature of Science: Relates the sub-topic to the overarching theme of Nature of Science.	
Understandings: • This section will provide specifics of the content requirements for each sub-topic. **Applications and skills:** • The content of this section gives details of how students are to apply the understandings. For example, these applications could involve demonstrating mathematical calculations or practical skills. **Guidance:** • This section will provide specifics and give constraints to the requirements for the understandings and applications and skills. • This section will also include links to specific sections in the data booklet.	**International-mindedness:** • Ideas that teachers can easily integrate into the delivery of their lessons. **Theory of knowledge:** • Examples of TOK knowledge questions. **Utilization:** (including syllabus and cross-curricular links) • Links to other topics within the chemistry guide, to a variety of real-world applications and to other Diploma Programme courses. **Aims:** • Links to the group 4 subjects aims.

Although it can all look quite daunting, the Guide is written for teachers and is to help them to deliver an interesting and educationally sound course. It is important to realize that the Diploma Guide provides guidance as to how the whole course should be delivered not just on what you will be examined on. In order to be fully prepared for both the internal and external assessment, you need to know that the exam tests what is listed under 'Understandings', 'Applications and skills' and 'Guidance'. This is what has been covered in Chapters 1–15 and if you have a sound understanding and knowledge of this content you will achieve a high score in the exams. It can also test the Nature of Science and international-mindedness but only in a very small way. Only about 1–3% of the exams will be testing these specifically. Some of the content under these two headings has been included in the relevant places in Chapters 1–15 and in some of the questions. It has not been highlighted specifically as there is not room in a book of this size to address everything under these two headings thoroughly. The Nature of Science and international-mindedness sections that follow in this chapter provide some general information and some specific examples that should enable you to respond to the very small number of questions on these two particular issues. Note that Essential ideas, TOK, Utilizations and Aims are not examined. However the Essential ideas and Utilization sections can be used to review what you have already learned.

Essential ideas (1)

ESSENTIAL IDEAS

The exam does not test the essential ideas as such, just the chemistry behind them. It can be a useful exercise to look at the essential ideas listed for each topic and check that you can relate the chemistry covered in these topics to each specific idea.

Stoichiometric relationships

Physical and chemical properties depend on the ways in which different atoms combine.

The mole makes it possible to correlate the number of particles with the mass that can be measured.

Mole ratios in chemical equations can be used to calculate reacting ratios by mass and gas volume.

Atomic structure

The mass of an atom is concentrated in its minute, positively charged nucleus.

The electron configuration of an atom can be deduced from its atomic number.

The quantized nature of energy transitions is related to the energy states of electrons in atoms and molecules.

Periodicity

The arrangement of elements in the periodic table helps to predict their electron configuration.

Elements show trends in their physical and chemical properties across periods and down groups.

The transition elements have characteristic properties; these properties are related to their all having incomplete d sub-levels. d-orbitals have the same energy in an isolated atom, but split into two sub-levels in a complex ion. The electric field of ligands may cause the d-orbitals in complex ions to split so that the energy of an electron transition between them corresponds to a photon of visible light.

Chemical bonding and structure

Ionic compounds consist of ions held together in lattice structures by ionic bonds.

Covalent compounds form by the sharing of electrons.

Lewis (electron dot) structures show the electron domains in the valence shell and are used to predict molecular shape.

The physical properties of molecular substances result from different types of forces between their molecules.

Metallic bonds involve a lattice of cations with delocalized electrons.

Larger structures and more in-depth explanations of bonding systems often require more sophisticated concepts and theories of bonding. Hybridization results from the mixing of atomic orbitals to form the same number of new equivalent hybrid orbitals that can have the same mean energy as the contributing atomic orbitals.

Energetics/thermochemistry

The enthalpy changes from chemical reactions can be calculated from their effect on the temperature of their surroundings. In chemical transformations energy can neither be created nor destroyed (the first law of thermodynamics). Energy is absorbed when bonds are broken and is released when bonds are formed.

The concept of the energy change in a single step reaction being equivalent to the summation of smaller steps can be applied to changes involving ionic compounds. A reaction is spontaneous if the overall transformation leads to an increase in total entropy (system plus surroundings). The direction of spontaneous change always increases the total entropy of the Universe at the expense of energy available to do useful work. This is known as the second law of thermodynamics.

Chemical kinetics

The greater the probability that molecules will collide with sufficient energy and proper orientation, the higher the rate of reaction.

Rate expressions can only be determined empirically and these limit possible reaction mechanisms. In particular cases, such as a linear chain of elementary reactions, no equilibria and only one significant activation barrier, the rate equation is equivalent to the slowest step of the reaction. The activation energy of a reaction can be determined from the effect of temperature on reaction rate.

Equilibrium

Many reactions are reversible. These reactions will reach a state of equilibrium when the rates of the forward and reverse reaction are equal. The position of equilibrium can be controlled by changing the conditions.

The position of equilibrium can be quantified by the equilibrium law. The equilibrium constant for a particular reaction only depends on the temperature.

Acids and bases

Many reactions involve the transfer of a proton from an acid to a base. The characterization of an acid depends on empirical evidence such as the production of gases in reactions with metals, the colour changes of indicators or the release of heat in reactions with metal oxides and hydroxides.

The pH scale is an artificial scale used to distinguish between acid, neutral and basic/alkaline solutions. The pH depends on the concentration of the solution. The strength of acids or bases depends on the extent to which they dissociate in aqueous solution.

Increased industrialization has led to greater production of nitrogen and sulfur oxides leading to acid rain, which is damaging our environment. These problems can be reduced through collaboration with national and intergovernmental organizations.

The acid–base concept can be extended to reactions that do not involve proton transfer. The equilibrium law can be applied to acid–base reactions. Numerical problems can be simplified by making assumptions about the relative concentrations of the species involved. The use of logarithms is also significant here.

pH curves can be investigated experimentally but are mathematically determined by the dissociation constants of the acid and base. An indicator with an appropriate end point can be used to determine the equivalence point of the reaction.

Redox processes

Redox (reduction–oxidation) reactions play a key role in many chemical and biochemical processes.

Voltaic cells convert chemical energy to electrical energy and electrolytic cells convert electrical energy to chemical energy.

Energy conversions between electrical and chemical energy lie at the core of electrochemical cells.

Essential ideas (2)

ESSENTIAL IDEAS (CONT.)

Organic chemistry

Organic chemistry focuses on the chemistry of compounds containing carbon. Structure, bonding and chemical reactions involving functional group interconversions are key strands in organic chemistry.

Key organic reaction types include nucleophilic substitution, electrophilic addition, electrophilic substitution and redox reactions. Reaction mechanisms vary and help in understanding the different types of reaction taking place. Organic synthesis is the systematic preparation of a compound from a widely available starting material or the synthesis of a compound via a synthetic route that often can involve a series of different steps. Stereoisomerism involves isomers that have different arrangements of atoms in space but do not differ in connectivity or bond multiplicity (i.e. whether single, double or triple) between the isomers themselves.

Measurement, data processing and analysis

All measurement has a limit of precision and accuracy, and this must be taken into account when evaluating experimental results. Graphs are a visual representation of trends in data. Analytical techniques can be used to determine the structure of a compound, analyse the composition of a substance or determine the purity of a compound. Spectroscopic techniques are used in the structural identification of organic and inorganic compounds.

Although spectroscopic characterization techniques form the backbone of structural identification of compounds, typically no one technique results in a full structural identification of a molecule.

Option A – Materials

Materials science involves understanding the properties of a material, and then applying those properties to desired structures. Metals can be extracted from their ores and alloyed for desired characteristics. ICP-MS/OES spectroscopy ionizes metals and uses mass and emission spectra for analysis. Catalysts work by providing an alternate reaction pathway for the reaction. Catalysts always increase the rate of the reaction and are left unchanged at the end of the reaction. Liquid crystals are fluids that have physical properties that are dependent on molecular orientation relative to some fixed axis in the material. Polymers are made up of repeating monomer units, which can be manipulated in various ways to give structures with desired properties. Chemical techniques position atoms in molecules using chemical reactions while physical techniques allow atoms/molecules to be manipulated and positioned to specific requirements. Although materials science generates many useful new products there are challenges associated with recycling of and high levels of toxicity of some of these materials.

Superconductivity is zero electrical resistance and expulsion of magnetic fields. X-ray crystallography can be used to analyse structures. Condensation polymers are formed by the loss of small molecules as functional groups from monomers join. Toxicity and carcinogenic properties of heavy metals are the result of their ability to form coordinated compounds, have various oxidation states and act as catalysts in the human body.

Option B – Biochemistry

Metabolic reactions involve a complex interplay between many different components in highly controlled environments. Proteins are the most diverse of the biopolymers responsible for metabolism and structural integrity of living organisms. Lipids are a broad group of biomolecules that are largely non-polar and therefore insoluble in water. Carbohydrates are oxygen-rich biomolecules that play a central role in metabolic reactions of energy transfer. Vitamins are organic micro-nutrients with diverse functions that must be obtained from the diet. Our increasing knowledge of biochemistry has led to several environmental problems, while also helping to solve others.

Analyses of protein activity and concentration are key areas of biochemical research. DNA is the genetic material that expresses itself by controlling the synthesis of proteins by the cell. Biological pigments include a variety of chemical structures with diverse functions which absorb specific wavelengths of light. Most biochemical processes are stereospecific and involve only molecules with certain configurations of chiral carbon atoms.

Option C – Energy

Societies are completely dependent on energy resources. The quantity of energy is conserved in any conversion but the quality is degraded. The energy of fossil fuels originates from solar energy, which has been stored by chemical processes over time. These abundant resources are non-renewable but provide large amounts of energy due to the nature of chemical bonds in hydrocarbons. The fusion of hydrogen nuclei in the Sun is the source of much of the energy needed for life on Earth. There are many technological challenges in replicating this process on Earth but it would offer a rich source of energy. Fission involves the splitting of a large unstable nucleus into smaller stable nuclei. Visible light can be absorbed by molecules that have a conjugated structure with an extended system of alternating single and multiple bonds. Solar energy can be converted to chemical energy in photosynthesis. Gases in the atmosphere that are produced by human activities are changing the climate as they are upsetting the balance between radiation entering and leaving the atmosphere.

Chemical energy from redox reactions can be used as a portable source of electrical energy. Large quantities of energy can be obtained from small quantities of matter. When solar energy is converted to electrical energy, the light must be absorbed and charges must be separated. In a photovoltaic cell, both of these processes occur in the silicon semiconductor, whereas these processes occur in separate locations in a dye-sensitized solar cell (DSSC).

Option D – Medicinal chemistry

Medicines and drugs have a variety of different effects on the functioning of the body. Natural products with useful medicinal properties can be chemically altered to produce more potent or safer medicines. Potent medical drugs prepared by chemical modification of natural products can be addictive and become substances of abuse. Excess stomach acid is a common problem that can be alleviated by compounds that increase the stomach pH by neutralizing or reducing its secretion. Antiviral medications have recently been developed for some viral infections while others are still being researched. The synthesis, isolation and administration of medications can have an effect on the environment.

Chiral auxiliaries allow the production of individual enantiomers of chiral molecules.Nuclear radiation, while dangerous owing to its ability to damage cells and cause mutations, can also be used to both diagnose and cure diseases. A variety of analytical techniques is used for detection, identification, isolation and analysis of medicines and drugs.

Nature of Science (1)

NATURE OF SCIENCE AND THEORY OF KNOWLEDGE

The Theory of Knowledge course considers eight different ways of knowing: reason, emotion, language, sense perception, intuition, imagination, faith and memory. In TOK classes you examine the strengths and weakness of four of these ways and apply them to different areas of knowledge – one of which is the Natural Sciences. Put simply, the Nature of Science is this particular part of TOK, i.e. TOK as applied to the area of knowledge that is Natural Science and, for this course, Chemistry in particular. In the IB Chemistry Guide (which you can obtain from your teacher) there is a seven page statement describing the Nature of Science. It is well worth reading this as it could be helpful to you when you come to write your TOK essay or give your TOK seminar. Essentially it covers five key points and lists sub-headings under each of these five points.

1. What is science and what is the scientific endeavour?
2. The understanding of science
3. The objectivity of science
4. The human face of science
5. Scientific literacy and the public understanding of science

As you go through the two-year chemistry course, it is instructive to see how specific chemistry topics relate to these points. Different sub-topics exemplify different aspects of these five points and some examples are given below to illustrate these. If you know and understand these you will be in a good position to answer the very few questions involving the Nature of Science on the chemistry exam papers. You will also be in a strong position when it comes to writing your TOK essay. This is because TOK examiners tend to give credit for relevant specific examples taken from subjects you study to back up your arguments. Too often they see students using the typical examples provided by TOK class teachers who of course cannot be experts in all the different disciplines you study.

EXAMPLES OF SOME KEY ASPECTS

Changing theories. Theories change to accommodate new information and understanding as chemistry develops. The different theories of acids provide a good example of this. In Roman times, acids were defined as sour substances, then as the oxide of a non-metal in water by Lavoisier in the 1780s, then as a substance that donates protons by Lowry–Brønsted in 1923 before being defined as a substance that accepts a pair of electrons by Lewis also in 1923. An even later definition by Usanovich in the 1930s defines an acid as a substance that accepts negative species or donates positive species.

Occam's razor. This basically states that simple explanations are, other things being equal, generally better than more complex ones. For example, collision theory is very simple but based on models of reacting species it explains kinetic theory and the factors affecting the rate of chemical reactions. Another example is the pH scale, which is a simple way of dealing with very small hydrogen ion concentrations.

Falsification. Popper in the 1950s maintained that it is impossible to prove something by doing an experiment as you would need to do an infinite number of experiments to cover every possible permutation. However if you try to disprove a theory you only need one successful experiment to disprove it. He maintained that a theory is only scientific if it is capable of being falsified. $pV = nRT$ is only true for an ideal gas, real gases do occupy some volume and do have some weak intermolecular forces of attraction. Lavoisier's theory of acids was falsified when scientists realized that HCN and HCl do not contain oxygen.

Paradigm shift. Thomas Kuhn proposed that scientific progress works through paradigms. This is an established model accepted by the scientific community. As more becomes known the paradigm has to accommodate the new knowledge. Eventually it becomes unwieldy and a new model becomes accepted – a paradigm shift. The classic old example in chemistry is phlogiston – a substance that was given off by everything when it burned. Even when Priestly discovered oxygen he called it 'dephlogisticated air' as substances readily gave up phlogiston to it. It took the genius of Lavoisier to explain that combustion occurs when substances combine with oxygen rather than give off a substance. A more modern paradigm in chemistry is simple covalent bonding using the octet rule. Sharing a pair of electrons (one from each atom) to form a share in an inert gas structure works for simple substances such as hydrogen, methane and carbon dioxide. It has to be stretched to include coordinate bonding and resonance hybrids but cannot explain substances such as SF_6 or NO. A new paradigm is molecular orbital theory or valence bond theory, which takes account of the fact that electrons are in different orbitals and energy levels and also explains why diatomic oxygen is paramagnetic. Another example of a paradigm shift is the change from the understanding that atoms are indivisible to the paradigm in which they can be broken into many different sub-atomic particles.

Serendipity. The accidental discovery of something useful when not looking for it. Legend has it that glass was discovered when Phoenician sailors cooked a meal on a sandy beach. Liquid crystals were discovered by Friedrich Reinitzer when he was doing experiments with cholesteryl benzoate, William Perkin discovered azo-dyes when he was trying to synthesize quinine and Alexander Fleming is credited with discovering penicillin when he was working with staphylococci bacteria. Other examples include the discovery of Teflon (poly(tetrafluoroethene)) and superglue.

Predictions. A good scientific theory enables scientists to make predictions. For example, theory explains that the lines in the visible emission spectrum of hydrogen are due to excited electrons dropping from higher energy levels to the $n = 2$ level. If this is true, one can predict that there should be another series of lines at higher energy corresponding to electrons dropping to the lower $n = 1$ level. This cannot be seen by the naked eye but the series is there if you use an ultraviolet spectrometer. Similarly there are more series at lower energy in the infrared region due to electrons dropping to the $n = 3$ and $n = 4$ levels, etc.

Nature of Science (2)

EXAMPLES OF SOME KEY ASPECTS (CONT.)

Models. Chemists use a variety of molecular models to represent the structure of molecules (e.g. ball and stick, space filling) and more sophisticated computer modelling to represent systems where there are many different variables such as climate change. Many potential drugs are first made virtually and modelled to see whether they might be effective before synthesizing those which look as if they may have some potential to test *in vitro* and then *in vivo*.

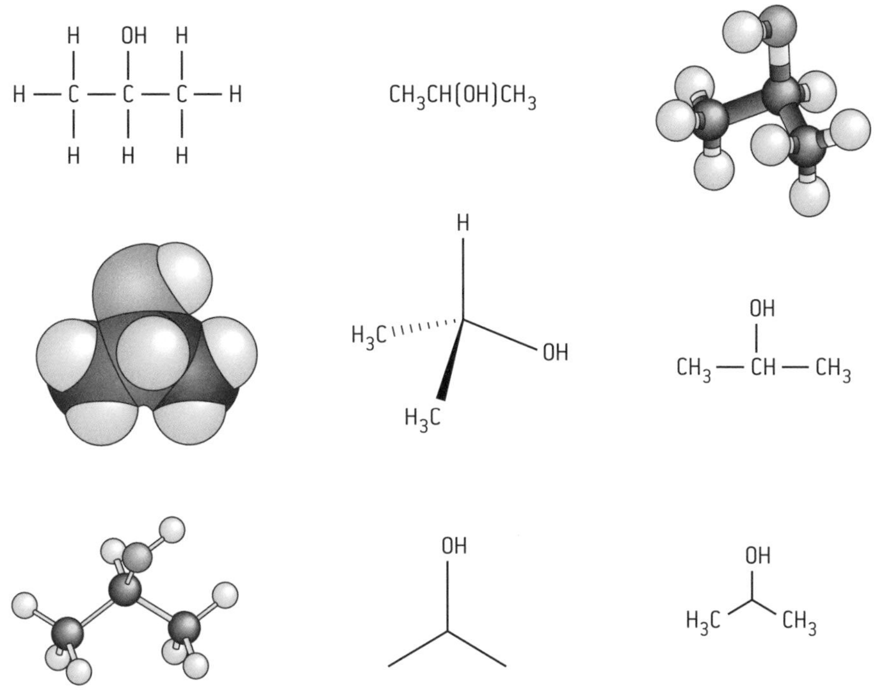

Some of the different models that can be used to represent the structure of propan-2-ol

Use of concepts. Chemists often use concepts to work out values that cannot be determined directly. Energy cycles are a good example of this because they are based on the first law of thermodynamics – energy can neither be created nor destroyed. It is easy to determine the enthalpies of combustion of carbon, hydrogen and methane practically. It is impossible to determine the enthalpy of formation of methane directly as carbon and hydrogen can react together to form many different compounds. However by using an energy cycle the value for the enthalpy of formation can be readily obtained indirectly.

Assumptions. Sometimes chemistry is based upon assumptions, even though they are not true, in order to provide a useful model. For example to determine the oxidation states of elements in a molecule such as ammonia, you assume it is ionic with the 'anion' being the element, in this case nitrogen, with the higher electronegativity so the oxidation state of nitrogen is -3 and the oxidation state of hydrogen $+1$. For the simple octet rule and to determine shapes using VSEPR theory, it is assumed that all valence electrons are the same; the fact that there are s and p electrons in different energy levels is ignored. Oxidation is commonly defined as the loss of electrons and yet when carbon burns in oxygen neither carbon nor oxygen loses electrons.

Ethical implications. The story of Fritz Haber who won the Nobel Prize for discovering how to fix nitrogen and thus provide artificial fertilizers to feed the world and yet who also worked on chlorine as a poison gas in the First World War is a classic illustration of the ethical problems facing chemists. Energy is needed by society but almost all the ways in which it is generated are also bad for society. Green chemistry is one way in which chemists try to act responsibly towards society and yet still make scientific advances and supply goods such as drugs and pesticides which are helpful to society.

Instrumentation. Modern instrumentation has revolutionized the way chemists work. For example, it is now possible to determine very small quantities of enhancing drugs in athletes who cheat, to catch criminals from DNA residues, and to manipulate the addition of precise amounts of metals during the production of alloys to achieve the desired properties. It has also enabled chemists to quickly and unambiguously assign structures to new compounds.

Language of chemistry. Chemists communicate to each other in very precise language. Sometimes the meaning of a word is different to its use in everyday English. For example, spontaneous in chemistry means that the reaction is able to do useful work, i.e. ΔG has a negative value. In everyday English spontaneous means without preplanning, i.e. 'off the cuff'. Some examples of other words with different meanings to chemists are strong, reduce, degenerate, weak, mole, phase and volatile. Chemists need to be aware that, unlike other chemists, the general public may misinterpret these words if they are used in the chemical sense. A strong drink or a weak acid means something different to a chemist than it does to everyone else. There are other ways in which the language of chemistry is precise. The IUPAC naming of organic compounds follows clearly defined rules and the oxidation numbers of elements is shown using Roman numerals in inorganic compounds, e.g. iron(III) chloride. Other language rules include writing physical constants in italics, e.g. equilibrium constants should be written as K_c or pK_a, and using lower case letters to write the names of elements and compounds in sentences even though their chemical symbols start with a capital letter.

International-mindedness

INTRODUCTION

A view often propagated is that it was the Greeks who first came up with the idea of an element. The four elements being earth, air, fire and water. In times of ancient Greece there was an overland trade route connecting the East to the West called the 'Silk Route'. In addition to goods, knowledge also travelled along this route and some now believe that the idea of elements was first developed in China or India one hundred years before the Greeks. Scientific knowledge has always tended to be freely available and the scientific method incorporates peer-review, open-mindedness and freedom of thought. The Chemistry Guide which discusses the international dimension claims that it transcends politics but one of the reasons why many of the contributions made by Eastern and Arab civilizations in the past have been forgotten, or at least unattributed, is due to the rise of Western culture. As a result of the Cold War, few in the West know the contributions made by many of the scientists in former Soviet Union countries as they tended to be ignored during the second half of the last century.

Chemistry clearly has a very large international dimension. Gases once released (whether polluting or not) do not remain in one country, and oil and other raw feedstocks are moved across oceans as well as across continents. In one way even the elements are international. Not only are some of them distributed unevenly on the planet but they can also have slightly different relative atomic masses depending on their origin. It has been known for a while that the atomic mass of lead can vary as it is a decay product from naturally occurring radioactive isotopes. With the advent of very sophisticated and accurate instrumentation it has been shown that the atomic mass of other elements can also vary by location. In 2010, IUPAC accepted that relative atomic masses for many elements should now be given as a range. For example the A_r of boron is now in the range 10.806 to 10.821 whereas previously (and still is in the IB data booklet) it was given as 10.81. Some examples of international-mindedness in chemistry are given below.

SOME INTERNATIONAL ASPECTS

IUPAC. The International Union of Pure and Applied Chemistry (IUPAC) was formed in 1919. It is an international, non-governmental body consisting of chemists from industry and academia that has the aim of fostering worldwide communication in chemistry.

Currently, about 1000 chemists throughout the world are engaged on a voluntary basis working for IUPAC. One of the greatest achievements of IUPAC has been to bring in a logical and accepted way of naming compounds.

International cooperation. There are many examples of chemists cooperating both with other chemists and with other scientists and international bodies. These include international attempts to deal with climate change. In 1987, the Montreal Protocol set in motion ways to deal with the threat to the ozone layer by banning the use of CFCs. This century the Kyoto Protocol paved the way for combatting global warming by reducing and minimizing the release of greenhouse gases. Other examples of international cooperation include the international space station and CERN in Geneva, both of which involve scientists from many different countries.

International symbols and units. Whatever their mother tongue chemists can communicate with each other through the use of a recognized system of symbols and units. Look at a Periodic Table written in a language unfamiliar to you. You will still be able to use it, as the elements have the same symbols and the units for their properties are also international. Similarly in any text book you will see chemical equations written in the same format. Chemists use the International System of Units (SI) derived from the French **Le Système international d'unités**. It is true that some countries still cling to their traditional units, the UK has miles and pints, the US has Fahrenheit and the French have millilitres but in scientific papers chemists will use metres (m) , kelvin (K) (or still sometimes °C), centimetres cubed (cm^3) and decimetres cubed (dm^3). Gone too are the days when pressure was measured in millimetres of mercury or atmospheres; now pascals (Pa or N m^2) are used internationally.

International legacy. It was Isaac Newton who said 'If I have seen further it is by standing on the shoulders of giants.' Chemists communicate mainly by publishing in peer-reviewed journals. When doing research they will make searches from journals published from around the world, many of which are now online. All chemists recognize the work that others have done or are doing and build upon it. Chemists also build on the international legacy of words. *Acidus* means sour in Latin and the Germans and Norwegians still use the words *sauer* and *syre* respectively for acid. *Alkali* is derived from the Arabic word for calcined ashes and *oxygene* is a Greek word meaning acid-forming.

International wealth. A few countries, such as Bhutan, define their wealth in terms of GNH (Gross National Happiness); most of the others define it in terms of GNP (Gross National Product). A country's wealth depends upon many factors but one of them is the availability of natural resources and the means to exploit them. This can change over time. Many states have grown rich on their oil deposits. Rare-earth metals, which are a very limited resource and spread unevenly around the world, are now in high demand as they are needed in the manufacture of many hi-tech products. They can be difficult to extract and are now being stockpiled or controlled by some countries even though they are used by many. Related to wealth is life-expectancy. Life expectancy has increased dramatically in the developed world due to the contributions from chemists in areas such as the provision of safe drinking water, good sanitation and the availability of effective drugs, pesticides, insecticides and artificial fertilizers.

Utilization

INTRODUCTION

The Utilization or Links section for each sub-topic makes suggestions as to how the topic relates to other topics within Chemistry and to other IB Diploma subjects, and gives examples of real-world applications. This clearly makes a lot of sense as Chemistry is not just an isolated academic subject with no relevance elsewhere. However, it is not something that is examined so it is for interest only. That said, the first reason is worth considering – how does this topic relate to other topics within Chemistry? The IB syllabus is conveniently divided into modules as it is easy and logical to set it out this way. Some other examination boards call these modules and do examine them separately. They are not separate entities though as Chemistry is a holistic subject. Topic 1 on stoichiometric relationships makes no sense if you do not know what an atom is. Organic chemistry depends upon knowledge of bonding, and acids and bases and redox chemistry are really just different aspects of equilibrium. A good teacher will teach holistically rather than in modules. Taking each topic in turn and thinking of actual examples as to how it can relate to all the other ten topics can be a really good way to review your knowledge and understanding at the end of the course before you take your final exams. One example is given below.

A HOLISTIC EXAMPLE

This example places Topic 4 in the centre and gives examples of how the chemistry covered in Topic 4 can be related in turn to all the other ten topics. You can place another one of the ten topics in the centre and make relationships between that and all the other topics and then repeat the exercise until each of the topics has had its turn in the middle. It's done here just for the Core but you can include the AHL as well if you are taking Higher Level.

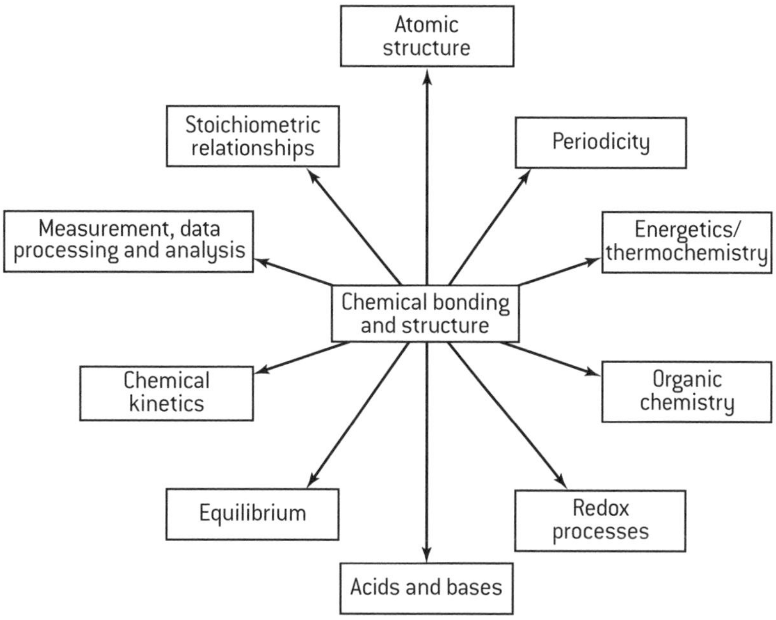

Topic 1. Different types of bonding affect the properties of substances including changes of state (1.1).

Topic 2. Ionic compounds are formed as the result of electron transfer. The electrons that are transferred result in the ions formed having the same electron arrangement as a noble gas (2.2).

Topic 3. In order to predict whether a compound formed between two elements will be covalent or ionic, their position in the periodic table and their electronegativities need to be taken into account (3.1 and 3.2).

Topic 5. To be able to explain the relationship between the number of bonds and bond strength it is necessary to be able to calculate the enthalpy changes associated with breaking and making bonds (5.3).

Topic 6. The rate of a reaction increases as the temperature is increased because the extra energy increases the likelihood of the shared pair of electrons in a covalent bond having the necessary activation energy to split apart when molecules collide (6.1).

Topic 7. Resonance structures such as those shown in benzene and the carbonate anion are not in dynamic equilibrium (7.1) with each other. Resonance structures are extreme forms and the true structure lies between the resonance structures.

Topic 8. When weak acids dissociate in aqueous solution (8.4) a covalent bond needs to be broken to form a hydronium ion, H_3O^+ and the acid radical.

Topic 9. The formation of ionic compounds (such as NaCl or MgO) from their elements involves the transfer of electrons so that it can always be classified as a redox reaction. The species that has lost electrons is oxidized and the species that has gained electrons is reduced (9.1).

Topic 10. Nucleophiles are electron-rich species that contain a non-bonding pair of electrons that they donate to an electron-deficient carbon (10.2). The donated pair of electrons form a coordinate bond.

Topic 11. The relative polarity of bonds can be predicted from electronegativity values. These values are not exact and there are different scales of measurement. Pauling's scale gives the values to one decimal place which for most values is two significant figures (11.1).

Study methods

This book has been written to provide you with all the information you need to gain the highest grade in Chemistry whether at SL or at HL. It is not intended as a 'teach yourself' book and is not a substitute for a good teacher nor for the practical work to support the theory. There is no magic solution that will compensate for a lack of knowledge or understanding but there are some pieces of advice that should ensure that you achieve to the best of your ability.

DURING THE COURSE

The IB course for both SL and HL is scheduled to last for two years, although some schools do attempt to cover the whole course in one year. There is a tendency for some students to take it easy in the first year as the final exams seem a long way off. Don't be tempted to do this as it will be hard to catch up later. Equally do not try to simply learn all the information given about each topic. The exam does not particularly test recall, more how to apply your knowledge in different situations. Although there are some facts that must be learned, much of Chemistry is logical and knowledge about the subject tends to come much more from understanding than from 'rote learning'. During each lesson concentrate on trying to understand the content. A good teacher will encourage you to do this by challenging you to think. At the end of the lesson or in the evening go over your notes, add to them or rearrange them to ensure you have fully understood everything. Read what this book has to say on the subject and read around the topic in other books or on the internet to increase your understanding. If there are parts you do not understand ask your teacher to explain them again. You can also benefit much by talking and working through problems with other students. You will only really know if you understand something if you have to explain it to someone else. You can test your understanding by attempting the problems at the end of each topic in this book.

Some of the early parts of the course involve basic calculations. Some students do find these hard initially. Persevere and see if you can identify exactly what the difficulty is and seek help. Most students find that as the course progresses and more examples are covered their confidence to handle numerical problems increases considerably. If you ensure that you do understand everything during the course then you will find that by the time it comes to the exam, learning the essential facts to support your understanding is much easier.

MATHEMATICAL SKILLS

One big advantage of the IB is that all students study maths so the mathematical skills required for Chemistry should not present a problem. Essentially they concern numeracy rather than complex mathematical techniques. Make sure that you are confident in the following areas.

- Perform basic functions: addition, subtraction, multiplication and division.
- Carry out calculations involving means, decimals, fractions, percentages, ratios, approximations and reciprocals.
- Use standard notation (e.g. 1.8×10^5).
- Use direct and indirect proportion.
- Solve simple algebraic equations.
- Plot graphs (with suitable scales and axes) and sketch graphs.
- Interpret graphs, including the significance of gradients, changes in gradient, intercepts and areas.
- Interpret data presented in various forms (e.g. bar charts, histograms, pie charts etc.).

USING YOUR CALCULATOR

Most calculators are capable of performing functions far beyond the demands of the course. When simple numbers are involved try to solve problems without using your calculator (you will need to do this for real in Paper 1). Even when the numbers are more complex try to estimate approximately what the answer will be before using the calculator. This should help to ensure that you do not accept and use a wrong answer because you failed to realize that you pushed the wrong buttons. Don't just give the 'calculator answer' but record the answer to the correct number of significant figures.

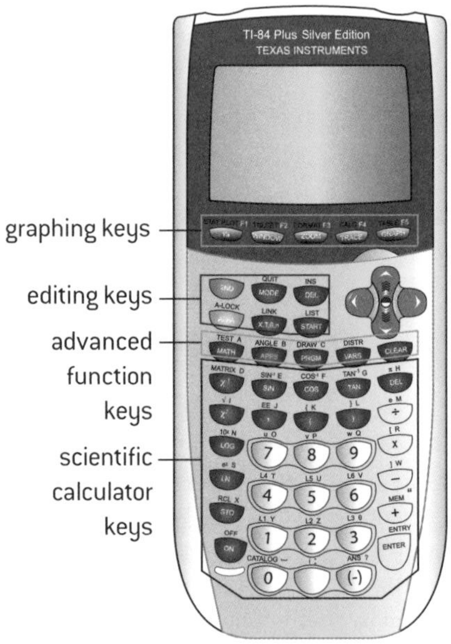

Make sure you know how to use your calculator to work out problems involving logarithms for pH and pK_a calculations. The examples given below are for a TI-84 plus.

To convert a hydrogen ion concentration of 1.8×10^{-5} mol dm^{-3} into pH.

$pH = -\log_{10} 1.8 \times 10^{-5}$

To obtain the value press the following keys in sequence.

(−) (LOG 1.8 × 10 ^ (−) 5) ENTER

This will give a value of 4.74.

To convert a pK_a value of 3.75 into a K_a value

Press

2nd 10^x (−) 3.75 ENTER

to give a value of 1.78×10^{-4}.

The final examinations

PREPARING FOR THE EXAMINATIONS

Hopefully for much of the course the emphasis has been on enjoying learning and understanding Chemistry rather than always worrying about grades. Towards the end of the course however it does make sense to prepare yourself for the final exam. Examiners are human and mark positively (i.e. they look to give credit rather than penalize mistakes). You have to help them by being clear in your answers and addressing the particular question(s) asked.

- **Know what it is you have to know.** Ask your teacher for a copy of the current programme for the Core and the one option you are taking. Higher Level students should also have a copy of the Additional Higher Level material. Go through the programme carefully and make sure you recognize and have covered all the points listed for each topic and sub-topic. Make sure you understand all the contents of this book as it applies to your particular level and choice of option and that you have worked through the relevant questions and answers at the end of each chapter.
- **Be familiar with key command terms.** Each question in the exam will normally contain a key command term. A list of all these terms and their precise meaning is given on the next page. If a question asks you to *describe* a reaction then a very different answer is required than if the question had asked you to *explain* a reaction. Examiners can only award marks for the correct answers to the question asked. Not paying careful attention to the correct command term may cost you marks unnecessarily.
- **Practice with past papers.** Most schools will give their students a 'mock' or 'trial' exam. This is helpful as it enables you to judge the correct amount of time to spend on each question. Make your mistakes in the mock exam and learn from them. Of course the IB questions are different each year but they do tend to follow a similar pattern. It helps to have seen similar questions before and know what level to expect.
- **Organize your notes.** As you review your work it is often helpful to rewrite your notes. Concentrate on just the key points – they should trigger your memory. This book already contains the important points in a fairly condensed form. Condense them even more to make your own set of review notes. Each time you review each topic try to condense the notes even more. By the time you are ready to take the exam all your personal review notes should ideally fit onto a single page!
- **Be familiar with using the IB data booklet.** You should get into the habit of using this throughout the course so that you are completely familiar with its contents and how to use them by the time of the exams.
- **Know the format of the exam papers.** Both HL and SL students take three exam papers. Papers 1 and 2 examine the core (and Additional HL) material. Paper 1 is multiple choice and you are not allowed a calculator or the IB data booklet. A periodic table is provided. Paper 2 contains short answer questions and longer response questions. All the questions are compulsory, i.e. you must attempt to answer all the questions. Paper 3 will normally be taken on the next day. You must attempt all the questions in Section A of Paper 3 which contains a data response question and short answer questions on experimental work. Section B contains questions on all the options. You are required to answer all the questions on one of the four options.
- **Know the dates of the exams.** Plan your review timetable carefully in advance. Remember that you will have exams in other subjects and that you may not have much time for a 'last minute' review.

TAKING THE EXAMINATIONS

- Try to ensure that the night before you are able to take some time to relax and get a good night's sleep.
- Take all you need with you to the examination room i.e. pens, pencils, ruler and a simple translating dictionary if English is not your first language. You will need your calculator for Paper 2 and Paper 3 – remember to include a spare battery.
- There is no reading time allowed for Paper 1. Work through the questions methodically. HL have 40 questions in 1 hour, SL have 30 questions in 45 minutes. If you get stuck on a question move on and then come back to it if you have time at the end. Make a note of those questions you are unsure about. You can then come back to these at the end rather than going through all of them again. Make sure you give one answer for each question. You are not penalized for wrong answers so if you run out of time make an educated guess rather than leave any questions unanswered.
- You will have five minutes' reading time before you are allowed to write your answers for both Paper 2 and Paper 3. You must attempt to answer all the questions on Paper 2 so use the time to gather your thoughts and get an overview of what is on the paper. Once you have selected the particular option you are answering on Paper 3 then again you must attempt to answer all the questions.
- Read each question very carefully. Make a mental note of the key command term so that you give the required answer.
- You must write your answers only within the required space as your answers are scanned and marked digitally. Write as legibly as you can. For questions involving calculations do not round up too early but make sure your final answer is given to the correct number of significant figures. Always include the correct units. If you change your mind then clearly cross out whatever you do not want to be marked. If you do need extra space put a note in the answer box that your answer is continued on a separate sheet of answer paper which is then attached to the booklet.
- Attempt all the required number of questions. If you do not attempt a question you can receive no marks. For sequential numerical questions even if you get the first part wrong continue as you will not be marked wrong twice for the same mistake. For this reason it is essential that you show your working. Do not answer more questions than required.
- Leave yourself time to read through what you have written to correct any mistakes.
- Ensure that you have filled in the front of the paper correctly including stating the number of the optional questions answered and the number of attached extra pages (if any) before leaving the examination room.

Command terms

Command terms are the imperative form of specific verbs that are used in examination questions. Although these are the terms that will normally be used, other terms may be used to direct you to produce an argument or answer in a specific way. You should understand the precise meaning of each term. They are divided into three different objectives. Objective 1 is the lowest and implies you just have to define or state the information. Objective 2 means you have to apply your knowledge of the topic in a straightforward situation. Objective 3 is the highest level and means that you will have to recognize the problem and select the appropriate method to solve it. Each of the three papers tests about 50% Objectives 1 and 2 and 50% Objective 3. It is worth noting that if you are also studying Physics or Biology then you will need to be careful as the command terms used for these two subjects are slightly different to the ones used for Chemistry.

OBJECTIVE 1

Classify	Arrange or order by class or category.
Define	Give the precise meaning of a word, phrase, concept or physical quantity.
Draw	Represent by means of a labelled, accurate diagram or graph, using a pencil. A ruler (straight edge) should be used for straight lines. Diagrams should be drawn to scale. Graphs should have points correctly plotted (if appropriate) and joined in a straight line or smooth curve.
Label	Add labels to a diagram.
List	Give a sequence of brief answers with no explanation.
Measure	Obtain a value for a quantity.
State	Give a specific name, value or other brief answer without explanation or calculation.

OBJECTIVE 2

Annotate	Add brief notes to a diagram or graph.
Apply	Use an idea, equation, principle, theory or law in relation to a given problem or issue.
Calculate	Obtain a numerical answer showing the relevant stages in the working.
Describe	Give a detailed account.
Distinguish	Make clear the differences between two or more concepts or items.
Estimate	Obtain an approximate value.
Formulate	Express precisely and systematically the relevant concept(s) or argument(s).
Identify	Provide an answer from a number of possibilities.
Outline	Give a brief account or summary.

OBJECTIVE 3

Analyse	Break down in order to bring out the essential elements or structure.
Comment	Give a judgment based on a given statement or result of a calculation.
Compare	Give an account of the similarities between two (or more) items or situations, referring to both (all) of them throughout.
Compare and contrast	Give an account of similarities and differences between two (or more) items or situations, referring to both (all) of them throughout.
Construct	Display information in a diagrammatic or logical form.
Deduce	Reach a conclusion from the information given.
Demonstrate	Make clear by reasoning or evidence, illustrating with examples or practical application.
Derive	Manipulate a mathematical relationship to give a new equation or relationship.
Design	Produce a plan, simulation or model.
Determine	Obtain the only possible answer.
Discuss	Offer a considered and balanced review that includes a range of arguments, factors or hypotheses. Opinions or conclusions should be presented clearly and supported by appropriate evidence.
Evaluate	Make an appraisal by weighing up the strengths and limitations.
Examine	Consider an argument or concept in a way that uncovers the assumptions and interrelationships of the issue.
Explain	Give a detailed account including reasons or causes.
Explore	Undertake a systematic process of discovery.
Interpret	Use knowledge and understanding to recognize trends and draw conclusions from given information.
Justify	Give valid reasons or evidence to support an answer or conclusion.
Predict	Give an expected result.
Show	Give the steps in a calculation or derivation.
Sketch	Represent by means of a diagram or graph (labelled as appropriate). The sketch should give a general idea of the required shape or relationship, and should include relevant features.
Solve	Obtain the answer(s) using algebraic and/or numerical and/or graphical methods.
Suggest	Propose a solution, hypothesis or other possible answer.

Internal Assessment (1)

INTRODUCTION

You are expected to spend 40 hours (SL) or 60 hours (HL) during the two years on the practical scheme of work. This is essentially time spent in the laboratory or on simulations. Chemistry is an experimental science and practical work is an important component of the course. Your teacher should devise a suitable practical programme for you to follow. Practical work can have many different aims. For example, it could be to improve your skills at different techniques, to reinforce the theoretical part of the course and to give you experience of planning your own investigations. Hopefully it will make studying Chemistry much more challenging and rewarding and also fun. Through the practical course you are expected to understand and implement safe practice and also to respect the environment.

INTERNAL ASSESSMENT – THE FACTS

The internal assessment component counts 20% towards the final mark with the external examinations counting 80%. The assessment is exactly the same for both SL and HL. Most of the work you do will not be assessed as the assessment will only be on **one** individual scientific investigation taking about 10 hours and its subsequent write-up, which should be about 6–12 pages long. This will be marked out of 24, which will then be scaled to a mark out of 20. It will be marked by your teacher but moderated externally by the IB. Although the other investigations do not count towards the 20% of the internal assessment mark they are tested to some extent on Section A on Paper 3. In order that students have some common understanding, there are six mandatory areas where practicals must be performed. However the teacher is free to determine which method to choose to cover these six areas as well as all the other practicals to cover other topics. The six areas are: Topic 1: Determining empirical formula from mass changes, Topics 1 and 8: Titration, Topic 1: Determining the molar mass of a gas, Topic 5: Enthalpy of a reaction by calorimetry, Topic 9: Reactions involving voltaic cells and Topic 10: Construction of 3-D models of organic molecules. In addition you are required to spend ten hours on the group 4 project but this does not count towards the final internal assessment mark.

GRADING OF INTERNAL ASSESSMENT

You will only gain good marks for your Individual Investigation if you address each criterion fully. The descriptors to gain the maximum mark for each criterion are given below.

Criterion	Max mark	Descriptor for maximum mark
Personal engagement	2	The evidence of personal engagement with the exploration is clear with significant independent thinking, initiative or creativity. The justification given for choosing the research question and/or the topic under investigation demonstrates personal significance, interest or curiosity. There is evidence of personal input and initiative in the designing, implementation or presentation of the investigation.
Exploration	6	The topic of the investigation is identified and a relevant and fully focused research question is clearly described. The background information provided for the investigation is entirely appropriate and relevant and enhances the understanding of the context of the investigation. The methodology of the investigation is highly appropriate to address the research question because it takes into consideration all, or nearly all, of the significant factors that may influence the relevance, reliability and sufficiency of the collected data. The report shows evidence of full awareness of the significant safety, ethical or environmental issues that are relevant to the methodology of the investigation (if appropriate).
Analysis	6	The report includes sufficient relevant quantitative and qualitative raw data that could support a detailed and valid conclusion to the research question. Appropriate and sufficient data processing is carried out with the accuracy required to enable a conclusion to the research question to be drawn that is fully consistent with the experimental data. The report shows evidence of full and appropriate consideration of the impact of measurement uncertainty on the analysis. The processed data is correctly interpreted so that a completely valid and detailed conclusion to the research question can be deduced.
Evaluation	6	A detailed conclusion is described and justified, which is entirely relevant to the research question and fully supported by the data presented. A conclusion is correctly described and justified through relevant comparison to the accepted scientific context. Strengths and weaknesses of the investigation, such as limitations of the data and sources of error, are discussed and provide evidence of a clear understanding of the methodological issues involved in establishing the conclusion. Realistic and relevant suggestions for the improvement and extension of the investigation have been discussed.
Communication	4	The presentation of the investigation is clear. Any errors do not hamper understanding of the focus, process and outcomes. The report is well-structured and clear: the necessary information on focus, process and outcomes is present and presented in a coherent way. The report is relevant and concise thereby facilitating a ready understanding of the focus, process and outcomes of the investigation. The use of subject-specific terminology and conventions is appropriate and correct. Any errors do not hamper understanding.

Internal Assessment (2)

MAXIMIZING YOUR INTERNAL ASSESSMENT MARKS

General points

- Before you undertake your individual investigation, familiarize yourself with the assessment criteria. Remember this is not an Extended Essay and you are not using the EE criteria.
- Look at examples of excellent past Individual Investigations (available from your teacher).
- Determine the title of your investigation and discuss it with your teacher. For ideas you could refer to interesting developments from experiments you have already done, look at the suggestions for Extended Essay titles which are given on page 179 or base it upon something you have read in a newspaper, journal or online.
- Decide whether you are going to generate your own primary data by experiment or whether you are going to use secondary data from other sources.
- Record all your work as you proceed including precise details of references.
- If you are doing a 'hands on' investigation, record precise details of all equipment used, e.g. a balance weighing to ±0.001 g, a thermometer measuring from −10 to +110 °C to an accuracy of ±0.1°C, a 25.00 cm^3 pipette measuring to ±0.04 cm^3, etc. Also record precise details of any chemicals used, e.g. copper(II) sulfate pentahydrate $CuSO_4.5H_2O(s)$ and if it is a solution include the concentration, e.g. 0.100 mol dm^{-3} NaOH(aq).
- Record all measurements accurately to the correct number of significant figures and include all units.
- Record all observations. Include colour changes, solubility changes, whether heat was evolved or taken in.
- Draw up a checklist to cover each criterion being assessed. As you write up your individual investigation check that each criterion is addressed fully.
- Remember that academic honesty is paramount. You must always acknowledge the ideas of other people.

Specific points for each criterion

Personal engagement

Ideally generate your own research question. Make sure that in your report you justify why it was chosen and why it is significant to you personally. Show clearly how you have designed the investigation and, if you are using an experimental method, how you have adapted it to fit your particular investigation.

Exploration

Whether you are generating primary data by experiment or whether you are using secondary data you must set it into context by discussing your research on what is already known about the topic under investigation. The research question must be fully focused and the methodology chosen must address all the variables whether dependent, independent or controlled. Safety and environmental concerns must be identified and addressed.

Analysis

Make sure that you have sufficient relevant raw data to address your research question and that you have processed it correctly. This means that, for example, for graphs you should have a minimum of five readings and the readings repeated for accuracy. The processing should include attention to uncertainties and error in the individual measurements and the total uncertainty associated with the final outcome recorded. Assumptions made should be clearly stated.

Evaluation

Make sure your conclusion is justified from the data gathered and is relevant to your research question. Put your conclusion into context by comparing it with literature sources if possible and work out the percentage error. Identify the strengths and weaknesses of your method and suggest sensible ways in which weaknesses could be improved.

Communication

Write your report concisely and make it intellectually neat. It should be well-structured and present the information and argument logically and clearly. Make sure you use correct chemical language and terminology with attention to the use of correct units, significant figures and decimal places. Images, graphs and tables should all be correctly labelled.

THE GROUP 4 PROJECT

The group 4 project is a collaborative activity whereby all the IB students in the school from the different group 4 subjects work together on a scientific or technological topic. The aim is to encourage an understanding of the relationships between the different scientific disciplines and the overarching nature of the scientific method. Collaboration between different schools in different regions is actively encouraged. There is considerable flexibility in how the project may proceed and different schools will approach it in different ways. However you are required to spend about ten hours in total on the group 4 project. In the planning stage you should decide on an overall topic with your fellow students and then, in small groups, decide how you will investigate a particular aspect of the chosen topic. During the action stage, which lasts for about six hours, you should investigate your topic. The investigation may be practically or theoretically based and may be just in chemistry or across all the scientific disciplines. You should collaborate with other students and in any practical work pay attention to safety, ethical and environmental considerations. Finally there is the evaluation stage which involves sharing your results, including your successes and failures, with all the other students. The emphasis for the group 4 project is on the collaborative experience of working with other students. It is the **process** not the **product** that is important. Although it is not assessed you will be required to write a short reflection on how you contributed towards the group 4 project.

Extended Essays (1)

WHAT IS AN EXTENDED ESSAY?

In order to fulfil the requirements of the IB all Diploma candidates must submit an Extended Essay in an IB subject of their own choice. The Essay is an in-depth study of a limited topic within a subject. The purpose of the Essay is to provide you with an opportunity to engage in independent research. Approximately 40 hours should be spent in total on the Essay. Each Essay must be supervised by a competent teacher. The length of the Essay is restricted to a maximum of 4000 words and it is assessed according to a carefully worded set of criteria. The marks awarded for the Extended Essay are combined with the marks for the Theory of Knowledge course to give a maximum of three bonus points.

EXTENDED ESSAYS IN CHEMISTRY

Although technically any IB Diploma student can choose to write their Essay in Chemistry it does help if you are actually studying Chemistry as one of your six subjects! Most Essays are from students taking Chemistry at Higher Level but there have been some excellent Essays submitted by Standard Level students. All Essays must have a sharply focused Research Question. Essays may be just library-based or also involve individual experimental work. Although it is possible to write a good Essay containing no experimental work it is much harder to show personal input and rarely do such Essays gain high marks. The experimental work is best done in a school laboratory although the word 'laboratory' can be interpreted in the widest sense and includes the local environment. It is usually much easier for you to control, modify, or redesign the simpler equipment found in schools than the more sophisticated (and expensive) equipment found in university or industrial research laboratories.

CHOOSING THE RESEARCH QUESTION

Choosing a suitable Research Question is really the key to the whole Essay. Some supervisors have a list of ready-made topics. The best Essays are almost always submitted by students who identify a particular area or chemistry problem that they are interested in and together with the supervisor formulate a precise and sharply focused research question. It must be focused. A title such as 'A study of analysis by chromatography' is far too broad to complete in 4000 words. A focused title might be 'An analysis of (a named red dye) present in (a specified number of) different brands of tomato ketchup by thin layer chromatography'. It is more usual to choose a topic and then decide which technique(s) might be used to solve the problem. An alternative way is to look at what techniques are available and see what problems they could address.

SOME DIFFERENT TECHNIQUES (TOGETHER WITH A RESEARCH QUESTION EXAMPLE) THAT CAN BE USED FOR CHEMISTRY EXTENDED ESSAYS

The list below shows some examples of how standard techniques or equipment available in a school laboratory can be used to solve some general Research Questions. For precise research questions these exemplars may need to be more sharply focused. Although one example has been provided for each technique many Research Questions will, of course, involve two or more of these techniques.

Redox titration
Do different (specified) varieties of seaweed contain different amounts of iodine?

Extension of a standard practical
What gas is evolved when zinc is added to $CuSO_4(aq)$ and what factors affect its formation?

Acid–base titration
How do storage time and temperature affect the vitamin C content of (specified) fruit juices?

Chromatography
Do strawberry jellies from (specified) different countries contain the same red dye(s)?

Calorimetry
How efficient is dried cow dung as a fuel compared to fossil fuels?

pH meter
Can (specified) different types of chewing gum affect the pH of the mouth and prevent tooth decay?

Steam distillation
What is the amount of aromatic oil that can be extracted from (a specified) plant species?

Electrochemistry
What is the relationship between concentration and the ratio of O_2:Cl_2 evolved during the electrolysis of NaCl(aq)?

Refinement of a standard practical
How can the yield be increased in the laboratory preparation of 1,3-dinitrobenzene?

Microwave oven
What is the relationship between temperature increase and dipole moment for (specified substances)?

Polarimetry
Is it possible to prepare the different enantiomers of butan-2-ol in a school laboratory?

Data logging probes
What is the rate expression for (a specified reaction)?

Visible spectrometry
What is the percentage of copper in different ores found in (specified area)?

Gravimetric analysis
Do (specified) 'healthy' pizzas contain less salt than normal pizzas?

Inorganic reactions
An investigation into the oxidation states of manganese – does Mn(V) exist?

Microscale/small scale
How can the residues from a typical IB school practical programme be reduced?

Extended Essays (2)

RESEARCHING THE TOPIC

Once the topic is chosen research the background to the topic thoroughly before planning the experimental work. Information can be obtained from a wide variety of sources: a library, the internet, personal contacts, questionnaires, newspapers, etc. Make sure that each time you record some information you make an accurate note of the source as you will need to refer to this in the bibliography. Treat information from the internet with care. If possible try to determine the original source. Articles in journals are more reliable as they have been vetted by experts in the field. Together with your supervisor plan your laboratory investigation carefully. Your supervisor should ensure that your investigation is safe, capable of producing results (even if they are not the expected ones) and lends itself to a full evaluation.

THE LABORATORY INVESTIGATION

Make sure you understand the chemistry that lies behind any practical technique before you begin. Keep a careful record of everything you do at the time that you are doing it. If the technique 'works' then try to expand it to cover new areas of investigation. If it does not 'work' (and most do not the first time) try to analyse what the problem is. Try changing some of the variables, such as increasing the concentration of reactants, changing the temperature or altering the pH. It may be that the equipment itself is faulty or unsuitable. Try to modify it. Use your imagination to design new equipment in order to address your particular problem (modern packaging materials from supermarkets can often be used imaginatively to great effect). Because of the time limitations it is often not possible to get reliable repeatable results but attempt to if you can. Remember that the written Essay is all that the external examiner sees so leave yourself plenty of time to write the Essay.

WRITING THE ESSAY

Before starting to write the Essay make sure you have read and understood the assessment criteria. Your school or supervisor will provide you with a copy. It may be useful to look at some past Extended Essays to see how they were set out. Almost all Essays are word-processed and this makes it easier to alter draft versions but they may be written by hand. You will not be penalized for poor English but you will be penalized for bad chemistry so make sure that you do not make simple word-processing errors when writing formulas, etc.

Start the Essay with a clear introduction and make sure that you set out the Research Question clearly and put it into context. The rest of the Essay should then be very much focused on addressing the Research Question. Some of the marks are gained simply for fulfilling the criteria (e.g. numbering the pages, including a list of contents, etc.). These may be mechanical but you will lose marks if you do not do them. Give precise details of any experimental techniques and set out the results clearly and with the correct units and correct number of significant figures. If you have many similar calculations then show the method clearly for one and set the rest out in tabular form. Numerical results should give the limits of accuracy and a suitable analysis of uncertainties should be included. Relate your results to the Research Question in your discussion and compare them with any expected results and with any secondary sources of information you can locate. State any assumptions you have made and evaluate the experimental method fully. Suggest possible ways in which the research could be extended if more time were available. Throughout the whole Essay show that you understand what it is that you are doing and demonstrate personal input and initiative.

When you have drafted the whole of the Essay give it to your supervisor for comment. Your final version should then take on board the feedback from your supervisor. Before handing in the final version go through the check list on the next page very carefully. If you can honestly answer "yes" to every question then your Essay will be at least satisfactory and hopefully much better than this.

REFLECTIONS ON PLANNING AND PROGRESS – FORM EE/RPPF

In addition to the finished Essay (the product) some of the assessment is also based on the Extended Essay process. During the course of your Extended Essay work you will have three formal meetings with your supervisor. These are for you to reflect on the planning and progress of the Essay. The first will take place near the start of the process when you will discuss approaches and strategies. The second meeting (the interim reflection session) will take place after you have attempted to refine the research question and have recorded some relevant evidence and data. The final session takes place after the final version of the Essay has been handed in and you will be asked to reflect on the whole process. After each meeting you must record your reflections on Form EE/RPPF. You should show how you have responded to any setbacks, and how you have engaged in the research process. Through your reflections (which should not exceed 500 words in total) you should also show that your work is authentic and demonstrate intellectual initiative and creativity. Your supervisor will also write their own comments at the end of the process and the completed Form EE/RPPF will be sent to the external assessor along with the final version of your Essay for assessment.

Extended Essays (3)

EXTENDED ESSAY CHECK LIST

In order to gain the maximum credit possible for your extended essay it is crucial that you can answer YES to the following questions before you finally submit the final version of your essay and the completed Reflections on Planning & Progress form.

The maximum number of marks available for each criterion is given in brackets.

A: Focus & method [6]

Is the research question outlined and clearly stated in the introduction? [ ]
Is the research question focused and capable of being addressed in 4000 words? [ ]
Is the purpose and focus of the RQ set into context of background knowledge and understanding of chemistry? [ ]
Have you shown that a range of appropriate different sources/methods were considered? [ ]
Have you provided evidence of how you selected effective and appropriate sources/method(s)? [ ]
Have you shown how the RQ was arrived at from the sources gathered? [ ]
Does the whole essay remain focused on the research question? [ ]

B: Knowledge & understanding [6]

Is correct chemical terminology used consistently and appropriately throughout the essay? [ ]
Have you shown that you understand the chemistry behind the sources/method(s) used? [ ]
Have the sources/method(s) chosen throughout the essay been used effectively to address the RQ? [ ]

C: Critical thinking [12]

Is the research appropriate to the RQ and its application consistently relevant? [ ]
Have you critically analysed the sources/method(s) used? [ ]
Have you developed an effective and reasoned argument from your research? [ ]
Is your argument well-structured? [ ]
Are all your conclusions effectively supported by the evidence and relevant to the RQ? [ ]
Have you analysed your research effectively and stated any limitations and/or counter arguments? [ ]

D: Formal presentation [4]

Is your essay within the 4000 word limit? [ ]
Does your essay include all the required elements (Title page, table of contents, introduction, discussion, conclusion & bibliography)? [ ]
Is all illustrative material (graphs, tables, chemical structures etc.) clearly and accurately labelled? [ ]
Does the bibliography include all, and only, those works that have been consulted? [ ]
Is the bibliography set out in a standard format that is consistently applied? [ ]
Is all the work of others clearly acknowledged? [ ]
Are all the pages numbered? [ ]
Are all pages double-spaced and formatted using Arial font size 12? [ ]
If you have included an appendix does it only contain information necessary to support the essay? [ ]

E: Engagement [6]

Have you engaged fully in discussions with your supervisor regarding the planning and progress of your essay? [ ]
Have you shown how your reflections enabled you to refine the research process? [ ]
Have you shown how you responded to setbacks and challenges during the research process? [ ]
Have you shown initiative, creativity and personal input throughout the research process? [ ]
Have you made suggestions as to how you can improve your own working practice? [ ]

Answers to questions

STOICHIOMETRIC RELATIONSHIPS Multiple choice (page 7)

1. D **2.** C **3.** B **4.** C **5.** D **6.** B **7.** D **8.** D **9.** A
10. B **11.** B **12.** A **13.** C **14.** B **15.** C **16.** A

STOICHIOMETRIC RELATIONSHIPS Short answers (page 8)

1. a) Amount of ethanoic anhydride $= \frac{15.0}{102.1} = 0.147$ mol **[1]**, amount of 2-hydroxybenzoic acid $= \frac{15.0}{138.13} = 0.109$ mol **[1]**, 2 mol of 2-hydroxybenzoic acid required to react with 1 mol of ethanoic anhydride so 2-hydroxybenzoic acid is the limiting reagent.**[1]**; b) M_r(aspirin) $= [(9 \times 12.01) + (8 \times 1.01) + (4 \times 16.00)] = 180.17$ **[1]**, 0.109 mol of 2-hydroxybenzoic acid produces 0.109 mol of aspirin so maximum mass of aspirin $= 0.109 \times 180.17 = 19.6$ g **[1]**; c) Percentage yield $= (\frac{13.7}{19.6}) \times 100 = 69.9\%$ **[1]**

2. a) $M_r(BaSO_4) = 137.33 + 32.07 + (4 \times 16.00) = 233.4$ **[1]**; amount of $BaSO_4 = \frac{9.336}{233.4} = 0.0400$ mol **[1]**; b) 0.0400 mol **[1]**; c) $M_r(M_2SO_4) = \frac{14.48}{0.0400} = 362$ **[1]**; d) $M_r(M_2SO_4) = 2 \times A_r(M) + 32.07 + (4 \times 16.00) = 362$ so A_r for M = 133 **[1]**. Since M has an A_r of 133 and forms a unipositive cation M is caesium. **[1]**

3. a) $Mg(s) + 2HCl(aq) \rightarrow MgCl_2(aq) + H_2(g)$ **[1]**; b) Amount of Mg $= \frac{(7.40 \times 10^{-2})}{24.31} = 3.04 \times 10^{-3}$ mol **[1]**, amount of HCl $= \left(\frac{15.0}{1000}\right) \times 2.00 = 3.00 \times 10^{-2}$ mol **[1]**, for all the HCl to react would require 1.50×10^{-2} mol of Mg so Mg is the limiting reagent **[1]**; c) i) Theoretical yield of $H_2 = 3.04 \times 10^{-3}$ mol **[1]**; ii) 1 mol of gas occupies 22700 cm^3 at 273 K, 1.00×10^5 Pa so at 293 K volume occupied by 3.04×10^{-3} mol of $H_2 = 3.04 \times 10^{-3} \times 22700 \times \frac{293}{273} = 74.1$ cm^3 (74.1 cm^3 can also be obtained by using $pV = nRT$) **[2]**; d) *Any two from:* hydrogen is not an ideal gas, the syringe sticks, the magnesium is impure, the hydrogen dissolves in the solution, uncertainties associated with the concentration of the acid. **[2]**

4. a) i) Ratio of Pb : C : H $= \left(\frac{64.052}{207.2}\right) : \left(\frac{29.703}{12.01}\right) : \left(\frac{6.245}{1.01}\right) = 0.309 : 2.47 : 6.18$ simplest ratio is 1 : 8 : 20 so empirical formula is PbC_8H_{20} **[3]**; ii) 1 mol of PbC_8H_{20} contains 1 mol of Pb so molecular formula is PbC_8H_{20} **[1]**; iii) $PbC_8H_{20} + 14O_2 \rightarrow PbO_2 + 8CO_2 + 10H_2O$ **[2]**; b) *Local:* carbon monoxide or volatile organics or nitrogen oxide or unburned hydrocarbons or particulates **[1]**, *global:* carbon dioxide or nitrogen oxide. **[1]**

5. a) Amount of $MnO_4^- = \frac{22.50}{1000} \times 2.152 \times 10^{-2} = 4.842 \times 10^{-4}$ mol **[2]**; b) Amount of $Fe^{2+} = 5 \times 4.842 \times 10^{-4} = 2.421 \times 10^{-3}$ mol **[2]**; c) Mass of Fe $= 2.421 \times 10^{-3} \times 55.85 = 0.1352$ g **[1]** so percentage of Fe in ore $= \frac{0.1352 \times 100}{0.36\ 82} = 36.72\%$. **[1]**

6. a) Amount of $Cu_2O = \frac{10000}{[(2 \times 63.55) + 16.00]} = 69.89$ mol **[1]**, amount of $Cu_2S = \frac{5000}{159.17} = 31.41$ mol **[1]**, Cu_2S requires 2×31.41 mol of Cu_2O to react so Cu_2S is the limiting reagent **[1]**; b) Max amount of Cu $= 6 \times 31.41 = 188.46$ mol **[1]**, maximum mass $= 188.46 \times 63.55 = 11976$ g $= 12.0$ kg **[1]**; c) Atom economy $= \frac{381.3}{445.37} \times 100 = 85.6\%$. **[3]**

7. *Any four from:* some of the product escaped when the lid was lifted, the Mg may have combined with nitrogen in the air, the Mg may not have been pure, not all the Mg may have burned, the crucible might have also reacted. **[4]**

8. a) C : H $= \frac{85.6}{12.01} : \frac{14.4}{1.01} = 7.13 : 14.3 = 1 : 2$, empirical formula is CH_2 **[2]**; b) i) $n = \frac{PV}{RT} = \frac{(1.00 \times 10^5 \times 0.405 \times 10^{-3})}{8.314 \times 273} = 1.784 \times 10^{-2}$ mol, $M = \frac{1.00}{0.01784} = 56.1$ g mol^{-1} **[2]**; ii) Empirical formula mass $= 12.01 + 1.01 \times 2 = 14.03$, $\frac{56.1}{14.03} = 4$ so molecular formula is C_4H_8 **[2]**; c) Carbon monoxide is produced which can combine irreversibly with the iron in haemoglobin in the blood or carbon particulates are formed which can cause problems with the lungs. **[2]**

ATOMIC STRUCTURE Multiple choice (page 14)

1. C **2.** A **3.** B **4.** A **5.** B **6.** A **7.** D **8.** D **9.** C
10. B **11.** D **12.** C **13.** A **14.** C **15.** A **16.** B

ATOMIC STRUCTURE Short answers (page 15)

1. a) The weighted mean mass of all the naturally occurring isotopes of the element, relative to one twelfth of the mass of a carbon-12 atom. **[1]**; b) $37x + 35(100 - x) = 100 \times 35.45 = 3545$, so $2x = 45$ and $x = 22.5$. Hence $^{37}Cl = 22.5\%$ and $^{35}Cl = (100 - 22.5) = 77.5\%$. **[2]**; c) i) $1s^22s^22p^63s^23p^5$ **[1]**; ii) $1s^22s^22p^63s^23p^6$ **[1]**; d) They do not differ as chemical properties are determined by the electron configuration and both isotopes have the same configuration. **[2]**

2. a) The weighted average of the sum of protons and neutrons for Co (Z = 27) is greater than for Ni (Z = 28) even though nickel atoms contain one more proton. **[1]**; b) 27 protons, 25 electrons **[1]**; c) i) $1s^22s^22p^63s^23p^64s^23d^7$ or $[Ar]4s^23d^7$ **[1]**; ii) $1s^22s^22p^63s^23p^63d^7$ or $[Ar]3d^7$ **[1]**

3. a) i) Atoms of the same element that contain the same number of protons but have a different number of neutrons in their nucleus. **[1]**; ii) 56 **[1]**; iii) It emits radiation which can potentially damage cells. **[1]**; b) i) Both contain 6 protons and 6 electrons, carbon-12 contains 6 neutrons and carbon-14 8 neutrons. **[2]**; ii) An electron **[1]**; iii) The length of time since death can be determined by looking at the ratio of ^{12}C to ^{14}C as it will increase at a uniform rate as the amount of ^{14}C halves every 5300 years. **[2]**

4. **[1]**

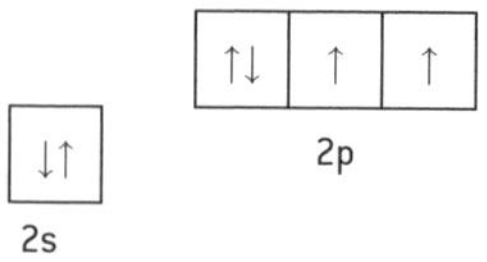

5.

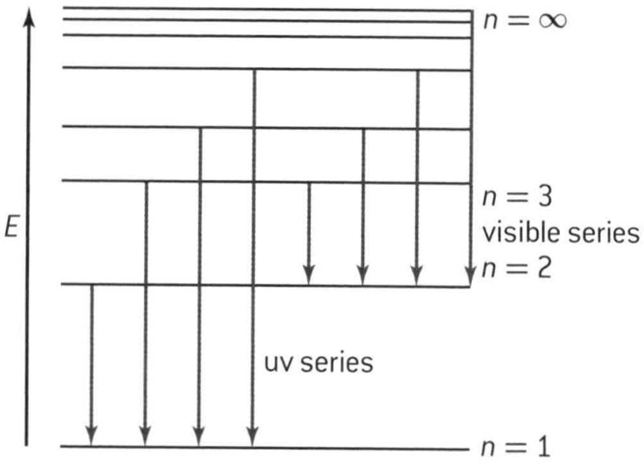

showing y-axis labelled as energy/E/labelling at least two energy levels;
showing a minimum of four energy levels/lines with convergence;
showing jumps to $n = 1$ for ultraviolet series;
showing jumps to $n = 2$ for visible light series; **[4]**

6. i) The electron configuration of argon, i.e. $1s^22s^22p^63s^23p^6$ **[1]**; ii) x = 1, y = 5 **[1]**;
iii) **[1]**

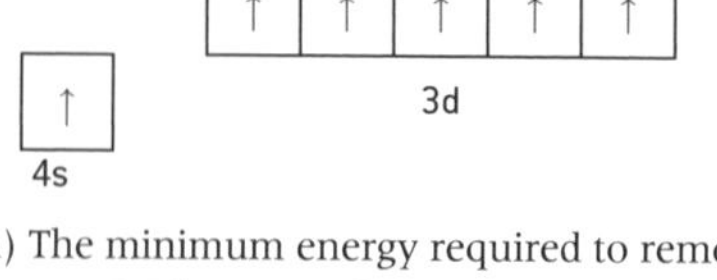

7. a) The minimum energy required to remove a mole of electrons **[1]** from a mole of gaseous atoms to form a mole of univalent cations in the gaseous state. **[1]**; b) i) Electrons are being added to the same main energy level but the number of protons in the nucleus increases thus attracting the electrons more strongly as the atomic radius decreases. **[2]**; ii) The outermost electron in sodium is the $3s^1$ electron which is much further from the nucleus than the 2p electrons in the outer energy level of neon. **[2]**; iii) The outer 2p sub-level in boron ($1s^22s^22p^1$) is higher in energy than the 2s outer sub-level in beryllium ($1s^2 2s^2$). **[2]**; iv) Phosphorus ([Ne]$3s^2 3p^3$) has three unpaired electrons, the pairing of two electrons in one of the 3p orbitals in oxygen ([Ne]$3s^2 3p^4$) causes repulsion between the electrons making it easier to remove one of them. **[2]**

8. Emission spectra consist of discrete lines representing transitions between the levels, if electrons could be anywhere the spectra would be continuous **[2]**; the convergence of lines in the spectra show that successive levels get closer in energy **[1]**; graphs of successive ionization energies give evidence for the number of electrons occupying each main energy level **[1]**; the irregularities in the graph of 1st ionization energies against atomic number (e.g. B and Be, and Mg and Al) give evidence for sub-levels. **[1]**

9. E for one electron $= \frac{(1312 \times 1000)}{(6.02 \times 10^{23})} = 2.179 \times 10^{-18}$ J **[1]**; $\nu = \frac{E}{h} = \frac{(2.179 \times 10^{-18})}{(6.63 \times 10^{-34})} = 3.29 \times 10^{15}\ s^{-1}$ **[1]**; $\lambda = \frac{c}{\nu} = \frac{(3.00 \times 10^{8})}{(3.29 \times 10^{15})} = 9.12 \times 10^{-8}$ m (or 91.2 nm) **[1]**

PERIODICITY Multiple choice (page 23)

1. B **2.** B **3.** D **4.** D **5.** A **6.** A **7.** A **8.** C **9.** C **10.** D
11. B **12.** D **13.** C **14.** C **15.** A **16.** A

PERIODICITY Short answers (page 24)

1. a) Elements in the same group have similar outer electron configurations, in this case ns^2np^2, **[1]** in the same period the elements have a different number of electrons in the same outer energy level. **[1]**; b) p-block **[1]**; c) The outer 3p electron in Si is higher in energy than the outer 2p electron in C so easier to remove. **[2]**; d) Attempt to pass electricity through the elements using a DC battery, leads and a multimeter. Diamond does not conduct (graphite is unusual for non-metal in that it does conduct), silicon, a metalloid is a semiconductor and tin a metal is a good conductor. **[3]**

2. a) The levels are split into sub-levels. The 4s sub-level is lower in energy than the 3d sub-level so fills before the 3d sub-level. **[2]**; b) The 4f sub-level is being filled for the lanthanides and the 5f sub-level for the actinides. **[2]**

3. a) From basic (Na_2O, MgO) through amphoteric (Al_2O_3) to acidic (all non-metal oxides) **[3]**; b) i) $Na_2O(s) + H_2O(l) \rightarrow 2NaOH(aq)$ **[1]**; ii) $P_4O_{10}(s) + 6H_2O(l) \rightarrow 4H_3PO_4(aq)$ **[1]**; $SO_3(g) + H_2O(l) \rightarrow H_2SO_4(aq)$ **[1]**; c) Both are acidic oxides and dissolve in rain water to form acid rain that reacts with calcium carbonate in building materials. e.g. $SO_2(g) + H_2O(l) \rightarrow H_2SO_3(aq)$, $3NO_2(g) + H_2O(l) \rightarrow 2HNO_3(aq) + NO(g)$, $CO_3^{2-}(s) + 2H^+(aq) \rightarrow H_2O(l) + CO_2(g)$ **[4]**

4. a) $1s^2 2s^2 2p^6 3s^2 3p^6$ **[1]**; b) Both have the same electron configuration but S has one more proton so has a greater attraction to the outer electrons. **[2]**; c) Br has more protons (35 compared to K's 19) so the outer energy level (which is the same for both) is attracted more strongly to the nucleus. **[2]**; d) Atomic radius is measured by dividing the distance between the two atoms in a molecule, Ne does not combine with another Ne atom. **[1]**

5. a) i) Atomic number **[1]**; ii) Increase in number of electrons in outer energy level, smaller atoms **[2]**, do not form bonds easily with other elements **[1]**; b) i) The energy (in kJ mol^{-1}) required to remove an electron from an atom in the gaseous state **[2]**; ii) Increasing number of protons causes decreasing radius and lower outer energy level. **[2]**; iii) The electron in the outer 2p sub-level in B ($1s^2 2s^2 2p^1$) is higher in energy than the electrons in the outer 2s sub-level in Be ($1s^2 2s^2$). **[2]**; c) Na is a metal and contains delocalized electrons, phosphorus consists of atoms covalently bonded in P_4 molecules in which there are no delocalized electrons. **[2]**

6. Less particulates/C/CO/VOCs and SO_2 produced **[2]**, particulates, CO, VOC and SO_2 toxic or SO_2 causes acid rain. **[1]**

7. a) Fe^{2+} acts as a Lewis acid and accepts one non-bonding pair of electrons from each water molecule which act as Lewis bases. **[2]**; b) Fe^{3} has one less electron than Fe^{2+} so the attraction of the nucleus on the remaining d electrons is stronger – this affects the splitting of the d orbitals in the complex ion. **[2]**; c) Iron metal contains unpaired electrons which align parallel to each other in domains irrespective of whether an external magnetic or electric field is present **[1]**, CN^- is higher in the spectrochemical series than H_2O so causes greater splitting of the d orbitals **[1]**, in $[Fe(H_2O)_6]^{2+}$ there are four unpaired electrons as each of the five d orbitals is occupied (2 electrons in one of the lower orbitals and 1 in each of the other four orbitals), $[Fe(CN)_6]^{4-}$ contains no unpaired electrons as the 6 electrons are spin-paired in the three lower d orbitals. **[2]**; d) As the reaction is exothermic increasing the temperature causes a lower yield **[1]** so a catalyst speeds it up at a lower temperature so reducing costs **[1]**, iron is abundant and a cheap metal to use as a catalyst. **[1]**

8. a) $[CuC_{10}H_{12}O_8N_2]^{4-} = 351.8\ g\ mol^{-1}$ **[2]**; b) 6, octahedral **[2]** c) It 'wraps' around the metal ions using six non-bonding pairs of electrons to form coordinate bonds with Cu^{2+} which is acting as a Lewis acid. **[3]**

9. a) 3 of the d orbitals lie between the axes and 2 lie along the axes. As the ligands approach they repel the d orbitals that lie along the axes more. More electron dense ligands cause greater repulsion/splitting. **[2]**; b) $[Cu(H_2O)_6]^{2+}$ absorbs orange light so transmits blue light, as the ligand changes to NH_3 ΔE increases so the absorbed colour shifts to a smaller wavelength (yellow) and the complementary transmitted colour is purple. **[3]**; c) Sc^{3+} contains no d electrons so no transitions between split d levels can occur. **[2]**

CHEMICAL BONDING AND STRUCTURE Multiple choice (page 36)

1. C **2.** A **3.** A **4.** D **5.** D **6.** A **7.** C **8.** B **9.** B **10.** C
11. C **12.** D **13.** A **14.** D **15.** B **16.** D

CHEMICAL BONDING AND STRUCTURE Short answers (page 37)

1. Electron domain geometry for all three species is tetrahedral as they all have 4 electron domains. **[2]** Molecular shapes:

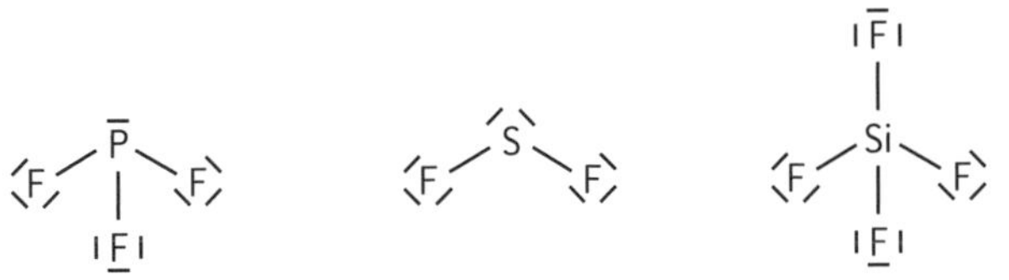

trigonal pyramid **[2]**; bent or V-shaped **[2]**; tetrahedral **[2]**

2. a) i) 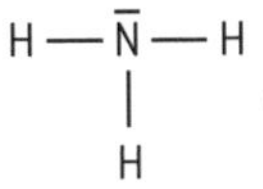

[1]; trigonal pyramid **[1]**; 107° **[1]**; the non-bonding pair repels more than the three bonding pairs **[1]**

ii) Boiling points increase going down the group **[1]**; M_r/number of electrons/molecular size increases down the group **[1]**; greater London dispersion forces/van der Waals' forces **[1]**; NH_3 has a higher boiling point than expected due to the hydrogen bonding between the molecules **[1]** b) **[3]**;

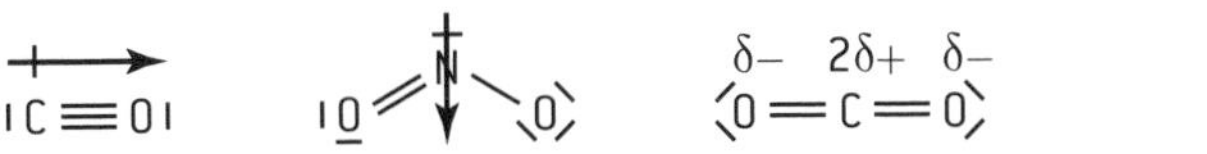

O is more electronegative than C in CO, NO_2 is bent and the C=O dipoles cancel out in the linear CO_2 **[2]**

3. a) 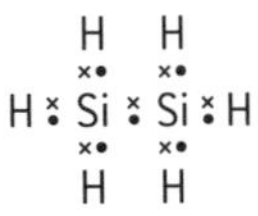

[1] b) 109.5° **[1]**, four equal electron domains around central C atom **[1]** c) C–H **[1]**, greater difference in electronegativity values **[1]**; d) Both molecules are non-polar **[1]** as they are symmetrical **[1]**; e) Stronger London dispersion forces **[1]** due to greater mass/more electrons. **[1]**

4. a) Methoxymethane **[1]** as the strongest attraction between molecules is dipole–dipole **[1]** whereas between the alcohol molecules there is stronger hydrogen bonding **[1]**;

b) Propan-1-ol is more soluble **[1]** as the non-polar hydrocarbon chain is shorter **[1]**; c) Graphite forms layers of flat hexagonal rings, each C atom bonded strongly to three other C atoms, layers held by weak attractive forces so can slide over each other with delocalized electrons between layers **[3]** diamond all C atoms strongly bonded to four other C atoms, giant tetrahedral structure with strong covalent bonds and no delocalized electrons. **[3]**

5. a) $1s^2 2s^2 2p^6 3s^2 3p^5$ **[1]** $[Ne]3s^2 3p^5$ **[1]**

b) 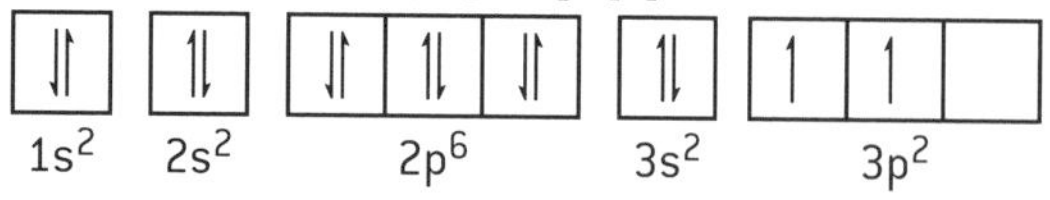

[2]

c) There is a large difference in electronegativity values between Cl and Na and only a small difference between Cl and Si so an electron is completely transferred when NaCl is formed and electrons are shared when $SiCl_4$ is formed. **[2]**

6. a) i) 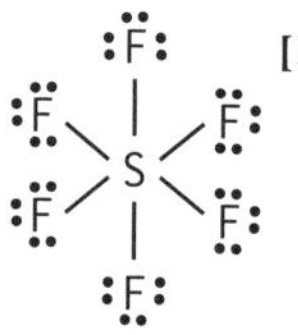

[1]; ii) S has readily available empty 3d orbitals which can be utilized, in O the d orbitals are too high in energy to be used. **[1]**; iii) O in ozone has 3 electron domains so trigonal planar with actual shape bent or V-shaped, S has 6 electron domains and 6 pairs of electrons so both are octahedral **[2]**; iv) Ozone: <120° (actual value 117°), SF_6 90° and 180° **[2]**

b) i) From data booklet $E_{C\text{-}F} = 492$ kJ mol^{-1} and $E_{C\text{-}Cl} = 324$ kJ mol^{-1}. $\lambda_{C\text{-}F} = \frac{hcL}{E} = \frac{(6.63 \times 10^{-34} \times 3.00 \times 10^{8} \times 6.02 \times 10^{23})}{(4.92 \times 10^{5})} = 2.43 \times 10^{-7}$ m = 243 nm, $\lambda_{C\text{-}Cl} = 370$ nm **[4]**; ii) Visible light does not have enough energy to break bonds but in ozone layer uv light can break C–Cl bond. **[2]**

7. Structure I O(LHS) = 6 − 6 − 1 = −1, N = 5 − 0 − 4 = +1, O(RHS) = 6 − 2 − 3 = +1 **[2]**, Structure II O = 6 − 4 − 2 = 0, N = 5 − 0 − 4 = +1 **[1]**, Structure II favoured as formal charges lower. **[1]**

8. a) Triple covalent bond (one of which is a dative/coordinate bond with electron pair donated by O to C) made up of one σ and two σ bonds **[2]**; b) Spread of π electrons over more than two nuclei giving equal bond strengths and greater stability **[3]**; c) Combination of two or more atomic orbitals to form new orbitals with lower energy **[1]** CO_2 sp, diamond sp^3, graphite sp^2, CO_3^{2-} sp^2 **[4]**; d) i) Molten NaCl conducts as ions are mobile in liquid state, SO_3 has neutral molecules and no mobile ions or electrons so it does not conduct electricity. **[2]**; ii) $Na_2O(s) + H_2O(l) \rightarrow 2NaOH(aq)$ **[1]**, $SO_3(l) + H_2O(l) \rightarrow H_2SO_4(aq)$, **[1]** both conduct as they react with water to form ions. **[1]**

ENERGETICS / THERMOCHEMISTRY Multiple choice (pages 44 and 45)

1. D **2.** B **3.** A **4.** A **5.** C **6.** D **7.** B **8.** C **9.** D **10.** A **11.** C **12.** C

ENERGETICS / THERMOCHEMISTRY Short answers (page 46)

1. a) 100 × 4.18 × 35.0 = 14 600 J = 14.6 kJ **[2]**; b) Amount of ethanol $= \frac{1.78}{46.08} = 0.0386$ mol **[1]**, $\Delta H_c = -\frac{14.6}{0.0386} = -378$ kJ mol^{-1} **[1]**; c) *Any two from* heat loss, incomplete combustion, heat absorbed by calorimeter not included. **[2]**

2. a) i) 100 g ethanol, 900 g octane **[1]**; ii) 2.17 mol of ethanol, 7.88 mol of octane **[1]**; iii) *E* from ethanol = (2.17 × 1367) = 2966 kJ, *E* from octane = (7.88 × 5470) = 43104 kJ, total energy = 4.61×10^4 kJ **[3]**; b) Greater as fewer intermolecular forces to break or vaporization is endothermic or gaseous fuel has greater enthalpy than liquid fuel. **[2]**

3. a) *Any two from* all heat transferred to copper sulfate solution, specific heat capacity of zinc negligible, density of solution same as density of pure water, specific heat capacity of solution same as for pure water **[2]**; b) i) 48.2 °C **[2]**; ii) Temperature decreases at uniform rate **[1]**; iii) 10.1 kJ **[1]** c) Amount of Zn = amount of $CuSO_4 = \frac{(1.00 \times 50.0)}{1000} = 5.00 \times 10^{-2}$ mol **[1]**; d) −201 kJ mol^{-1} **[1]**

4. a) ½ equation 1 − ½ equation 2 − equation 3 = + 137 kJ **[4]**; b) Positive as number of moles of gas increasing **[2]**; c) At low T, $\Delta H^{\ominus}$ is positive and $\Delta G^{\ominus}$ is positive, at high T, factor $T\Delta S^{\ominus}$ predominates and $\Delta G^{\ominus}$ is negative **[2]**; d) Energy in due to C–C and 2C–H = 1174 kJ mol^{-1}, energy out due to C=C and H–H = 1050 kJ mol^{-1}, ΔH = +124 kJ mol^{-1} **[3]**; **(e)** bond enthalpy values are average values. **[1]**

5. a) At 298 $\Delta G^{\ominus} = \Delta H^{\ominus} - T\Delta S^{\ominus} = 210 - (298 \times 0.216) = +146$ kJ, positive value so non-spontaneous **[2]**; b) When $\Delta G^{\ominus} = 0$, $T = \frac{210}{0.216} = 972$ K **[2]**

CHEMICAL KINETICS Multiple choice (pages 51 and 52)

1. A **2.** A **3.** D **4.** B **5.** A **6.** B **7.** D **8.** D **9.** C **10.** D **11.** A **12.** A **13.** D **14.** C

CHEMICAL KINETICS Short answers (page 53)

1. a) i)

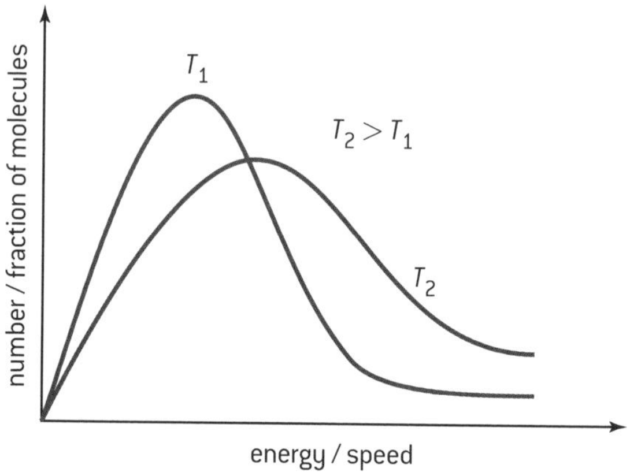

[2];

ii) Minimum energy required for the reaction to proceed **[1]**; iii) Increases rate of reaction by lowering the activation energy **[2]**

b) i)

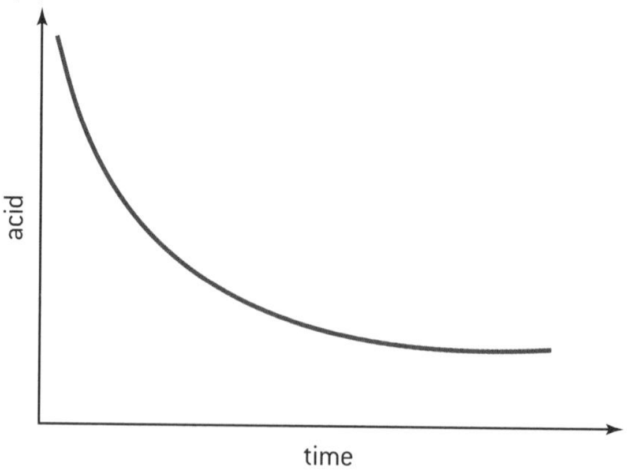

[1];

ii) Slope decreases **[1]**; iii) Rate decreases as fewer collisions per unit time. **[2]**

2. a) Reaction is complete so no more O_2 evolved **[1]** b) Rate = gradient of the tangent to the graph at 120 s = 0.017 mm s^{-1} **[3]**

c) i)

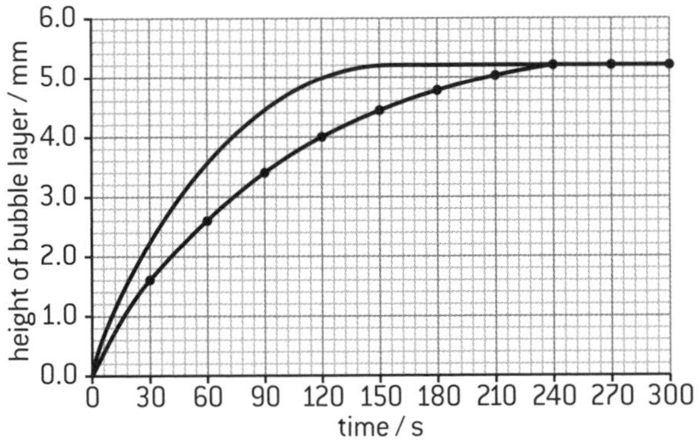

[1];

ii) Catalyst provides an alternative pathway with a lower activation energy **[1]** so more molecules have energy greater than or equal to activation energy. **[1]**

3. a) Doubling $[H_2]$ doubles the rate **[1]**; b) 2 **[1]**, (using experiments 1 and 4) as [NO] is halved, rate goes down to $\frac{1}{4}$ **[1]**; c) rate = $k[H_2(g)][NO(g)]^2$ **[1]**; d) Experiment 3 rate = 1.2×10^{-2} mol dm^{-3} s^{-1} **[1]**, Experiment 5 rate = 2.5×10^{-4} mol $dm^{-3}s^{-1}$ **[1]**; e) $k = \frac{4.0 \times 10^{-3}}{(2.0 \times 10^{-3})(4.0 \times 10^{-3})^2} = 1.3 \times 10^5$ **[1]**, mol^{-2} dm^6 s^{-1} **[1]**;

f) e.g. $NO + NO \rightleftharpoons N_2O_2$ (fast), $N_2O_2 + H_2 \rightarrow H_2N_2O_2$ (slow), $H_2N_2O_2 + H_2 \rightarrow 2H_2O + N_2$(slow) **[2]**;

g) A substance that increases the rate and is in a different phase to the reactants. **[1]**;

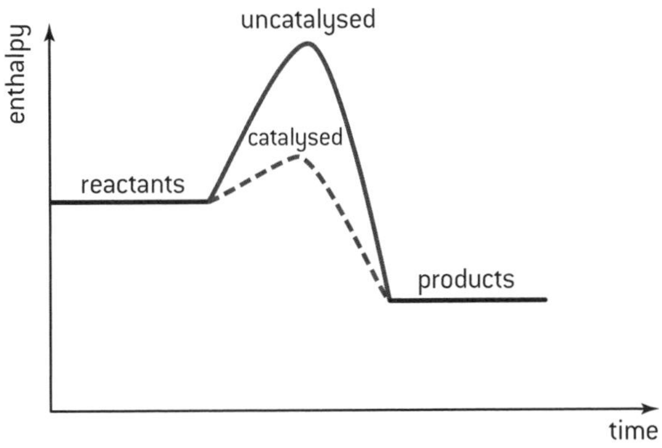

[2]

4. a) i) Rate = $k[NO(g)]^2[Cl_2(g)]$ **[1]**; ii) rate of reaction will decrease by a factor of 4, no effect on the rate constant **[2]**;
b) Above 775 K: rate = $k[NO_2(g)][CO(g)]$ **[1]**, Below 775 K: rate = $k[NO_2(g)]^2$ **[1]**; c) Zero-order reaction **[1]**, all concentrations are 1.0 mol dm^{-3} **[1]**

EQUILIBRIUM Multiple choice (page 57)

1. C **2.** B **3.** D **4.** B **5.** C **6.** A **7.** D **8.** C **9.** D
10. C **11.** D **12.** A

EQUILIBRIUM Short answers (page 58)

1. a) $K_c = \frac{[C_2H_5OH(g)]}{[C_2H_4(g)] \times [H_2O(g)]}$ **[1]**; b) i) Favours reactants (shifted to the left) **[1]**; ii) ΔH negative **[1]** as forward reaction is exothermic **[1]**, heat is absorbed when equilibrium moves to left. **[1]**;
c) Rate of reaction increased **[1]**, increased collision frequency **[1]**, equilibrium shifted to right **[1]** fewer moles of gas on right. **[1]**

2. a) $K_c = \frac{[NH_3(g)]^2}{[N_2(g)] \times [H_2(g)]^3}$ **[1]**; b) Shifts to the right (products) **[1]**, 4 mol decreases to 2 mol **[1]**; c) K_c decreases **[1]**, forward reaction is exothermic **[1]**; d) Lowers the activation energy of both forward and backward reaction equally. **[1]**

3. a) $K_c = \frac{[NO(g)]^4 \times [H_2O(g)]^6}{[NH_3(g)]^4 \times [O_2(g)]^5}$ **[1]**; b) i) Right **[1]**; ii) Right **[1]**; iii) Right **[1]**; iv) No change **[1]**

c) Minimum energy needed by colliding particles to react. **[1]**

4. a) Reactants and products in same phase **[1]**, rate of forward reaction = rate of reverse reaction**[1]**; b) $K_c = \frac{[HI(g)]^2}{[H_2(g)] \times [I_2(g)]}$ **[1]**;
c) No change to position of equilibrium or to K_c **[2]**

d) Reaction is exothermic **[1]**; e) $K_c' = \frac{1}{160} = 6.25 \times 10^{-2}$ **[1]**

5. a) $Br_2(g) \rightleftharpoons Br_2(l)$ **[1]**, $Br_2(g) + H_2(g) \rightleftharpoons 2HBr(g)$ **[1]**;
b) i) Increase in volume of liquid and no change in colour of vapour **[1]**; ii) Shift to the right (towards products) **[1]**;
iii) no effect as same amount of moles on both sides **[2]**;
c) i) $K_c = \frac{[HBr(g)]^2}{[H_2(g)] \times [Br_2(g)]}$ **[1]**; ii) no effect**[1]**

6. a) $\Delta S^\ominus = (\sum S^\ominus \textit{ products}) - (\sum S^\ominus \textit{ reactants}) = (2 \times 192) - [193 + (3 \times 131)] = -202$ J K^{-1} mol^{-1} **[2]**; b) $\Delta G^\ominus = \Delta H^\ominus - T\Delta S^\ominus = (-92 \times 1000) - 298 \times (-202) = -31\,804$ J (31.8 kJ) **[2]**; c) As the temperature increases $T\Delta S^\ominus$ becomes greater and eventually $\Delta G^\ominus$ becomes positive and the reaction becomes non-spontaneous **[2]**; d), $\Delta G = -RT\ lnK$, $-31804 = -8.31 \times 298 \times \ln K$, $K = e^{12.84} = 3.62 \times 10^5$ **[3]**;
e) $K_c = \frac{0.060^2}{[0.17 \times 0.11^3]} = 16$ **[3]**; f) The value for K_c is lower at temperature T_2 so T_2 must be higher than 298 K as the reaction is exothermic **[2]**; g) Increases the yield as 4 moles of gas on left decreases to 2 moles of gas on right **[2]**; h) More expensive to build plant to withstand higher pressure than 200 atm, 500 K gives reasonable yield in a reasonable time, a higher temperature would lower the yield and a lower temperature would slow the rate **[2]**; i) To increase the rate at which the position of equilibrium is reached. **[1]**

ACIDS AND BASES Multiple choice (page 68)

1. A **2.** B **3.** D **4.** C **5.** B **6.** A **7.** B **8.** D **9.** D **10.** A
11. B **12.** C **13.** C **14.** B **15.** A **16.** B **17.** D

ACIDS AND BASES Short answers (page 69)

1. a) i) An electron pair donor **[1]**; ii) HCl/Cl^- and H_3O^+/H_2O **[2]**;
b) i) A strong acid is fully dissociated, a weak acid is only partially dissociated **[2]**; ii) $CH_3COOH(aq) + NH_3(aq) \rightleftharpoons CH_3COO^- NH_4^+(aq)$ **[1]**; iii) Both react to give the same amount of hydrogen **[1]** and form a salt **[1]** but the hydrochloric acid will react faster **[1]** as the hydrogen ion concentration is greater. **[1]**

2. a) i) $[H^+(aq)] = 1.00 \times 10^{-2}$ mol dm^{-3} **[1]**, pH = 2 **[1]**;
ii) 3 **[2]** b) $H_2SO_4(aq) + 2NaOH(aq) \rightarrow Na_2SO_4(aq) + 2H_2O(l)$ **[1]**;
c) i) One drop will cause the colour of the solution to change from pink to colourless **[2]**; ii) Amount of $OH^-(aq)$ in 25.0 cm^3 of NaOH(aq) = $\frac{25.0}{1000} \times 1.00 \times 10^{-4} = 2.50 \times 10^{-6}$ mol. Amount of 5.00×10^{-5} H_2SO_4 required = 1.25×10^{-6} mol, volume = 25.0 cm^3 **[2]**

3. a) $HCl(aq) + NaHCO_3(aq) \rightarrow NaCl(aq) + H_2O(l) + CO_2(g)$ **[2]**;
b) i) $CO_3^{2-}(aq)$ **[1]**; ii) $H_2CO_3(aq)$ or $H_2O(l)/CO_2(g)$ **[1]**;
c) $2HCl(aq) + CuO(s) \rightarrow CuCl_2(aq) + H_2O(l)$ **[1]**, $2HCl(aq) + Na_2CO_3(aq) \rightarrow 2NaCl(aq) + H_2O(l) + CO_2(g)$ **[1]**

4. a) Rain water contains dissolved $CO_2(g)$ which can give it a pH as low as 5.6 so acid rain which contains other dissolved acids must have a pH below 5.6. **[2]**; b) $N_2(g) + O_2(g) \rightarrow 2NO(g)$ **[1]**, $2NO(g) + O_2(g) \rightarrow 2NO_2(g)$ **[1]** then either $2NO_2(g) + H_2O(l) \rightarrow HNO_3(aq) + HNO_2(aq)$ or $4NO_2(g) + O_2(g) + 2H_2O(l) \rightarrow 4HNO_3(aq)$ **[2]**; c) The $CaCO_3$ reacts with the acid to form soluble salts, $CaCO_3(s) + 2H^+(aq) \rightarrow Ca^{2+}(aq) + CO_2(g) + H_2O(l)$ **[2]**;
d) The hydroxide ions neutralize the hydrogen ions to form water, $OH^- + H^+(aq) \rightarrow H_2O(l)$ **[2]**

5. a) $H_2O(l) \rightleftharpoons H^+(aq) + OH^-(aq)$, endothermic as K_w increases with an increase in temperature **[2]**; b) At 90 °C $K_w \approx 38 \times 10^{-14}$, $[H^+(aq)] = [OH^-(aq)] = (38 \times 10^{-14})^{1/2} = 6.16 \times 10^{-7}$ mol dm^{-3} **[2]**, pH = 6.2 **[1]**

6. a) A nucleophile contains a non-bonded pair of electrons, which it can donate to form a bond **[1]**; b) i) CN^- **[1]**;
ii) The CN^- donates a pair of electrons to the atom which acts as a Lewis acid to form an intermediate. The oxygen ion in the intermediate behaves as a Lewis base donating a pair of electrons to the proton (Lewis acid) to form the product **[3]**;
c) Each ligand donates a pair of electrons to the transition metal ion so acts as a Lewis base. The transition metal ion is a good Lewis acid as it has a high charge density. **[3]**

7. a) i)$C_2H_5COOH(aq) + H_2O(l) \rightleftharpoons C_2H_5COO^-(aq) + H_3O^+(l)$ **[1]**, $K_a = \frac{[C_2H_5COO^-(aq)][H^+(aq)]}{[C_2H_5COOH(aq)]}$ **[1]**; ii) $[H^+(aq)] = (2.00 \times 10^{-3} \times 10^{-4.87})^{½} = 1.64 \times 10^{-4}$ mol dm^{-3}, pH = 3.8 **[3]**;
iii) The temperature is 25 °C as pK_a is measured at 25 °C **[1]**, the concentration of the acid at equilibrium is the same as the undissociated acid**[1]**; b) i) $Na^+(aq)$, $C_2H_5COO^-(aq)$, $C_2H_5COOH(aq)$, $H^+(aq)$ or $H_3O^+(aq)$, $OH^-(aq)$, $H_2O(l)$ **[2]**;
ii) Additional OH^- ions react with H^+ ions to form water but more of the acid dissociates to replace the H^+ ions so the concentration of H^+ ions remains almost constant. **[2]**

8. a) Pink **[1]**; b) Not suitable as the p$K_a \approx 9.5$ it will not change colour in the range 4–6 where there is a sudden change in pH at the end point for a strong acid/weak base titration **[2]**;
c) There is no sudden change in pH for a weak acid/weak base titration so no indicator will be accurate to one drop. **[1]**

REDOX PROCESSES Multiple choice (page 77)

1. C **2.** D **3.** A **4.** B **5.** B **6.** A **7.** B **8.** A **9.** D **10.** A
11. A **12.** C **13.** A **14.** C

REDOX PROCESSES Short answers (page 78)

1. a) $5Fe^{2+}(aq) + MnO_4^-(aq) + 8H^+(aq) \rightarrow Mn^{2+}(aq) + 5Fe^{3+}(aq) + 4H_2O(l)$ **[2]**; b) $Fe^{2+}(aq)$ **[1]**; c) $\frac{22.50}{1000} \times 2.152 \times 10^{-2} = 4.842 \times 10^{-4}$ mol **[2]**; d) $5 \times 4.842 \times 10^{-4} = 2.421 \times 10^{-2}$ mol **[2]**;
e) $\left(\frac{55.85 \times 2.421 \times 10^{-2}}{3.682 \times 10^{-1}}\right) \times 100 = 36.72\%$ **[2]**

2. a) The electrons flow from the Mg half-cell to the Fe half-cell showing that Mg is reducing $Fe^{2+}(aq)$. **[2]**; b) Fe **[1]**;
c) i) $Fe^{2+}(aq) + 2e^- \rightarrow Fe(s)$ **[1]**; ii) $Mg(s) + Fe^{2+}(aq) \rightarrow Fe(s) + Mg^{2+}(aq)$ **[2]**; d) Greater as Cu is below Fe in the activity series so Cu^{2+} is more readily reduced by Mg. **[2]**

3. a) i) A cell showing the container, liquid, electrodes and power supply (as shown on page 74), (+) electrode: Cl_2, (-) electrode Na(l) **[2]**; ii) $2Cl^-(aq) \rightarrow Cl_2(g) + 2e^-$, $Na^+(l) + e^- \rightarrow Na(l)$, $2NaCl(l) \rightarrow 2Na(l) + Cl_2(g)$ **[2]**; b) There are no free electrons and the ions are not able to move **[1]**; c) Al forms a protective layer or does not rust or is less dense. **[1]**

4. a) Oxidation states change: I from 0 to -1 and S from +2 to +2.5 **[2]**; b) In SO_4^{2-} the oxidation state of S is +6 and O is -2. Since an O atom has been replaced by an S atom the two S atoms are +6 and -2 in $S_2O_3^{2-}$ but all S atoms are assumed to be the same when calculating their oxidation state in an ion so the average is +2 **[2]**; In $S_4O_6^{2-}$ where the oxidation state of S works out to be 2.5 the man-made concept of oxidation states breaks down as the S atoms cannot have an oxidation state that is not a whole number and suggests that not all four S atoms are in the same chemical environment **[3]**; c) Addition of oxygen and increase in oxidation state both fit as definitions of oxidation but removal of hydrogen and loss of electrons do not fit as the carbon atom is surrounded by eight shared outer electrons in both diamond itself and in carbon dioxide. **[4]**

5. a) $Cl_2(g) + 2Ag(s) \rightarrow 2Ag^+(aq) + 2Cl^-(aq)$ **[2]**; b) from the chlorine half-cell to the silver half-cell **[1]**; c) 0.56 V **[1]**;
d) $\Delta G^{\ominus} = -nFE^{\ominus} = -2 \times 96500 \times 0.56 = 1.08 \times 10^5$ J, energy produced = 108 kJ. **[2]**

6. a) i) (+) (mainly) $O_2(g)$, (-) $H_2(g)$ and ii) (+) (mainly) $Cl_2(g)$, (-) $H_2(g)$ **[3]**; b) Both give $H_2(g)$ and $O_2(g)$ at the (-) and (+) electrodes respectively in the ratio of 2:1. **[3]**; c) Make the spoon the (-) electrode (cathode) and pass a current of electricity through a solution of silver ions (e.g. $AgNO_3$ (aq)). **[2]**

ORGANIC CHEMISTRY Multiple choice (page 97)

1. B **2.** C **3.** C **4.** D **5.** D **6.** A **7.** B **8.** D **9.** A
10. D **11.** C **12.** B **13.** A **14.** B **15.** C **16.** A **17.** A

ORGANIC CHEMISTRY Short answers (page 98)

1. a) Boiling points increase **[1]**, increasing size leads to greater contact/surface area **[1]**, and greater London dispersion forces **[1]**; b) *Any two from*: same general formula, successive members differ by CH_2, same functional group, similar chemical properties, gradual change in physical properties **[2]**;

c) **A:**

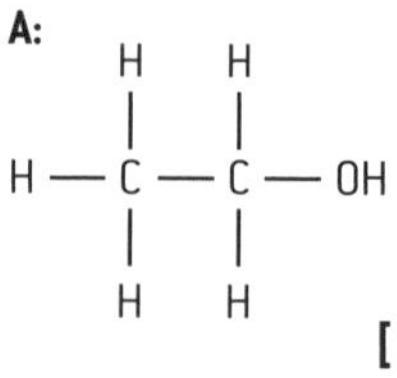

[1]

B:

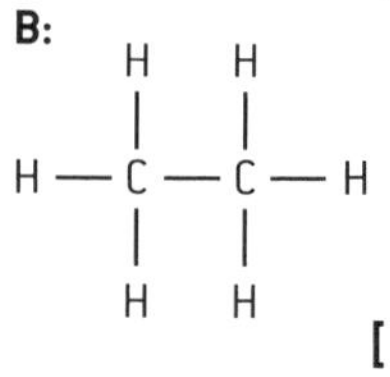

[1]

C:

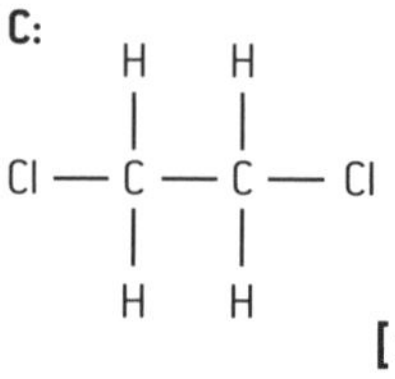

[1]

D:

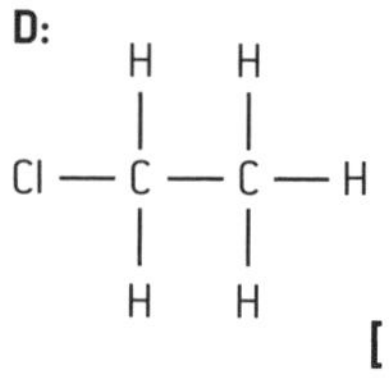

[1]

C is 1, 2-dichlorethane **[1]**

d) Add bromine water **[1]**, pentane no change/stays brown, pent-1-ene decolourizes the bromine water **[1]**; e) **E** contains two H atoms and one R group bonded to C so primary **[2]**, **F** contains two R groups so secondary **[2]**; f) *Initiation:* $Cl_2 \rightarrow 2Cl\cdot$ in ultraviolet light **[1]**, *propagation*: $Cl\cdot + CH_4 \rightarrow CH_3\cdot + HCl$ and $CH_3\cdot + Cl_2 \rightarrow CH_3Cl + Cl\cdot$ **[2]**, *termination: any one from:* $Cl\cdot + Cl\cdot \rightarrow Cl_2$/$CH_3\cdot + Cl\cdot \rightarrow CH_3Cl$/$CH_3\cdot + CH_3\cdot \rightarrow C_2H_6$ **[1]**

2. a) i) *Any one of* C1 to C6 **[1]**, ii) C7 **[1]**, iii) C8 or C9 **[1]**;
b) C8 to O3 is a single bond, C8 to O4 is a double bond **[1]**, less shared electron pairs so single bond is longer than double bond **[1]**;
c) C1 to C6 is part of benzene ring so bond length between single and double bond and shorter than single bond between C1 and C7 **[2]**; d) i) Reflux with an excess of acidified potassium dichromate(VI) solution **[2]**; ii) Warm with a carboxylic acid in the presence of concentrated sulfuric acid. **[2]**

3.

a)

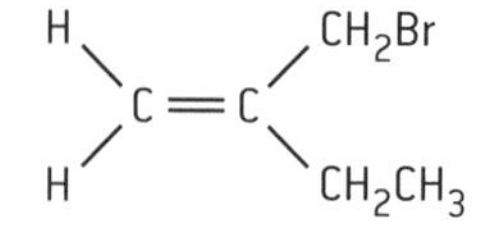

[1]

b)

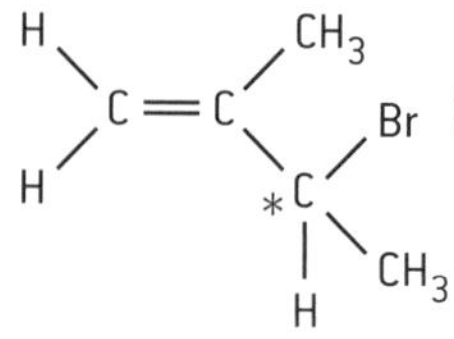

[1]

c)

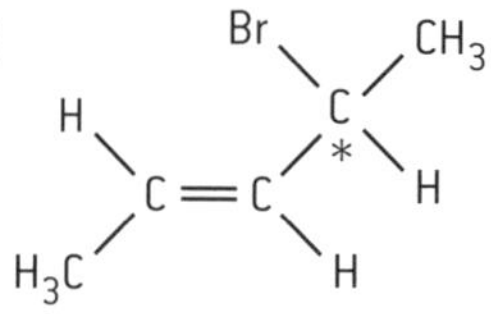

[1]

d)

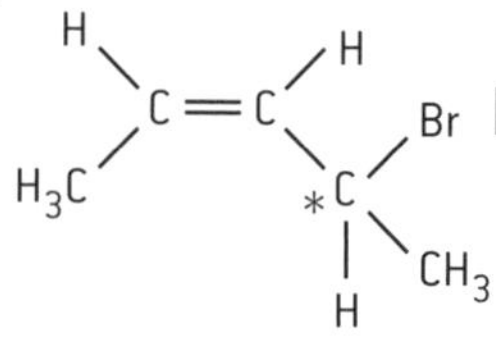

[1]

e)

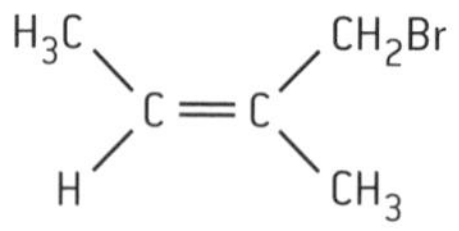

[1]

f)

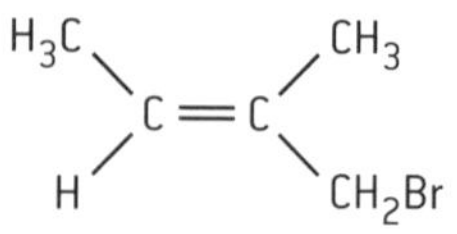

[1]

Note that other correct answers are possible, e.g. $(CH_3)_2C=C(CH_3)Br$ for a).

4. a) Convert half of the propan-1-ol into propanoic acid by refluxing with excess acidified potassium dichromate(VI) solution **[1]**, $3CH_3CH_2CH_2OH + 2Cr_2O_7^{2-} + 16H^+ \rightarrow 3CH_3CH_2COOH + 4Cr^{3+} + 11H_2O$ **[1]**; warm the propanoic acid with the remaining propan-1-ol and a few drops of concentrated sulfuric acid as a catalyst **[1]**, $CH_3CH_2COOH + CH_3CH_2CH_2OH \rightleftharpoons CH_3CH_2COOCH_2CH_2CH_3 + H_2O$ **[1]**; b) Add hydrogen bromide to propene (Markovnikov addition) **[1]**, $CH_3CH{=}CH_2 + HBr \rightarrow CH_3CHBrCH_3$ **[1]**; warm with aqueous sodium hydroxide **[1]**, $CH_3CHBrCH_3 + NaOH \rightarrow CH_3CH(OH)CH_3 + NaBr$ **[1]**;
heat with acidified potassium dichromate(VI) solution **[1]**, $3CH_3CH(OH)CH_3 + Cr_2O_7^{2-} + 8H^+ \rightarrow 3CH_3COCH_3 + 2Cr^{3+} + 7H_2O$ **[1]**;

c) Add hydrogen with a nickel catalyst at 180 °C **[1]**, $C_3H_6 + H_2 \rightarrow C_3H_8$ **[1]**; chlorine and ultraviolet light **[1]**, $CH_3CH_2CH_3 + Cl_2 \rightarrow CH_3CH_2CH_2Cl + HCl$ **[1]**; warm with aqueous sodium hydroxide solution **[1]**, $CH_3CH_2CH_2Cl + NaOH \rightarrow CH_3CH_2CH_2OH + NaCl$ **[1]**; warm with acidified potassium dichromate(VI) solution and distil the product as it is formed **[1]**, $3CH_3CH_2CH_2OH + Cr_2O_7^{2-} + 8H^+ \rightarrow 3CH_3CH_2CHO + 2Cr^{3+} + 7H_2O$ **[1]**

5. a) The carbon atom has a small positive charge (δ^+) as the C–Br bond is polar due to bromine being more electronegative than carbon this attracts the OH^- nucleophile. **[1]**; b) Rate = k [R–Br] [OH^-] **[1]**, S_N2 **[1]**;

c)

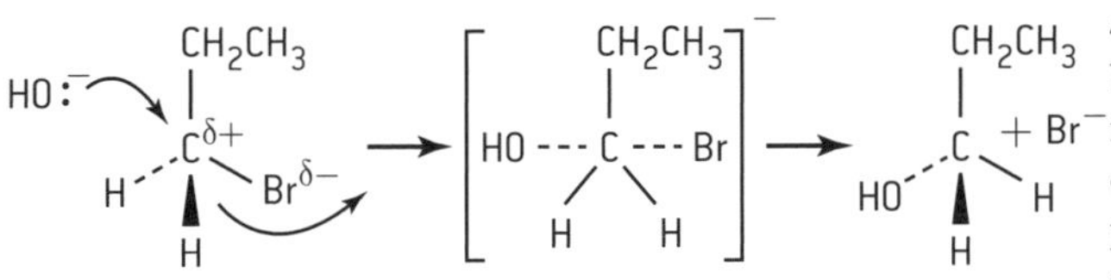

[4]

d) Slower rate **[1]** as the C–Cl bond is stronger than the C–Br bond so harder to break. **[1]**

6. H_2SO_4 acts as a catalyst **[1]**. It protonates the nitric acid to form the NO_2^+ electrophile **[1]**. $HNO_3 + H_2SO_4 \rightarrow H_2NO_3^+ + HSO_4^-$ **[1]**, $H_2NO_3^+ \rightarrow H_2O + NO_2^+$ **[1]**. After the NO_2^+ has reacted with benzene to form a positive intermediate the HSO_4^- ion extracts a proton to form nitrobenzene and regenerate the sulfuric acid **[1]**, $C_6H_6NO_2^+ + HSO_4^- \rightarrow C_6H_5NO_2 + H_2SO_4$ **[1]**

MEASUREMENT, DATA PROCESSING AND ANALYSIS

Multiple choice (pages 107 and 108)

1. B **2.** C **3.** B **4.** A **5.** D **6.** D **7.** C **8.** A **9.** B **10.** C **11.** A **12.** B **13.** D **14.** C **15.** A **16.** D **17.** C **18.** D

MEASUREMENT, DATA PROCESSING AND ANALYSIS

Short answers (page 109)

1. a) Mass: 2, temperature: 4, pressure: 3 **[1]**; b) $\frac{0.072 \times 10^3}{22.99 + (3 \times 14.01)}$ = 1.1 mol **[2]**; c) mass: $\left(\frac{0.001}{0.072}\right) \times 100 = 1.4\%$, temperature: $\left(\frac{0.50}{20.00}\right) \times 100 = 2.5\%$, pressure: $\left(\frac{1}{106}\right) \times 100 = 0.94\%$ **[1]**; $n(N_2) = \frac{3}{2} \times 1.1 = 1.65$ mol [1], $V = \frac{nRT}{P} = \frac{(1.65 \times 8.31 \times 293)}{106} \times 10^3 = 3.8 \times 10^{-2}$ m^3 = 38 dm^3 (note answer to 2 significant figures as mass only given to 2 sf) **[2]**; total uncertainty = 1.4 + 2.5 + 0.94 = 4.8%, volume = 38 ± 4.8% dm^3 = 38 ± 2 dm^3 **[2]**.

2. C: $\frac{15.40}{12.01}$: H: $\frac{3.24}{1.01}$: I: $\frac{81.36}{126.9}$ gives C: 1.28 : H: 3.21 : I: 0.641 i.e. C_2H_5I **[1]**; b) i) Mass spectrum shows molecular mass is 156 so molecular formula is C_2H_5I **[1]**; b) ii) IHD = $\frac{(2 \times 2 + 2 - 6)}{2} = 0$ **[1]**; b) iii) 127 due to I^+ and 29 due to $C_2H_5^+$ **[2]**; iv) iodoethane **[1]**; c) Hydrogen atoms in CH_3CH_2 are in two different environments with three H atoms in one and two H atoms in the other so two peaks in the ratio of 3:2. **[2]**

3. Pentan-2-one: 4 peaks in ratio of 3:2:2:3 **[2]**; pentan-3-one: 2 peaks in ratio of 3:2. **[2]**

4. a) As a reference **[1]**; b) quartet **[1]**; c) 3:3:2 **[1]**; d) $RCOOCH_2R$ **[1]**, no protons due to phenyl group present in spectrum or other structure has hydrogen atoms in only two different chemical environments **[1]**; e) A: C–H **[1]**, B: C=O **[1]**, C: C–O **[1]**; f) 88: $C_4H_8O_2^+$ **[1]**, 73: $C_3H_5O_2^+$ or $(M - CH_3)^+$ **[1]**, 59: $C_2H_3O_2^+$ or $(M - C_2H_5)^+$ **[1]**; g) ethyl ethanoate **[1]**, $CH_3COOC_2H_5$. **[1]**

OPTION A – MATERIALS (page 123)

1. a) Iron oxide can be reduced relatively easily by carbon monoxide **[1]**. Molten sodium chloride required the discovery of electricity to extract sodium as Na is much higher than Fe in the activity series **[1]**; b) Paramagnetism is associated with unpaired electrons whereas diamagnetism occurs when all the electrons are spin paired **[2]**; c) *Any two from* change in tensile strength / melting point / density / malleability / brittleness. **[2]**

2. a) 1000 kg = $\frac{1.00 \times 10^6}{26.98}$ = 3.706 × 10^4 mol of Al **[1]**; since $Al^{3+}(l) + 3e^- \rightarrow Al(s)$ amount of charge required = 3 × 3.706 × 10^4 = 1.112 × 10^5 F **[1]**; time = $\frac{(1.112 \times 10^5 \times 9.65 \times 10^4)}{2.00 \times 10^5}$ = 5.365 × 10^4 s = 14.9 h **[1]**; b) Carbon dioxide, which is a greenhouse gas, released by burning fossils fuels to produce the electricity **[1]**, transport of ore and product **[1]** and from oxidation of carbon electrodes. **[1]**

3. a) They are polar **[1]**; b) *Any two from* graphite / cellulose / silk / Kevlar / DNA / biphenyl nitriles / soap **[1]**; c) *Thermotropic*: pure substances **and** exhibit liquid crystal properties in a certain temperature range **[1]**; *lyotropic*: solutions **and** exhibit liquid-crystal properties in a certain concentration range. **[1]**

4. a) *Homogeneous*: catalyst in the same phase as the reactants, proceed by formation of an intermediate compound **[2]**; *heterogeneous:* catalyst in a different phase to reactants, work by adsorption of reactants onto surface **[2]**; b) *Any three from* cost / selectivity / susceptibility to poisoning / environmental impact / efficiency **[3]**; c) To increase the surface area making it more efficient. **[1]**

5. a)

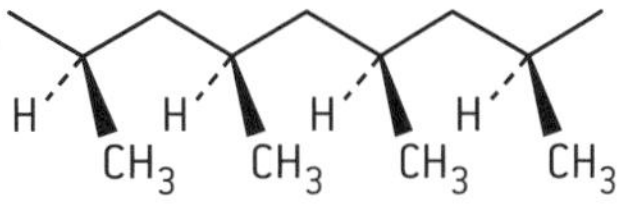

[2]

b) More crystalline **[1]**, tougher **[1]**; c) LDPE has more side chains (branching) whereas HDPE has little branching so the chains can fit more closely together **[2]**. Since all the $CH_3CH{=}CH_2$ molecules are converted into the polymer the atom economy is 100%. **[1]**

6. a) In CVD a carbon-containing gas in the plasma phase is decomposed in the presence of an inert carrier gas, if oxygen was present the carbon would be oxidized to carbon dioxide. **[2]**; b) Nanotubes contain delocalized electrons. **[1]**

7. a)

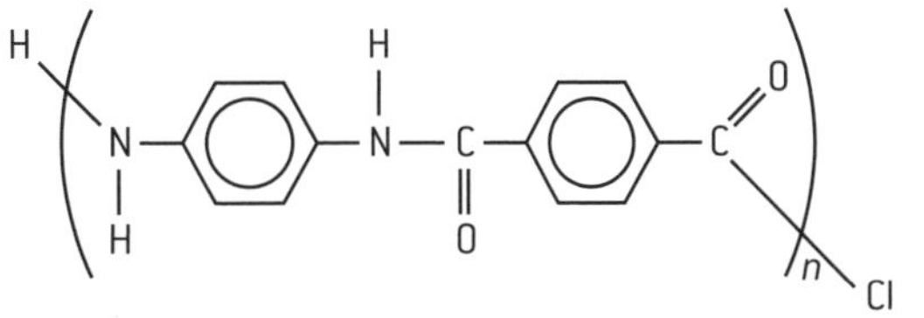

[2]

b) It is able to form cross links due to hydrogen bonding giving it a very ordered three-dimensional structure **[2]**; c) The acid can protonate the N atoms so that they are unable to form hydrogen bonds and hence it destroys the 3-D structure (strong acid can also break down the amide links within the chains). **[2]**

8. a) K_{sp} = [Cd^{2+}(aq)] × [OH^-(aq)]^2 and [OH^-(aq)] = 2[Cd^{2+}(aq)] **[1]**; K_{sp} = 7.20 × 10^{-15} = [Cd^{2+}(aq)] × (2[Cd^{2+}(aq)])^2 = 4[Cd^{2+}(aq)]^3 **[1]**; Hence [Cd^{2+}(aq)] = [$Cd(OH)_2$(aq)] = $\left(\frac{7.20}{4} \times 10^{-15}\right)^{\frac{1}{3}}$ = $(1.80 \times 10^{-15})^{\frac{1}{3}}$ = 1.22 × 10^{-5} mol dm^{-3} **[1]**; b) Since K_{sp} = [Cd^{2+}(aq)] × [OH^-(aq)]^2 adding more hydroxide ions must reduce the concentration of cadmium ions to keep K_{sp} constant **[2]**; c) Ethane-1,2-diamine contains two amino groups both of which contain a non-bonding pairs of electrons which can form coordinate bonds with a transition metal ion making it a bidentate ligand. **[2]**

9. a) 8 × $\frac{1}{8}$ = one **[1]**; b) 6 **[1]**. Using the equation $\lambda = 2d\sin\theta$ where X-rays with wavelength λ produce constructive interference when diffracted at an angle θ to the surface of the crystal to give the distance, d. **[2]**

OPTION B – BIOCHEMISTRY (page 140)

1. a) i) H_3N^+–$CH(CH_2CH_2SCH_3)$–COOH **[1]**; ii) H_3N^+–$CH(CH_2CH_2SCH_3)$–COO^- **[1]**; iii) H_2N–$CH(CH_2CH_2SCH_3)$–COO^- **[1]**; b) H_2N–$CH(CH_2CH_2SCH_3)$–CONH–$CH(CH_3)$–COOH **[1]**; H_2N–$CH(CH_3)$–CONH–$CH(CH_2CH_2SCH_3)$–COOH **[1]**; c) 6 **[1]**; d) Paper chromatography **[1]**; hydrolyse the protein with hot hydrochloric acid to release the separate amino acids **[1]**; spot the sample on paper together with samples of known amino acids above the level of a suitable solvent and let the solvent rise up the paper **[1]**; develop the chromatogram using ninhydrin and compare the R_f values of the spots. **[1]**

2. a) Lipids (or fats) **[1]**; b) Vegetable **[1]** as all three fatty acid residues are unsaturated **[1]**; c) Liquid **[1]** as the structure around the C=C double bonds prevents close packing **[1]**; d) *Any two from* energy storage or source / insulation / cell membrane. **[2]**

3. a) i) Alkene **[1]** hydroxyl **[1]**; ii) Vitamin A is fat soluble **[1]** as it has a long non-polar hydrocarbon chain **[1]** and vitamin C is water soluble as it contains many −OH groups which can hydrogen bond with water **[1]**; iii) *Vitamin A:* night blindness/ xerophthalmia **[1]**, *Vitamin C:* bleeding of gums etc. **[1]**, scurvy **[1]**; b) Amount of $I_2 = \frac{14.2}{(2 \times 126.9)} = 0.056$ mol **[1]**, each molecule of oil contains 4 C=C double bonds. **[2]**

4. a) i) Using a dye/ninhydrin **[1]**; ii) glutamic acid **[1]**; isoelectric point is below pH of buffer **[1]**; it becomes negatively charged **[1]**; iii) it forms a zwitterion with no overall charge **[1]**; b) $H_2NCH_2COOH + H^+ \rightleftharpoons H_3N^+CH_2COOH$ **[1]**, $H_2NCH_2COOH + OH^- \rightleftharpoons H_2NCH_2COO^- + H_2O.$ **[1]**

5. a) Form coloured compounds **[1]**, form complex ions **[1]**; b) i) 55% − 25% = 30% **[1]**; ii) CO_2 and lactic acid are acidic so lower the pH **[1]**, haemoglobin is less able to bind to oxygen at lower pH. **[1]**

6. a) It contains a chiral carbon atom **[1]**; b) $K_a = 1.38 \times 10^{-4}$ = $[H^+] \times [CH(CH_3)(OH)COO^-] / [CH(CH_3)(OH)COOH] \approx$ $[H^+]^2/0.100$ **[1]**; $[H^+] = (1.38 \times 10^{-5})^{\frac{1}{2}} = 0.00371$ mol dm^{-3} **[1]**, pH = 2.43 **[1]**; c) M_r(lactic acid) = (3 × 12.01) + (6 × 1.01) + (3 × 16.00) = 90.1 and 2.00 g of NaOH = $\frac{2}{(22.99 + 16.00 + 1.01)}$ = 0.0500 mol **[1]**, $K_a = 1.38 \times 10^{-4} = \frac{[H^+] \times [CH(CH_3)(OH)COO^-]}{[CH(CH_3)(OH)COOH]}$ and pH = 4.00 so $\frac{[CH(CH_3)(OH)COO^-]}{[CH(CH_3)(OH)COOH]} = \frac{1.38 \times 10^{-4}}{10^{-4}} = 1.38$ **[1]**, $[CH(CH_3)(OH)COO^-] = [NaOH] = 0.1$ so $[CH(CH_3)(OH)COOH]$ = 0.0725 mol dm^{-3} **[1]**, mass required in 500 cm^3 = $[(\frac{0.725}{2})$ + 0.0500 (to react with the NaOH)] × 90.1 = 7.77 g. **[1]**

OPTION C – ENERGY (page 153)

1. a) The combination of two light nuclei to form a heavier nucleus **[1]**; b) The mass defect **[1]** is converted into energy according to $E = mc^2$ **[1]**; c) The problems of controlling and maintaining the plasma at very high temperatures have not been overcome **[2]**; X = n, Y = Pu, Z = e$^-$, a = 92, b = 1, c = 0, d = 0, e = − 1. **[4]**

2. a) $^{238}_{94}Pu \rightarrow ^{234}_{92}U + ^{4}_{2}He$ **[2]**; b) The time an individual atom takes to decay is not fixed **[1]**, half-life is the average time for many different atoms **[1]**; 2→1→0.5→0.25→0.125→0.0625 = 5 half-lives = 440 years. **[2]**

3. a) *Cracking:* breaking larger hydrocarbons into smaller ones **[1]**, *reforming:* making branched or cyclic hydrocarbons from straight chain hydrocarbons **[1]**; b) C_2H_4: = $(\frac{1411}{28.04}) \times 10^3 = 5.03 \times 10^4$ kJ kg^{-1} **[1]**, C_8H_{18}: = $(\frac{5470}{114.26}) \times 10^3 = 4.79 \times 10^3$ kJ kg^{-1} **[1]**; c) It combusts as efficiently as a mixture of 5% heptane and 95% 2,2,4-trimethylpentane **[2]**; d) Not so easy to transport as liquid gasoline, more useful as a feedstock for plastics. **[2]**

4. a) Extensive conjugation **[1]**; b) $6CO_2 + 6H_2O \rightarrow C_6H_{12}O_6 + 6O_2$ **[2]**; c) i) Too viscous, as molar mass is too high. **[2]**

ii)
$$H_2C-O-COR \qquad H_2C-O-COH$$
$$| \qquad\qquad\qquad\qquad |$$
$$HC-O-COR+ 3C_2H_5OH \rightleftharpoons HC-O-COH + 3C_2H_5-O-CO-R$$
$$| \qquad\qquad\qquad\qquad |$$
$$H_2C-O-COR \qquad H_2C-O-COH$$
[2]

iii) To move the position of equilibrium to the right to increase the amount of biodiesel formed. **[1]**

5. a) Greenhouse gases allow the passage of incoming shortwave radiation but absorb some of the longer wavelength reflected radiation from the Earth and re-radiate it back to the Earth **[2]**. Unlike N_2 and O_2, the bond vibrations of CO_2 and H_2O involve a

change in dipole moment so they can absorb infrared radiation **[2]**; b) Increased use of fossils fuels **[1]**, there is an equilibrium between CO_2 in the atmosphere and dissolved CO_2 in sea-water **[1]**, more CO_2 in water forms more carbonic acid which weakly dissociates to H^+ ions **[1]**; c) *Any two from*: move toward more renewable forms of energy, use carbon credits, insulate homes better, increase carbon capture. **[2]**

6. a) $E = +0.34 - \frac{(8.31 \times 298)}{(2 \times 96500)} \ln 0.01 = 0.40$ V **[2]**; b) 0.40 − (− 0.76) = 1.16 V **[2]**; c) The standard cell will have a lower emf of 1.10 V **[1]**, but will last longer, i.e. do more total work **[1]**, as more materials are present. **[1]**

7. a) Total mass of 92 protons and 146 neutrons = (92 × 1.672622 × 10^{-27}) + (146 × 1.674927 × 10^{-27}) = 3.98420566 × 10^{-25} kg. Mass of one atom of ^{238}U = $\frac{238.050789 \times 10^{-3}}{6.02 \times 10^{23}}$ = 3.954332043 × 10^{-25} kg, mass defect = (3.954332043 × 10^{-25} − 3.98420566 × 10^{-25}) = 2.987 × 10^{-27} kg **[1]**, E = 2.987 × 10^{-27} × (3.00 × 10^8)2 = 2.69 × 10^{-10} J (2.69 × 10^{-13} kJ) **[1]**; b) it is converted into UF$_6$ which is a gas **[1]**, this can **either** then be allowed to effuse and the lighter ^{235}UF$_6$ will effuse at a faster rate than ^{238}UF$_6$ **or** by gas centrifugation where the heavier ^{238}UF$_6$ moves to the outside of the container **[2]**; c) $\lambda = \frac{\ln 2}{69} = 1.00 \times$ 10^{-2} year^{-1} **[1]**, $t = \frac{\ln(\frac{100\%}{1.00\%})}{1.00 \times 10^{-2}}$ = 461 years. **[1]**

8. a) *Any three from:* simpler to manufacture, semi-flexible, semi-transparent, cheaper, absorb visible light over a larger range of wavelengths **[3]**; b) The electrolyte reduces the oxidized form of the dye and is itself oxidized to triiodide ions at the anode **[1]**, 3I$^-$ → I$_3^-$ + 2e$^-$ **[1]**, at the cathode electrons are received from the external circuit and triiodide ions are reduced to iodide ions **[1]**, I$_3^-$ + 2e$^-$ → 3I$^-$ **[1]**; c) It acts as a semiconductor and covalently bonds to the dye **[2]**, nanoparticles increase the surface area to attach the dye. **[1]**

OPTION D – MEDICINAL CHEMISTRY (page 166)

1. a) The therapeutic index for animals uses LD_{50} and it is not reasonable to test a drug to kill 50% of the human population so toxic dose is used instead **[1]**; b) *Therapeutic window*: the range of dosages between the minimum amounts of the drug that produce the desired effect and a medically unacceptable adverse effect **[1]**; *therapeutic index* is equal to TD_{50} divided by ED_{50} **[1]**; c) Morphine may be taken to ease pain in which case constipation etc. is a side-effect **[2]**; or taken for diarrhoea in which case side effects such as nausea, sleepiness etc. may be experienced **[1]**; d) *tolerance:* require more of the drug to achieve the same effect **[1]**; *addiction:* going without the drug causes withdrawal symptoms. **[1]**

2. a) Increases the resistance to penicillinase enzyme or alters the effectiveness of the penicillin **[1]**; b) Some bacteria may remain unaffected **[1]**; c) Increases the risk of bacteria becoming resistant **[1]**; d) Due to ring strain the beta-lactam group breaks easily **[1]** to form two parts that bond to an enzyme in the bacterium **[1]** so that it is unable to make cell walls. **[1]**

3. a) Esterification (also condensation) **[1]**; b) 2-hydroxybenzoic acid will have one peak whereas aspirin will have two as it has two functional groups containing C=O **[2]**;

c)
$$H_2N-\langle\!\!\langle\;\;\rangle\!\!\rangle-OH$$ **[2]**;

d) *Aspirin: any one from:* stomach bleeding, Reye's syndrome, allergic reactions **[1]**, *paracetamol:* overdose can permanently damage kidneys or liver. **[1]**

4. a) i) Antivirals **[1]**, influenza **[1]**; ii) Zanamivir **[1]**, it contains many polar hydroxyl groups **[1]**; b) i) Proton pump inhibitor **[1]**, *any one from* heartburn, stomach ulcer, gastric reflux, indigestion **[2]**; ii) They are mirror images/enantiomers/ optical isomers. **[1]**

5. a) The different enantiomers can have different biological properties, one form may be beneficial, the other form may be

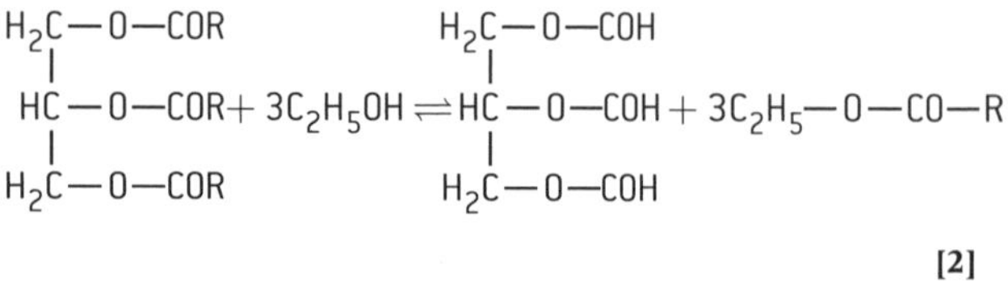

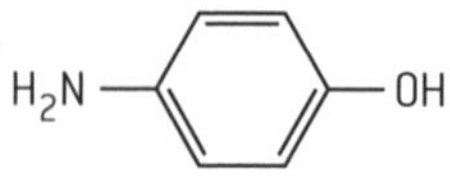

harmful **[2]**; b) A chiral auxiliary is itself an enantiomer **[1]**, it is bonded to the reacting molecule to create the stereochemical conditions necessary to follow a certain geometric path **[1]**, once the desired enantiomer is formed the auxiliary is removed **[1]**.

c)

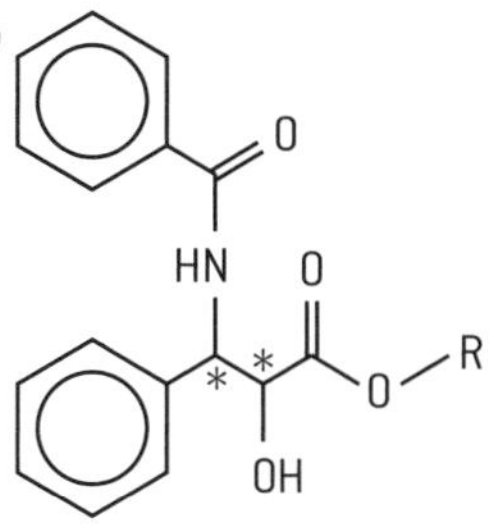

[2]

6. a) i) $^{212}_{82}Pb \rightarrow\ ^{208}_{80}Hg +\ ^{4}_{2}He$ **[2]**. ii) TAT involves taking the alpha emitter directly to the tumour **[1]**, BNCT involves targeting the tumour with boron-10, a non-radioactive isotope **[1]**, then bombarding it with neutrons so that it then emits radiation on the tumour **[1]**; b) i) $^{131}_{53}I \rightarrow\ ^{131}_{54}Xe +\ ^{0}_{-1}e^-$ **[2]**; ii) $\lambda = \frac{0.693}{8.02} =$ 0.08641, $\ln(\frac{1}{x}) = 0.08641 \times 31.0 = 2.679$ **[1]**, $x = \frac{1}{14.57} = 0.0686$, so 6.86% remaining **[1]**; iii) it has a shorter half-life **[1]**, it stays in the body for a shorter time. **[1]**

Origin of individual questions

STOICHIOMETRIC RELATIONSHIPS
Multiple choice: 1. N11SLP1(1) **11.** N11HLP1(3) **12.** N11SLP1(4) **13.** N10HLP1(2) **15.** N10SLP1(4)

Short answer: 3. M12TZ2SLP2(2) **4.** M12TZ1SLP2(2) **5.** Adapted from N09SLP2(1) **8.** M05SLP2(2)

ATOMIC STRUCTURE
Multiple choice: 3. N99SLP1(5) **4.** M11TZ1HLP1(6) **5.** N11SLP1(5) **6.** M12TZ2HLP1(5) **8.** N11HLP1(6) **15.** M98HLP1(8) **16.** N10HLP1(5)

Short answer: 5. M10TZ2SLP2(2) **6.** M11TZ2HLP2(3a)

PERIODICITY
Multiple choice: 6. N10SLP1(9) **7.** N10HLP1(7) **12.** M00HLP1(8)

Short answer: 5. Adapted from M10TZ2SLP2(5) **6.** N09SLP2(4c)

CHEMICAL BONDING AND STRUCTURE
Multiple choice: 1. N11SLP1(9) **2.** N11SLP1(12) **4.** N10HLP1(11) **5.** M99SLP1(12) **6.** N10SLP1(13) **7.** N10HLP(10) **13.** N10HLP1(13)

Short answer: 1. N09SLP2(2) **2.** M09TZ2SLP2(7b) **3.** M12TZ1SLP2(5a) **4.** M12TZ2SLP2(5b) **8.** M11TZ1HLP2(6c,d,e & f)

ENERGETICS/THERMOCHEMISTRY
Multiple choice: 3. N11SLP1(14) **4.** M99SLP1(16) **5.** N10HLP1(15) **7.** N99SLP1(17) **9.** N10HLP1(17) **10.** M12TZ1HLP1(16) **11.** N10HLP1(18) **12.** M12TZ1HLP1(17)

Short answer: 1. M10TZ1SLP2(6a) **2.** N10SLP2(1a–d) **3.** M11TZ1SLP2(4a&b)

CHEMICAL KINETICS
Multiple choice: 1. N10SLP1(17) **2.** N10HLP1(19) **3.** M99SLP1(19) **4.** N09SLP1(18) **5.** N09SLP1(19) **6.** M12TZ2SLP1(17) **7.** M12TZ2SLP1(18) **8.** M09TZ2SLP1(17) **9.** M09TZ1SLP1(18) **10.** M11TZ2SLP1(17) **11.** M12TZ1HLP1(20) **12.** M12TZ2HLP1(19) **13.** M12TZ2HLP1(19) **14.** N10HLP1(20)

Short answer: 1. N07SLP2(7b&c) **2.** M12TZ2SLP2(1a,b &c) **3.** M08TZ2HLP2(1) **4.** N09HLP2(6a–c)

EQUILIBRIUM
Multiple choice: 2. M12TZ2SLP1(19) **3.** N11SLP1(20) **4.** M09TZ2SLP1(19) **5.** M09TZ1SLP1(21) **6.** M10TZ1SLP1(20) **7.** N10SLP1(20) **9.** M10TZ2HLP1(23) **11.** N98HLP1(27)

Short answer: 1. M08TZ2SLP2(3) **2.** M07SLP2(4) **3.** N11SLP2(6a) **4.** M11TZ2SLP2(6) **5.** M12TZ1HLP2(6a,b&c)

REDOX PROCESSES
Short answer: 1. N09SLP2(1) **3.** N11SLP2(3)

ORGANIC CHEMISTRY
Short answer: 1. M11TZ1SLP2(6)

Option A – MATERIALS
Short answer: 3. M13TZ1SLP3(C3)

Option B – BIOCHEMISTRY:

4. M11TZ2SLP3(B2)

Index

Page numbers in *italics* refer to question sections.

Periodic table for use with the IB

atomic number
element
relative atomic mass

	1	2	3	4	5	6	7	8	9	10	11	12	13	14	15	16	17	18
1	1 H 1.01																	2 He 4.00
2	3 Li 6.94	4 Be 9.01											5 B 10.81	6 C 12.01	7 N 14.01	8 O 16.00	9 F 19.00	10 Ne 20.18
3	11 Na 22.99	12 Mg 24.31											13 Al 26.98	14 Si 28.09	15 P 30.97	16 S 32.07	17 Cl 35.45	18 Ar 39.95
4	19 K 39.10	20 Ca 40.08	21 Sc 44.96	22 Ti 47.87	23 V 50.94	24 Cr 52.00	25 Mn 54.94	26 Fe 55.85	27 Co 58.93	28 Ni 58.69	29 Cu 63.55	30 Zn 65.38	31 Ga 69.72	32 Ge 72.63	33 As 74.92	34 Se 78.96	35 Br 79.90	36 Kr 83.90
5	37 Rb 85.47	38 Sr 87.62	39 Y 88.91	40 Zr 91.22	41 Nb 92.91	42 Mo 95.96	43 Tc [98]	44 Ru 101.07	45 Rh 102.91	46 Pd 106.42	47 Ag 107.87	48 Cd 112.41	49 In 114.82	50 Sn 118.71	51 Sb 121.76	52 Te 127.60	53 I 126.90	54 Xe 131.29
6	55 Cs 132.91	56 Ba 137.33	57 * La 138.91	72 Hf 178.49	73 Ta 180.95	74 W 183.84	75 Re 186.21	76 Os 190.23	77 Ir 192.22	78 Pt 195.08	79 Au 196.97	80 Hg 200.59	81 Tl 204.38	82 Pb 207.2	83 Bi 208.98	84 Po [209]	85 At [210]	86 Rn [222]
7	87 Fr [223]	88 Ra [226]	89 * Ac* [227]	104 Rf [267]	105 Db [268]	106 Sg [269]	107 Bh [270]	108 Hs [269]	109 Mt [278]	110 Ds [281]	111 Rg [281]	112 Cn [285]	113 Uut [286]	114 Uuq [289]	115 Uup [288]	116 Uuh [293]	117 Uus [294]	118 Uuo [294]

*	58 Ce 140.12	59 Pr 140.91	60 Nd 144.24	61 Pm [145]	62 Sm 150.36	63 Eu 151.96	64 Gd 157.25	65 Tb 158.93	66 Dy 162.50	67 Ho 164.93	68 Er 167.26	69 Tm 168.93	70 Yb 173.05	71 Lu 174.97
**	90 Th 232.04	91 Pa 231.04	92 U 238.03	93 Np [237]	94 Pu [244]	95 Am [243]	96 Cm [247]	97 Bk [247]	98 Cf [251]	99 Es [252]	100 Fm [257]	101 Md [258]	102 No [259]	103 Lr [262]